**Advances in Experimental Medicine and Biology**

Volume 892

*Advances in Experimental Medicine and Biology* presents multidisciplinary and dynamic findings in the broad fields of experimental medicine and biology. The wide variety in topics it presents offers readers multiple perspectives on a variety of disciplines including neuroscience, microbiology, immunology, biochemistry, biomedical engineering and cancer research. *Advances in Experimental Medicine and Biology* has been publishing exceptional works in the field for over 30 years and is indexed in Medline, Scopus, EMBASE, BIOSIS, Biological Abstracts, CSA, Biological Sciences and Living Resources (ASFA-1), and Biological Sciences. The series also provides scientists with up to date information on emerging topics and techniques.
2013 Impact Factor: 2.012.

More information about this series at http://www.springer.com/series/5584

José Ramos • Hana Sychrová • Maik Kschischo
Editors

# Yeast Membrane Transport

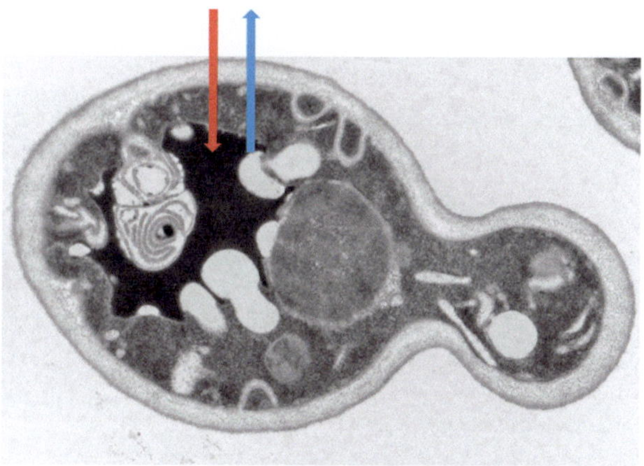

Springer

*Editors*
José Ramos
Department of Microbiology
University of Córdoba
Córdoba, Spain

Maik Kschischo
Department of Mathematics and Technology
University of Applied Sciences
Koblenz, Remagen, Germany

Hana Sychrová
Department of Membrane Transport
Institute of Physiology
The Czech Academy of Sciences
Prague, Czech Republic

ISSN 0065-2598 ISSN 2214-8019 (electronic)
Advances in Experimental Medicine and Biology
ISBN 978-3-319-25302-2 ISBN 978-3-319-25304-6 (eBook)
DOI 10.1007/978-3-319-25304-6

Library of Congress Control Number: 2015960745

Springer Cham Heidelberg New York Dordrecht London


Printed on acid-free paper

Springer International Publishing AG Switzerland is part of Springer Science+Business Media (www.springer.com)

# Preface

All cells need to communicate with their environment, which means exchange information and material across their membranes. To ensure the uptake of nutrients and other necessary compounds or to extrude toxic metabolites, metabolic waste, or chemical signals, cells use transport systems consisting of one or more proteins embedded in the cell membranes. The significance of transport in organisms and their cells was summarized by Daniel C. Tosteson who wrote in his book *Membrane Transport: People and Ideas* (Tosteson 1989): "I conjure a picture of a living organism as an extraordinarily complex, self-replicating, self-assembling, dynamic, open, physicochemical system maintained near the steady state by the continual entry and exit of matter and energy." Though the indispensability and complexity of cell membrane transport have attracted researchers for more than 100 years, research on membrane transport has made continuous progress in the last decades and remains an active field of scientific investigation. Yeasts, and mainly the model organism *Saccharomyces cerevisiae*, are among the favorite cells for researchers studying all aspects of transport systems and their mechanisms. However, an important amount of knowledge has been obtained when employing the so-called non-conventional and pathogenic yeasts, especially because of their peculiarities and, in some cases, specific transport systems. The transporters mediating uptake of nutrients, fluxes of cations and anions, or the extrusion of toxic compounds across the plasma membrane are discussed together with the transport systems localized in the membranes of cell organelles. Each chapter summarizes our current knowledge on important transport processes in yeasts, and as it is based on the experience of the experts working in the field for many years, it provides both a general overview of the main transport characteristics for a specific substrate or a group of substrates and unique details that only an expert working in the field is able to transmit to the reader.

Córdoba, Spain — José Ramos
Remagen, Germany — Maik Kschischo
Prague, Czech Republic — Hana Sychrová

# Contents

# Chapter 1
# Membrane Transport in Yeast, An Introduction

**Maik Kschischo, José Ramos, and Hana Sychrová**

**Abstract** Research on membrane transport has made continuous progress in the last decades and remains an active field of scientific investigation. In the case of yeast, most of the research has been conducted for the model organism *Saccharomyces cerevisiae,* but also the so-called non-conventional yeasts are being studied, especially because of their peculiarities and, in some cases, specific transport systems. This book is based on the experience of several experts summarizing the current knowledge about important substrate transport processes in yeast. Each chapter provides both a general overview of the main transport characteristics of a specific substrate or group of substrates and the unique details that only an expert working in the field is able to transmit to the reader.

**Keywords** Yeast • Membrane • Transport

## 1.1 Importance of Membrane Transport Processes

A cell's existence depends on its membrane. Although membranes were first considered just as physical boundaries, their function as a permeability barrier was quickly realized. Later on, the crucial sensing functions of membranes to facilitate information exchange between the external and internal milieu were discovered. In addition to the plasma membrane, eukaryotic cells contain systems of internal

M. Kschischo
Department of Mathematics and Technology, University of Applied Sciences Koblenz, RheinAhrCampus, Remagen, Germany
e-mail: kschischo@rheinahrcampus.de

J. Ramos (✉)
Department of Microbiology, University of Córdoba, Córdoba, Spain
e-mail: mi1raruj@uco.es

H. Sychrová
Department of Membrane Transport, Institute of Physiology, The Czech Academy of Sciences, Prague, Czech Republic
e-mail: Hana.Sychrova@fgu.cas.cz

 
J. Ramos et al. (eds.), *Yeast Membrane Transport*, Advances in Experimental Medicine and Biology 892, DOI 10.1007/978-3-319-25304-6_1

membranes which form different membrane-enclosed compartments within the cell. The main constituents of membranes are lipids, proteins and carbohydrates, but the relative portion can vary between different organisms and cell types. Although the concept of a universal membrane structure is usually accepted it is also well known that membrane systems differ widely. Examples, at different levels of complexity can be found in the outer membrane of gram negative bacteria, in archaea or even in the composition of the membrane of a cell growing under different external situations.

Living organisms can survive under ever changing environmental conditions. However all vital processes within the cell can only function for a certain range of substrate and solvent concentrations. Adaptation to stressful conditions requires cells to sense both changes in environmental conditions and changes of their own intracellular state. One fundamental mechanism for this adaptation is regulated transport across biological membranes. Importantly, this includes orchestrated uptake or release of substances across the plasma membrane as well as transport across intracellular membranes.

Most of the solutes are taken across the membranes by transport systems. Now we know that even some substances such as water or glycerol, that were believed to diffuse readily through membranes, are also transported by specific transport proteins. Daniel C Tosteson (1989) wrote in his book "Membrane Transport: people and ideas" (1989): *"I conjure a picture of a living organism as an extraordinarily complex, self-replicating, self-assembling, dynamic, open, physicochemical system maintained or near the steady state by the continual entry and exit of matter and energy"*. For these purposes biological membranes and membrane transport are essential.

## 1.2 Why Yeast?

Yeasts are eukaryotic micro-organisms classified in the kingdom Fungi. The 1,500 species currently described are estimated to be only a small fraction of all yeast species (Kurtzman et al. 2010). These species do not form a single taxonomic or phylogenetic grouping. They form two separate phyla, the Ascomycota and the Basidiomycota. The term "yeast" is often taken as a synonym for *Saccharomyces cerevisiae,* which is one of the most thoroughly researched eukaryotic microorganisms. The reasons of using *S. cerevisiae* are multiple. Besides sharing the complex internal cell structure of plants and animals, it is mainly its fast growth, a broad choice of genetic tools to add new genes or delete its own genes through homologous recombination, and the complete sequencing and annotation of its genome as the first one amongst eukaryotic cells (Goffeau et al. 1996). Importantly, the broad use of yeast cells in research is also driven by biotechnological applications with a growing economic impact.

## 1.3 A Bit of "History"

Yeasts belong to the earliest domesticated organisms. Archaeologists confirmed their use in food production several thousand years ago. At the end of the seventeenth century they were most probably the first microorganisms observed by Antonie van Leeuwenhoek in his microscope, and in the eighteenth century they helped Antoine-Laurent de Lavoisier to formulate the fundamental physical law of mass conservation. Nevertheless, only in the middle-of the nineteenth century, Louis Pasteur proved that alcoholic fermentation was conducted by living yeasts and not by a chemical catalyst. Several excellent reviews have summarized the history of research on yeast membrane transport (Barnett 2003; Barnett and Davson 2008; Eddy and Barnett 2007).

The advent of new measurement techniques like the $^{14}C$- labelled organic compounds (such as sugars and amino acids) in the first half of the twentieth century initiated a boom in research on membrane transport in living cells including yeasts. In the second half of the twentieth century, several tens of yeast transporters were described kinetically; in many cases the mutants lacking various transport activities have been isolated. It was in the 1980s when the first genes encoding yeast transporters were cloned and sequenced. The sequencing of the yeast genome triggered a second important breakthrough in studying cell transport systems. It enabled a comprehensive *in silico* prediction of all yeast transporters (De Hertogh et al. 2002; Nelissen et al. 1997) and made cloning or deletion of their genes easy. Later, a comparative analysis of sequenced genomes from different yeast species showed that approximately 10 % of a yeast's genome correspond to membrane transporters (De Hertogh et al. 2006). A detailed analysis of genes encoding putative transporters revealed that many important transport proteins are highly conserved in yeast species.

## 1.4 Mechanistic Processes Involved in Membrane Transport

As chemoorganotrophs, yeasts use organic compounds as a source of energy and do not require sunlight to grow. Carbon is obtained mostly from hexose sugars, such as glucose and fructose, or disaccharides such as sucrose and maltose. As a source of nitrogen, yeast cells use either inorganic ammonium cations or various amino acids and other nitrogen-containing organic compounds. To ensure the uptake of all necessary nutrients, export of vast metabolites or toxic compounds, and maintenance of ion and pH homeostasis, yeast cells employ more than 600 transport systems mediating the transport of solutes across the plasma and organelle membranes with various mechanisms and affinities. In yeast cells, almost all known types of transport mechanisms exist, and this book describes in detail examples for most of them. Besides **simple diffusion** across the membrane lipids (a mechanism for the transport of gasses or small lipophilic molecules), yeast cells have many

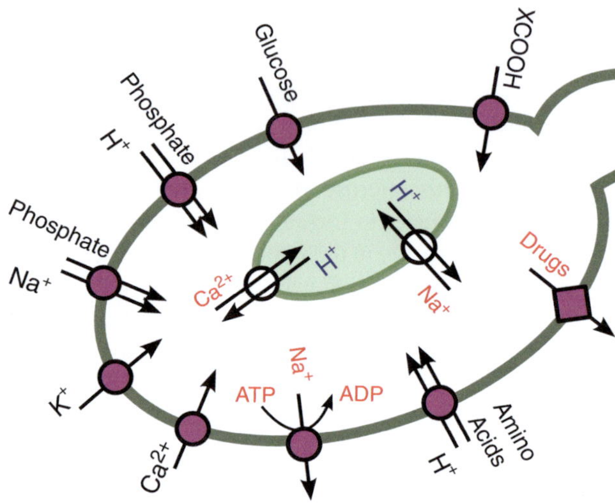


**Fig. 1.1** A schematic picture of a yeast cell with several organelles and fluxes of different important substrates

systems operating by **facilitated diffusion.** This mechanism includes transporters and channels (Saier et al. 2009); many of them are used to acquire sugars (Chap. 6) or to maintain the homeostasis of cations (Chaps. 8 and 11).

**Primary active transport mechanisms** are essential to create and maintain the gradients of protons and cations across cell membranes (Chap. 3) and play an important role in the detoxification of the cytosol (Chaps. 13 and 15). The yeast genome encodes almost one hundred ATPase genes whose products consume the energy of ATP hydrolysis. Among them, the plasma-membrane Pma1 and vacuolar Vma ATPases have a crucial role. They pump protons out of the cytosol and thereby help to maintain the intracellular pH in an optimal range. Simultaneously they create the necessary gradients of protons that serve as a source of energy for **secondary active transporters.**

The proton gradient generated by the essential Pma1 $H^+$-ATPase (Serrano et al. 1986) at the plasma membrane serves as the source of energy for secondary active **symporters** and **antiporters,** which provide cells with the majority of nutrients including amino acids (Chap. 4), phosphates (Chap. 10), sulphates (Chap. 11) or eliminate surplus cations (Chaps. 8 and 10) (Fig. 1.1).

The proton gradient across the vacuolar membrane as well as that across membranes of organelles is generated by the Vma $H^+$-ATPase of the secretory pathway. The free energy stored in the proton gradient is consumed by symporting or antiporting transport systems to supply organelles with various nutrients and metabolites, and to ensure organelle homeostasis of pH and cations, e.g. (Arino et al. 2010; Horak 1997). Another important process used in yeast as well as in many other eukaryotic cells is the consumption of the proton gradient in mitochondria by ATP-synthase to synthesize ATP.

Altogether, the existence of several hundreds of transporters with various mechanisms, substrate specificities and different ways of regulation of their synthesis and degradation in a single cell justifies again the role of yeast as a model system for membrane transport in eukaryotic cells. The reader will find many examples in this book, where research in yeast was crucial for the characterization of transporters from higher eukaryotes related either to human disease or to the effective production of crops (see Chap. 14 or 8 just as two examples).

## 1.5 A Comment on Intracellular Membranes

For many years, most researchers focused their work on the plasma membrane. However, we should pay attention to the role of intracellular membranes in the organellar compartmentalization of substrates. Many characteristics of organelle membrane transport mechanisms are still unknown, despite of being crucially important for cellular physiology. Some research groups have worked on this aspect. Two examples illustrating the complexity of these attempts are given by the vacuolar amino acid transporters and the mitochondrial carrier family.

The vacuole is the main compartment for regulating the cytosolic amounts of amino acids, as it is required for effective protein synthesis. In *S. cerevisiae*, a numerous group of amino acid transporters is located at the tonoplast. The transport of the amino acids is coupled to the extrusion of protons by the vacuolar $H^+$-ATPase. Most of the corresponding genes involved are known and also homologous genes for these transporters have been identified in higher eukaryotic organisms (Sekito et al. 2008).

The metabolic activities performed by mitochondria require the exchange of molecules between the cytosol and the mitochondrial matrix space. Members of the Mitochondrial Carrier Family (MCF), involved in these activities, have been found in many different organisms including yeast, plants or humans. These proteins mediate the transport of several metabolites, cofactors or nucleotides across the mitochondrial inner membrane. More than 30 carriers of the family have been identified in *S. cerevisiae* (del Arco and Satrustegui 2005) and new carriers are still being identified.

## 1.6 Are There Still New Interesting Discoveries in Yeast?

Substances transported across membranes interact with various biochemical reaction pathways. The resulting changes of the cell's biochemical and physiological state feed back to the regulation of membrane transport. Clearly, a holistic understanding of biological transport requires a systems biology approach integrating traditional data and experiments with large scale omics data, e.g. gene expression and metabolic measurements. Bridging the gap between these global approaches and detailed experimental studies is not always easy. Several chapters in this book meet this challenge and illustrate what can be gained in the future from such a two sided perspective. One example is the regulation of pH (Chap. 3), where such an integrative approach has contributed to an improved (but still incomplete) understanding of the relationship between pH, nutrient signaling and growth control. Another example is provided in Chap. 11: The old, but largely forgotten observation (Goodman and Rothstein 1957; Schmidt et al. 1949) that cation homeostasis and phosphate transport are closely related was recently rediscovered by expression analysis and then elucidated in detail by a series of ingenious experiments. Another direction for the future is the combination of mathematical modelling and experimental data analysis (Chaps. 8 and 12) to encompass the complexity of the multicausal and nonlinear dynamics of transport and regulation.

## 1.7 Basic Versus Applied Research in Membrane Transport

Without any doubt, the investigation of membrane transport processes has been a matter of basic or fundamental research. But, in addition to being fundamental for life, membrane transport processes play a key role in many practical or applied fields (see (Johnson 2013) and references therein). Various areas of yeast biotechnology such as Biocontrol, Biocatalysis or Biomedical Research (see Chaps. 14 and 15) are directly or indirectly related to membrane transport. One example is heterologous protein production in yeasts, which has been utilized since the 1980s and thus can already be considered a classical biotechnological process. Since then, the field has experienced an enormous growth driven by its biotechnological potential. A crucial point in heterologous protein production is the secretion through the plasma membrane.

The cloning, expression, processing, and secretion of human proinsulin in *Saccharomyces cerevisiae* in the 1980s was a breakthrough in heterologous protein expression in yeast. Subsequent studies showed that *S. cerevisiae* effectively synthesized and secreted many mammalian and human proteins including α-interferon, epidermal growth factor, human serum albumin, hepatitis B surface antigen, β-endorphin, and prochymosin. *S. cerevisiae* and *Kluyveromyces pastoris* have been most extensively developed for heterologous protein production. More than 500 heterologous proteins have been produced in *K. pastoris*, including plant, animal and human proteins, as well as membrane-associated proteins which are often difficult to functionally express in many other systems. *K. pastoris* has the exceptional ability to produce high quantities of proteins, glycosylate them in patterns similar to that of humans, efficiently secrete them into the medium, and avoid aggregation and proteolysis during recovery for many proteins (Johnson 2013 and references therein).

The second example is the attempt to link a traditional yeast fermentative process such as brewing with bioremediation of real industrial effluents containing heavy metals (for a review see Soares and Soares 2013). *S. cerevisiae* yeast cells are a by-product from the brewing industry, but they are still a relatively underutilized raw material. The application of brewing yeast cells in the bioremediation of industrial effluents loaded with heavy metals constitutes another possibility for the valorization of this type of biomass. Living yeast cells can remove metals from a solution by two main mechanisms: a passive process, called biosorption and a metabolism-dependent process, known as bioaccumulation. Biosorption is a physicochemical process wherein the metal interacts with and is accumulated in the surface of yeast cells. Bioaccumulation, however, only occurs in metabolic active cells and is a slow process; it implies intracellular accumulation, associated with the passage of the metals through pores or channels of the cell membrane by the primary action of the plasma membrane $H^+$-ATPase. Bioaccumulation is affected by temperature and the presence of metabolic inhibitors. More work is required to optimize conditions and to define metal transporters working during the bioaccumulation process of a specific metal to be removed. In a green process, theoretically, the regeneration of the biomass and the simultaneous reclamation of the valuable metals are desirable. With this aim, different eluents have been tested such as diluted inorganic and organic acids, bases and chelating. Alternatively, biomass loaded with metals can be dissolved with strong acids or incinerated.

## 1.8 Overview of the Contributions

The relationships between the cell wall and the plasma membrane and how they affect transport are poorly known. Jean-Marie François reviews in Chap. 2 the state of the art in the field. In addition to its “mechanical” properties, the cell wall keeps an intimate connection with the plasma membrane that we are starting to uncover. The author analyses the structure and properties of the cell wall and illustrates with a few examples the contribution of the cell wall to transport processes.

Cytosolic and organelle pH is a fundamental parameter of cellular physiology and proton transport is coupled to the transport of many other substrates and to the membrane potential. Progress towards a systems level understanding of pH control is reviewed by Patricia M. Kane in Chap. 3. She also summarizes very interesting recent findings that dynamic pH changes play a pivotal role in nutrient signaling and growth control.

Amino acids are important nutrients for many fungi and have been largely studied for the model system *Saccharomyces cerevisiae* and for the filamentous fungus *Aspergillus nidulans.* As reviewed in Chap. 4 by the group of Bruno Andre, structural modeling and mutational analysis have contributed to the current understanding of the key molecular elements facilitating amino acid recognition and transport as well as amino-acid sensing in relation to their subcellular location and substrate specificity range.

Water can diffuse across the lipid bilayer but can also be transported through specialized water channels named aquaporins. In Chap. 5, the groups of Maria C. Loureiro-Dias and Graça Soveral have prepared a collaborative report on recent advances in the field of water transport and aquaporins. *Pichia pastoris* Aqy1 can be considered as a model to study water transporters in yeast, as it is the only fungal aquaporin structure solved so far and essential details of its mechanism of action and structure have been revealed during the last years. Additionally, different methods to assess water transport in yeast and information related to the state of the art in *Saccharomyces cerevisiae* are provided.

The yeast *Saccharomyces cerevisiae* can be found in ecosystems with changing sugar concentrations. Therefore it is not surprising that it has developed not only different systems for sugar transport but also complex regulatory mechanisms to efficiently cope with varying conditions. The HXT family of transporters and its regulation is a fascinating example of adaptation to the environment. Linda Bisson et al. (Chap. 6) summarize the state of the art of sugar transport and transcriptional and post-translational regulation of the process, including information about the role of prions in the control of transport. Finally glycerol transport and the corresponding proteins involved in glycerol fluxes are also discussed.

All living cells require calcium. The "CRaZy" calcium cycle is reviewed in Chap. 7 by Eduardo Espeso. In a concise and clear paper, the calcium transporters/channels and the possible mechanisms involved in the fluxes of this divalent cation are described. In addition the role of calcium as a second messenger and the regulatory processes devoted to maintain tightly regulated levels of free intracellular calcium are analyzed. The functions of the protein phosphatase calcineurin and the transcription factor Crz1 are discussed in detail.

Chapter 8 is devoted to "Potassium and sodium transport in yeast". In the first and more extensive part of the chapter, Lynne Yenush reviews in detail the mechanisms and proteins involved in alkali cation homeostasis including genes and processes that modulate influx and efflux processes. The author analyses the well-known plasma membrane potassium and sodium transporters but also those present in intracellular organelles that have been less studied. The second (Physiological consequences and cellular responses to alterations in potassium and sodium concentrations) and the third (Extrapolations and Applications) section are of special interest, since they provide a holistic view of the subject. Consequences of perturbations in extra- or intracellular potassium and sodium levels, mathematical modeling approaches and possible applications ranging from medicine to industrial processes are commented.

The group of Margarida Casal summarizes the transport of carboxylic acids, an ubiquitous class of organic compounds acting as substrates or products of fermentative processes. The transport depends largely on the environmental pH and reacts to environmental changes including carbon source availability and acid stress conditions. Chapter 9 discusses the functionally characterized plasma membrane carboxylic acids transporters Jen1, Ady2, Fps1 and Pdr12 in the yeast *Saccharomyces cerevisiae* and their homologues in filamentous fungi and bacteria.

Efficient phosphate uptake is crucial for cell growth and proliferation, as Pi is required for the generation of high-energy phosphate esters as ATP. It is also indispensable for synthesis of many molecules such as DNA, RNA, lipids, and is involved in cell signaling and regulation of protein activities. Dieter Samyn and Bengt Persson (Chap. 10) focus on known phosphate transporters and on the sophisticated way how the yeast cells regulate their expression and activity in response to their needs and to changing environment.

Transport, consumption and metabolism of different nutrients and substrates are influenced by monovalent cation (mainly $H^+$, $K^+$ and $Na^+$) homeostasis. David Canadell and Joaquín Ariño have compiled the available information on the subject in Chap. 11. They prepared a combination of historical background and exciting new findings on how cation homeostasis is linked to phosphate, nitrogen, sulfate or sugar transport.

Mathematical models can facilitate the understanding of membrane transport, provide testable predictions and can be used to prioritize experiments. Chapter 12 by Mathias Kahm and Maik Kschischo explains the specific characteristics of mathematical models coupling membrane transport, membrane potential and pH with the transport of cations. The principle and structure of such models is illustrated for the example of potassium transport. This forms the basis for a review of two recent mathematical models of ion regulation in *S. cerevisiae,* which illustrate, how modelling can be used to support experimental research on membrane transport.

Chapter 13 has been prepared in collaboration between the groups of Ana Plemenitaš and Nina Gunde Cimerman in Ljubljana. That contribution covers transport processes in extremophilic yeast/fungi from hypersaline habitats. Special emphasis is placed on the study of alkali cation transporters in the ascomycetous yeast *Hortaea werneckii* and *Aureobasidium* spp. and the basidiomycetous *Wallemia ichthyophaga.* On the basis of the analysis of their sequenced genomes, the existence of different strategies to survive and thrive under hypersaline conditions in these organisms is proposed. The last section of the chapter briefly covers other substrate transporters identified in *Aureobasidium* and *Wallemia*. Transporters such as aquaporins or those belonging to the Major Facilitator Superfamily are important determinants of the halophilic behaviour of these interesting organisms.

Rajendra Prasad and collaborators summarize our knowledge on drug resistance of yeast pathogens and the role of various transporters in it in Chaps. 14 and 15. Though that there has been an expansion in the number of antifungal drugs in the last years, the treatment of fungal disease remains unsatisfactory, mainly due to a dramatically increasing appearance of strains resistant to treatment with classical antifungal drugs, which are very often targeted to the cell-wall and plasma-membrane components. The so called multiple drug resistance (MDR) is indeed a multifactorial phenomenon wherein a resistant cell possesses several mechanisms which contribute to display reduced susceptibility to not only single drug in use but also show collateral resistance to several drugs. Chapter 14 brings an overview of the current understanding of the antifungal drugs in use, their mechanism of action and the emerging possible novel antifungal drugs with great promise, whereas Chap. 15 focuses on the two main types of transporters involved in drug detoxification, ATPases and antiporters.

## References

Arino J, Ramos J, Sychrova H (2010) Alkali metal cation transport and homeostasis in yeasts. Microbiol Mol Biol Rev 74:95–120

Barnett JA (2003) Beginnings of microbiology and biochemistry: the contribution of yeast research. Microbiol SGM 149:557–567

Barnett JA, Davson H (2008) A history of research on yeasts 13. Active transport and the uptake of various metabolites. Yeast 25:689–731

De Hertogh B, Carvajal E, Talla E, Dujon B, Baret P, Goffeau A (2002) Phylogenetic classification of transporters and other membrane proteins from *Saccharomyces cerevisiae*. Funct Integr Genomics 2:154–170

De Hertogh B, Hancy F, Goffeau A, Baret PV (2006) Emergence of species-specific transporters during evolution of the hemiascomycete phylum. Genetics 172:771–781

del Arco A, Satrustegui J (2005) New mitochondrial carriers: an overview. Cell Mol Life Sci 62:2204–2227

Eddy AA, Barnett JA (2007) A history of research on yeasts II. The study of solute transport: the first 90 years, simple and facilitated diffusion. Yeast 24:1023–1059

Goffeau A, Barrell BG, Bussey H, Davis RW, Dujon B, Feldmann H, Galibert F, Hoheisel JD, Jacq C, Johnston M et al (1996) Life with 6000 genes [see comments]. Science 274(546):563–547

Goodman J, Rothstein A (1957) The active transport of phosphate into the yeast cell. J Gen Physiol 40:915–923

Horak J (1997) Yeast nutrient transporters. Biochim Biophys Acta 1331:41–79

Johnson EA (2013) Biotechnology of non-*Saccharomyces* yeasts—the ascomycetes. Appl Microbiol Biotechnol 97:503–517

Kurtzman CP, Fell JW, Boekhout T (2010) The yeasts, a taxonomic study, 5th edn. Elsevier, Amsterdam/Oxford

Nelissen B, De Wachter R, Goffeau A (1997) Classification of all putative permeases and other membrane plurispanners of the major facilitator superfamily encoded by the complete genome of Saccharomyces cerevisiae. FEMS Microbiol Rev 21:113–134

Saier MH Jr, Yen MR, Noto K, Tamang DG, Elkan C (2009) The transporter classification database: recent advances. Nucleic Acids Res 37:D274–D278

Schmidt G, Hecht L, Thannhauser SJ (1949) The effect of potassium ions on the absorption of orthophosphate and the formation of metaphosphate by bakers' yeast. J Biol Chem 178:733–742

Sekito T, Fujiki Y, Ohsumi Y, Kakinuma Y (2008) Novel families of vacuolar amino acid transporters. IUBMB Life 60:519–525

Serrano R, Kielland-Brandt MC, Fink GR (1986) Yeast plasma membrane ATPase is essential for growth and has homology with ($Na^+ + K^+$), $K^+$- and $Ca^{2+}$-ATPases. Nature 319:689–693

Soares EV, Soares HMVM (2013) Cleanup of industrial effluents containing heavy metals: a new opportunity of valorising the biomass produced by brewing industry. Appl Microbiol Biotechnol 97:6667–6675

Tosteson DC (1989) Membrane transport: people and ideas. Oxford University Press, New York

# Chapter 2
# Cell Surface Interference with Plasma Membrane and Transport Processes in Yeasts

**Jean Marie Francois**

**Abstract** The wall of the yeast *Saccharomyces cerevisiae* is a shell of about 120 nm thick, made of two distinct layers, which surrounds the cell. The outer layer is constituted of highly glycosylated proteins and the inner layer is composed of β-glucan and chitin. These two layers are interconnected through covalent linkages leading to a supramolecular architecture that is characterized by physical and chemical properties including rigidity, porosity and biosorption. The later property results from the presence of highly negative charged phosphate and carboxylic groups of the cell wall proteins, allowing the cell wall to act as an efficient barrier to metals ions, toxins and organic compounds. An intimate connection between cell wall and plasma membrane is indicated by the fact that changes in membrane fluidity results in change in cell wall nanomechanical properties. Finally, cell wall contributes to transport processes through the use of dedicated cell wall mannoproteins, as it is the case for Fit proteins implicated in the siderophore-iron bound transport and the Tir/Dan proteins family in the uptake of sterols.

**Keywords** *Yeasts* • Cell surface • Cell wall • Porosity • Biosorption • Membrane • Metal ions • Transport

## 2.1 Introduction

For essential molecules to enter the yeast cell, they must penetrate the cell wall, cross a periplasmic space before to have access to transporters or permeases embedded in the plasma membrane. The yeast cell wall can be viewed as a sieve with fairly large holes through which molecules of any kind can diffuse freely, if they are not too big. This function of the cell wall in selective entrance of molecules was already

J.M. Francois (✉)
Université de Toulouse; INSA, UPS, INP; LISBP, 135 Avenue de Rangueil, F-31077 Toulouse, France

INRA, UMR792 Ingénierie des Systèmes Biologiques et des Procédés, F-31400 Toulouse, France

CNRS, UMR5504, F-31400 Toulouse, France
e-mail: fran_jm@insa-toulouse.fr



J. Ramos et al. (eds.), *Yeast Membrane Transport*, Advances in Experimental Medicine and Biology 892, DOI 10.1007/978-3-319-25304-6_2

appreciated more than 50 years ago by several researchers including Conway & Downey, Gerhardt and colleagues (see an historical research on yeast transport reviewed by (Eddy and Barnett 2007)). However, this view must be complimented with the fact that the cell wall also acts as ion exchangers being highly negatively charged by phosphate and carboxyl groups present in cell wall proteins. These ionic charges furthermore contribute to the biosorption properties of the wall which is proposed as an efficient metal ions detoxifier, thus opening opportunities in the exploitation of the increasing amount of yeast biomass produced by biorefinery industries. The purpose of this chapter is to make an account on the structure-function of cell wall, on the intimate connection of the cell wall with the plasma membrane and on the interference of the cell wall as a first selective barrier for entrance of essential nutrients, minerals and other biomolecules required for growth. While the model eukaryote *Saccharomyces cerevisiae* will be the focus of this review, relevant information with other yeasts such as *C. albicans* will be also presented to further support the critical function of the cell wall envelope in transports processes.

## 2.2 An Overview of the Structure and Biophysical Properties of Yeast Cell Wall

The wall of the yeast *Saccharomyces cerevisiae* is an armour of about 120 nm thick surrounding the cell as determined from transmission electron micrographs (Yamaguchi et al. 2011) (see Fig. 2.1) and by Atomic Force Microscopy (AFM) (Dupres et al. 2010). It occupies about 17 % of the total cell volume and accounts for 20–30 % of the dry mass of the vegetative cell (Yamaguchi et al. 2011; Aguilar-Uscanga and Francois 2003; Nguyen et al. 1998). As depicted in Fig. 2.1, two distinct layers can be distinguished from electron microscopy of thin sections through the cell wall. The outer layer is electron-dense, has brush-like surface (Osumi 1998; Hagen et al. 2004; Kapteyn et al. 1999) and is constituted of highly

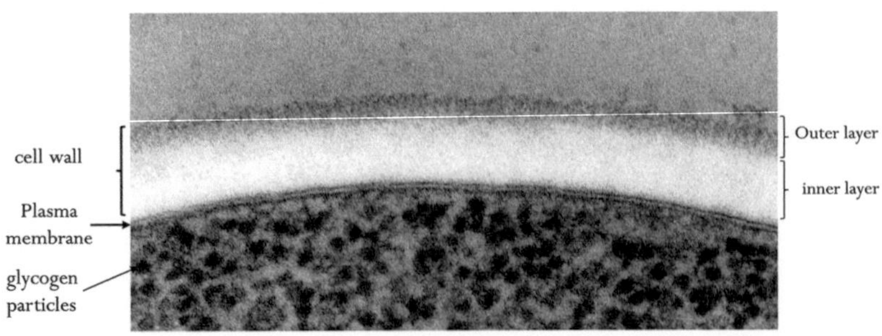


**Fig. 2.1** Electron microscopy image of the cell wall and membrane of the yeast *S. cerevisiae* (Courtesy from Dr M. Yamaguchi, Chiba University, Japan)

glycosylated proteins that can be removed by proteolysis (Zlotnik et al. 1984). The thickness of this outer layer is estimated to around 34 nm ± 3 nm (Yamaguchi et al. 2011). The more electron transparent inner layer is microfibrillar and β-glucanase digestible, with an estimated thickness of about 86 ± 14 nm (Yamaguchi et al. 2011). Detailed biochemical analyses over the last 30 years employing fractionation procedures combined with chemical and enzymatic methods revealed the presence of 4 different polysaccharides that compose the wall (Fleet 1991; Magnelli et al. 2002; Dallies et al. 1998; Schiavone et al. 2014). The mannan is a complex oligosaccharide comprising 10 to more than 50 mannose units linked in α-(1,2); α-(1,3); α-(1,5) and α-(1,6), which is attached to proteins by either Asn (large manno-oligosaccharides for N-glycosylation) or Ser/thr residues (short manno-oligosaccharides to make the O-glycosylation). These mannoproteins constitute the outer layer of the wall. The β-glucan are composed of β-1,3-glucan which are linear chains of around 1500 glucose units linked in β-(1,3) and of β-(1,6) glucan composed of 140–350 residues of glucose linked by a β-(1,6) linkage. The fourth polysaccharide is chitin which is a polymer of 100–190 *N*-acetylglucosamine units linked by β-(1,4) linkages. The inner layer of the cell wall is mainly composed of β-glucan and chitin. The relative proportion of the four wall polysaccharide can vary according to cultures conditions, process fermentation, strain background, etc., but it is generally admitted that 50 % of the cell wall dry mass is made of β-glucan from which 10 to 15 % are β-1,6-glucan, 40 % of mannans, 3–5 % of chitin and the remaining mass corresponds to proteins (Dallies et al. 1998; Aguilar-Uscanga and Francois 2003; Schiavone et al. 2014; Magnelli et al. 2002; Nguyen et al. 1998). These polysaccharides are not simply juxtaposed, as would suggest the electron micrograph but they are assembled to each other by covalent linkages to generate a supramolecular architecture in which the chains of β-(1,3)-glucose residues are branched to β-(1,6)-glucan, forming a fibrillar network, that serves as backbone to which are linked chitin, β-(1,6)-glucan and mannoproteins. In addition, some cell wall mannoproteins (CWP) are linked to the inner layer through β-(1,6)-glucan via a remnant of a GPI anchor (Klis et al. 2006; Lesage and Bussey 2006). In addition to these GPI-CWP, the cell wall contain two other types of glycosylated proteins, namely the PIR-proteins that are ester-linked to the β- 1,3 glucan matrix by mild-alkali/ β-1,3 glucanase sensitive links (Mrsa et al. 1997; Ecker et al. 2006) and the disulphide-linked proteins which are retained in the cell wall through disulphide bound to other PIR or GPI proteins and which can be released by sulfhydryl reagents such as mercaptoethanol or dithiothreitol (Cappellaro et al. 1994, 1998). The molecular architecture of the cell wall from the yeast *S. cerevisiae* is schematically represented in Fig. 2.2. It is estimated that 1200 genes, up to about a quarter of the genome of the *S. cerevisiae*, have some role in the construction and maintenance of a normal wall. However, most of them have an indirect impact (De Groot et al. 2001), and the number of genes encoding enzymes that are directly involved in the biogenesis and remodelling of the wall and non-enzymatic wall proteins is now estimated to less than 200 (Orlean 2012).

The wall gives to the yeast its morphology during budding growth, pseudo-hyphae formation, mating and sporulation. Four major physiological functions

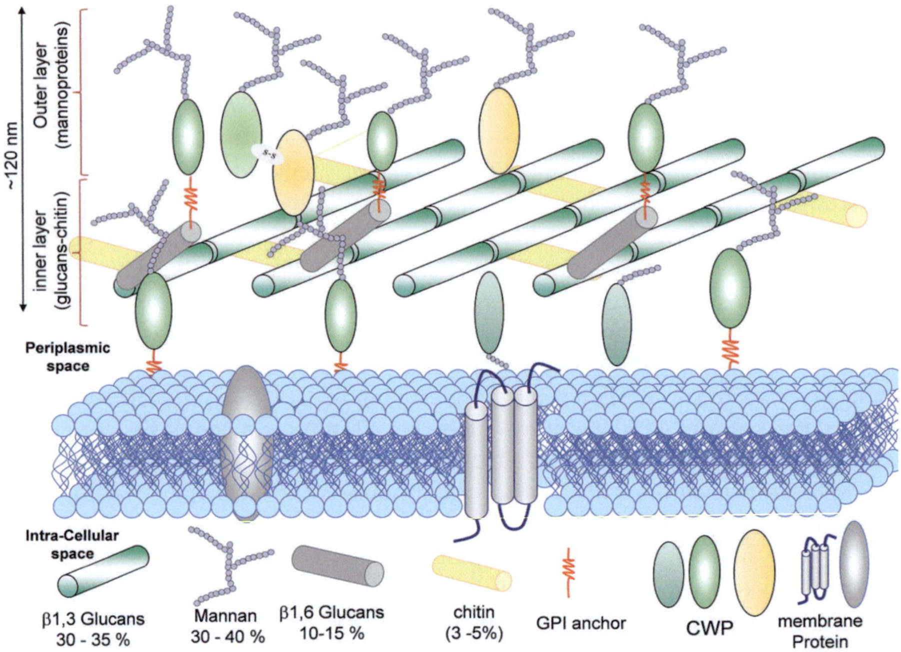


**Fig. 2.2** Schematic organization of the cell wall organisation in the yeast *Saccharomyces cerevisiae*. *CWP* means cell wall proteins, and coloration marks their distinction: from *left* to *right*, non-covalently linked CWP with periplasmic location such as invertase, *GPI*-anchored cell wall protein and *PIR*- cell wall proteins

have been recognized for the cell wall, namely stabilization and internal osmotic conditions, protection against stresses, maintenance of the cell shape and a scaffold for cell wall proteins. The latter property likely determines the limit of wall permeability for macromolecules (see below) and hence shielding the inner layer of polysaccharides from attack by foreign proteins. Collectively, these functions account for constant remodeling of the cell wall molecular structure in response to environmental constraints that can be physics, chemicals or biologicals. This cell wall remodeling process is mainly under the control of the cell wall integrity (CWI) signaling pathway that transmits the signals from the cell surface sensors to a MAP kinases cascade (Levin 2011), with the main consequence to reorganize cell wall architecture through changes in the content of carbohydrate polymers, in the cross-links between these polymers and in a transient redistribution of the cell wall repair machinery to the site of the cell wall injuries (Klis et al. 2006; Lesage and Bussey 2006). Although the CWI can be activated by a wide variety of stimuli, including antifungal drugs, cell wall degrading enzymes, etc., these stimuli lead to increased membrane tension. For instance, cells with a weakened cell wall caused by genetic means or cultures conditions tend to swollen because their walls are less capable to withstand the turgor pressure. In addition, and not surprisingly, the cell wall construction is temporally and spatially controlled, with the expression of many cell wall related genes varying in a cell-cycle dependent fashion (Spellman et al. 1998), and genes encoding orthologous proteins expressed at different location in the cell, such as Cwp1 and Cwp2 that are transcribed at the same time but the former being specifically located in the birth scar (Smits et al. 2006). A cell wall integrity checkpoint has been described which operates at the vicinity of the G2/M arrest (Suzuki et al. 2004). However, this checkpoint is clearly distinguishable from other cell cycle checkpoints because it occurs after completion of DNA replication, and it functions differently from the morphogenesis checkpoint. Nevertheless, the molecular machinery that makes up the signaling module of this checkpoint is still unclear (Negishi and Ohya 2010).

Complimentary to these molecular and biochemical data, nanomechanical properties of the cell wall have been recently explored thanks to the development of the Atomic Force Microscopy in liquid (Dufrene 2001; Dague et al. 2007). AFM is a tool that can be used as a microscope to directly visualize the cell surface topology and as a force spectroscopy tool to measure the mechanical properties of the cell (termed elasticity and expressed as Young modulus, (Stenson et al. 2011)). Using this tool, it was found that the nanomechanical properties of the yeast cell wall could not be ascribed to a specific cell wall component but are mainly dependent on the intrinsic molecular organization of the wall, with cross-links of chitin to β-glucan playing an important role in the cell wall elasticity (Dague et al. 2010). AFM is also the appropriate tool to visualize and quantify the physical, morphological and structural changes that can take place at the cell surface of yeast in response to various types of stresses. Two cases study illustrate this powerful biophysical approach. While there is an abundant literature on the heat shock response in yeast at the cellular and molecular levels (reviewed in (Verghese et al. 2012)), the physical consequences of this stress was only recently appreciated using the AFM technology. A temperature shift of a yeast culture from 30 to 42 °C uncovered the formation of circular rings that take their origin at a single location on the cell surface and reached 2–3 μm diameter in less than 1 h after the heat stress. This morphological process was found to be dependent on genes required for the budding process such as *BNI1* and *CHS3,* and was also under the control of the CWI signaling pathway. These genetic data led to the suggestion that the formation of these circular rings arise from a defective bud scar or bud emergence site during the temperature stress (Pillet et al. 2014). A second case study was to explore the response of yeast cells to ethanol shock. The tolerance of the yeast *S. cerevisiae* to its main fermentation product ethanol has been the concern of extensive researches for many years, with the aim to understand the mechanism of its toxicity (reviewed in (Stanley et al. 2010; Ma and Liu 2010)). In spite of a wealth of biochemical and molecular data on the response of yeast to ethanol stress, there is almost no physical information describing the biophysical effect that ethanol may exert at the single cell level. Using AFM, we found that the cell wall elasticity of yeast cells exposed to 9 % ethanol was dramatically reduced, in spite of the fact that the content and proportion of the cell wall polysaccharides in these ethanol-treated yeast cells remained unchanged (C. Elzstein, C. Formosa, M. Schiavone, E Dague and Jean François, submitted to publication). Conversely, cell elasticity was found to strongly increase in response to hydric perturbations caused by high salt, high glycerol concentration or air-drying which are known to induce a pronounced cell dehydration (Dupont et al. 2014).

Collectively, these observations suggest that perturbations of cell wall structure result in stretching of the plasma membrane which must have some consequences on transport processes. Also, the tight linkage of the plasma membrane with the cell wall via integral, most likely GPI- anchored-glycoproteins, may explain that change in cell wall elasticity can be caused by alteration of physical properties (e.g. swelling or shrinkage) of the plasma membrane (Schaber et al. 2010; Dupont et al. 2014). This suggestion was already proposed to explain the activation of the CWI by chlorpromazine, an amphipathic molecule that causes membrane stretch by asymmetric insertion into the plasma membrane (Kamada et al. 1995). This interaction between membrane and cell wall is also evidenced by the response of yeast to organic solvents which induce activation of the PDR (pleiotropic drug resistance) and the CWI pathways (Nishida et al. 2014). While these authors suggested that both pathways are independently activated by the organic solvents, an alternative proposition is that the CWI activation results from the membrane damage caused by these toxic compounds. This intimate connection between these two cellular structures also account for the involvement of the HOG pathway in the cell wall organization (Kapteyn et al. 2001; Del et al. 2008; Schaber et al. 2010). For further complementary information on the cell wall biogenesis, the molecular regulation of cell wall assembly, as well as on the advances in cell wall physical properties using AFM technology, the reader is referred to many excellent reviews that recently appeared on these subjects (Orlean 2012; Levin 2011; Free 2013; Muller and Dufrene 2011; Francois et al. 2013).

## 2.3 Cell Wall Porosity and Impact for Material Exchange

Porosity is an important property of the cell wall because it limits the secretion of high molecular mass compounds such as homologous and heterologous proteins (de Nobel and Barnett 1991). The limited permeability of the cell wall contributes to water retention at the inner region adjacent to the plasma membrane, leading to the 'periplasmic space' that is hardly seen in any electron microscopy picture (see Fig. 2.1). This retention of water is due to high degree of glycosylation of cell wall proteins together with the presence of negatively charged phosphates groups on their carbohydrates side chains. In addition, this periplasmic space is the receptacle of many secreted proteins, such as invertase (Carlson and Botstein 1982), alpha-galactosidase (or melibiase) (Sumner-Smith et al. 1985) and acid trehalase (Parrou et al. 2005), which cleave sucrose, melibiose and trehalose, respectively into hexose monomers that are taken up by sugar transporters (Boles and Hollenberg 1997). At variance to siderophores binding/uptake proteins encoded by *FIT1*, *FIT2* and *FIT3* genes (Protchenko et al. 2001) (see below) that are retained to the cell wall matrix through a GPI anchor, these periplasmic proteins do not show any specific linkage with cell wall polymers. Maintenance of the cell wall porosity is thus critical to retain these periplasmic enzymes.

The porosity of the yeast cell wall has been initially estimated by the penetration of inert polymer solutes varying in molecular size into isolated wall material or directly on intact yeast cells. Plotting the percentage of wall space penetrated these polymer solutes of known concentration and size gave a regression line with an inflexion a point corresponding to $M_w$ of about 4500 for isolated wall material and 760 for intact cell (Scherrer et al. 1974). This difference was accounted for by the partial disorganization of the outer layer of the isolated cell wall (Scherrer et al. 1977), suggesting an important role of the layer in cell wall porosity. This suggestion was confirmed later by Zlotnik *et al.* (Zlotnik et al. 1984) when they showed that the horseradish peroxidase (with Mr 40000) causes cell lysis after removal of cell wall mannoproteins by Z-protease. Similarly, cell lysis by protease free β- glucanase is largely facilitated in the presence of β-mercaptoethanol likely because of the removal of disulfide bound proteins at the outer layer of the cell wall. However, the small openings in the outer part of the cell wall obtained by Gerhardt and colleagues was not in agreement with either release of bigger molecules such as invertase ($M_r > 200{,}000$) or some proteases ($M_r > 50{,}000$) or penetration of cytochrome c (Mw 12,500), serum bovine albumin (Mw 67000). Later on, De Nobel and colleagues reconsidered this aspect of cell wall porosity and set up a new assay that was based on DEAE-Dextran-induced and polycations-induced leakage of UV-absorbing compounds from yeast (most likely purines nucleotides). As the polycations with a small hydrodynamic radius caused leakage independently of cell wall porosity whereas DEAE-dextran with a higher hydrodynamic radius caused only limited cell leakage due to limited passage through the cell wall, the ratio between the two measurements was taken as an estimation of the wall porosity. Using this approach, these authors estimated a permeation of the wall for molecules with a hydrodynamic radius of up to 5.8 nm corresponding to a globular protein with $M_r$ of 400,000 (de Nobel et al. 1990; de Nobel and Barnett 1991). In addition, it was shown that the cell wall permeability is a cell-cycle related event, showing higher level during the initial stages of bud growth and decreasing sharply at nuclear migration, coincidentally with optimal expression of β-glucanase-extractable mannoproteins (de Nobel et al. 1991). These β-glucanase extractable proteins are mostly glycosylphosphatidylinoisitol- (GPI) –modified proteins that are retained to the inner layer of β-1,3-glucan through a β-1,6-glucan chain, with the latter being in turn connected to GPI (Klis et al. 2006) (Fig. 2.2). In particular, *CWP1* and *CWP2* genes which belong to the TIR family and encode two major mannoproteins of the outer layer of the cell wall in maintaining cell wall stability (Orlean 2012) are considered to play an important function in the cell wall porosity since their deletion markedly increased the cell wall permeability of yeast to several compounds including also genotoxic agents (Zhang et al. 2008). Cell wall porosity was found to be strongly altered upon deletion of *KNR4* gene which codes for a intrinsically disordered protein (Durand et al. 2008) the role of which is to connect cell wall assembly to cell growth (Martin-Yken et al. 2002), as witnessed by the hyper permeability of the *knr4Δ* mutant to the chromophore 5-bromo-4-chloro-3-indolyl-beta-D-galactopyranoside (X-Gal blue) (Martin et al. 1999). This high permeability associated with the loss of *KNR4* could be due to the oversecretion of cell wall mannoproteins in the culture medium (Penacho et al. 2012).

## 2.4 Biosorbant and Selective Molecular Sieve Properties of the Yeast Cell Wall

In the environmental context of industrialized countries, accelerating pollution by toxic metals, metalloids, radionuclides and organometals has influenced research towards the biotechnological potential of utilizing microorganisms for metal removal and recovery from the biosphere (Blackwell et al. 1995). Conversely, the widespread in nature of metals and metalloids that can locally reach fairly high concentration led microorganism to develop cellular protection systems to survive in such polluted environments (Wysocki and Tamas 2010). Yeasts possess an acknowledged potential for accumulating a range of metal ions and protect themselves against toxicity caused by essential (e.g. zinc, iron, cobalt, manganese, magnesium) as well as non–essential metals for life (e.g. chromium, lead, uranium, cadmium, silver, mercury in their ionic form). The metal uptake process is essentially biphasic consisting of a metabolism-independent (or passive) and metabolism-dependent (or active) step. The first passive step is rapid, typically only a few minutes in duration and is independent to temperature, metabolic energy, and metabolic inhibitors. Biosorption is the coined term to describe this passive process of metal binding by either dead biomass, cells fragments or active /living cells. When performed with living cells, biosorption corresponds to the fast, reversible and metabolically-independent adsorption step that is usually followed by a much slower, metabolically-dependent –irreversible- bioaccumulation mechanism. Thus, the biosorption process with living biomass can be impacted also by the physiology of the cells and the physico-chemical conditions of the culture conditions (Fomina and Gadd 2014).

The cell wall constituents play a key role in metal ions sequestration. Such constituents possess numerous functional groups, including carboxylate, phosphate, hydroxide, amine, sulfate and sulfhydryl, which are brought about by the outer - mannanoproteins- layer of the wall (Brady et al. 1994). However, the adsorptive capacity of cell wall to these heavy metals is not solely determined by its charges composition but also by the structural organization of the entire protein-carbohydrates and the degree of dissociation of the negatively charged groups (Blackwell et al. 1995). While this is not the purpose here to present a comprehensive review on biosorption processes which can be found elsewhere (Pacheco et al. 2011; Volesky 2007; Fomina and Gadd 2014), one can pointed out that the cell wall is an essential barrier protecting yeast towards toxicity caused by heavy metals such as cadmium (Cd), mercury (Hg) and lead (Pb) that are non-essential for biological functions with a capacity of 0.1–0.5 mmol metals adsorbed/g biomass (Volesky and Holan 1995; Vieira and Volesky 2000). This remarkable biosorption property of yeast cell wall is receiving industrial interest notably in treatment of wastewaster as an opportunity to valorize the yeast biomass produced from industrial fermentations (Soares and Soares 2013). Also, the biosorption capacity of cell wall is not limited to metals ions but can be attractive as binder to other potential toxic compounds such as mycotoxins (Yiannikouris et al. 2013). More recently, the involvement of cell wall as a molecular sieve to protect yeast from toxicity by hop iso-α-acids (which are responsible of the bitter flavor of the beer) by reducing its intracellular accessibility was proposed, based on a strong upregulation of a set of genes encoding cell wall mannoproteins, and the finding that the mannosylation process is necessary to ensure yeast hop resistance (Hazelwood et al. 2010).

## 2.5 Inference of Cell Wall in Metal Ions Uptake and Ionome Homeostasis

Metal mobilization from soils, minerals or other substrates can be achieved by protonolysis, complexation by excreted metabolites and Fe(III)-binding siderophores or redox mobilization (Gadd 2010). Iron acquisition has been the focus of detailed study in microbial system because this metal ion is mostly present in the environment as very poorly soluble ferric oxyhydroxides ($Ks < 10^{-38}M$) while it is requested at μM concentration in the cell to support growth. The acquisition of iron is thus a critical, a tightly regulated and a multifaceted process which ensures adequate supply of iron while avoiding deleterious effects associated with iron hyperaccumulation (Kornitzer 2009). To solve the bioavailability problem of iron, microorganisms have elaborated three modes of iron solubilization and assimilation, namely (i) acidification of the environment; (ii) reduction of ferric iron to the more soluble ferrous form and (iii) secretion of soluble iron-chelating molecules. Many organisms and plants synthetize and secrete siderophores which are heterogeneous organic compounds of small molecular weight (<15,000) that bind ferric ion with strong affinity. While the yeast *S. cerevisiae* does not synthetize nor secrete siderophores, it can take up and utilize the iron bound to siderophores secreted by a variety of other species. For these iron-containing compounds to gain access to the plasma membrane uptake systems, the compounds must first penetrate the cell wall and cross the periplasmic space. Using a cDNA microarray approach, Philpott & collaborators identified three genes designated *FIT1*, *FIT2*, *FIT3* (for facilitor of iron transport) that were highly expressed under iron poor cultures as well as in strain bearing the *AFT1-1$^{up}$* allele which mimics iron deprivation conditions (Protchenko et al. 2001). These genes encode cell wall mannoproteins that are attached to the β-glucan layer through a GPI- anchor. The encoded Fit proteins are very rich in serine and threonine residues (40–50 % serine and threonine) and bear several short repeat of 6–7 amino acids sequence. These proteins likely function to trap the siderophores-bound iron in the cell wall and facilitate the uptake of these complexes via the Arn1-4 siderophores-transporter present in the plasma membrane, although direct evidence for this uptake facilitation remains to be established. Nevertheless, the deletion of *FIT* genes led to defects in the uptake of siderophore-iron complexes. In addition, the loss of function of these *FIT* genes resulted in the activation of Aft1-dependent iron regulon, indicating that the Fit proteins contribute to the storage of iron at the cell surface. In line with this suggestion was the finding that the amount of iron associated with the cell wall and periplasmic space is significantly reduced in a mutant defective in Fit proteins (Protchenko et al. 2001). The importance of this siderophore transport system in iron homeostasis is further supported by the upregulation of the *FIT* genes under high copper conditions (Cankorur-Cetinkaya et al. 2013) or in evolutionary engineered strains resistant to high levels of nickel and cobalt (Alkim et al. 2013; Kucukgoze et al. 2013). Because of the proximity of these metals to iron in the Mendeleev periodic table, it is suggested that nickel and cobalt can compete for uptake of iron leading to iron deficiency and therefore to the induction of iron transporter system caused by these metal ions overload. However, this hypothesis is not in accordance with the finding that the intracellular level of iron is apparently not altered in strains challenged with excess of copper, nickel or cobalt (Alkim et al. 2013; Kucukgoze et al. 2013). Another hypothesis was proposed based on the fact that these metal ions provoke an oxidative stress response characterized by a rise of reactive oxygen species (ROS). Given the role of Fe-dependent catalase in decomposing $H_2o_2$, one possibility is an enhanced requirement for iron in mounting a response to increased oxidative burden to match ROS disproportionation to $H_2o_2$ by the Sod1 enzyme (Nevitt et al. 2012). This hypothesis is still not completely satisfactory because neither *CTT1* nor *SOD1* are found to be upregulated in response to excess of these metal ions (Alkim et al. 2013; Cankorur-Cetinkaya et al. 2013).

Like *S. cerevisiae*, other yeasts including *S. pombe*, *C. glabrata* and *C. albicans* do not synthetize and secrete siderophores. However, they do possess well-defined siderophores uptake systems comprising not only the membrane localized Arn1-4 siderophore-iron transporters (Cyert and Philpott 2013) but also cell wall attachment proteins. Homologous of the S. *cerevisiae FIT1, 2, 3* exists in the yeast *S. pombe* but in contrast to the former yeast, these genes are not induced by iron deprivation in *S. pombe*. However, a genome wide transcriptomic study of this yeast under iron limitation revealed the strong induction of *Shu1* gene that encodes a predicted cell surface protein bearing a potential glycosylphosphatidylinositol (GPI) – attachment sequence (Rustici et al. 2007). The role of the encoded protein in iron homeostasis has not been investigated yet. A systematic analysis of *C glabrata* mutants' collection for iron homeostasis identified the cell wall structural protein CgCcw14 as implicated in the maintenance of intracellular iron content and in the adherence to epithelial cells. This protein is 60 % similar to *S. cerevisiae* cell wall glycoprotein Ccw14 (Mrsa et al. 1999) and displays the so-called CFEM domain sequence (common in fungal extracellular membranes). This domain is characterized by eight cysteine residues with conserved spacing and was shown to be required for attachment of the protein to the cell wall (de Groot et al. 2008). The finding that a deletion of *CgCCW14* results in higher intracellular iron content could argue that this protein may regulate iron uptake at the cell wall or periplasmic space. The commensal fungus *C. albicans* residing on the skin and on mucosal surfaces has also developed sophisticated strategy to scavenge iron from its surrounding. Thus, besides the high affinity reductive and non-reductive pathways commonly identified in fungal cells (Kornitzer 2009), it was recently shown that *C. albicans* is able to extract iron from hemoglobin, the largest iron pool in its host's body, using a relay network of extracellular-anchored proteins that facilitates heme – iron uptake by a still unknown mechanism. The system is composed of Rtb5 which is found at the cell periphery and Pga7 which is buried in the cell wall. This differential localization together with the fact that Pga7 has higher affinity to heme than Rtb5 favors a channeling that could explain how heme – binding proteins embedded in the cell wall can mediate the uptake of heme, which is afterwards internalize through the plasma membrane by endocytosis (Kuznets et al. 2014). It is important to notice that this relay network system is likely not present in *S. cerevisiae* because this yeast cannot utilize haemoglobin-bound iron for growth. Finally, a wall proteome study of *C. albicans* subjected to iron restriction uncovered strong increase in the levels of GPI-modified Als3 which serves as a ferritin receptor, and several CFEM-domain containing proteins including Csa1, Pga7, Pga10 and Rtb5. Moreover, it is suggested that the CFEM domains can bind directly ferrous and ferric iron, making them a promising antifungal target, possibly relevant to other pathogenic fungi.

## 2.6 Inference of Cell Wall with Membrane Fluidity and Transport Processes from Genome Scale Analyses

DNA microarray (Schena et al. 1995) is a powerful tool to tract differentially transcribing genes that are important for cell fitness, survival or response to a given stimulus. Complimentary to this approach, is the screening of the nearly complete set of yeast open reading frame haploid knockouts strains generated by the *Saccharomyces* Genome Deletion project (Winzeler et al. 1999), which allows to identify genes implicated in nutrient homeostasis. These genome scale analyses were scrutinized to highlight cell wall related genes that may impact on membrane fluidity and import pathways essential for nutritional requirement and growth. A striking interaction of cell wall with transport process recognized a long time ago is the sterol influx which is restricted to anaerobiosis, while under the aerobic condition, the uptake of sterol is prevented leading to the so-called aerobic sterol exclusion, the mechanism of which is not yet completely understood (Raychaudhuri and Prinz 2006). However, the potential role of the cell wall in the sterol influx came in light with the characterization of two hypoxic transcriptional regulators, namely *UPC2* and *SUT1*. It was shown that their de-regulation, as a result of *upc2-1* gain of function or as constitutive *SUT1* expression allowed uptake of sterol in aerobic condition (Wilcox et al. 2002; Alimardani et al. 2004). Genome-wide transcriptional analysis of these mutant strains brought to light two genes families the expression of which was dramatically increased (Table 2.1). The first family is composed of the *TIR1, 2, 3 & 4* and *DAN1* genes, which encodes O-mannosylated proteins that are linked to the inner layer of the cell wall through a glycosyl-phosphatidyl inositol (GPI) anchor. The other identified genes family encodes Pau proteins which are referred to seripauperins, based on their serine-poor composition. They are small proteins of about 120–125 amino acids in length, showing 85–100 % identify and share a domain of about 90 residues with the N-terminal region of the Tir/Dan proteins (Abramova et al. 2001). Contrary to the latter, the Pau proteins lack the long C-terminal serine/threonine stretch and the GPI anchoring signal, thus leading to the prediction that these proteins can be secreted in the medium, although experimental evidence is still lacking (Rachidi et al. 2000; Viswanathan et al. 1994). These two genes families are strongly expressed under anaerobiosis (Kwast et al. 2002; Lai et al. 2005), induced by low temperature or cold shock (Sahara et al. 2002; Murata et al. 2006), as well as during wine fermentation that likely coincided with total oxygen depletion from the culture (Rossignol et al. 2003) (see Table 2.1). Concomitantly with this upregulation, changes in expression of genes involved in sterols, phospholipids, sphingolipids and vesicles trafficking were also noticed (Kwast et al. 2002; Rossignol et al. 2003; Schade et al. 2004; Murata et al. 2006), suggesting a profound remodeling of the cell wall and of the plasma membrane. These changes are likely necessitated to accommodate with reorganization of carbon and energy metabolism associated with reduced or lack of oxygen for the synthesis of sterols, unsaturated fatty acids and other membrane components and enzymes. The sterol influx is a typical model by which a cell wall protein, namely Dan1, contributes with an ABC plasma membrane encoded by *AUS1* to drive sterol import in yeast cell. Yet, it is not proven that any of these actors correspond to a sterol transporter. In addition, the sterol uptake in a *STU1* overexpressing strain is 10 times higher than in a strain in which both *DAN1* and *AUS1* are overexpressed, suggesting other participants, either from cell wall or cell membrane or both, in the sterol import (Alimardani et al. 2004). Also, the abundance of Pau protein in cell wall and their actual function in the cell remain a mystery. They may be required for sterols import since the inactivation of ergosterol synthesis pathway strongly induces the expression of many *PAU* genes (Mnaimneh et al. 2004) and a computer-aided analysis of heterogeneous genome-wide data predicted functional annotation of *PAU* genes into the sterol metabolism. Moreover, the decrease of membrane fluidity by addition of dimethylsufoxide (DMSO) or high glycerol level blocks hypoxic induction of *PAU* genes (Hickman et al. 2011), which further supports the idea of an interplay between cell wall porosity and membrane fluidity. Overall, such remodeling activity can be expected to alter cell wall porosity to accommodate the transport and processing of essential nutrients that are required under these specific, low oxygen or oxygen deprivation conditions.

**Table 2.1** Listing of differentially expressed genes encoding cell wall related proteins in response to sterol uptake activation, mimicking also by hypoxic/anaerobiosis conditions

| ORF | Gene name | Sterol uptake(1) | *upc2-1* strain(2) | Glucose $N_2$vsAir(3) | Galactose $N_2$vsAir(4) | Wine fermentation(5) | Low temperature(6) | Gene function according to SGD(7) |
|---|---|---|---|---|---|---|---|---|
| *YER011W* | *TIR1* | nf | + | + | + | ++ | ++ | Cell wall mannoprotein of the Srp1p/Tip1p family |
| *YOR010C* | *TIR2* | ++++ | + | ++++ | + | ++ | ++ | Cell wall mannoprotein of the Srp1p/Tip1p family |
| *YIL011w* | *TIR3* | +++ | ++ | ++++ | +++ | +++ | nf | Cell wall mannoprotein of the Srp1p/Tip1p family |
| *YOR009W* | *TIR4* | ++++ | +++ | ++++ | +++ | +++ | nf | Cell wall mannoprotein of the Srp1p/Tip1p family |
| *YJR150C* | *DAN1* | ++++ | +++ | ++++ | +++ | +++ | nf | Cell wall mannoprotein, expressed under anaerobic conditions, repressed in aerobic growth |
| *YLR037c* | *DAN2/PAU23* | ++ | + | ++++ | +++ | nd | nf | Cell wall mannoprotein with similarity to TIR family, and member of the seripauperin multigene family |
| *YBR301w* | *DAN3/PAU24* | + | nf | +++ | + | nd | nf | Member of the seripauperin multigene family |
| *YJR151C* | *DAN4* | + | +++ | ++++ | ++ | nd | nf | Cell wall mannoprotein with similarity to TIR family |
| *YBR067c* | *TIP1* | ++ | ++ | + | nf | nd | ++ | major cell wall mannoprotein with possible lipase activity |
| *YJL223C* | *PAU1* | nf | ++ | ++++ | ++ | +++ | ++ | Member of the seripauperin multigene family |
| *YEL049W* | *PAU2* | ++ | + | +++ | ++ | ++++ | ++ | Member of the seripauperin multigene family |
| *YCR104W* | *PAU3* | nf | + | +++ | + | ++++ | ++ | Member of the seripauperin multigene family |
| *YLR461W* | *PAU4* | nf | + | +++ | ++ | ++++ | ++ | Member of the seripauperin multigene family |
| *YFL020C* | *PAU5* | nf | + | +++ | +++ | ++++ | +++ | Member of the seripauperin multigene family |
| *YNR076W* | *PAU6* | ++ | + | ++++ | ++ | nd | ++ | Member of the seripauperin multigene family |

(continued)

**Table 2.1** (continued)

| ORF | Gene name | Sterol uptake[1] | *upc2-1* strain[2] | Glucose $N_2$vsAir[3] | Galactose $N_2$vsAir[4] | Wine fermentation[5] | Low temperature[6] | Gene function according to SGD[7] |
|---|---|---|---|---|---|---|---|---|
| *YAR020C* | *PAU7* | nf | + | +++ | ++ | nd | ++ | Member of the seripauperin multigene family |
| *YAL068C* | *PAU8* | ++ | + | +++ | ++ | nd | ++ | Member of the seripauperin multigene family |
| *YDR542W* | *PAU9* | | + | ++++ | +++ | +++ | nf | Member of the seripauperin multigene family |
| *YGL261C* | *PAU11* | ++ | + | ++ | ++ | +++ | nf | Member of the seripauperin multigene family |
| *YGR294w* | *PAU12* | nf | + | ++ | ++ | +++ | nf | Member of the seripauperin multigene family |
| *YHL046C* | *PAU13* | nf | + | ++ | ++ | +++ | nd | Member of the seripauperin multigene family |
| *YIL176C* | *PAU14* | ++ | + | +++ | ++ | ++++ | ++ | Member of the seripauperin multigene family |
| *YIR041W* | *PAU15* | nf | + | +++ | ++ | +++ | nd | Member of the seripauperin multigene family |
| *YKL224C* | *PAU16* | + | + | +++ | nf | ++++ | nd | Member of the seripauperin multigene family |
| *YLL025W* | *PAU17* | + | nf | +++ | ++ | ++++ | nd | Member of the seripauperin multigene family |
| *YLL064C* | *PAU18* | + | + | +++ | ++ | ++++ | nd | Member of the seripauperin multigene family |
| *YMR325W* | *PAU19* | nf | + | ++++ | +++ | ++++ | nd | Member of the seripauperin multigene family |
| *YOL161C* | *PAU20* | nf | + | ++++ | ++ | nd | nd | Member of the seripauperin multigene family |
| *YOR394W* | *PAU21* | nf | nf | +++ | nf | +++ | nd | Member of the seripauperin multigene family |
| *YPL282C* | *PAU22* | nf | nf | +++ | nf | nd | nd | Member of the seripauperin multigene family |
| *YOL155c* | *HPF1* | nd | nd | ++ | ++ | nf | nd | mannoprotein with putative glucan alpha-1,4-glucosidase activity |
| *YIL169c* | *HPF1'* | nd | nd | ++ | ++ | nd | nd | protein similar to YOL155C |
| *YKL096W* | *CWP1* | – | nd | ++ | nf | nd | nd | Cell wall mannoprotein linked to a beta-1,3- and beta-1,6-glucan |
| *YKL096W-A* | *CWP2* | nd | nd | – | nf | nd | nd | Cell wall mannoprotein linked to a beta-1,3- and beta-1,6-glucan |
| *YHR143W* | *DSE2* | nd | nd | nf | nf | nd | nd | Daughter cell-specific secreted protein with similarity to glucanases |
| *YNR067C* | *ENG1* | ++ | nd | nf | + | nd | nd | Daughter cell-specific secreted protein with similarity to glucanases |

| | | | | | | | | |
|---|---|---|---|---|---|---|---|---|
| *YER150W* | *SPI1* | nd | nd | ++ | nf | ++ | nd | GPI-anchored cell wall protein involved in weak acid resistance |
| *YOR092W* | *ECM3* | nd | nd | ++ | nf | nd | nd | protein of unknown function; may play a role in cell wall biosynthesis, |
| *YLR436C* | *ECM30* | nd | nd | ++ | nf | nd | nd | protein of unknown function; may play a role in cell wall biosynthesis, |
| *YGR279C* | *SCW4* | nd | nd | ++ | + | nd | nd | cell wall protein with similarity to glucanases |
| *YMR305C* | *SCW10* | nd | nd | – | nf | nd | nd | cell wall protein with similarity to glucanases |
| *YGL028C* | *SCW11* | + | nd | + | nf | nd | nd | cell wall protein with similarity to glucanases |
| *YEL040W* | *CRH2* | + | nd | – | nf | nd | nd | Chitin transglycosylase that functions in the transfer of chitin to beta(1–6) and beta(1–3) glucans in the cell wall; |
| *YGR032W* | *FKS2* | nd | nd | ++ | ++ | + | nd | Catalytic subunit of 1,3-beta-glucan synthase |

++++ = upregulated >15 fold; +++ = upregulated 5–10 fold; ++ = upregulated 3–5 fold, + = upregulated 1.5–3 fold increase, − = downregulated >3 fold

Data are from (1) Alimardani et al. (2004), (2) Wilcox et al. (2002), (3) Kwast et al. (2002), (4) Lai et al. (2005), (5) Rossignol et al. (2003), (6) Murata et al. (2006), (7) http://www.yeastgenome.org/

## 2.7 Summing Up

The yeast cell wall is a complex extracellular structure that is not solely an armour protecting the cell against internal and external pressure and thus warrants the cell shape, but it is an essential and efficient barrier to several molecules including metals ions, toxins, proteins, etc. that can impede the proper function of the plasma membrane in *e.g.* transport and signalisation. It is thus not surprising that an intimate connection between these two structures exists, although the molecular details of this connection, namely the nature and number of the molecular linkages are not yet known. What is also intriguing is the role of cell wall in the sterol uptake and metabolism that is restricted to anaerobic condition. It should be worth to investigate the biophysical properties of cell wall from yeast culture under strict anaerobiosis. Also, the function of several cell wall proteins, notably the *DAN/TIR* and the *PAU* multigene family in membrane fluidity remains to be investigated.

**Acknowledgments** The author is grateful to his close collaborators and to several colleagues for stimulating discussion and for providing unpublished data. Research in my laboratory has been supported by grants from EU (6thFP, Fungcellwall project), European Science Foundation (ESF, Fuminomics project), Agence Nationale de la Recherche (ANR contract NT05-2_42127 & R-11-JSV5-001-01), Region Midi Pyrénées (grant n° 10051296) and industrial biotechnologies.

## References

Abramova N, Sertil O, Mehta S, Lowry CV (2001) Reciprocal regulation of anaerobic and aerobic cell wall mannoprotein gene expression in *Saccharomyces cerevisiae*. J Bacteriol 183:2881–2887

Aguilar-Uscanga B, Francois JM (2003) A study of the yeast cell wall composition and structure in response to growth conditions and mode of cultivation. Lett Appl Microbiol 37:268–274

Alimardani P, Regnacq M, Moreau-Vauzelle C, Ferreira T, Rossignol T, Blondin B, Berges T (2004) SUT1-promoted sterol uptake involves the ABC transporter Aus1 and the mannoprotein Dan1 whose synergistic action is sufficient for this process. Biochem J 381:195–202

Alkim C, Benbadis L, Yilmaz U, Cakar ZP, Francois JM (2013) Mechanisms other than activation of the iron regulon account for the hyper-resistance to cobalt of a *Saccharomyces cerevisiae* strain obtained by evolutionary engineering. Metallomics 5:1043–1060

Blackwell KJ, Singleton I, Tobin JM (1995) Metal cation uptake by yeast: a review. Appl Microbiol Biotechnol 43:579–584

Boles E, Hollenberg CP (1997) The molecular genetics of hexose transport in yeasts. FEMS Microbiol Rev 21:85–111

Brady D, Stoll AD, Starke L, Duncan JR (1994) Chemical and enzymatic extraction of heavy metal binding polymers from isolated cell walls of *Saccharomyces cerevisiae*. Biotechnol Bioeng 44:297–302

Cankorur-Cetinkaya A, Eraslan S, Kirdar B (2013) Transcriptional remodelling in response to changing copper levels in the Wilson and Menkes disease model of *Saccharomyces cerevisiae*. Mol Biosyst 9:2889–2908

Cappellaro C, Baldermann C, Rachel R, Tanner W (1994) Mating type-specific cell-cell recognition of *Saccharomyces cerevisiae*: cell wall attachment and active sites of a- and alpha-agglutinin. EMBO J 13:4737–4744

Cappellaro C, Mrsa V, Tanner W (1998) New potential cell wall glucanases of *Saccharomyces cerevisiae* and their involvement in mating. J Bacteriol 180:5030–5037

Carlson M, Botstein D (1982) Two differentially regulated mRNAs with different 5′ ends encode secreted with intracellular forms of yeast invertase. Cell 28:145–154

Cyert MS, Philpott CC (2013) Regulation of cation balance in *Saccharomyces cerevisiae*. Genetics 193:677–713

Dague E, Gilbert Y, Verbelen C, Andre G, Alsteens D, Dufrene YF (2007) Towards a nanoscale view of fungal surfaces. Yeast 24:229–237

Dague E, Bitar R, Ranchon H, Durand F, Yken HM, Francois JM (2010) An atomic force microscopy analysis of yeast mutants defective in cell wall architecture. Yeast 27:673–684

Dallies N, Francois J, Paquet V (1998) A new method for quantitative determination of polysaccharides in the yeast cell wall. Application to the cell wall defective mutants of Saccharomyces cerevisiae. Yeast 14:1297–1306

De Groot P, Ruiz C, Vazquez de Aldana CR, Duenas E, Cid VJ, Del Rey F, Rodriguez-Pena JM, Perez P, Andel A, Caubin J, Arroyo J, Garcia JC, Gil C, Molina M, Garcia LJ, Nombela C, Klis F (2001) A genomic approach for the identification and classification of genes involved in cell wall formation and its regulation in *Saccharomyces cerevisiae*. Comp Funct Genomics 2:124–142

de Groot PW, Kraneveld EA, Yin QY, Dekker HL, Gross U, Crielaard W, de Koster CG, Bader O, Klis FM, Weig M (2008) The cell wall of the human pathogen Candida glabrata: differential incorporation of novel adhesin-like wall proteins. Eukaryot Cell 7:1951–1964

de Nobel JG, Barnett JA (1991) Passage of molecules through yeast cell walls: a brief essay-review. Yeast 7:313–323

de Nobel JG, Klis FM, Munnik T, Priem J, Van den Ende H (1990) An assay of relative cell wall porosity in *Saccharomyces cerevisiae*, Kluyveromyces lactis and Schizosaccharomyces pombe. Yeast 6:483–490

de Nobel JG, Klis FM, Ram A, Van UH, Priem J, Munnik T, Van den Ende H (1991) Cyclic variations in the permeability of the cell wall of *Saccharomyces cerevisiae*. Yeast 7:589–598

Del VV, Casagrande V, Bianchi MM, Piccinni E, Frontali L, Militti C, Fardeau V, Devaux F, Di SC, Presutti C, Negri R (2008) Role of Hog1 and Yaf9 in the transcriptional response of *Saccharomyces cerevisiae* to cesium chloride. Physiol Genomics 33:110–120

Dufrene YF (2001) Application of atomic force microscopy to microbial surfaces: from reconstituted cell surface layers to living cells. Micron 32:153–165

Dupont S, Rapoport A, Gervais P, Beney L (2014) Survival kit of *Saccharomyces cerevisiae* for anhydrobiosis. Appl Microbiol Biotechnol 98:8821–8834

Dupres V, Dufrene YF, Heinisch JJ (2010) Measuring cell wall thickness in living yeast cells using single molecular rulers. ACS Nano 4:5498–5504

Durand F, Dagkessamanskaia A, Martin-Yken H, Graille M, van Tilbeurgh H, Uversky VN, Francois JM (2008) Structure-function analysis of Knr4/Smi1, a newly member of intrinsically disordered proteins family, indispensable in the absence of a functional PKC1-SLT2 pathway in *Saccharomyces cerevisiae*. Yeast 25:563–576

Ecker M, Deutzmann R, Lehle L, Mrsa V, Tanner W (2006) Pir proteins of *Saccharomyces cerevisiae* are attached to beta-1,3-glucan by a new protein-carbohydrate linkage. J Biol Chem 281:11523–11529

Eddy AA, Barnett JA (2007) A history of research on yeasts 11. The study of solute transport: the first 90 years, simple and facilitated diffusion(1). Yeast 24:1023–1059

Fleet GH (1991) Cell Walls. In: Rose AH, Harrison JS (eds) The yeast, 2nd edn. Academic, London, pp 199–277

Fomina M, Gadd GM (2014) Biosorption: current perspectives on concept, definition and application. Bioresour Technol 160:3–14

Francois JM, Formosa C, Schiavone M, Pillet F, Martin-Yken H, Dague E (2013) Use of atomic force microscopy (AFM) to explore cell wall properties and response to stress in the yeast *Saccharomyces cerevisiae*. Curr Genet 59:187–196

Free SJ (2013) Fungal cell wall organization and biosynthesis. Adv Genet 81:33–82

Gadd GM (2010) Metals, minerals and microbes: geomicrobiology and bioremediation. Microbiology 156:609–643

Hagen I, Ecker M, Lagorce A, Francois JM, Sestak S, Rachel R, Grossmann G, Hauser NC, Hoheisel JD, Tanner W, Strahl S (2004) Sed1p and Srl1p are required to compensate for cell wall instability in *Saccharomyces cerevisiae* mutants defective in multiple GPI-anchored mannoproteins. Mol Microbiol 52:1413–1425

Hazelwood LA, Walsh MC, Pronk JT, Daran JM (2010) Involvement of vacuolar sequestration and active transport in tolerance of *Saccharomyces cerevisiae* to hop iso-alpha-acids. Appl Environ Microbiol 76:318–328

Hickman MJ, Spatt D, Winston F (2011) The Hog1 mitogen-activated protein kinase mediates a hypoxic response in *Saccharomyces cerevisiae*. Genetics 188:325–338

Kamada Y, Jung US, Piotrowski J, Levin DE (1995) The protein kinase C-activated MAP kinase pathway of *Saccharomyces cerevisiae* mediates a novel aspect of the heat shock response. Genes Dev 9:1559–1571

Kapteyn JC, Van den Ende H, Klis FM (1999) The contribution of cell wall proteins to the organization of the yeast cell wall. Biochem Biophys Acta 1426:373–383

Kapteyn JC, ter Riet B, Vink E, Blad S, de Nobel H, van den EH, Klis FM (2001) Low external pH induces HOG1-dependent changes in the organization of the *Saccharomyces cerevisiae* cell wall. Mol Microbiol 39:469–479

Klis FM, Boorsma A, de Groot PW (2006) Cell wall construction in *Saccharomyces cerevisiae*. Yeast 23:185–202

Kornitzer D (2009) Fungal mechanisms for host iron acquisition. Curr Opin Microbiol 12: 377–383

Kucukgoze G, Alkim C, Yilmaz U, Kisakesen HI, Gunduz S, Akman S, Cakar ZP (2013) Evolutionary engineering and transcriptomic analysis of nickel-resistant *Saccharomyces cerevisiae*. FEMS Yeast Res 13:731–746

Kuznets G, Vigonsky E, Weissman Z, Lalli D, Gildor T, Kauffman SJ, Turano P, Becker J, Lewinson O, Kornitzer D (2014) A relay network of extracellular heme-binding proteins drives C. albicans iron acquisition from hemoglobin. PLoS Pathog 10:e1004407

Kwast KE, Lai LC, Menda N, James DT III, Aref S, Burke PV (2002) Genomic analyses of anaerobically induced genes in *Saccharomyces cerevisiae*: functional roles of Rox1 and other factors in mediating the anoxic response. J Bacteriol 184:250–265

Lai LC, Kosorukoff AL, Burke PV, Kwast KE (2005) Dynamical remodeling of the transcriptome during short-term anaerobiosis in *Saccharomyces cerevisiae*: differential response and role of Msn2 and/or Msn4 and other factors in galactose and glucose media. Mol Cell Biol 25:4075–4091

Lesage G, Bussey H (2006) Cell wall assembly in *Saccharomyces cerevisiae*. Microbiol Mol Biol Rev 70:317–343

Levin DE (2011) Regulation of cell wall biogenesis in *Saccharomyces cerevisiae*: the cell wall integrity signaling pathway. Genetics 189:1145–1175

Ma M, Liu ZL (2010) Mechanisms of ethanol tolerance in *Saccharomyces cerevisiae*. Appl Microbiol Biotechnol 87:829–845

Magnelli P, Cipollo JF, Abeijon C (2002) A refined method for the determination of *Saccharomyces cerevisiae* cell wall composition and beta-1,6-glucan fine structure. Anal Biochem 301:136–150

Martin H, Dagkessamanskaia A, Satchanska G, Dallies N, Francois J (1999) *KNR4*, a suppressor of *Saccharomyces cerevisiae* cwh mutants, is involved in the transcriptional control of chitin synthase genes. Microbiology 145:249–258

Martin-Yken H, Dagkessamanskaia A, Talibi D, Francois J (2002) KNR4 is a member of the PKC1 signalling pathway and genetically interacts with BCK2, a gene involved in cell cycle progression in *Saccharomyces cerevisiae*. Curr Genet 41:323–332

Mnaimneh S, Davierwala AP, Haynes J, Moffat J, Peng WT, Zhang W, Yang X, Pootoolal J, Chua G, Lopez A, Trochesset M, Morse D, Krogan NJ, Hiley SL, Li Z, Morris Q, Grigull

J, Mitsakakis N, Roberts CJ, Greenblatt JF, Boone C, Kaiser CA, Andrews BJ, Hughes TR (2004) Exploration of essential gene functions via titratable promoter alleles. Cell 118:31–44
Mrsa V, Seidl T, Gentzsch M, Tanner W (1997) Specific labelling of cell wall proteins by biotynilation. Identification of four covalently linked 0-mannosylated proteins of *Saccharomyces cerevisiae*. Yeast 13:1145–1154
Mrsa V, Ecker M, Strahl-Bolsinger S, Nimtz M, Lehle L, Tanner W (1999) Deletion of new covalently linked cell wall glycoproteins alters the electrophoretic mobility of phosphorylated wall components of *Saccharomyces cerevisiae*. J Bacteriol 181:3076–3086
Muller DJ, Dufrene YF (2011) Atomic force microscopy: a nanoscopic window on the cell surface. Trends Cell Biol 21:461–469
Murata Y, Homma T, Kitagawa E, Momose Y, Sato MS, Odani M, Shimizu H, Hasegawa-Mizusawa M, Matsumoto R, Mizukami S, Fujita K, Parveen M, Komatsu Y, Iwahashi H (2006) Genome-wide expression analysis of yeast response during exposure to 4 degrees C. Extremophiles 10:117–128
Negishi T, Ohya Y (2010) The cell wall integrity checkpoint: coordination between cell wall synthesis and the cell cycle. Yeast 27:513–519
Nevitt T, Ohrvik H, Thiele DJ (2012) Charting the travels of copper in eukaryotes from yeast to mammals. Biochim Biophys Acta 1823:1580–1593
Nguyen TH, Fleet GH, Rogers PL (1998) Composition of the cell walls of several yeast species. Appl Microbiol Biotechnol 50:206–212
Nishida N, Jing D, Kuroda K, Ueda M (2014) Activation of signaling pathways related to cell wall integrity and multidrug resistance by organic solvent in *Saccharomyces cerevisiae*. Curr Genet 60:149–162
Orlean P (2012) Architecture and biosynthesis of the *Saccharomyces cerevisiae* cell wall. Genetics 192:775–818
Osumi M (1998) The ultrastructure of yeast: cell wall structure and formation. Micron 29:207–233
Pacheco PH, Gil RA, Cerutti SE, Smichowski P, Martinez LD (2011) Biosorption: a new rise for elemental solid phase extraction methods. Talanta 85:2290–2300
Parrou JL, Jules M, Beltran G, Francois J (2005) Acid trehalase in yeasts and filamentous fungi: Localization, regulation and physiological function. FEMS Yeast Res 5:503–511
Penacho V, Blondin B, Valero E, Gonzalez R (2012) Flocculation and transcriptional adaptation to fermentation conditions in a recombinant wine yeast strain defective for KNR4/SMI1. Biotechnol Prog 28:327–336
Pillet F, Lemonier S, Schiavone M, Formosa C, Martin-Yken H, Francois JM, Dague E (2014) Uncovering by atomic force microscopy of an original circular structure at the yeast cell surface in response to heat shock. BMC Biol 12:6
Protchenko O, Ferea T, Rashford J, Tiedeman J, Brown PO, Botstein D, Philpott CC (2001) Three cell wall mannoproteins facilitate the uptake of iron in *Saccharomyces cerevisiae*. J Biol Chem 276:49244–49250
Rachidi N, Martinez MJ, Barre P, Blondin B (2000) *Saccharomyces cerevisiae* PAU genes are induced by anaerobiosis. Mol Microbiol 35:1421–1430
Raychaudhuri S, Prinz WA (2006) Uptake and trafficking of exogenous sterols in *Saccharomyces cerevisiae*. Biochem Soc Trans 34:359–362
Rossignol T, Dulau L, Julien A, Blondin B (2003) Genome-wide monitoring of wine yeast gene expression during alcoholic fermentation. Yeast 20:1369–1385
Rustici G, van Bakel H, Lackner DH, Holstege FC, Wijmenga C, Bahler J, Brazma A (2007) Global transcriptional responses of fission and budding yeast to changes in copper and iron levels: a comparative study. Genome Biol 8:R73
Sahara T, Goda T, Ohgiya S (2002) Comprehensive expression analysis of time-dependent genetic responses in yeast cells to low temperature. J Biol Chem 277:50015–50021
Schaber J, Adrover MA, Eriksson E, Pelet S, Petelenz-Kurdziel E, Klein D, Posas F, Goksor M, Peter M, Hohmann S, Klipp E (2010) Biophysical properties of *Saccharomyces cerevisiae* and their relationship with HOG pathway activation. Eur Biophys J 39:1547–1556

Schade B, Jansen G, Whiteway M, Entian KD, Thomas DY (2004) Cold adaptation in budding yeast. Mol Biol Cell 15:5492–5502
Schena M, Shalon D, Davis RW, Brown PO (1995) Quantitative monitoring of gene expression patterns with a complementary DNA microarray. Science 270:467–470
Scherrer R, Louden L, Gerhardt P (1974) Porosity of the yeast cell wall and membrane. J Bacteriol 118:534–540
Scherrer R, Berlin E, Gerhardt (1977) Density, porosity, and structure of dried cell walls isolated from Bacillus megaterium and *Saccharomyces cerevisiae*. J Bacteriol 129:1162–1164
Schiavone M, Vax A, Formosa C, Martin-Yken H, Dague E, Francois JM (2014) A combined chemical and enzymatic method to determine quantitatively the polysaccharide components in the cell wall of yeasts. FEMS Yeast Res 14:933–947
Smits GJ, Schenkman LR, Brul S, Pringle JR, Klis FM (2006) Role of cell cycle-regulated expression in the localized incorporation of cell wall proteins in yeast. Mol Biol Cell 17:3267–3280
Soares EV, Soares HM (2013) Cleanup of industrial effluents containing heavy metals: a new opportunity of valorising the biomass produced by brewing industry. Appl Microbiol Biotechnol 97:6667–6675
Spellman PT, Sherlock G, Zhang MQ, Iyer VR, Anders K, Eisen MB, Brown PO, Botstein D, Futcher B (1998) Comprehensive identification of cell cycle-regulated genes of the yeast *Saccharomyces cerevisiae* by microarray hybridization. Mol Biol Cell 9:3273–3297
Stanley D, Bandara A, Fraser S, Chambers PJ, Stanley GA (2010) The ethanol stress response and ethanol tolerance of *Saccharomyces cerevisiae*. J Appl Microbiol 109:13–24
Stenson JD, Hartley P, Wang C, Thomas CR (2011) Determining the mechanical properties of yeast cell walls. Biotechnol Prog 27:505–512
Sumner-Smith M, Bozzato RP, Skipper N, Davies RW, Hopper JE (1985) Analysis of the inducible MEL1 gene of Saccharomyces carlsbergensis and its secreted product, alpha-galactosidase (melibiase). Gene 36:333–340
Suzuki M, Igarashi R, Sekiya M, Utsugi T, Morishita S, Yukawa M, Ohya Y (2004) Dynactin is involved in a checkpoint to monitor cell wall synthesis in *Saccharomyces cerevisiae*. Nat Cell Biol 6:861–871
Verghese J, Abrams J, Wang Y, Morano KA (2012) Biology of the heat shock response and protein chaperones: budding yeast (*Saccharomyces cerevisiae*) as a model system. Microbiol Mol Biol Rev 76:115–158
Vieira RH, Volesky B (2000) Biosorption: a solution to pollution? Int Microbiol 3:17–24
Viswanathan M, Muthukumar G, Cong YS, Lenard J (1994) Seripauperins of *Saccharomyces cerevisiae*: a new multigene family encoding serine-poor relatives of serine-rich proteins. Gene 148:149–153
Volesky B (2007) Biosorption and me. Water Res 41:4017–4029
Volesky B, Holan ZR (1995) Biosorption of heavy metals. Biotechnol Prog 11:235–250
Wilcox LJ, Balderes DA, Wharton B, Tinkelenberg AH, Rao G, Sturley SL (2002) Transcriptional profiling identifies two members of the ATP-binding cassette transporter superfamily required for sterol uptake in yeast. J Biol Chem 277:32466–32472
Winzeler EA, Shoemaker DD, Astromoff A, Liang H, Anderson K, Andre B, Bangham R, Benito R, Boeke JD, Bussey H, Chu AM, Connelly C, Davis K, Dietrich F, Dow SW, El Bakkoury M, Foury F, Friend SH, Gentalen E, Giaever G, Hegemann JH, Jones T, Laub M, Liao H, Davis RW (1999) Functional characterization of the *S. cerevisiae* genome by gene deletion and parallel analysis. Science 285:901–906
Wysocki R, Tamas MJ (2010) How *Saccharomyces cerevisiae* copes with toxic metals and metalloids. FEMS Microbiol Rev 34:925–951
Yamaguchi M, Namiki Y, Okada H, Mori Y, Furukawa H, Wang J, Ohkusu M, Kawamoto S (2011) Structome of *Saccharomyces cerevisiae* determined by freeze-substitution and serial ultrathin-sectioning electron microscopy. J Electron Microsc (Tokyo) 60:321–335

Yiannikouris A, Kettunen H, Apajalahti J, Pennala E, Moran CA (2013) Comparison of the sequestering properties of yeast cell wall extract and hydrated sodium calcium aluminosilicate in three in vitro models accounting for the animal physiological bioavailability of zearalenone. Food Addit Contam Part A Chem Anal Control Expo Risk Assess 30:1641–1650
Zhang M, Liang Y, Zhang X, Xu Y, Dai H, Xiao W (2008) Deletion of yeast CWP genes enhances cell permeability to genotoxic agents. Toxicol Sci 103:68–76
Zlotnik H, Fernandez MP, Bowers B, Cabib E (1984) *Saccharomyces cerevisiae* mannoproteins form an external cell wall layer that determines wall porosity. J Bacteriol 159:1018–1026

# Chapter 3
# Proton Transport and pH Control in Fungi

**Patricia M. Kane**

**Abstract** Despite diverse and changing extracellular environments, fungi maintain a relatively constant cytosolic pH and numerous organelles of distinct lumenal pH. Key players in fungal pH control are V-ATPases and the P-type proton pump Pma1. These two proton pumps act in concert with a large array of other transporters and are highly regulated. The activities of Pma1 and the V-ATPase are coordinated under some conditions, suggesting that pH in the cytosol and organelles is not controlled independently. Genomic studies, particularly in the highly tractable *S. cerevisiae*, are beginning to provide a systems-level view of pH control, including transcriptional responses to acid or alkaline ambient pH and definition of the full set of regulators required to maintain pH homeostasis. Genetically encoded pH sensors have provided new insights into localized mechanisms of pH control, as well as highlighting the dynamic nature of pH responses to the extracellular environment. Recent studies indicate that cellular pH plays a genuine signaling role that connects nutrient availability and growth rate through a number of mechanisms. Many of the pH control mechanisms found in *S. cerevisiae* are shared with other fungi, with adaptations for their individual physiological contexts. Fungi deploy certain proton transport and pH control mechanisms not shared with other eukaryotes; these regulators of cellular pH are potential antifungal targets. This review describes current and emerging knowledge proton transport and pH control mechanisms in *S. cerevisiae* and briefly discusses how these mechanisms vary among fungi.

**Keywords** *S. cerevisiae* • Proton pump • Cytosolic pH • Organelle acidification • pH sensing and growth • pH signaling • V-ATPase • Pma1

P.M. Kane (✉)
Department of Biochemistry and Molecular Biology, SUNY Upstate Medical University, Syracuse, NY 13219, USA
e-mail: kanepm@upstate.edu



J. Ramos et al. (eds.), *Yeast Membrane Transport*, Advances in Experimental Medicine and Biology 892, DOI 10.1007/978-3-319-25304-6_3

## 3.1 Introduction

Tight control of cytosolic and organelle pH is critical for viability in all eukaryotic cells, including fungi. Protein conformation is exquisitely sensitive to pH. Most enzymes have a pH optimum for activity, and this property is exploited to provide pH-dependent, organelle-specific activity regulation. The pH gradient across the inner mitochondrial membrane drives ATP synthesis, and pH gradients across membranes can drive nutrient import coupled to transport of ions and other solutes. In recent years, the importance of pH as an intracellular signal has begun to come into focus. Given the diverse and critical functions directly linked to pH control, it is not surprising that genomic studies indicate that cytosolic pH can quantitatively control growth rate in *S. cerevisiae*.

However, many fungi encounter particularly challenging environments for pH control. They tolerate wide ranges of extracellular pH and must adapt to wide-ranging concentrations of other ions that facilitate or challenge pH control. Strains such as *S. cerevisiae* that frequently utilize aerobic glycolysis rapidly acidify their surroundings and generate copious amounts of organic acids. As a result, fungi have robust mechanisms for pH control and $H^+$-transport, incorporating both mechanisms common to all eukaryotes and specialized factors that facilitate adaptation to more extreme conditions. Interestingly, pH control in yeast is of considerable practical interest as well, as weak acids such as sorbate are widely used as preservatives to inhibit fungal growth. Thus, pH control in fungi can be viewed both as remarkably adaptable and as an Achilles heel.

This review outlines current knowledge of fungal proton transport and pH control, focusing initially on *S. cerevisiae*, where a wide array of genomic, biochemical and cell biological tools are available to address the proton transporters and underlying mechanisms of pH control. pH regulation in other filamentous fungi, and particularly pathogens, will then be compared to *S. cerevisiae*, and the possibilities for designing antifungal agents that target pH control mechanisms will then be addressed.

## 3.2 The Physiological Context of pH Homeostasis in *S. cerevisiae*

In *S. cerevisiae*, as in other organisms, pH control mechanisms are adapted to the cells' physiological context. Under conditions of abundant glucose, their preferred carbon source, *S. cerevisiae* cells will undergo rapid fermentative growth, producing ethanol, $CO_2$ and organic acids through glycolysis (reviewed in Johnston 1999; Conrad et al. 2014). Cells grown in glucose rapidly acidify their medium and require robust mechanisms to maintain cytosolic pH during growth, and cytosolic pH decreases as cells reach stationary phase (reviewed in Orij et al. 2011). Although *S. cerevisiae* is quite tolerant of ethanol, ethanol production ultimately limits growth, and this limitation may reflect a combination of plasma membrane permeabilization at high alcohol concentrations, which compromises nutrient uptake, and a resulting inability to control cytosolic pH. Interestingly, recent experiments have indicated that ethanol tolerance can be substantially increased by preventing extracellular acidification during fermentation and including excess $K^+$ in the medium (Lam et al. 2014). These modifications promote activity of the plasma membrane proton pump, and highlight the central importance of maintaining pH gradients and plasma membrane potential for cell viability and growth. It should be noted that under glucose-rich conditions, there is very little oxidative phosphorylation in *S. cerevisiae*, and transcription of proteins in the respiratory chain and the ATP synthase is strongly repressed (Johnston and Carlson 1992). Consistent with a limited role for oxidative phosphorylation as a source of ATP under fermentative conditions, respiratory inhibitors like antimycin A have little effect on cytosolic pH in cells grown in glucose (Orij et al. 2009). *S. cerevisiae* can also grow on non-fermentable carbon sources such as glycerol and ethanol, and in fact, will shift to metabolism of ethanol as a carbon source during prolonged growth when glucose is exhausted (Johnston and Carlson 1992). During growth on non-fermentable carbon sources, synthesis of the enzymes required for oxidative phosphorylation is derepressed (Johnston and Carlson 1992), and overall growth is generally slower.

Superimposed on the requirement for cytosolic pH control is a requirement for precise control of organellar pH (Kane 2006). All cells have a number of organelles, including vacuoles/lysosomes, endosomes, and the Golgi apparatus that maintain an acidic lumenal pH relative to the cytosol (reviewed in Mellman et al. 1986; Casey et al. 2010). The internal pH of these organelles is tuned to their functions: for example, vacuolar proteases have optimal activity at acidic pH and the affinity of various receptor-ligand complexes is tuned to compartment pH. In contrast, mitochondria are alkaline relative to the cytosol, consistent with the requirements for a membrane potential across the mitochondrial inner membrane and for a pH gradient able to drive ATP synthesis during oxidative phosphorylation (Orij et al. 2011). Under conditions where cytosolic pH control is challenged, the impact on organelle pH must also be considered. An overview of the cellular pH gradients in cells at log phase in glucose is depicted in Fig. 3.1.

In virtually all cells, pH control is achieved primarily by the balanced activity of three types of molecules: pumps, exchangers, and buffers, coupled to multiple layers of regulation (Casey et al. 2010). Proton pumps, including Pma1 at the plasma membrane and V-ATPases on organelle membranes in fungi, couple hydrolysis of ATP to proton transport and thus are key players in establishing pH gradients (Orij et al. 2011). Exchangers can exploit the energy stored in pH gradients to transport ions and solutes against a gradient or to assist in pH control using the gradient of another ion, and may represent the source of the "proton leak" that helps to determine final pH in various organelles (Kondapalli et al. 2014). Buffers, particularly the robust phosphate buffering system in *S. cerevisiae* (Poznanski et al. 2013), protect cells and organelles from short-term pH transients, but cannot withstand long-term shifts without assistance from proton transporters (Casey et al. 2010).

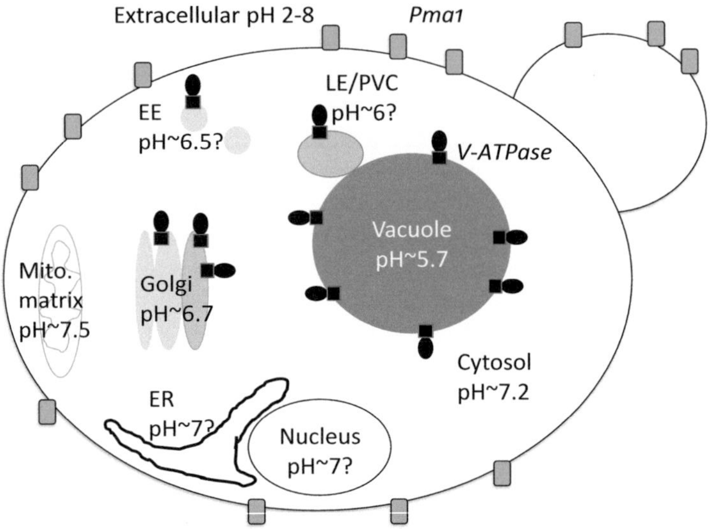


**Fig. 3.1 Compartment pH and pH gradients in glucose-grown *S. cerevisiae*.** Approximate pH values for various yeast organelles are shown. Organelles are abbreviated as follows: *EE* early endosome, *LE/PVC* late endosome/prevacuolar compartment, *ER* endoplasmic reticulum, *Mito.* mitochondrion. Approximate pH values that have been measured in log-phase cells growing in glucose (Orij et al. 2009; Martinez-Munoz and Kane 2008; Tarsio et al. 2011; Brett et al. 2005a; Braun et al. 2010) are shown; for compartments where organelle-specific pH measurements have not yet been made, expected pH values are indicated with a question mark. The V-ATPase is shown in black and Pma1 as grey rectangles. It should be noted that although cytosolic pH is quite constant, organelle pH is quite sensitive to growth conditions and thus will vary with buffer conditions and medium content as described in the text

## 3.3 The Plasma Membrane $H^+$-pump Pma1 and Organellar V-ATPases: Central Players in Cellular pH Control

### *3.3.1 Pma1 Structure, Function, and Genetics*

Pma1 is a single-subunit P-type $H^+$-ATPase belonging to the same family as the ubiquitous $Na^+/K^+$-ATPase of mammalian cells (Serrano et al. 1986). It is the most abundant protein of the *S. cerevisiae* plasma membrane and the major determinant of plasma membrane potential, as a result of its electrogenic transport of $H^+$ without counterions (Serrano 1991). It is believed to be the primary determinant of cytosolic pH, and is a major consumer of cellular ATP (Serrano et al. 1986). Pma1 has ten transmembrane domains, cytosolic N- and C-termini, and a large intracellular loop between the 4th and 5th transmembrane helices (Kuhlbrandt et al. 2002). Aspartate 378 of *S. cerevisiae PMA1* resides in the large intracellular loop and forms the characteristic β-aspartyl-phosphate intermediate during each catalytic cycle (Ambesi et al. 2000; Pedersen et al. 2007). Pma1 homologues are found in all fungi, as well as in plants. Although there is no high-resolution structure of any fungal Pma1, the *Neurospora* Pma1 was modeled by incorporating insights from the related SERCA $Ca^{2+}$-ATPase structure into an 8 A map of the *Neurospora* proton pump from electron microscopy (Kuhlbrandt et al. 2002). Subsequently, an X-ray crystal structure of the related *Arabidopsis* Pma1 in complex with the non-hydrolyzable ATP analog AMP-PNP was solved in 2007 (Pedersen et al. 2007). Consistent with the predictions of the model, the plant Pma1 structure revealed that, like previously characterized P-type pumps (Toyoshima et al. 2000; Olesen et al. 2007; Morth et al. 2007), Pma1 folds into the three cytosolic domains, designated P for phosphorylation, N for nucleotide binding, and A for actuator, despite rather low sequence identity with the previously crystallized pumps (Pedersen et al. 2007). In other P-type ATPases, ATP binding and hydrolysis drives large conformational changes between the three cytosolic domains that are communicated to the transmembrane helices to drive directional ion transport (Moller et al. 2005). It is very likely similar mechanisms are present in Pma1. The plant Pma1 structure also provided insights into the possible path of proton transport, involving conserved amino acids in the transmembrane helices (Pedersen et al. 2007; Morth et al. 2011). Although many P-type ATPases exhibit counterion transport, there is no evidence of any other ion being transported by Pma1. A model for the *S. cerevisiae* Pma1 was generated using the Phyre2 server (Kelley and Sternberg 2009). The highest confidence model was based on the *Neurospora* model (Kuhlbrandt et al. 2002) (*N. crassa* Pma1 has a higher degree of sequence identity with the *S. cerevisiae* pump that the *A. thaliana* enzyme) and is shown in Fig. 3.2a.

Consistent with its central role in pH control and bioenergetics at the plasma membrane, *PMA1* is an essential gene in *S. cerevisiae* (Serrano et al. 1986). Mutations that partially compromise Pma1 function generally result in reduced growth rate, particularly at acidic extracellular pH, as well as resistance to hygromycin (McCusker et al. 1987). Hygromycin resistance has been attributed to depolarization of the plasma membrane as a result of loss of electrogenic proton transport through Pma1 and has been used to isolate Pma1 mutants (Perlin et al. 1988, 1989). Certain *pma1* mutations also result in a distinctive multi-budded phenotype (McCusker et al. 1987). The central role of Pma1 in fungal bioenergetics and its absence from higher eukaryotes also suggested that it might be a target for antifungal development (Seto-Young et al. 1997).

*S. cerevisiae* has a paralog of *PMA1* that likely arose during the whole genome duplication. The open reading frame of *PMA2* is highly conserved, but *PMA2* is expressed at much lower levels than *PMA1* and has different catalytic properties (Supply et al. 1993a). It can partially compensate for the lethality of a *PMA1* deletion if overexpressed from the *PMA1* promoter (Supply et al. 1993b), but even under these conditions, only low concentrations reach the plasma membrane, and phenotypes such as sensitivity to low pH are observed. Unlike *PMA1*, deletion of *PMA2* gives few phenotypes, and thus its overall physiological function is not clear. The *pma2*Δ mutant was identified as having greatly decreased filamentation in a genomic screen for genes required for filamentous growth, but the mechanistic implications of this have not been examined (Jin et al. 2008).

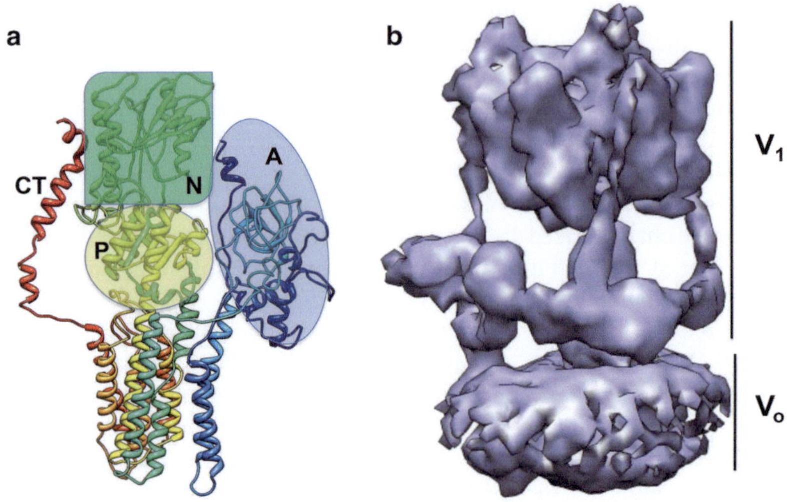


**Fig. 3.2 Structural views of the yeast Pma1 and V-ATPase proton pumps.** (**a**) The sequence of *S. cerevisiae PMA1* was submitted to the PHYRE2 server (Bennett-Lovsey et al. 2008) for modeling. The highest confidence model is shown and was based on the *Neurospora crassa* plasma membrane $H^+$-ATPase (PDB 1MHS). The model is visualized using UCSF Chimera (Pettersen et al. 2004), and the approximate locations of the cytosolic actuator (A), nucleotide-binding (N), and phosphorylation (P) domains are indicated based on comparison to the *A. thaliana* plasma membrane $H^+$-ATPase structure (Pedersen et al. 2007). The regulatory C-terminal tail is also indicated. (**b**) The 11A structure of the *S. cerevisiae* V-ATPase (Benlekbir et al. 2012) (EMDB 5476) was visualized using UCSF Chimera. The cytosolic $V_1$ and membrane $V_o$ domains are indicated

## *3.3.2 V-ATPases: Structure, Function, and Genetics*

V-ATPases are multisubunit proton pumps that acidify organelles such the vacuole/lysosomes, endosomes, and Golgi apparatus of all eukaryotic cells (Kane 2006; Forgac 2007). These pumps are very highly conserved among eukaryotes, and consist of a complex of peripheral membrane subunits containing sites of ATP hydrolysis (the $V_1$ subcomplex) attached to a complex of integral membrane proteins that comprise the proton pore (the $V_o$ subcomplex). Although there is no high resolution structure of any assembled V-ATPase, an 11 A structure from cryo-EM (Benlekbir et al. 2012) is shown in Fig. 3.2b. The pump is oriented with the $V_1$ complex toward the cytosol to allow ATP-driven proton transport into organelles. Although V-ATPases can reside at the cell surface of certain mammalian cells (Forgac 2007), they appear to be exclusively intracellular in fungi.

V-ATPases have an evolutionary relationship to F-type ATP synthases and archaeal ATPases and ATP synthases (Kibak et al. 1992; Gruber et al. 2001, 2014). All of these enzymes consist of peripheral and integral membrane subcomplexes that separate the ATP-binding and proton translocation functions and communicate

ATP binding state to the proton pore through long-range conformational changes. All three types of proton pumps exhibit rotary catalysis, which requires that conformational changes in the catalytic subunits with ATP binding and hydrolysis be transmitted to central rotor stalk in the center of the catalytic complex (Svergun et al. 1998; Owegi et al. 2006; Balakrishna et al. 2014). This central stalk is capable of turning a membrane rotor complex consisting of several protonatable proteolipid subunits. Proton transport then occurs at the interface between the proteolipid subunits and a single larger membrane subunit ($V_o$ subunit a) (Forgac 2007). Productive rotary catalysis requires not only a rotor, but also stator connections that maintain stable connections between the hexamer of catalytic and regulatory subunits and the $V_o$ a subunit. Eukaryotic V-ATPases have the most complex stator subunit arrangement of any of the rotary proton pumps with three peripheral stator stalks (Benlekbir et al. 2012; Kitagawa et al. 2008), while F-type and archaebacterial ATPases have one and two respectively (Rubinstein et al. 2003; Lau and Rubinstein 2012; Esteban et al. 2008). In addition, while the stator stalks of F-type and archaebacterial ATPases are tightly associated with the membrane sector (Dunn et al. 2000; Dickson et al. 2006; Yokoyama et al. 2003), the three peripheral stalks of V-ATPases (consisting of subunits E and G) associate tightly with the peripheral $V_1$ sector (Kitagawa et al. 2008; Ohira et al. 2006). Two bridging subunits, $V_1$ subunits H and C, provide the primary contacts between the $V_1$ and $V_o$ sectors (Benlekbir et al. 2012; Zhang et al. 2008; Oot and Wilkens 2012). This more complex arrangement in the stator subunits may support the regulatory reversible disassembly of V-ATPases (see below). The highest levels of sequence identity between F- and V-type ATPases are found in the catalytic and regulatory ATP-binding subunits of V-ATPases, and the proteolipid subunits, while other subunits show structural similarity, but limited sequence identity (Kibak et al. 1992). In contrast, sequence identities between V-ATPase subunits of different organisms tend to be very high (Forgac 2007).

Higher eukaryotes have multiple isoforms for many of the V-ATPase subunits that impart both tissue- and organelle-specific regulation on V-ATPases (Forgac 2007). In general, fungi show a more limited spectrum of isoforms. *S. cerevisiae* has a single set of two subunit isoforms for the largest membrane subunit. These two isoforms, Vph1 and Stv1, have a distinct cellular distribution with Vph1 residing primarily in the vacuole at steady state, and Stv1 cycling between Golgi and endosomes, with most Stv1 visualized in the Golgi (Manolson et al. 1994; Kawasaki-Nishi et al. 2001a). Overexpression of *STV1* can suppress the phenotypes of a *vph1*Δ mutation and Vph1 is believed to mask the phenotype of *STV1* deletion (Manolson et al. 1994). However, V-ATPases containing Vph1 and Stv1 have distinct biochemical properties, suggesting that they are not fully functionally redundant (Kawasaki-Nishi et al. 2001b).

Loss of V-ATPase activity is conditionally lethal in fungi, but lethal in higher eukaryotes (Nelson and Nelson 1990; Sun-Wada et al. 2000). Deletion of any of the single copy V-ATPase subunit genes (*vma* mutants) or deletion of both *STV1* and *VPH1* generates a very characteristic $Vma^-$ growth phenotype. This phenotype is characterized by slow growth under all conditions, with optimal growth at extracellular pH 5 (Nelson and Nelson 1990). *vma* mutants are unable to grow at alkaline extracellular pH or under high calcium concentrations (Ohya et al. 1991), and the mutants are sensitive to a wide array of heavy metals and drugs ((Parsons et al. 2004); reviewed in (Kane 2007)). *vma* mutants have proven to be very valuable in characterizing V-ATPase function *in vivo*, and the viability of loss-of-function mutants has made *S. cerevisiae* a major organism for studies of V-ATPase structure and mechanism.

### *3.3.3 Regulation of Pma1*

In *S. cerevisiae*, Pma1 and V-ATPases function in a physiological context characterized by variable availability of carbon sources and wide-ranging extracellular pH. Like many cancer cells, *S. cerevisiae* grows very rapidly under conditions of aerobic glycolysis when glucose is abundant (Vander Heiden et al. 2009). This generates large amounts of organic acid and exerts significant pH stress on cells (Orij et al. 2011). With their exposure to the cytosol and dependence on cytosolic ATP, it is not surprising that Pma1 and the V-ATPase respond to certain common environmental factors such as glucose levels and intracellular and extracellular pH. Glucose metabolism activates both pumps, consistent with a role in removing metabolically generated protons from the cytosol. Their responses to pH are more complex, but mutations in either class of pump can generate pH-dependent lethality, consistent with their central role in cellular pH homeostasis. The functions of these two classes of pumps are increasingly recognized as interconnected, although their individual regulatory mechanisms are quite different. These distinct modes of regulation will be outlined first, followed by evidence for coordination of Pma1 and V-ATPase activity.

Reversible activation of Pma1 in response to glucose was first reported by Serrano in 1983 (Serrano 1983). Incubation of yeast cells in glucose resulted in a 10-fold activation of ATPase activity that was rapidly reversed by glucose removal. Glucose activation was accompanied by a decrease in $K_m$ for ATP and a shift toward neutral pH optimum for activity. The biochemical basis for this regulation has been a subject of debate for some time. A number of P-type ATPases are regulated by autoinhibitory interactions between their C-terminal tail and the catalytic site, with activation of the enzyme arising from relief of these interactions. Deletion analysis of the Pma1 C-terminal tail implicated this region of the yeast pump in glucose regulation, since deletion of the last 11 amino acids resulted in glucose-independent activation (Portillo et al. 1989). Mutation of two amino acids in the deleted region, S911A and T912A, abolished glucose activation of Pma1 and compromised growth of cells on glucose (Eraso and Portillo 1994). Significantly, a number of second site suppressors of the glucose-dependent growth defects mapped to mutations near the catalytic site (Eraso and Portillo 1994), consistent with the proposed autoinhibitory interaction. Chang and Slayman provided convincing evidence of phosphorylation and dephosphorylation of Pma1 at the plasma membrane in response to glucose addition and deprivation, respectively (Chang and Slayman 1991). This led to the hypothesis that phosphorylation of the C-terminal tail was involved in relief of autoinhibition, and attention focused on S911 and T912 of Pma1, which are highly conserved among fungi, along with another highly conserved serine S899 (Portillo et al. 1991; Lecchi et al. 2005; Eraso et al. 2006; Lecchi et al. 2007). A phosphomimetic mutation at S899, S899D, partially mimicked the effects of glucose addition, decreasing the $K_m$ for ATP even in the absence of glucose, but did not change the $V_{max}$ in the absence of glucose (Eraso and Portillo 1994). Ptk2, a member of a family of fungal kinases, is required for full glucose activation of Pma1 (Goossens et al. 2000), and Ptk2 was later shown to generate S899-dependent phosphorylation of a Pma1 C-terminal peptide in vivo (Eraso et al. 2006). However, phosphoproteomic analysis of the Pma1 C-terminal tail has provided the most definitive data to date that phosphorylation of S911 and T912 are the primary mediators of glucose activation of Pma1 (Lecchi et al. 2007). A Pma1 peptide containing amino acids 896–918 was isolated from differentially labeled glucose-starved and glucose-fermenting yeast cells by LC-MS. Comparison of the phosphorylation state of this peptide indicated that glucose-starved cells contained predominantly singly phosphorylated peptide on T912, while glucose-fermenting cells exhibit a major shift to a species with both S911 and T912 phosphorylated. In contrast, no phosphorylation of S899 was observed under either condition in these experiments. Taken together, these data support an autoinhibition of Pma1 under conditions of glucose starvation that occurs through interactions between a singly phosphorylated C-terminal tail and the catalytic domain. This inhibition is relieved upon glucose addition through phosphorylation of S911 (Lecchi et al. 2007). It is still possible that other sites, such as S899, are involved in fine-tuning this regulation under different conditions. Consistent with a complex regulatory response to glucose, multiple protein kinases have been implicated in glucose regulation of Pma1 (Goossens et al. 2000; Portillo 2000), and the primary kinases responsible for phosphorylation of S911 and T912 have yet to be conclusively defined.

The *Arabidopsis* $H^+$-ATPase structure provides limited insights into how autoinhibition might work, even though some form of direct or indirect autoinhibition involving the C-terminal tail is likely to occur in the plant proton pumps as well (Pedersen et al. 2007; Portillo 2000). The pump was crystallized in a detergent-activated form, and most of the C-terminal tail was not resolved. Nevertheless, it was suggested that the C-terminal regulatory domain might inhibit activity by restricting movements of the A-domain (Pedersen et al. 2007; Morth et al. 2011). The model for the *S. cerevisiae* enzyme in Fig. 3.2a, shows the C-terminus interacting with the N-domain. In fact, any interaction with the C-terminal tail that restricts inter-domain movement could provide autoinhibition.

Pma1 is also regulated by both intracellular and extracellular pH (Eraso and Gancedo 1987; Carmelo et al. 1997). Depending on the composition of the growth medium, yeast cells can generate (and tolerate) an extracellular pH as low as pH 2–3 as they grow. Pma1 is activated in response to decreases in intracellular pH as long as glucose is present (Ullah et al. 2012; Holyoak et al. 1996). However, as glucose is depleted and organic acids such as acetic acid are produced, cytosolic pH falls from neutral or slightly alkaline in actively growing cells to approximately pH 6 in stationary phase cells (Orij et al. 2012). The glucose requirement for Pma1 activation at low pH reflects the need for a continued supply of cytosolic ATP. However, the mechanisms of Pma1 activation in response to low cytosolic pH are not exactly the same as those seen with glucose activation. Specifically, at low cytosolic pH, the $K_m$ for ATP decreases without the corresponding increase in $V_{max}$ or shift in pH optimum that is seen in glucose activation of Pma1 (Eraso and Gancedo 1987).

As an electrogenic pump, Pma1 activity is also sensitive to plasma membrane potential, and this accounts for the significant interdependence of $K^+$ and $H^+$ transport (Perlin et al. 1988; Calahorra et al. 1998). The Trk1 and Trk2 $K^+$ transporters play a major role in this interdependence. *trk1*Δ mutants share phenotypes such as sensitivity to extracellular acidification with mutants that have reduced Pma1 activity, suggesting that loss of Trk1 limits Pma1 activity (Yenush et al. 2002). $K^+$ addition promotes $H^+$ efflux from yeast cells treated with glucose to activate Pma1 (Martinez-Munoz and Kane 2008). Yenush et al. uncovered a very interesting pH-dependent regulation of Trk1, which highlights the interdependence of these two systems (Yenush et al. 2005). Trk1 is inhibited by the Ppz1 and Ppz2 protein phosphatases, and the activity of these phosphatases is in turn inhibited by regulatory protein Hal3. Interactions between Hal3 and Ppz1 proved to be pH-dependent. Specifically, at low cytosolic pH, Hal3 associated more strongly with Ppz1, relieving Trk1 inhibition, while at pH 7.5, Hal3 was released and Trk1 inhibition by Ppz1 was promoted. This regulation would appear to correlate well with pH-dependent regulation of Pma1, since at low cytosolic pH, Trk1 would be better able to neutralize the membrane potential generated by an activated Pma1, while at higher pH, less Trk1 activity would be necessary in the presence of a less active $H^+$ pump. Transport of monovalent cations is described in more detail elsewhere in this volume.

### 3.3.4 Regulation of V-ATPases

Like Pma1, V-ATPases are also regulated by both glucose and pH. Glucose regulation of V-ATPases involves the process of reversible disassembly of peripheral $V_1$ subunits from the membrane-bound $V_o$ sector (reviewed in (Kane and Smardon 2003)). Loss of $V_1$- $V_o$ association upon acute glucose deprivation in yeast was first observed by immunoprecipitation of assembled V-ATPase complexes, free (i.e. not bound to $V_o$) $V_1$ complexes, and free (i.e. not bound to $V_1$) $V_o$ complexes with subunit-specific monoclonal antibodies (Kane 1995). These experiments showed that in glucose-replete cells, the majority of $V_1$ complexes were bound to $V_o$, but upon acute glucose depletion, the $V_1$ and $V_o$ sectors were predominantly separated. Interestingly, one of the subunits involved in bridging $V_1$ and $V_o$, $V_1$ subunit C, was released from both $V_1$ and $V_o$ sectors. Vacuoles isolated from glucose-deprived cells also have reduced levels of $V_1$ subunits and reduced ATPase activity, consistent with disassembly representing a mechanism of V-ATPase inhibition. Disassembly of $V_1$ from $V_o$ is fully reversible, even in the absence of protein synthesis, and repeated cycles of V-ATPase assembly and disassembly in living cells were observed using a microfluidics format (Dechant et al. 2010). The physiological relevance of this mechanism in controlling organelle acidification is supported by the observation that vacuolar pH rises after even a brief (10–15 min.) period of glucose deprivation, but rapidly decreases upon glucose readdition (Martinez-Munoz and Kane 2008; Tarsio et al. 2011). Recent experiments with GFP-tagged proteins have suggested that many $V_1$ subunits may stay in proximity to the vacuolar membrane during periods of glucose deprivation, with only $V_1$ subunit C assuming a fully cytosolic localization (Tabke et al. 2014). It is possible that the extensive dissociation of $V_1$ observed by multiple other approaches (Kane 1995; Dechant et al. 2010; Seol et al. 2001; Bond and Forgac 2008) represents a disruption of a very fragile $V_1$-$V_o$ association in glucose deprived cells or that $V_1$ subunits are retained in proximity to the membrane by some other mechanism, but it is clear that the V-ATPase is not functional under these conditions. Interestingly, both ATP hydrolysis in dissociated $V_1$ sectors and proton pumping through the free $V_o$ complex are inhibited under conditions of disassembly (Parra et al. 2000; Zhang et al. 1992). Reversible disassembly was reported in the tobacco hornworm *M. sexta* and the yeast *S. cerevisiae* at about the same time (Kane 1995; Sumner et al. 1995). Since then, reversible disassembly of V-ATPases has been observed in a number of higher eukaryotes, suggesting it is a general mode of V-ATPase assembly (Trombetta et al. 2003; Sautin et al. 2005).

The mechanism of glucose-induced reversible disassembly is not well-understood. Neither glucose nor glucose-6-phosphate is required to promote assembly, since the initial steps of glycolysis can be bypassed (Parra and Kane 1998). A number of glycolytic enzymes, including aldolase and phosphofructokinase, appear to associate with the V-ATPase and would thus be positioned well for transmitting a glucose signal (Lu et al. 2003; Su et al. 2003; Chan and Parra 2014). Consistent with such a role, Chan and Parra observed that loss of the phosphofructokinase subunit Pfk2 resulted in reduced reassembly of the V-ATPase after glucose deprivation and readdition (Chan and Parra 2014). Bond and Forgac provided evidence implicating the ras-cyclic AMP pathway in V-ATPase disassembly (Bond and Forgac 2008). Specifically, they observed a failure of V-ATPases to disassemble in *ira1*△ and *ira2*△ mutants. These mutant fails to downregulate ras signaling in glucose-poor conditions (Conrad et al. 2014). Constitutive activation of either ras or cAMP-dependent protein kinase A also suppressed disassembly of the V-ATPase. These results suggest that protein kinase A is upstream of V-ATPase assembly and serves as an activator of the V-ATPase. However, Dechant et al. have presented conflicting results indicating that protein kinase A activation is downstream of V-ATPase assembly (Dechant et al. 2010). At the level of the V-ATPase itself, no post-translational modifications in the yeast enzyme have been definitively associated with reversible disassembly, although there is evidence of phosphorylation of the $V_1$ C subunit in insect cells under conditions of reassembly (Voss et al. 2007, 2009). The connection between

V-ATPase assembly and activity and glucose levels remains an important and incompletely understood question, and recent evidence linking the V-ATPase to nutritional sensing and growth control have only increased its importance (Dechant et al. 2010, 2014; Zoncu et al. 2011) (see Sect. 3.5).

V-ATPases are also regulated in response to extracellular pH. Diakov and Kane demonstrated increased V-ATPase activity when cells were grown in medium buffered to pH 7 compared to cells grown in medium buffered to pH 5 (Diakov and Kane 2010a). The elevated ATPase activity was accompanied by higher levels of $V_1$ subunits associated with the membrane in isolated vacuolar membranes, suggesting that V-ATPase activation in response to pH also occurs in at the level of $V_1$-$V_o$ sector assembly. Disassembly in response to glucose deprivation was largely suppressed at extracellular pH 7, consistent with a stabilization of the intact V-ATPase at high pH (Dechant et al. 2010; Diakov and Kane 2010b). It is interesting to note that while both the V-ATPase and Pma1 can be activated in response to pH, Pma1 is activated at low extracellular pH and the V-ATPase is activated at high extracellular pH. This may have implications for the balanced activity of the two pumps as described below.

Sodium ions generally do not really play the central bioenergetic role in fungi that they do in mammalian cells. In general, they are treated as toxins by the cell and are exported across the plasma membrane or sequestered (reviewed in Arino et al. 2010). High extracellular concentrations of NaCl induce stress responses in yeast (Posas et al. 2000; Szopinska et al. 2011). Vacuolar uptake of $Na^+$ ions is part of the immediate response to salt stress, and occurs by $Na^+/H^+$ exchange (Nass and Rao 1999; Silva and Geros 2009). V-ATPases establish the required $H^+$ gradient and are strongly activated in response to salt stress (Silva and Geros 2009; Li et al. 2012). This activation also occurs through increased assembly of $V_1$ and $V_o$ subunits. Interestingly, the salt activation of the V-ATPase is almost completely dependent on the presence of the signaling phosphoinositide phospholipid PI(3,5)$P_2$ (Li et al. 2014). This lipid of the vacuole and late endosome increases up to 20-fold in response to salt stress (Duex et al. 2006), and recent data indicate that it interacts directly with the $V_o$ sector of the V-ATPase and promotes stable $V_1$-$V_o$ assembly (Li et al. 2014). Notably, lack of PI(3,5)$P_2$ has no effect on reversible disassembly in response to glucose, suggesting that multiple signaling mechanisms regulate the level of V-ATPase assembly and activity (Li et al. 2014).

### *3.3.5 Coordination of V-ATPase and Pma1 Activity*

For many years V-ATPase and Pma1 activities were treated as largely independent. Pma1 was regarded as the major regulator of cytosolic pH, and V-ATPases were assigned an independent role in organelle acidification. Understanding the relationship between cytosolic and vacuolar pH control requires coordinated measurement of these parameters, as well as their responses to relevant regulators, in living cells. Cytosolic and vacuolar pH have been measured in whole yeast cells by a number of methods such as $^{31}$P-NMR, and such measurements have suggested a coordinated response of cytosolic and vacuolar pH to weak acid stress (Carmelo et al. 1997; Beauvoit et al. 1991). However, the advent of ratiometric fluorescent pH measurements for use *in vivo* have made these measurements much more accessible. In *S. cerevisiae*, vacuolar pH can be conveniently measured with BCECF-AM, which accumulates in vacuoles in vivo (Plant et al. 1999; Ali et al. 2004; Diakov et al. 2013). Ratiometric measurement of pH is possible from ~pH 5 to 7, and normalized measurements of fluorescence intensity at 490 nm to cell density have been used to measure vacuolar pH below pH 5 (Ali et al. 2004; Brett et al. 2011). Cytosolic pH can be measured with the ratiometric pH-sensitive GFP, pHluorin, which gives a linear response to pH from 6 to 8, covering the range of cytosolic pH typically seen in vivo (Orij et al. 2009; Diakov et al. 2013; Brett et al. 2005a). Using these ratiometric fluorescence methods, vacuolar and cytosolic pH responses have been measured under a variety of growth conditions and in a large number of mutant strains (see below). Targeted versions of pHluorin have also been developed and used to measure Golgi and mitochondrial pH in *S. cerevisiae* (Orij et al. 2009; Braun et al. 2010).

In wild-type cells, the responses of cytosolic and vacuolar pH to glucose addition are consistent with the glucose activation of Pma1 and the V-ATPase. Cells briefly (15–30 min) deprived of glucose show a slightly elevated vacuolar pH that is reduced upon glucose addition (Martinez-Munoz and Kane 2008; Tarsio et al. 2011). The involvement of the V-ATPase in this reduction is supported by parallel measurements in *vma* mutants, which reveal that vacuolar pH is higher before glucose addition, and increased, instead of decreased, upon glucose addition (Martinez-Munoz and Kane 2008; Tarsio et al. 2011). Cytosolic pH responses measured via pHluorin were very similar to those observed previously by other methods (Orij et al. 2009; Martinez-Munoz and Kane 2008). In log-phase cells, cytosolic pH is decreased even by a brief glucose deprivation, and readdition of glucose results in a transient acidification followed by rapid alkalinization to neutral or slightly alkaline pH. Measurement of cytosolic pH in *vma* mutants revealed that cytosolic pH responses are significantly slower in the *vma* mutants than in wild-type cells, and cytosolic pH is somewhat lower (Martinez-Munoz and Kane 2008; Tarsio et al. 2011). This result suggested that loss of V-ATPase activity has an unexpected impact on cytosolic pH and possibly on Pma1 activity. Consistent with loss of V-ATPase activity affecting the activity of Pma1, the rate of glucose-activated proton export from cells proved to be much lower in the *vma* mutants (Martinez-Munoz and Kane 2008; Tarsio et al. 2011).

A number of different mechanisms could coordinate activities of the vacuolar and plasma membrane pumps. However, Perzov et al. (2000) had observed that Pma1 behaved differently in differential centrifugation and subcellular fractionation in the *vma* mutants and proposed that Pma1 was trapped in the endoplasmic reticulum of the mutants. Hirata et al. reported localization of Pma1 to the vacuole in the *vma* mutants, but the significance of this was not clear (Hirata and Takatsuki 2001). These data suggested that the levels of Pma1 at the plasma membrane might be reduced in the *vma* mutants. Immunofluorescence microscopy confirmed that Pma1 was partially localized to intracellular compartments including the interior of the vacuole in *vma* mutants, particularly in strains that were also deficient for vacuolar proteases, suggesting that Pma1 might be targeted to the vacuole for degradation (Martinez-Munoz and Kane 2008). Huang and Chang provided evidence that a number of plasma membrane transporters, including Pma1, might be mislocalized to the vacuole as a result of loss of organelle acidification in the secretory pathway (Huang and Chang 2011). This is very likely to be one mechanism of coordination under conditions of long-term loss of V-ATPase activity. However, Smardon et al. demonstrated that acute loss of V-ATPase activity also results in a rapid ubiquitination and endocytosis of Pma1 from the plasma membrane in wild-type cells (Smardon and Kane 2014). This endocytosis requires the ubiquitin ligase Rsp5, homologue of the mammalian NEDD4, and an adaptor protein Rim8. Importantly, double mutants containing both a V-ATPase deletion and a loss of function mutation in *RSP5* or *RIM8* grow very poorly. Furthermore, double mutants containing both a V-ATPase deletion and a mutation that prevents endocytosis are inviable (Munn and Riezman 1994), and treatment of an endocytosis mutant with the V-ATPase inhibitor concanamycin A resulted in rapid loss of cell integrity (Smardon and Kane 2014). These data indicate that the rapid endocytosis of Pma1 upon loss of V-ATPase function likely represents a compensatory response rather than an error in trafficking. It is possible that balancing levels of Pma1 and V-ATPase activity is important for maintenance of overall pH homeostasis.

Further support for coordination of Pma1 and V-ATPase levels was recently obtained through studies of aging cells. Pma1 is an exceptionally long-lived protein that appears to be retained predominantly by mother cells during yeast cell division (Thayer et al. 2014). As a result of this, Pma1 progressively accumulates on the surface of aging cells, but there is no evidence of any parallel increase in V-ATPase levels. Loss of vacuolar acidification as yeast cells age had been reported and appeared to be associated with loss of mitochondrial function (Hughes and Gottschling 2012). Henderson et al. (2014) tested whether reducing Pma1 activity via a *pma1-105* mutation would restore vacuolar acidification in aged cells and "rejuvenate" these cells. They found that vacuolar acidification, as indicated by increased quinacrine uptake, was restored in the *pma1-105* mutant. Furthermore, they provided evidence of higher cytosolic pH in cortical regions of wild-type mother cells containing excess Pma1 than in daughter cells containing limited Pma1, suggesting that excess Pma1 in mothers alkalinizes the cytosol (Henderson et al. 2014). Taken together, these data highlight the importance of balancing the activities of Pma1 and the V-ATPase, and indicate that failure to maintain this balance has functional implications for overall pH homeostasis and lifespan (Henderson et al. 2014). One question that remains is why the Rim8/Rsp5 pathway for internalization of Pma1 is not activated in aging cells as they begin to lose the Pma1/V-ATPase balance. In addition, the signal to Rim8/Rsp5 for loss of V-ATPase activity, which might be compromised in aging cells, is not known. Further experiments are needed to address these questions.

### 3.3.6 Other Regulators of Cellular pH in Yeast

Although the V-ATPase and Pma1 may be regarded as primary drivers of cellular pH gradients, they are not the sole regulators. Several other transporters have been implicated in control of cytosolic and/or organelle pH, and additional players are emerging from genomic screens (see below). In this section, a few of the better characterized players in pH control will be described.

In mammalian cells, $Na^+/H^+$ exchangers (NHE1, NHE2, and NHE3 proteins) act as the primary exporters of $H^+$, using the $Na^+$ gradient established by the $Na^+/K^+$-ATPase (Casey et al. 2010). There are also $Na^+(K^+)/H^+$ exchangers in organelles (NHE6-9) that are implicated in organelle pH control (Kondapalli et al. 2014). Yeast cells have homologues of both classes of exchanger, which exhibit both functional similarities and differences from their mammalian homologues (Banuelos et al. 1998; Brett et al. 2005b). The yeast Nhx1 protein resides in late endosomes/pre-vacuolar compartments and bears homology to the intracellular exchangers (Nass and Rao 1998). These intracellular exchangers are likely to transport both $Na^+$ and the more abundant cytosolic $K^+$ (Brett et al. 2005a), and thus have the potential to modulate pH by coupling export of $H^+$ from organelles to uptake of $Na^+$ or $K^+$ into organelles. Consistent with such a role, yeast *nhx1*$\Delta$ mutants were shown to have a more acidic vacuolar pH than wild-type cells when subjected to acid stress (Ali et al. 2004; Brett et al. 2005b), but also have a more acidic cytosol, for reasons than are less clear (Brett et al. 2005a). Interestingly, Nhx1 effects on organelle pH seem to be tied to intracellular trafficking, supporting the idea that lumenal pH is sensed as part of organelle identity (Brett et al. 2005a). Plasma membrane exchangers play a key role in alkali metal tolerance, as they use the $H^+$ gradient across the plasma membrane to support export of $Na^+$ (Banuelos et al. 1998). Under some circumstances, they have also been shown to influence cellular pH (Sychrova et al. 1999).

## 3.4 Genomic Perspectives on $H^+$-Transport and pH Control

### 3.4.1 pH Measurements Across Deletion Mutant Arrays

The development of ordered *S. cerevisiae* deletion mutant arrays (Winzeler et al. 1999) has allowed a system-level understanding of pH control that was not previously possible. Vacuolar and cytosolic pH have been compared across thousands of individual deletion strains under a defined set of conditions (medium composition, extracellular pH, and growth phase) and deviations in pH from the wild type strain have implicated sets of genes and processes in maintaining pH (Orij et al. 2012; Brett et al. 2011). In addition, arrays of deletion strains have been tested for sensitivity to pH challenges such as weak acid stress (Ullah et al. 2012; reviewed in Mira et al. 2010) and alkaline extracellular pH (Serrano et al. 2004). This type of experiment has provided insights into the requirements for survival under conditions of varied pH and the cellular functions most sensitive to pH change. A major strength of these approaches is that they have identified novel intersections between pH control and other processes, as well as new players that would never have been tested by more targeted approaches. One potential weakness is that cellular pH is a very dynamic parameter that is responsive to multiple environmental factors ranging from growth phase of the cells to pH and buffering of the medium (Orij et al. 2012; Diakov and Kane 2010a); as a result, conditions for screens must be carefully controlled, and the results cannot be automatically extrapolated to other growth conditions. Nevertheless, such screens have provided a wealth of new information and new candidates for pH control and pH responses.

### 3.4.2 A Systems-Level View of Cytosolic pH Control

Orij et al. introduced a cytosolic pHluorin into the deletion mutant array in order to identify critical determinants of cytosolic pH control (Orij et al. 2012). Remarkably, in their screen of more that 4200 deletion mutants they determined that no single mutation had a measured cytosolic pH more than 0.3 pH units lower or 0.5 pH units higher than the wild-type average. The screen was conducted in defined monosodium glutamate medium buffered to pH 5 with sodium citrate, and care was taken to ensure that the cultures were well-aerated in fresh medium and maintained in log phase growth. These parameters are critical, as they determined that lack of aeration and conditioning of the medium, presumably as a result of production of dissolved $CO_2$ during metabolism, had significant effects on cytosolic pH. The 73 mutants identified as having low cytosolic pH and the 104 identified as having high cytosolic pH were enriched for mutants in vacuolar proteins, including the V-ATPase, and for mitochondrial functions. The strong enrichment for mitochondrial mutants was surprising, given that pH was measured under glucose fermenting conditions, where respiration has little or no contribution to ATP generation. In addition, several categories such as ribosomal proteins and aminoacyl tRNA ligases were significantly enriched in the mutants with altered pH, even though these mutations appear to affect very general cellular functions associated with growth rate. One of the most interesting associations uncovered in this screen was a very tight correlation between cytosolic pH and growth rate (Orij et al. 2009, 2012). The hypothesis that cytosolic pH controls growth rate was tested in a mutant with lower Pma1 activity where cytosolic pH could be readily adjusted; in this strain, growth rate followed cytosolic pH closely (Orij et al. 2012). Further analysis revealed a small set of mutations that appear to uncouple cytosolic pH from growth rate. These mutants define proteins with a potential pH signaling role and include *kcs1*△ and *plc1*△, two mutants implicated in inositol pyrophosphate metabolism (Auesukaree et al. 2005). These mutants had not previously been linked specifically with pH control, and thus represent an excellent example of the power of such genomic screens to reveal new candidates.

### 3.4.3 A Systems-Level View of Vacuolar pH Control

Brett et al. surveyed 4600 strains of the deletion mutant array for loss of vacuolar pH control using the pH-sensitive probe BCECF-AM. In their screen, cells were grown for 19 h. in APG medium, a synthetic medium containing arginine, phosphate, and a variety of salts (Brett et al. 2011). It should be noted that these conditions of screening are very different than those used for the genomic cytosolic pH screen described above, which limits comparison of the two screens. The authors screened the deletion collection at extracellular pH 2.7, 4, and 7, allowing an assessment of the response of vacuolar pH to altered extracellular pH. Under these conditions, the median vacuolar pH for wild-type cells was the same at pH 2.7 and 4 (pH = 5.27–5.28), but increased (to pH 5.83) in medium buffered pH 7. 107 mutants displayed acidic or alkaline vacuoles relative to wild-type cells in more than one external pH condition. This collection of mutants was significantly enriched in functional categories of transporters, membrane organization and biogenesis, and membrane trafficking. As expected, the V-ATPase mutants were prominent among strains with vacuolar alkalinization, and there is evidence that several of the other mutants, including mutants affecting ergosterol biosynthesis (Zhang et al. 2010), likely alter vacuolar pH by virtue of effects on the V-ATPase. Intriguing patterns of vacuolar pH perturbation emerged among the trafficking mutants, with mutations implicated in defective retrograde trafficking frequently resulting in a more alkaline vacuolar pH and mutations in anterograde trafficking pathways generating vacuolar acidification (Brett et al. 2011). This trend deserves further exploration, as it may indicate that regulated transport of specific pH regulators is a critical determinant of organelle pH. No correlation between growth rate and vacuolar pH control was observed in this genomic screen, suggesting that cytosolic pH may be a dominant factor in determining overall growth. However, the very different conditions in the two screens may make such conclusions premature.

### 3.4.4 Cellular Responses to pH Stress

Whole genome transcriptional studies of cellular responses to stress, including pH stress, were reported within a few years of the completion of the *S. cerevisiae* genome. Causton et al. examined the acute transcriptional response upon shifting cells from pH 6 to acidic (pH 4) or alkaline (pH 7.9) conditions (Causton et al. 2001). They observed significant overlap between genes induced by these stresses and by other apparently unrelated stresses such as heat shock, peroxide, and osmotic stress, and characterized this set of genes as a "common environmental response" to environmental challenges. Many of these genes, including several induced by acid or alkali were targets of the Msn2 and Msn4 transcription factors (Causton et al. 2001). In this study, only a handful of genes showed inverse responses to acid and alkali that might be indicative of pH-responsive genes. These genes included *PDR12*, which is induced under acid conditions and repressed under alkali; *PDR12* encodes an ABC transporter important for weak acid resistance (Causton et al. 2001). The potassium transporter *TRK2*, a paralogue of *TRK1*, showed a similar pattern of expression. In contrast, *PHO89*, encoding the major phosphate importer active at alkaline pH (Serra-Cardona et al. 2014), was upregulated at alkaline pH and repressed upon a shift to low pH (Causton et al. 2001). This control was later shown to involve activation of the calcineurin-dependent transcription factor Crz1p and repressors regulated by the Rim pathway and Snf1p (Serra-Cardona et al. 2014).

Subsequent experiments, often combining microarray analysis of transcriptional responses with phenotypic screening for growth of the deletion mutant array under high or low extracellular pH, have expanded understanding of the transcriptional networks activated at high and low pH. In addition, screening of the deletion mutant array has helped to identify participants in pH responses that are not regulated at the level of transcription.

There have been multiple genomic screens for sensitivity to different weak acids (reviewed in Mira et al. 2010). These screens are motivated by the commercial and therapeutic importance of permeant weak acids, as well as their prominent role in normal fungal metabolism. Permeant weak acids are commercially important as fungal growth inhibitors (Mollapour et al. 2004). Several weak acids are also used therapeutically, but their "off-target" cellular impacts are not fully understood (Mira et al. 2010). In addition, fungi naturally produce organic acids during fermentative growth (Kawahata et al. 2006). Weak acids are usually uncharged at low extracellular pH and can cross the membrane and dissociate in the higher pH of the cytosol. This is stressful to the cells, as dissociation of the acid generates both protons, potentially reducing intracellular pH, and a counterion. Pma1 is activated by reduced cytosolic pH (Section 3.3), but activation of Pma1 into a futile cycle, where protonated weak acids re-enter immediately after proton export, can deplete cellular energy stores (Ullah et al. 2012). In addition, the counterion can also be toxic to the cells (Ullah et al. 2012), and different counterions may require distinct responses. Mira et al. (2010) have reviewed genomic studies addressing mechanisms of adaptation to a number of different weak acids in *S. cerevisiae.* Interestingly, there appear to be both general responses to weak acid stress and specific responses to individual acids, correlated with the structure and hydrophobicity of the counterion. Msn2 and Msn4 mediate a general environmental stress response, including a response to multiple weak acids (Causton et al. 2001; Gasch et al. 2000); their targets include chaperones and proteins involved in energy metabolism and ergosterol biosynthesis. In contrast, the transcription factor War1 responds more specifically to sorbate, and is required for upregulation of *PDR12*, a plasma membrane transporter implicated in sorbate export (Schuller et al. 2004).

Screens for mutants hypersensitive to multiple weak acids have also identified proteins that respond either generally or specifically to weak acid stress. For example, multiple experiments indicate that Pma1, the V-ATPase and ergosterol biosynthesis play a central role in resistance against weak acid stress (Ullah et al. 2012; Mira et al. 2010). However, beyond these few common players, the mutants showing hypersensitivity even to closely related weak acids such as acetic and propionic acid vary significantly (Mira et al. 2010). Thus, cells seem to detect and respond to both general features of weak acid stress and specific features of the acid's structure or cellular impact. In addition, the response to weak acid stress appears to depend on both transcriptional activation and activation of existing transporters such as Pma1.

Alkaline pH is also stressful to *S. cerevisiae*. The pH gradient across the plasma membrane is lost when extracellular pH approaches the cytosolic pH, and nutrient and ion uptake can be disrupted. Lamb et al. (Lamb et al. 2001) compared transcripts at pH 4 and 8 and identified multiple alkaline-inducible genes implicated in phosphate, copper, and iron uptake, as well as cell wall and membrane maintenance. The *ENA1* gene, encoding a $Na^+$-ATPase, is induced alkaline conditions (Lamb et al. 2001; Garciadeblas et al. 1993) and encodes a pump capable of exporting toxic $Na^+$ in the absence of a $H^+$ gradient (Haro et al. 1991). Subsequent work has highlighted the importance of the Rim101 and Crz1 transcription factors in the transcriptional response to alkaline stress (Viladevall et al. 2004) and implicated the conserved Rim pathway, the $Ca^{2+}$/calmodulin-dependent protein phosphatase calcineurin, and Snf1 kinase in transmitting alkaline pH signals to the transcriptional apparatus (Platara et al. 2006).

The Rim pathway is quite conserved among fungi and is critical for virulence of pathogenic fungi (see below). A working model for the pathway is depicted in Fig. 3.3. In *S. cerevisiae*, the Rim pathway involves plasma membrane pH sensor proteins, Dfg16 and Rim21, plasma membrane protein Rim9, and the alpha arrestin Rim8 that acts downstream of the plasma membrane sensors. Activation of

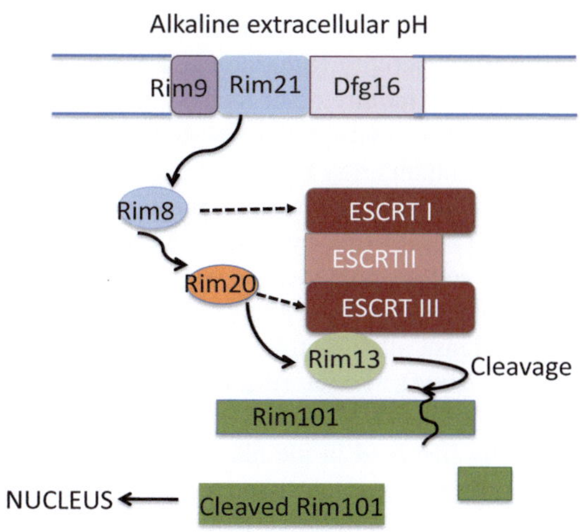


**Fig. 3.3 Working model for the *S. cerevisiae* Rim alkaline response pathway**. Rim21, Dfg16, and Rim9 are plasma membrane proteins that are implicated in the initial step of ambient pH sensing. Rim8 and Rim20 represent subsequent steps in the pathway, and both have interactions with proteins of the ESCRT complex that are required for signaling. In the final steps, the Rim13 protease is activated and cleaves Rim101, which is then transported to the nucleus. Further details are provided in the text (The diagram is adapted from Cornet and Gaillardin 2014)

this upstream "sensing module" of the Rim pathway appears to signal to Rim20, which permits activation of the Rim13 protease and cleavage of Rim101. Rim101 is a transcriptional repressor that represses Nrg1 transcription (Lamb and Mitchell 2003). Nrg1, in turn represses several alkaline response genes, including *ENA1*, and this repression is relieved by the action of Rim101 on Nrg1. In addition, multiple interactions have been reported between Rim proteins, particularly Rim20 and Rim8, and the proteins of the ESCRT pathway, which are involved in formation of the multivesicular body (Herrador et al. 2010; Xu et al. 2004). Importantly, deletion of ESCRT proteins prevents proteolytic activation of Rim101 (Xu et al. 2004). The Rim20 protein is recruited from the cytosol to endosomes by alkaline extracellular pH, and this recruitment is dependent on the ESCRT proteins as well as proteins upstream of Rim20 in the Rim pathway (Rim21, Rim8, and Rim9) (Boysen and Mitchell 2006). These data strongly suggest that endosome metabolism is connected to pH signaling (Boysen and Mitchell 2006). However, recent data has suggested that the ESCRT proteins can be recruited to the plasma membrane in response to alkaline pH and that Rim pathway activation could occur at the plasma membrane (Obara and Kihara 2014). The site of Rim101 activation is one of several questions that are not yet clear with regard to the Rim pathway. Rim8 can be ubiquitinated (Herrador et al. 2010), but it is not clear whether Rim8 ubiquitination, or ubiquitination of the plasma membrane pH sensors, is involved in signaling to Rim101 (Herrador et al. 2010, 2013; Obara and Kihara 2014). Surprisingly, there is also evidence that the Rim pathway can pay a role in the response to acid stress (Mira et al. 2009). These issues are interesting in the larger context of pH sensing and because of the physiological importance of the Rim pathway in other fungi. As described in Sect. 3.3.5, Rim8 also is required to promote Pma1 endocytosis in response to loss of V-ATPase activity (Smardon and Kane 2014). However, Pma1 endocytosis does not appear to require other Rim pathway components, suggesting that Rim8 is a multifunctional protein (A. Smardon and P. Kane, unpublished).

Screening of a deletion mutant array for reduced growth at alkaline extracellular pH identified over 100 hypersensitive mutants (Serrano et al. 2004). As expected, mutants lacking the V-ATPase were hypersensitive, but a number of mutants compromised in iron and copper homeostasis were also hypersensitive. Only the high affinity copper transporter *CTR1* and a low affinity iron/copper/zinc transporter (*FET4*) were found to increase alkaline tolerance when expressed from multicopy plasmids (Serrano et al. 2004). These results suggested that iron and copper limitation may be a primary cause of alkaline pH sensitivity, although other ions such as $Ca^{2+}$ are certainly involved in the alkaline pH response (Viladevall et al. 2004) and multisubunit proteins cannot be overexpressed by this approach. Notably, certain aspects of the transcriptional response to alkaline pH also parallel the cellular response to low glucose (Casamayor et al. 2012). Taken together, these results highlight that alkaline extracellular pH generates deficiencies in multiple nutrients. Interestingly, mutations in the Rim pathway components were not identified in this screen (Serrano et al. 2004), which was conducted at pH 7.2 and 7.5, and thus assessed the requirements for growth under mild alkaline stress.

## 3.5 pH as a Growth Signal in *S. cerevisiae*

Cytosolic pH is tightly connected to the metabolic status of the cell, and thus is positioned to serve a signaling role. Although the genomic studies described above indicate that *S. cerevisiae* strains maintain remarkably similar cytosolic pH values under comparable growth conditions, it is also clear that both cytosolic and vacuolar pH can vary significantly under different growth conditions, and that pH changes with growth as nutrients are exhausted (Orij et al. 2012). Correlations between cytosolic pH and growth rate have been observed for some time, with acidification of the cytosol generally correlated with slow growth and more alkaline cytosolic pH associated with rapid growth. Elevation of intracellular pH through expression of Pma1 in cultured fibroblasts was even shown to impart tumorigenic properties on these cells (Perona et al. 1990), leading to a hypothesis that elevated cytosolic pH promotes uncontrolled growth. Observation of synchronized *S. cerevisiae* cultures by $^{31}$P-NMR suggested that there was cytosolic alkalinization as the cells crossed into G1 (Gillies et al. 1981), linking cytosolic pH to cell cycle progression. Recently, several studies have begun to provide a mechanistic basis for observations connecting cytosolic pH and cell growth.

Young et al. described a mechanism for coordination of membrane biosynthesis by cytosolic pH (Young et al. 2010). Acidic cytosolic pH, generated by multiple methods, triggered release of the Opi1 transcriptional repressor from its binding to phosphatidic acid in the ER membrane and permitted its transit into the nucleus. Nuclear Opi1 is able to repress transcription of multiple phospholipid biosynthetic genes (Carman and Henry 2007), thus limiting membrane biosynthesis under conditions of low cytosolic pH. These experiments (Young et al. 2010) provided new insights into how cells might coordinate synthesis of membrane precursors with nutrient availability and growth rate. One very interesting aspect of this work was the finding that cytosolic pH changes as small as 0.3 pH units (from pH 7.1 type to 6.8) were sufficient to generate large-scale release of Opi1 and large changes in transcription. This work makes it clear that pH signaling can occur even in the face of the relatively stable cytosolic pH observed in wild type cells.

Dechant et al. have provided further evidence of pH signaling as a central component of nutrient sensing and growth control (Dechant et al. 2010, 2014). Consistent with the results of Orij et al. (2012), Dechant et al. found that growth rate and cell size correlate with cytosolic pH, with reduced glucose associated with low cytosolic pH, slower growth, and smaller cell size (Dechant et al. 2014). Depletion of Pma1 from daughter cells appeared to result in cell cycle arrest in early G1. This arrest could be reversed by increased extracellular pH, suggesting that the G1 arrest arose from cytosolic acidification. Cytosolic acidification also resulted in inactivation of Ras, consistent with low cytosolic pH acting as a signal for glucose limitation. Interestingly, this work and previous work from the same lab indicate that the V-ATPase plays a central role in this mode of pH signaling. Consistent with the results of Diakov and Kane, Dechant et al. found that V-ATPase disassembly upon glucose deprivation was suppressed at high pH, suggesting that cytosolic pH helps to signal V-ATPase assembly (Dechant et al. 2010; Diakov and Kane 2010a). They proposed that glucose levels are sensed, at least in part, through maintenance of neutral or slightly alkaline cytosolic pH, and suggested that the assembled V-ATPase is required to transmit the pH signal to downstream effectors such as protein kinase A and Ras (Dechant et al. 2010, 2014). The small GTPase Arf1 was shown to interact with the Stv1-containing (Golgi/endosome) form of the yeast V-ATPase, and mutational studies placed the V-ATPase upstream of Arf1 and Arf1 upstream of Ras (Dechant et al. 2014). One puzzling aspect of the Stv1-Arf1 interaction is that Stv1-containing V-ATPases in the Golgi and endosome do not appear to undergo much glucose-induced reversible disassembly (Qi and Forgac 2007), suggesting that increased assembly in response to elevated cytosolic pH is unlikely to signal through Stv1-containing complexes. However, although the picture is not yet complete, these data place pH control, and the V-ATPase and Pma1 specifically, at the center of glucose signaling, with Pma1 controlling cytosolic pH, and the V-ATPase acting as a signaling intermediate between glucose levels, cytosolic pH and the critical nutrient sensors PKA and Ras. Interestingly, the V-ATPase has also been implicated in growth control through sensing of amino acids in the TORC1 pathway (Zoncu et al. 2011). This pathway requires interactions between the Vph1-containing V-ATPase at the vacuole, the GTPase exchange protein Ragulator, and the small GTPase Gtr1p (a yeast Rag protein) (Zoncu et al. 2011; Dechant et al. 2014; Bar-Peled and Sabatini 2014). Cytosolic pH has not been specifically implicated in this pathway, but amino acid transporters, many of which require the V-ATPase-generated proton gradient for activity, are involved, suggesting a role for vacuolar pH as well as the V-ATPase itself (Wang et al. 2015; Jewell et al. 2015).

## 3.6 Proton Transport and pH Control in Other Fungi

Fungi other than *S. cerevisiae* encode homologues of many of the same transporters and regulators involved in pH control and response to pH stress, but may deploy these molecules in different ways depending on their metabolic demands and the pH challenges encountered in their environment. Additional pH regulators are also encoded in specific fungi to complement the core pH regulatory machinery and to provide additional defenses against specific environmental challenges. A complete comparison of proton transport and pH control between fungi is far beyond the scope of this review, but we will provide a brief comparison of several fungi with *S. cerevisiae*, and discuss prospects for targeting pH homeostasis in the development of therapeutic antifungal agents.

### 3.6.1 V-ATPase and Pma1 in *Neurospora crassa*

Some of the earliest information about structure and function of the fungal V-ATPase and Pma1 were obtained in the non-pathogenic, filamentous fungus *Neurospora crassa*. The first fungal V-ATPase subunit genes were cloned from *N. crassa* (Bowman et al. 1988a, b), and much of the early biochemical characterization of V-ATPases was also done in this fungus, including electron microscopic studies that supported the resemblance to the ATP synthase (Dschida and Bowman 1992). The sensitivity of fungal V-ATPases to bafilomycins and concanamycins, highly specific inhibitors that have proven to be essential tools, was established in *N. crassa*, and the inhibitor binding site was elucidated in this organism (Bowman and Bowman 2002; Bowman et al. 2004, 1988c; Drose et al. 1993). A number of structural studies of Pma1 were also initiated in *N. crassa*, and the model combining electron microscopic maps with information from the high resolution structures $Ca^{2+}$ pumps (Kuhlbrandt et al. 2002; Auer et al. 1998) provided fundamental mechanistic insights and remains one of the most complete models of a fungal Pma1.

In addition, work in *N. crassa* provided important insights into the cellular roles of V-ATPase in filamentous fungi (reviewed in (Bowman and Bowman 2000)). Both inhibition of the V-ATPase by concanamycin and genetic disruption of the catalytic subunit resulted in gross morphological alterations, as well as loss of growth at neutral pH (Bowman et al. 1997, 2000). When V-ATPase activity was compromised, very short and highly branched hyphae were produced. Interestingly, when mutants resistant to concanamycin were isolated, their mutations mapped not to V-ATPase genes but instead to the *pma-1* gene (Bowman et al. 1997). This provided the first genetic evidence for crosstalk between the vacuolar and plasma membrane proton pumps. Insights into the role of V-ATPases obtained in *Neurospora* helped to set the stage for exploring V-ATPases as a therapeutic target in pathogenic filamentous fungi.

### 3.6.2 pH Control in Pathogenic Fungi

Outside their hosts, fungal pathogens such as *Candida albicans* show a preference for abundant glucose and acidic pH, much like *S. cerevisiae*. However, within human hosts, the pathogens must tolerate neutral to slightly alkaline conditions in the bloodstream and frequent cycles of glucose deprivation. In other niches such as the oral and genital tracts, varied pH and other rapidly changing environmental conditions are also present. Fungi must adapt to varied host conditions to maintain virulence (reviewed in Davis 2009). *C. albicans* undergoes pronounced morphological changes in response to extracellular pH, with a yeast form predominating at low extracellular pH and a filamentous morphology emerging at higher extracellular pH. The ability to form hyphae appears to be critical for virulence (reviewed in Hayek et al. 2014).

*C. albicans* has a plasma membrane proton pump and V-ATPase that play a central role in pH homeostasis and are very similar to the *S. cerevisiae* pumps (Monk et al. 1993; Poltermann et al. 2005; Rane et al. 2013; Raines et al. 2013). Both proton pumps are critical for growth and virulence (Poltermann et al. 2005; Rane et al. 2013; Raines et al. 2013; Monk et al. 1995; Patenaude et al. 2013), and have been explored as anti-fungal targets (Hayek et al. 2014; Monk et al. 1995; Kulkarny et al. 2014). Pma1 activity is upregulated in preparation for filamentous growth (Monk et al. 1993), and loss of cytosolic pH control is associated with loss of filamentation (Stewart et al. 1989). Similar to the phenotypes of V-ATPase mutants in *N. crassa, C. albicans* V-ATPase mutants fail to grow at high pH and are defective in hyphae formation (Poltermann et al. 2005). Interestingly, the azole drugs currently in pharmacological use as antifungals may act by inhibition of V-ATPase activity (Zhang et al. 2010). These drugs target the fungal sterol ergosterol, and depletion of ergosterol reduces V-ATPase activity (Zhang et al. 2010). Future development of new antifungals targeting V-ATPases will depend on either identifying features of the V-ATPase that distinguish them from mammalian V-ATPases or targeting regulators, like ergosterol, that are not found in humans. Although V-ATPases are very highly conserved, inhibitors that distinguish between mammalian and fungal V-ATPases have been described (Hayek et al. 2014; Bowman et al. 2003), suggesting that development of antifungals that directly target the V-ATPase is possible.

Many pathogenic fungi rely on pathways analogous to the Rim pathway of *S. cerevisiae* for adaptation to extracellular pH (Davis 2009; Penalva et al. 2008, 2014). Compromising these pathways can reduce virulence, so they are being actively explored as antifungal targets (reviewed in Cornet and Gaillardin 2014). The Pal pathway in *Aspergillus nidulans* has been extensively characterized (Herranz et al. 2005) and shares many features with the *S. cerevisiae* Rim pathway. PalH and PalI serve as plasma membrane pH sensors, PalF shows homology to the alpha-arrestin Rim8, PalA (orthologue of Rim20) interacts with ESCRT proteins, and PacC (orthologue of Rim101) serves as the final transcriptional regulator of the pathway (reviewed in Penalva et al. 2008, 2014). The transcriptional activities of PacC in *Aspergillus* have also been well-studied and indicate a more complex regulation than that reported for Rim101 in *S. cerevisiae*, including an involvement in both acidic and alkaline pH responses (Penalva and Arst 2002). In *C. albicans*, Rim101 is required for virulence, but some aspects of Rim101 function can be partially bypassed by overexpression of Rim101 targets important for cell wall structure (Nobile et al. 2008). In addition, *C. albicans* Rim101 upregulates genes required for iron acquisition at alkaline pH (Bensen et al. 2004), suggesting that iron limitation is a consequence of high ambient pH in both pathogenic and non-pathogenic fungi.

Calcineurin is also an important contributor to growth of pathogenic fungi at high ambient pH, where it can act in parallel with the Rim pathway (Kullas et al. 2007). Deletion of the calcineurin catalytic subunit gene in *C. albicans* led to growth defects at alkaline pH, although it did not affect filamentation (Bader et al. 2006). Calcineurin is also required for *C. albicans* virulence, at least in certain host niches (reviewed in Steinbach et al. 2007). Calcineurin inhibitors have been used for severe systemic fungal infections (Steinbach et al. 2007), but it is not clear whether the effectiveness of these inhibitors is specifically due to the effects on ambient pH responses.

Taken together, the evidence cited above indicates that there are both similarities and differences between *S. cerevisiae*, non-pathogenic fungi like *N. crassa*, and pathogenic fungi such as *C. albicans* and *Aspergillus*. The core proton transport machinery, including the V-ATPase and plasma membrane proton pump are very similar. Certain phenotypes that accompany loss of function, such as failure to grow at alkaline pH in V-ATPase mutants, are also similar, suggesting parallel physiological roles. Because of these similarities, *S. cerevisiae* continues to be a valuable model for understanding the proton pumps, their regulation, and the consequences of their inhibition. However, it is also clear that there are aspects of pH regulation in filamentous, and particularly pathogenic, fungi that are not captured in studies of *S. cerevisiae*. These include the connections between dimorphic growth and pH regulation, which cannot be fully recapitulated in *S. cerevisiae*. Furthermore, the adaptation of pathogenic fungi to different environmental niches in their hosts can include an adjustment to ambient pH, but likely also involves sensing of other factors. Given these differences in fungal physiology and lifestyle, it is important that studies of pH transport and regulation continue to be extended to multiple fungi.

## 3.7 Conclusions and Future Directions

pH control is a critical requirement for growth of all organisms. Fungi have adopted a number of common strategies to address the challenges of maintaining pH control in the face of rapid metabolism and a changing extracellular environment. These strategies include a central role for organelle and plasma membrane proton pumps, with increasing evidence suggesting that the function of these two different types of pumps is actively coordinated in order to maintain overall pH homeostasis. Structural studies of these conserved pumps have provided significant insights into their mechanisms, but there is still no high resolution structure of any fungal Pma1 or assembled V-ATPase. Such structures promise to provide new mechanistic insights. These pumps tend to be adapted to optimal activity at low pH and high glucose, but multiple layers of regulation allow them to adjust activity in response to changing environmental conditions. Despite this capacity for regulation, extracellular acidification (particularly in the presence of permeant weak acids) and alkalinization of the extracellular environment are significant stresses. The response to these stresses invokes a combination of transcriptional activation to generate additional adaptive capacity and post-translational activation of constitutively expressed proteins that is still being deciphered. Beyond pH-specific stress responses, recent work has highlighted the importance of cellular pH as a potential cellular signal, capable of determining overall growth rate and driving cells into "life or death" decisions. Despite extensive work, the key question of the molecular basis of cellular pH sensing in these pathways is not understood.

The advent of whole genome approaches has greatly advanced the field of pH homeostasis and allowed identification of new players in regulation. Combination of *S. cerevisiae* deletion mutant arrays with readily introduced or genetically encoded fluorescent pH sensors has already proven to be exceptionally valuable. Some system-wide approaches, such as proteomics, have barely been applied to studies of pH homeostasis, and also promise to provide new insights into signaling pathways important for adjustment to ambient pH. In summary, the toolbox for addressing mechanisms of proton transport and pH control has rapidly expanded in recent years, but there is still much to learn before we have a true systems-level understanding of fungal pH homeostasis. The previous fruitfulness of the *S. cerevisiae* model system and the established importance of pH control in fungal pathogens will help to drive these studies forward.

**Acknowledgments** Work in the Kane lab is funded by NIH R01 GM50322. Many thanks to the investigators who provided the decades of work leading to our current understanding of fungal pH transport and pH homeostasis and apologies to all whose work I was unable to cite.

## References

Ali R, Brett CL, Mukherjee S, Rao R (2004) Inhibition of sodium/proton exchange by a Rab-GTPase-activating protein regulates endosomal traffic in yeast. J Biol Chem 279(6):4498–4506

Ambesi A, Miranda M, Petrov VV, Slayman CW (2000) Biogenesis and function of the yeast plasma-membrane H(+)-ATPase. J Exp Biol 203(Pt 1):155–160

Arino J, Ramos J, Sychrova H (2010) Alkali metal cation transport and homeostasis in yeasts. Microbiol Mol Biol Rev 74(1):95–120. doi:74/1/95 [pii]10.1128/MMBR.00042-09

Auer M, Scarborough GA, Kuhlbrandt W (1998) Three-dimensional map of the plasma membrane H+-ATPase in the open conformation. Nature 392(6678):840–843. doi:10.1038/33967

Auesukaree C, Tochio H, Shirakawa M, Kaneko Y, Harashima S (2005) Plc1p, Arg82p, and Kcs1p, enzymes involved in inositol pyrophosphate synthesis, are essential for phosphate regulation and polyphosphate accumulation in Saccharomyces cerevisiae. J Biol Chem 280(26):25127–25133. doi:M414579200 [pii]10.1074/jbc.M414579200

Bader T, Schroppel K, Bentink S, Agabian N, Kohler G, Morschhauser J (2006) Role of calcineurin in stress resistance, morphogenesis, and virulence of a Candida albicans wild-type strain. Infect Immun 74(7):4366–4369. doi:74/7/4366 [pii]10.1128/IAI.00142-06

Balakrishna AM, Basak S, Manimekalai MS, Gruber G (2014) Crystal structure of subunits D and F in complex give insight into energy transmission of the eukaryotic V-ATPase from Saccharomyces cerevisiae. J Biol Chem. doi:M114.622688 [pii]10.1074/jbc.M114.622688

Banuelos MA, Sychrova H, Bleykasten-Grosshans C, Souciet JL, Potier S (1998) The Nha1 antiporter of Saccharomyces cerevisiae mediates sodium and potassium efflux. Microbiology 144(Pt 10):2749–2758

Bar-Peled L, Sabatini DM (2014) Regulation of mTORC1 by amino acids. Trends Cell Biol 24(7):400–406. doi:S0962-8924(14)00036-1 [pii]10.1016/j.tcb.2014.03.003

Beauvoit B, Rigoulet M, Raffard G, Canioni P, Guerin B (1991) Differential sensitivity of the cellular compartments of Saccharomyces cerevisiae to protonophoric uncoupler under fermentative and respiratory energy supply. Biochemistry (Mosc) 30(47):11212–11220

Benlekbir S, Bueler SA, Rubinstein JL (2012) Structure of the vacuolar-type ATPase from Saccharomyces cerevisiae at 11-A resolution. Nat Struct Mol Biol 19(12):1356–1362. doi:nsmb.2422 [pii]10.1038/nsmb.2422

Bennett-Lovsey RM, Herbert AD, Sternberg MJ, Kelley LA (2008) Exploring the extremes of sequence/structure space with ensemble fold recognition in the program Phyre. Proteins 70(3):611–625. doi:10.1002/prot.21688

Bensen ES, Martin SJ, Li M, Berman J, Davis DA (2004) Transcriptional profiling in Candida albicans reveals new adaptive responses to extracellular pH and functions for Rim101p. Mol Microbiol 54(5):1335–1351. doi:MMI4350 [pii]10.1111/j.1365-2958.2004.04350.x

Bond S, Forgac M (2008) The Ras/cAMP/protein kinase A pathway regulates glucose-dependent assembly of the vacuolar (H+)-ATPase in yeast. J Biol Chem 283(52):36513–36521. doi:M805232200 [pii]10.1074/jbc.M805232200

Bowman EJ, Bowman BJ (2000) Cellular role of the V-ATPase in Neurospora crassa: analysis of mutants resistant to concanamycin or lacking the catalytic subunit A. J Exp Biol 203(Pt 1):97–106

Bowman BJ, Bowman EJ (2002) Mutations in subunit C of the vacuolar ATPase confer resistance to bafilomycin and identify a conserved antibiotic binding site. J Biol Chem 277(6):3965–3972

Bowman BJ, Allen R, Wechser MA, Bowman EJ (1988a) Isolation of genes encoding the Neurospora vacuolar ATPase. Analysis of vma-2 encoding the 57-kDa polypeptide and comparison to vma-1. J Biol Chem 263(28):14002–14007

Bowman EJ, Tenney K, Bowman BJ (1988b) Isolation of genes encoding the Neurospora vacuolar ATPase. Analysis of vma-1 encoding the 67-kDa subunit reveals homology to other ATPases. J Biol Chem 263(28):13994–14001

Bowman EJ, Siebers A, Altendorf K (1988c) Bafilomycins: a class of inhibitors of membrane ATPases from microorganisms, animal cells, and plant cells. Proc Natl Acad Sci U S A 85(21):7972–7976

Bowman EJ, O'Neill FJ, Bowman BJ (1997) Mutations of pma-1, the gene encoding the plasma membrane H+-ATPase of Neurospora crassa, suppress inhibition of growth by concanamycin A, a specific inhibitor of vacuolar ATPases. J Biol Chem 272(23):14776–14786

Bowman EJ, Kendle R, Bowman BJ (2000) Disruption of vma-1, the gene encoding the catalytic subunit of the vacuolar H(+)-ATPase, causes severe morphological changes in Neurospora crassa. J Biol Chem 275(1):167–176

Bowman EJ, Gustafson KR, Bowman BJ, Boyd MR (2003) Identification of a new chondropsin class of antitumor compound that selectively inhibits V-ATPases. J Biol Chem 278(45):44147–44152. doi:10.1074/jbc.M306595200M306595200 [pii]

Bowman EJ, Graham LA, Stevens TH, Bowman BJ (2004) The bafilomycin/concanamycin binding site in subunit c of the V-ATPases from Neurospora crassa and Saccharomyces cerevisiae. J Biol Chem 279(32):33131–33138

Boysen JH, Mitchell AP (2006) Control of Bro1-domain protein Rim20 localization by external pH, ESCRT machinery, and the Saccharomyces cerevisiae Rim101 pathway. Mol Biol Cell 17(3):1344–1353. doi:E05-10-0949 [pii]10.1091/mbc.E05-10-0949

Braun NA, Morgan B, Dick TP, Schwappach B (2010) The yeast CLC protein counteracts vesicular acidification during iron starvation. J Cell Sci 123(Pt 13):2342–2350. doi:jcs.068403 [pii]10.1242/jcs.068403

Brett CL, Tukaye DN, Mukherjee S, Rao R (2005a) The yeast endosomal Na+K+/H+ exchanger Nhx1 regulates cellular pH to control vesicle trafficking. Mol Biol Cell 16(3):1396–1405

Brett CL, Donowitz M, Rao R (2005b) Evolutionary origins of eukaryotic sodium/proton exchangers. Am J Physiol Cell Physiol 288(2):C223–C239

Brett CL, Kallay L, Hua Z, Green R, Chyou A, Zhang Y, Graham TR, Donowitz M, Rao R (2011) Genome-wide analysis reveals the vacuolar pH-stat of Saccharomyces cerevisiae. PLoS One 6(3):e17619. doi:10.1371/journal.pone.0017619

Calahorra M, Martinez GA, Hernandez-Cruz A, Pena A (1998) Influence of monovalent cations on yeast cytoplasmic and vacuolar pH. Yeast 14(6):501–515

Carman GM, Henry SA (2007) Phosphatidic acid plays a central role in the transcriptional regulation of glycerophospholipid synthesis in Saccharomyces cerevisiae. J Biol Chem 282(52):37293–37297. doi:R700038200 [pii]10.1074/jbc.R700038200

Carmelo V, Santos H, Sa-Correia I (1997) Effect of extracellular acidification on the activity of plasma membrane ATPase and on the cytosolic and vacuolar pH of Saccharomyces cerevisiae. Biochim Biophys Acta 1325(1):63–70
Casamayor A, Serrano R, Platara M, Casado C, Ruiz A, Arino J (2012) The role of the Snf1 kinase in the adaptive response of Saccharomyces cerevisiae to alkaline pH stress. Biochem J 444(1):39–49. doi:BJ20112099 [pii]10.1042/BJ20112099
Casey JR, Grinstein S, Orlowski J (2010) Sensors and regulators of intracellular pH. Nat Rev Mol Cell Biol 11(1):50–61. doi:nrm2820 [pii]10.1038/nrm2820
Causton HC, Ren B, Koh SS, Harbison CT, Kanin E, Jennings EG, Lee TI, True HL, Lander ES, Young RA (2001) Remodeling of yeast genome expression in response to environmental changes. Mol Biol Cell 12(2):323–337
Chan CY, Parra KJ (2014) Yeast phosphofructokinase-1 subunit Pfk2p is necessary for pH homeostasis and glucose-dependent V-ATPase reassembly. J Biol Chem. doi:M114.569855 [pii]10.1074/jbc.M114.569855
Chang A, Slayman CW (1991) Maturation of the yeast plasma membrane [H+]ATPase involves phosphorylation during intracellular transport. J Cell Biol 115(2):289–295
Conrad M, Schothorst J, Kankipati HN, Van Zeebroeck G, Rubio-Texeira M, Thevelein JM (2014) Nutrient sensing and signaling in the yeast Saccharomyces cerevisiae. FEMS Microbiol Rev 38(2):254–299. doi:10.1111/1574-6976.12065
Cornet M, Gaillardin C (2014) pH signaling in human fungal pathogens: a new target for antifungal strategies. Eukaryot Cell 13(3):342–352. doi:EC.00313-13 [pii]10.1128/EC.00313-13
Davis DA (2009) How human pathogenic fungi sense and adapt to pH: the link to virulence. Curr Opin Microbiol 12(4):365–370. doi:S1369-5274(09)00055-1 [pii]10.1016/j.mib.2009.05.006
Dechant R, Binda M, Lee SS, Pelet S, Winderickx J, Peter M (2010) Cytosolic pH is a second messenger for glucose and regulates the PKA pathway through V-ATPase. EMBO J 29(15):2515–2526. doi:emboj2010138 [pii]10.1038/emboj.2010.138
Dechant R, Saad S, Ibanez AJ, Peter M (2014) Cytosolic pH regulates cell growth through distinct GTPases, Arf1 and Gtr1, to promote Ras/PKA and TORC1 activity. Mol Cell 55(3):409–421. doi:S1097-2765(14)00484-5 [pii]10.1016/j.molcel.2014.06.002
Diakov TT, Kane PM (2010a) Regulation of vacuolar proton-translocating ATPase activity and assembly by extracellular pH. J Biol Chem 285(31):23771–23778. doi:M110.110122 [pii]10.1074/jbc.M110.110122
Diakov TT, Kane PM (2010b) Regulation of V-ATPase activity and assembly by extracellular pH. J Biol Chem. doi:M110.110122 [pii]10.1074/jbc.M110.110122
Diakov TT, Tarsio M, Kane PM (2013) Measurement of vacuolar and cytosolic pH in vivo in yeast cell suspensions. J Vis Exp Apr 19(74). doi:10.3791/50261
Dickson VK, Silvester JA, Fearnley IM, Leslie AG, Walker JE (2006) On the structure of the stator of the mitochondrial ATP synthase. EMBO J 25(12):2911–2918
Drose S, Bindseil KU, Bowman EJ, Siebers A, Zeeck A, Altendorf K (1993) Inhibitory effect of modified bafilomycins and concanamycins on P- and V-type adenosinetriphosphatases. Biochemistry (Mosc) 32(15):3902–3906
Dschida WJ, Bowman BJ (1992) Structure of the vacuolar ATPase from Neurospora crassa as determined by electron microscopy. J Biol Chem 267(26):18783–18789
Duex JE, Nau JJ, Kauffman EJ, Weisman LS (2006) Phosphoinositide 5-phosphatase Fig 4p is required for both acute rise and subsequent fall in stress-induced phosphatidylinositol 3,5-bisphosphate levels. Eukaryot Cell 5(4):723–731
Dunn SD, McLachlin DT, Revington M (2000) The second stalk of Escherichia coli ATP synthase. Biochim Biophys Acta 1458(2–3):356–363
Eraso P, Gancedo C (1987) Activation of yeast plasma membrane ATPase by acid pH during growth. FEBS Lett 224(1):187–192
Eraso P, Portillo F (1994) Molecular mechanism of regulation of yeast plasma membrane H(+)-ATPase by glucose. Interaction between domains and identification of new regulatory sites. J Biol Chem 269(14):10393–10399

Eraso P, Mazon MJ, Portillo F (2006) Yeast protein kinase Ptk2 localizes at the plasma membrane and phosphorylates in vitro the C-terminal peptide of the H+-ATPase. Biochim Biophys Acta 1758(2):164–170

Esteban O, Bernal RA, Donohoe M, Videler H, Sharon M, Robinson CV, Stock D (2008) Stoichiometry and localization of the stator subunits E and G in Thermus thermophilus H+-ATPase/synthase. J Biol Chem 283(5):2595–2603. doi:M704941200 [pii]10.1074/jbc.M704941200

Forgac M (2007) Vacuolar ATPases: rotary proton pumps in physiology and pathophysiology. Nat Rev Mol Cell Biol 8(11):917–929. doi:nrm2272 [pii]10.1038/nrm2272

Garciadeblas B, Rubio F, Quintero FJ, Banuelos MA, Haro R, Rodriguez-Navarro A (1993) Differential expression of two genes encoding isoforms of the ATPase involved in sodium efflux in Saccharomyces cerevisiae. Mol Gen Genet 236(2–3):363–368

Gasch AP, Spellman PT, Kao CM, Carmel-Harel O, Eisen MB, Storz G, Botstein D, Brown PO (2000) Genomic expression programs in the response of yeast cells to environmental changes. Mol Biol Cell 11(12):4241–4257

Gillies RJ, Ugurbil K, den Hollander JA, Shulman RG (1981) 31P NMR studies of intracellular pH and phosphate metabolism during cell division cycle of Saccharomyces cerevisiae. Proc Natl Acad Sci U S A 78(4):2125–2129

Goossens A, de La Fuente N, Forment J, Serrano R, Portillo F (2000) Regulation of yeast H(+)-ATPase by protein kinases belonging to a family dedicated to activation of plasma membrane transporters. Mol Cell Biol 20(20):7654–7661

Gruber G, Wieczorek H, Harvey WR, Muller V (2001) Structure-function relationships of A-, F- and V-ATPases. J Exp Biol 204(Pt 15):2597–2605

Gruber G, Manimekalai MS, Mayer F, Muller V (2014) ATP synthases from archaea: the beauty of a molecular motor. Biochim Biophys Acta 1837(6):940–952. doi:S0005-2728(14)00091-7 [pii]10.1016/j.bbabio.2014.03.004

Haro R, Garciadeblas B, Rodriguez-Navarro A (1991) A novel P-type ATPase from yeast involved in sodium transport. FEBS Lett 291(2):189–191. doi:0014-5793(91)81280-L [pii]

Hayek SR, Lee SA, Parra KJ (2014) Advances in targeting the vacuolar proton-translocating ATPase (V-ATPase) for anti-fungal therapy. Front Pharmacol 5:4. doi:10.3389/fphar.2014.00004

Henderson KA, Hughes AL, Gottschling DE (2014) Mother-daughter asymmetry of pH underlies aging and rejuvenation in yeast. Elife 3:e03504. doi:10.7554/eLife.03504

Herrador A, Herranz S, Lara D, Vincent O (2010) Recruitment of the ESCRT machinery to a putative seven-transmembrane-domain receptor is mediated by an arrestin-related protein. Mol Cell Biol 30(4):897–907. doi:MCB.00132-09 [pii]10.1128/MCB.00132-09

Herrador A, Leon S, Haguenauer-Tsapis R, Vincent O (2013) A mechanism for protein monoubiquitination dependent on a trans-acting ubiquitin binding domain. J Biol Chem 288(23):16206–16211. doi:C113.452250 [pii]10.1074/jbc.C113.452250

Herranz S, Rodriguez JM, Bussink HJ, Sanchez-Ferrero JC, Arst HN Jr, Penalva MA, Vincent O (2005) Arrestin-related proteins mediate pH signaling in fungi. Proc Natl Acad Sci U S A 102(34):12141–12146. doi:0504776102 [pii]10.1073/pnas.0504776102

Hirata R, Takatsuki A (2001) Role of organelle acidification in intracellular protein transport. RIKEN Rev 41:90–91

Holyoak CD, Stratford M, McMullin Z, Cole MB, Crimmins K, Brown AJ, Coote PJ (1996) Activity of the plasma membrane H(+)-ATPase and optimal glycolytic flux are required for rapid adaptation and growth of Saccharomyces cerevisiae in the presence of the weak-acid preservative sorbic acid. Appl Environ Microbiol 62(9):3158–3164

Huang C, Chang A (2011) pH-dependent cargo sorting from the Golgi. J Biol Chem 286(12):10058–10065. doi:M110.197889 [pii]10.1074/jbc.M110.197889

Hughes AL, Gottschling DE (2012) An early age increase in vacuolar pH limits mitochondrial function and lifespan in yeast. Nature 492(7428):261–265. doi:nature11654 [pii]10.1038/nature11654

Jewell JL, Kim YC, Russell RC, Yu FX, Park HW, Plouffe SW, Tagliabracci VS, Guan KL (2015) Metabolism. Differential regulation of mTORC1 by leucine and glutamine. Science 347(6218):194–198. doi:science.1259472 [pii]10.1126/science.1259472

Jin R, Dobry CJ, McCown PJ, Kumar A (2008) Large-scale analysis of yeast filamentous growth by systematic gene disruption and overexpression. Mol Biol Cell 19(1):284–296. doi:E07-05-0519 [pii]10.1091/mbc.E07-05-0519

Johnston M (1999) Feasting, fasting and fermenting. Glucose sensing in yeast and other cells. Trends Genet 15(1):29–33

Johnston M, Carlson M (1992) Regulation of carbon and phosphate utilization. The molecular and cellular biology of the yeast Saccharomyces. Gene Expr 2:193–281

Kane PM (1995) Disassembly and reassembly of the yeast vacuolar H(+)-ATPase in vivo. J Biol Chem 270(28):17025–17032

Kane PM (2006) The where, when, and how of organelle acidification by the yeast vacuolar H+-ATPase. Microbiol Mol Biol Rev 70(1):177–191

Kane PM (2007) The long physiological reach of the yeast vacuolar H+-ATPase. J Bioenerg Biomembr 39(5–6):415–421. doi:10.1007/s10863-007-9112-z

Kane PM, Smardon AM (2003) Assembly and regulation of the yeast vacuolar H+-ATPase. J Bioenerg Biomembr 35(4):313–321

Kawahata M, Masaki K, Fujii T, Iefuji H (2006) Yeast genes involved in response to lactic acid and acetic acid: acidic conditions caused by the organic acids in Saccharomyces cerevisiae cultures induce expression of intracellular metal metabolism genes regulated by Aft1p. FEMS Yeast Res 6(6):924–936

Kawasaki-Nishi S, Bowers K, Nishi T, Forgac M, Stevens TH (2001a) The amino-terminal domain of the vacuolar proton-translocating ATPase a subunit controls targeting and in vivo dissociation, and the carboxyl-terminal domain affects coupling of proton transport and ATP hydrolysis. J Biol Chem 276(50):47411–47420

Kawasaki-Nishi S, Nishi T, Forgac M (2001b) Yeast V-ATPase complexes containing different isoforms of the 100-kDa a-subunit differ in coupling efficiency and in vivo dissociation. J Biol Chem 276(21):17941–17948

Kelley LA, Sternberg MJ (2009) Protein structure prediction on the Web: a case study using the Phyre server. Nat Protoc 4(3):363–371. doi:nprot.2009.2 [pii]10.1038/nprot.2009.2

Kibak H, Taiz L, Starke T, Bernasconi P, Gogarten JP (1992) Evolution of structure and function of V-ATPases. J Bioenerg Biomembr 24(4):415–424

Kitagawa N, Mazon H, Heck AJ, Wilkens S (2008) Stoichiometry of the peripheral stalk subunits E and G of yeast V1-ATPase determined by mass spectrometry. J Biol Chem 283(6):3329–3337

Kondapalli KC, Prasad H, Rao R (2014) An inside job: how endosomal Na(+)/H(+) exchangers link to autism and neurological disease. Front Cell Neurosci 8:172. doi:10.3389/fncel.2014.00172

Kuhlbrandt W, Zeelen J, Dietrich J (2002) Structure, mechanism, and regulation of the Neurospora plasma membrane H+-ATPase. Science 297(5587):1692–1696. doi:10.1126/science.10725741072574 [pii]

Kulkarny VV, Chavez-Dozal A, Rane HS, Jahng M, Bernardo SM, Parra KJ, Lee SA (2014) Quinacrine inhibits Candida albicans growth and filamentation at neutral pH. Antimicrob Agents Chemother 58(12):7501–7509. doi:AAC.03083-14 [pii]10.1128/AAC.03083-14

Kullas AL, Martin SJ, Davis D (2007) Adaptation to environmental pH: integrating the Rim101 and calcineurin signal transduction pathways. Mol Microbiol 66(4):858–871. doi:MMI5929 [pii]10.1111/j.1365-2958.2007.05929.x

Lam FH, Ghaderi A, Fink GR, Stephanopoulos G (2014) Biofuels. Engineering alcohol tolerance in yeast. Science 346(6205):71–75. doi:346/6205/71 [pii]10.1126/science.1257859

Lamb TM, Mitchell AP (2003) The transcription factor Rim101p governs ion tolerance and cell differentiation by direct repression of the regulatory genes NRG1 and SMP1 in Saccharomyces cerevisiae. Mol Cell Biol 23(2):677–686

Lamb TM, Xu W, Diamond A, Mitchell AP (2001) Alkaline response genes of Saccharomyces cerevisiae and their relationship to the RIM101 pathway. J Biol Chem 276(3):1850–1856. doi:10.1074/jbc.M008381200M008381200 [pii]

Lau WC, Rubinstein JL (2012) Subnanometre-resolution structure of the intact Thermus thermophilus H+-driven ATP synthase. Nature 481(7380):214–218. doi:nature10699 [pii]10.1038/nature10699

Lecchi S, Allen KE, Pardo JP, Mason AB, Slayman CW (2005) Conformational changes of yeast plasma membrane H+-ATPase during activation by glucose: role of Threonine-912 in the carboxy-terminal tail. Biochemistry (Mosc) 44(50):16624–16632. doi:10.1021/bi051555f

Lecchi S, Nelson CJ, Allen KE, Swaney DL, Thompson KL, Coon JJ, Sussman MR, Slayman CW (2007) Tandem phosphorylation of Ser-911 and Thr-912 at the C terminus of yeast plasma membrane H+-ATPase leads to glucose-dependent activation. J Biol Chem 282(49):35471–35481

Li SC, Diakov TT, Rizzo JM, Kane PM (2012) Vacuolar H+-ATPase works in parallel with the HOG pathway to adapt Saccharomyces cerevisiae cells to osmotic stress. Eukaryot Cell 11(3):282–291. doi:EC.05198-11 [pii]10.1128/EC.05198-11

Li SC, Diakov TT, Xu T, Tarsio M, Zhu W, Couoh-Cardel S, Weisman LS, Kane PM (2014) The signaling lipid PI(3,5)P(2) stabilizes V(1)-V(o) sector interactions and activates the V-ATPase. Mol Biol Cell 25(8):1251–1262. doi:mbc.E13-10-0563 [pii]10.1091/mbc.E13-10-0563

Lu M, Sautin YY, Holliday LS, Gluck SL (2003) The glycolytic enzyme aldolase mediates assembly, expression and activity of V-ATPase. J Biol Chem 279:8732–8739

Manolson MF, Wu B, Proteau D, Taillon BE, Roberts BT, Hoyt MA, Jones EW (1994) STV1 gene encodes functional homologue of 95-kDa yeast vacuolar H(+)-ATPase subunit Vph1p. J Biol Chem 269(19):14064–14074

Martinez-Munoz GA, Kane P (2008) Vacuolar and plasma membrane proton pumps collaborate to achieve cytosolic pH homeostasis in yeast. J Biol Chem 283(29):20309–20319. doi:M710470200 [pii]10.1074/jbc.M710470200

Martinez-Munoz GA, Pena A (2005) In situ study of K+ transport into the vacuole of Saccharomyces cerevisiae. Yeast 22(9):689–704

McCusker JH, Perlin DS, Haber JE (1987) Pleiotropic plasma membrane ATPase mutations of Saccharomyces cerevisiae. Mol Cell Biol 7(11):4082–4088

Mellman I, Fuchs R, Helenius A (1986) Acidification of the endocytic and exocytic pathways. Annu Rev Biochem 55:663–700

Mira NP, Lourenco AB, Fernandes AR, Becker JD, Sa-Correia I (2009) The RIM101 pathway has a role in Saccharomyces cerevisiae adaptive response and resistance to propionic acid and other weak acids. FEMS Yeast Res 9(2):202–216. doi:FYR473 [pii]10.1111/j.1567-1364.2008.00473.x

Mira NP, Teixeira MC, Sa-Correia I (2010) Adaptive response and tolerance to weak acids in Saccharomyces cerevisiae: a genome-wide view. OMICS 14(5):525–540. doi:10.1089/omi.2010.0072

Mollapour M, Fong D, Balakrishnan K, Harris N, Thompson S, Schuller C, Kuchler K, Piper PW (2004) Screening the yeast deletant mutant collection for hypersensitivity and hyper-resistance to sorbate, a weak organic acid food preservative. Yeast 21(11):927–946. doi:10.1002/yea.1141

Moller JV, Nissen P, Sorensen TL, le Maire M (2005) Transport mechanism of the sarcoplasmic reticulum Ca2+-ATPase pump. Curr Opin Struct Biol 15(4):387–393. doi:S0959-440X(05)00122-3 [pii]10.1016/j.sbi.2005.06.005

Monk BC, Niimi M, Shepherd MG (1993) The Candida albicans plasma membrane and H(+)-ATPase during yeast growth and germ tube formation. J Bacteriol 175(17):5566–5574

Monk BC, Mason AB, Abramochkin G, Haber JE, Seto-Young D, Perlin DS (1995) The yeast plasma membrane proton pumping ATPase is a viable antifungal target. I. Effects of the cysteine-modifying reagent omeprazole. Biochim Biophys Acta 1239(1):81–90

Morth JP, Pedersen BP, Toustrup-Jensen MS, Sorensen TL, Petersen J, Andersen JP, Vilsen B, Nissen P (2007) Crystal structure of the sodium-potassium pump. Nature 450(7172):1043–1049

Morth JP, Pedersen BP, Buch-Pedersen MJ, Andersen JP, Vilsen B, Palmgren MG, Nissen P (2011) A structural overview of the plasma membrane Na+, K+-ATPase and H+-ATPase ion pumps. Nat Rev Mol Cell Biol 12(1):60–70. doi:nrm3031 [pii]10.1038/nrm3031
Munn AL, Riezman H (1994) Endocytosis is required for the growth of vacuolar H(+)-ATPase-defective yeast: identification of six new END genes. J Cell Biol 127(2):373–386
Nass R, Rao R (1998) Novel localization of a Na+/H+ exchanger in a late endosomal compartment of yeast. Implications for vacuole biogenesis. J Biol Chem 273(33):21054–21060
Nass R, Rao R (1999) The yeast endosomal Na+/H+ exchanger, Nhx1, confers osmotolerance following acute hypertonic shock. Microbiology 145(Pt 11):3221–3228
Nelson H, Nelson N (1990) Disruption of genes encoding subunits of yeast vacuolar H(+)-ATPase causes conditional lethality. Proc Natl Acad Sci U S A 87(9):3503–3507
Nobile CJ, Solis N, Myers CL, Fay AJ, Deneault JS, Nantel A, Mitchell AP, Filler SG (2008) Candida albicans transcription factor Rim101 mediates pathogenic interactions through cell wall functions. Cell Microbiol 10(11):2180–2196. doi:CMI1198 [pii]10.1111/j.1462-5822.2008.01198.x
Obara K, Kihara A (2014) Signaling events of the Rim101 pathway occur at the plasma membrane in a ubiquitination-dependent manner. Mol Cell Biol 34(18):3525–3534. doi:MCB.00408-14 [pii]10.1128/MCB.00408-14
Ohira M, Smardon AM, Charsky CM, Liu J, Tarsio M, Kane PM (2006) The E and G subunits of the yeast V-ATPase interact tightly and are both present at more than one copy per V1 complex. J Biol Chem 281:22752–22760
Ohya Y, Umemoto N, Tanida I, Ohta A, Iida H, Anraku Y (1991) Calcium-sensitive cls mutants of Saccharomyces cerevisiae showing a Pet- phenotype are ascribable to defects of vacuolar membrane H(+)-ATPase activity. J Biol Chem 266(21):13971–13977
Olesen C, Picard M, Winther AM, Gyrup C, Morth JP, Oxvig C, Moller JV, Nissen P (2007) The structural basis of calcium transport by the calcium pump. Nature 450(7172):1036–1042
Oot RA, Wilkens S (2012) Subunit interactions at the V1-Vo interface in yeast vacuolar ATPase. J Biol Chem 287(16):13396–13406. doi:M112.343962 [pii]10.1074/jbc.M112.343962
Orij R, Postmus J, Ter Beek A, Brul S, Smits GJ (2009) In vivo measurement of cytosolic and mitochondrial pH using a pH-sensitive GFP derivative in Saccharomyces cerevisiae reveals a relation between intracellular pH and growth. Microbiology 155(Pt 1):268–278. doi:155/1/268 [pii]10.1099/mic.0.022038-0
Orij R, Brul S, Smits GJ (2011) Intracellular pH is a tightly controlled signal in yeast. Biochim Biophys Acta 1810(10):933–944. doi:S0304-4165(11)00060-2 [pii]10.1016/j.bbagen.2011.03.011
Orij R, Urbanus ML, Vizeacoumar FJ, Giaever G, Boone C, Nislow C, Brul S, Smits GJ (2012) Genome-wide analysis of intracellular pH reveals quantitative control of cell division rate by pH(c) in Saccharomyces cerevisiae. Genome Biol 13(9):R80. doi:gb-2012-13-9-r80 [pii]10.1186/gb-2012-13-9-r80
Owegi MA, Pappas DL, Finch MW Jr, Bilbo SA, Resendiz CA, Jacquemin LJ, Warrier A, Trombley JD, McCulloch KM, Margalef KL, Mertz MJ, Storms JM, Damin CA, Parra KJ (2006) Identification of a domain in the V0 subunit d that is critical for coupling of the yeast vacuolar proton-translocating ATPase. J Biol Chem 281(40):30001–30014
Parra KJ, Kane PM (1998) Reversible association between the V1 and V0 domains of yeast vacuolar H+-ATPase is an unconventional glucose-induced effect. Mol Cell Biol 18(12):7064–7074
Parra KJ, Keenan KL, Kane PM (2000) The H subunit (Vma13p) of the yeast V-ATPase inhibits the ATPase activity of cytosolic V1 complexes. J Biol Chem 275(28):21761–21767
Parsons AB, Brost RL, Ding H, Li Z, Zhang C, Sheikh B, Brown GW, Kane PM, Hughes TR, Boone C (2004) Integration of chemical-genetic and genetic interaction data links bioactive compounds to cellular target pathways. Nat Biotechnol 22(1):62–69
Patenaude C, Zhang Y, Cormack B, Kohler J, Rao R (2013) Essential role for vacuolar acidification in Candida albicans virulence. J Biol Chem 288(36):26256–26264. doi:M113.494815 [pii]10.1074/jbc.M113.494815
Pedersen BP, Buch-Pedersen MJ, Morth JP, Palmgren MG, Nissen P (2007) Crystal structure of the plasma membrane proton pump. Nature 450(7172):1111–1114

Penalva MA, Arst HN Jr (2002) Regulation of gene expression by ambient pH in filamentous fungi and yeasts. Microbiol Mol Biol Rev 66(3):426–446

Penalva MA, Tilburn J, Bignell E, Arst HN Jr (2008) Ambient pH gene regulation in fungi: making connections. Trends Microbiol 16(6):291–300. doi:S0966-842X(08)00090-5 [pii]10.1016/j.tim.2008.03.006

Penalva MA, Lucena-Agell D, Arst HN Jr (2014) Liaison alcaline: Pals entice non-endosomal ESCRTs to the plasma membrane for pH signaling. Curr Opin Microbiol 22C:49–59. doi:S1369-5274(14)00125-8 [pii]10.1016/j.mib.2014.09.005

Perlin DS, Brown CL, Haber JE (1988) Membrane potential defect in hygromycin B-resistant pma1 mutants of Saccharomyces cerevisiae. J Biol Chem 263(34):18118–18122

Perlin DS, Harris SL, Seto-Young D, Haber JE (1989) Defective H(+)-ATPase of hygromycin B-resistant pma1 mutants from Saccharomyces cerevisiae. J Biol Chem 264(36):21857–21864

Perona R, Portillo F, Giraldez F, Serrano R (1990) Transformation and pH homeostasis of fibroblasts expressing yeast H(+)-ATPase containing site-directed mutations. Mol Cell Biol 10(8):4110–4115

Perzov N, Nelson H, Nelson N (2000) Altered distribution of the yeast plasma membrane H+-ATPase as a feature of vacuolar H+-ATPase null mutants. J Biol Chem 275(51):40088–40095

Pettersen EF, Goddard TD, Huang CC, Couch GS, Greenblatt DM, Meng EC, Ferrin TE (2004) UCSF Chimera – a visualization system for exploratory research and analysis. J Comput Chem 25(13):1605–1612. doi:10.1002/jcc.20084

Plant PJ, Manolson MF, Grinstein S, Demaurex N (1999) Alternative mechanisms of vacuolar acidification in H(+)-ATPase-deficient yeast. J Biol Chem 274(52):37270–37279

Platara M, Ruiz A, Serrano R, Palomino A, Moreno F, Arino J (2006) The transcriptional response of the yeast Na(+)-ATPase ENA1 gene to alkaline stress involves three main signaling pathways. J Biol Chem 281(48):36632–36642. doi:M606483200 [pii]10.1074/jbc.M606483200

Poltermann S, Nguyen M, Gunther J, Wendland J, Hartl A, Kunkel W, Zipfel PF, Eck R (2005) The putative vacuolar ATPase subunit Vma7p of Candida albicans is involved in vacuole acidification, hyphal development and virulence. Microbiology 151(Pt 5):1645–1655

Portillo F (2000) Regulation of plasma membrane H(+)-ATPase in fungi and plants. Biochim Biophys Acta 1469(1):31–42

Portillo F, de Larrinoa IF, Serrano R (1989) Deletion analysis of yeast plasma membrane H+-ATPase and identification of a regulatory domain at the carboxyl-terminus. FEBS Lett 247(2):381–385

Portillo F, Eraso P, Serrano R (1991) Analysis of the regulatory domain of yeast plasma membrane H+-ATPase by directed mutagenesis and intragenic suppression. FEBS Lett 287(1–2):71–74

Posas F, Chambers JR, Heyman JA, Hoeffler JP, de Nadal E, Arino J (2000) The transcriptional response of yeast to saline stress. J Biol Chem 275(23):17249–17255

Poznanski J, Szczesny P, Ruszczynska K, Zielenkiewicz P, Paczek L (2013) Proteins contribute insignificantly to the intrinsic buffering capacity of yeast cytoplasm. Biochem Biophys Res Commun 430(2):741–744. doi:S0006-291X(12)02254-1 [pii]10.1016/j.bbrc.2012.11.079

Qi J, Forgac M (2007) Cellular environment is important in controlling V-ATPase dissociation and its dependence on activity. J Biol Chem 282(34):24743–24751. doi:M700663200 [pii]10.1074/jbc.M700663200

Raines SM, Rane HS, Bernardo SM, Binder JL, Lee SA, Parra KJ (2013) Deletion of vacuolar proton-translocating ATPase V(o)a isoforms clarifies the role of vacuolar pH as a determinant of virulence-associated traits in Candida albicans. J Biol Chem 288(9):6190–6201. doi:M112.426197 [pii]10.1074/jbc.M112.426197

Rane HS, Bernardo SM, Raines SM, Binder JL, Parra KJ, Lee SA (2013) Candida albicans VMA3 is necessary for V-ATPase assembly and function and contributes to secretion and filamentation. Eukaryot Cell 12(10):1369–1382. doi:EC.00118-13 [pii]10.1128/EC.00118-13

Rubinstein JL, Walker JE, Henderson R (2003) Structure of the mitochondrial ATP synthase by electron cryomicroscopy. EMBO J 22(23):6182–6192. doi:10.1093/emboj/cdg608

Sautin YY, Lu M, Gaugler A, Zhang L, Gluck SL (2005) Phosphatidylinositol 3-kinase-mediated effects of glucose on vacuolar H+-ATPase assembly, translocation, and acidification of intracellular compartments in renal epithelial cells. Mol Cell Biol 25(2):575–589

Schuller C, Mamnun YM, Mollapour M, Krapf G, Schuster M, Bauer BE, Piper PW, Kuchler K (2004) Global phenotypic analysis and transcriptional profiling defines the weak acid stress response regulon in Saccharomyces cerevisiae. Mol Biol Cell 15(2):706–720. doi:10.1091/mbc.E03-05-0322E03-05-0322 [pii]

Seol JH, Shevchenko A, Deshaies RJ (2001) Skp1 forms multiple protein complexes, including RAVE, a regulator of V-ATPase assembly. Nat Cell Biol 3(4):384–391

Serra-Cardona A, Petrezselyova S, Canadell D, Ramos J, Arino J (2014) Coregulated expression of the Na+/phosphate Pho89 transporter and Ena1 Na+-ATPase allows their functional coupling under high-pH stress. Mol Cell Biol 34(24):4420–4435. doi:MCB.01089-14 [pii]10.1128/MCB.01089-14

Serrano R (1983) In vivo glucose activation of the yeast plasma membrane ATPase. FEBS Lett 156(1):11–14

Serrano R (1991) Transport across yeast vacuolar and plasma membranes. In: Broach JR, Pringle JR, Jones EW (eds) The molecular and cellular biology of the yeast Saccharomyces: genome dynamics, protein synthesis, and energetics, vol 1. Cold Spring Harbor Laboratory Press, Cold Spring Harbor, pp 523–585

Serrano R, Kielland-Brandt MC, Fink GR (1986) Yeast plasma membrane ATPase is essential for growth and has homology with (Na+ + K+), K+- and Ca2+-ATPases. Nature 319(6055):689–693

Serrano R, Bernal D, Simon E, Arino J (2004) Copper and iron are the limiting factors for growth of the yeast Saccharomyces cerevisiae in an alkaline environment. J Biol Chem 279(19):19698–19704. Epub 12004 Mar 19601

Seto-Young D, Monk B, Mason AB, Perlin DS (1997) Exploring an antifungal target in the plasma membrane H(+)-ATPase of fungi. Biochim Biophys Acta 1326(2):249–256. doi:S0005-2736(97)00028-X [pii]

Silva P, Geros H (2009) Regulation by salt of vacuolar H+-ATPase and H+-pyrophosphatase activities and Na+/H+ exchange. Plant Signal Behav 4(8):718–726

Smardon AM, Kane PM (2014) Loss of vacuolar H+-ATPase activity in organelles signals ubiquitination and endocytosis of the yeast plasma membrane proton pump Pma1p. J Biol Chem 289(46):32316–32326. doi:M114.574442 [pii]10.1074/jbc.M114.574442

Steinbach WJ, Reedy JL, Cramer RA Jr, Perfect JR, Heitman J (2007) Harnessing calcineurin as a novel anti-infective agent against invasive fungal infections. Nat Rev Microbiol 5(6):418–430. doi:nrmicro1680 [pii]10.1038/nrmicro1680

Stewart E, Hawser S, Gow NA (1989) Changes in internal and external pH accompanying growth of Candida albicans: studies of non-dimorphic variants. Arch Microbiol 151(2):149–153

Su Y, Zhou A, Al-Lamki RS, Karet FE (2003) The a-subunit of the V-type H+-ATPase interacts with phosphofructokinase-1 in humans. J Biol Chem 278(22):20013–20018. doi:10.1074/jbc.M210077200M210077200 [pii]

Sumner JP, Dow JA, Earley FG, Klein U, Jager D, Wieczorek H (1995) Regulation of plasma membrane V-ATPase activity by dissociation of peripheral subunits. J Biol Chem 270(10):5649–5653

Sun-Wada G, Murata Y, Yamamoto A, Kanazawa H, Wada Y, Futai M (2000) Acidic endomembrane organelles are required for mouse postimplantation development. Dev Biol 228(2):315–325

Supply P, Wach A, Goffeau A (1993a) Enzymatic properties of the PMA2 plasma membrane-bound H(+)-ATPase of Saccharomyces cerevisiae. J Biol Chem 268(26):19753–19759

Supply P, Wach A, Thines-Sempoux D, Goffeau A (1993b) Proliferation of intracellular structures upon overexpression of the PMA2 ATPase in Saccharomyces cerevisiae. J Biol Chem 268(26):19744–19752

Svergun DI, Konrad S, Huss M, Koch MH, Wieczorek H, Altendorf K, Volkov VV, Gruber G (1998) Quaternary structure of V1 and F1 ATPase: significance of structural homologies and diversities. Biochemistry (Mosc) 37(51):17659–17663

Sychrova H, Ramirez J, Pena A (1999) Involvement of Nha1 antiporter in regulation of intracellular pH in Saccharomyces cerevisiae. FEMS Microbiol Lett 171(2):167–172. doi:S0378-1097(98)00597-7 [pii]

Szopinska A, Degand H, Hochstenbach JF, Nader J, Morsomme P (2011) Rapid response of the yeast plasma membrane proteome to salt stress. Mol Cell Proteomics 10(11):M111.009589. doi:M111.009589 [pii]10.1074/mcp.M111.009589

Tabke K, Albertmelcher A, Vitavska O, Huss M, Schmitz HP, Wieczorek H (2014) Reversible disassembly of the yeast V-ATPase revisited under in vivo conditions. Biochem J 462(1):185–197. doi:BJ20131293 [pii]10.1042/BJ20131293

Tarsio M, Zheng H, Smardon AM, Martinez-Munoz GA, Kane PM (2011) Consequences of loss of Vph1 protein-containing vacuolar ATPases (V-ATPases) for overall cellular pH homeostasis. J Biol Chem 286(32):28089–28096. doi:M111.251363 [pii]10.1074/jbc.M111.251363

Thayer NH, Leverich CK, Fitzgibbon MP, Nelson ZW, Henderson KA, Gafken PR, Hsu JJ, Gottschling DE (2014) Identification of long-lived proteins retained in cells undergoing repeated asymmetric divisions. Proc Natl Acad Sci U S A 111(39):14019–14026. doi:1416079111 [pii]10.1073/pnas.1416079111

Toyoshima C, Nakasako M, Nomura H, Ogawa H (2000) Crystal structure of the calcium pump of sarcoplasmic reticulum at 2.6 A resolution. Nature 405(6787):647–655

Trombetta ES, Ebersold M, Garrett W, Pypaert M, Mellman I (2003) Activation of lysosomal function during dendritic cell maturation. Science 299(5611):1400–1403

Ullah A, Orij R, Brul S, Smits GJ (2012) Quantitative analysis of the modes of growth inhibition by weak organic acids in Saccharomyces cerevisiae. Appl Environ Microbiol 78(23):8377–8387. doi:AEM.02126-12 [pii]10.1128/AEM.02126-12

Vander Heiden MG, Cantley LC, Thompson CB (2009) Understanding the Warburg effect: the metabolic requirements of cell proliferation. Science 324(5930):1029–1033. doi:324/5930/1029 [pii]10.1126/science.1160809

Viladevall L, Serrano R, Ruiz A, Domenech G, Giraldo J, Barcelo A, Arino J (2004) Characterization of the calcium-mediated response to alkaline stress in Saccharomyces cerevisiae. J Biol Chem 279(42):43614–43624. Epub 42004 Aug 43606

Voss M, Vitavska O, Walz B, Wieczorek H, Baumann O (2007) Stimulus-induced phosphorylation of vacuolar H(+)-ATPase by protein kinase A. J Biol Chem 282(46):33735–33742. doi:M703368200 [pii]10.1074/jbc.M703368200

Voss M, Blenau W, Walz B, Baumann O (2009) V-ATPase deactivation in blowfly salivary glands is mediated by protein phosphatase 2C. Arch Insect Biochem Physiol 71(3):130–138. doi:10.1002/arch.20310

Wang S, Tsun ZY, Wolfson RL, Shen K, Wyant GA, Plovanich ME, Yuan ED, Jones TD, Chantranupong L, Comb W, Wang T, Bar-Peled L, Zoncu R, Straub C, Kim C, Park J, Sabatini BL, Sabatini DM (2015) Metabolism. Lysosomal amino acid transporter SLC38A9 signals arginine sufficiency to mTORC1. Science 347(6218):188–194. doi:science.1257132 [pii]10.1126/science.1257132

Winzeler EA, Shoemaker DD, Astromoff A, Liang H, Anderson K, Andre B, Bangham R, Benito R, Boeke JD, Bussey H, Chu AM, Connelly C, Davis K, Dietrich F, Dow SW, El Bakkoury M, Foury F, Friend SH, Gentalen E, Giaever G, Hegemann JH, Jones T, Laub M, Liao H, Davis RW et al (1999) Functional characterization of the S. cerevisiae genome by gene deletion and parallel analysis. Science 285(5429):901–906

Xu W, Smith FJ Jr, Subaran R, Mitchell AP (2004) Multivesicular body-ESCRT components function in pH response regulation in Saccharomyces cerevisiae and Candida albicans. Mol Biol Cell 15(12):5528–5537. doi:10.1091/mbc.E04-08-0666E04-08-0666 [pii]

Yenush L, Mulet JM, Arino J, Serrano R (2002) The Ppz protein phosphatases are key regulators of K+ and pH homeostasis: implications for salt tolerance, cell wall integrity and cell cycle progression. EMBO J 21(5):920–929. doi:10.1093/emboj/21.5.920

Yenush L, Merchan S, Holmes J, Serrano R (2005) pH-Responsive, posttranslational regulation of the Trk1 potassium transporter by the type 1-related Ppz1 phosphatase. Mol Cell Biol 25(19):8683–8692

Yokoyama K, Nagata K, Imamura H, Ohkuma S, Yoshida M, Tamakoshi M (2003) Subunit arrangement in V-ATPase from Thermus thermophilus. J Biol Chem 278(43):42686–42691

Young BP, Shin JJ, Orij R, Chao JT, Li SC, Guan XL, Khong A, Jan E, Wenk MR, Prinz WA, Smits GJ, Loewen CJ (2010) Phosphatidic acid is a pH biosensor that links membrane biogenesis to metabolism. Science 329(5995):1085–1088. doi:329/5995/1085 [pii]10.1126/science.1191026

Zhang J, Myers M, Forgac M (1992) Characterization of the V0 domain of the coated vesicle (H+)-ATPase. J Biol Chem 267(14):9773–9778

Zhang Z, Zheng Y, Mazon H, Milgrom E, Kitagawa N, Kish-Trier E, Heck AJ, Kane PM, Wilkens S (2008) Structure of the yeast vacuolar ATPase. J Biol Chem 283(51):35983–35995

Zhang YQ, Gamarra S, Garcia-Effron G, Park S, Perlin DS, Rao R (2010) Requirement for ergosterol in V-ATPase function underlies antifungal activity of azole drugs. PLoS Pathog 6(6):e1000939. doi:10.1371/journal.ppat.1000939

Zoncu R, Bar-Peled L, Efeyan A, Wang S, Sancak Y, Sabatini DM (2011) mTORC1 senses lysosomal amino acids through an inside-out mechanism that requires the vacuolar H-ATPase. Science 334(6056):678–683. doi:334/6056/678 [pii]10.1126/science.1207056

# Chapter 4
# Function and Regulation of Fungal Amino Acid Transporters: Insights from Predicted Structure

**Christos Gournas, Martine Prévost, Eva-Maria Krammer, and Bruno André**

**Abstract** Amino acids constitute a major nutritional source for probably all fungi. Studies of model species such as the yeast *Saccharomyces cerevisiae* and the filamentous fungus *Aspergillus nidulans* have shown that they possess multiple amino acid transporters. These proteins belong to a limited number of superfamilies, now defined according to protein fold in addition to sequence criteria, and differ in subcellular location, substrate specificity range, and regulation. Structural models of several of these transporters have recently been built, and the detailed molecular mechanisms of amino acid recognition and translocation are now being unveiled. Furthermore, the particular conformations adopted by some of these transporters in response to amino acid binding appear crucial to promoting their ubiquitin-dependent endocytosis and/or to triggering signaling responses. We here summarize current knowledge, derived mainly from studies on *S. cerevisiae* and *A. nidulans*, about the transport activities, regulation, and sensing role of fungal amino acid transporters, in relation to predicted structure.

**Keywords** Amino acid transporters • Yeast • Endocytosis • Signaling

## 4.1 Introduction

Thirty years ago were reported the first two amino acid sequences of yeast amino acid transporters, the histidine permease Hip1 and the arginine permease Can1 (Hoffmann 1985; Tanaka and Fink 1985). Since then, many other amino acid permeases have been molecularly characterized in yeast and other fungi, and this effort has been greatly boosted by genome sequencing data revealing, for each species concerned, the entire catalog of putative amino acid transporters (Van Belle

C. Gournas • B. André (✉)
Molecular Physiology of the Cell, Université Libre de Bruxelles (ULB), IBMM, 6041 Gosselies, Belgium
e-mail: bran@ulb.ac.be

M. Prévost • E.-M. Krammer
Structure and Function of Biological Membranes, Université Libre de Bruxelles (ULB), Campus Plaine, 1050 Brussels, Belgium

 
J. Ramos et al. (eds.), *Yeast Membrane Transport*, Advances in Experimental Medicine and Biology 892, DOI 10.1007/978-3-319-25304-6_4

and André 2001). For many of these proteins, however, the precise substrate specificity range, kinetic parameters, and even subcellular location remain only partially known. Several amino acid permeases have provided powerful model systems for addressing the general question of how plasma-membrane transporters are regulated according to environmental changes. Such studies have revealed, for instance, that modification of these proteins by ubiquitin (Ub) plays a major role in controlling their intracellular traffic (Haguenauer-Tsapis and André 2004). Furthermore, amino acid permeases and derivatives having lost their transport activity have been shown to function as sensors that elicit signaling in response to detection of external amino acids (Boles and André 2004; Conrad et al. 2014). More recently, resolution of the atomic structure of a bacterial amino acid transporter similar to fungal amino acid permeases, namely the AdiC arginine/agmatine antiporter, opened the way to building structural models of several fungal amino acid permeases. These studies, combined with extensive mutational analyses, have revealed the first key molecular elements of what seems to be a common mechanism for amino acid recognition and translocation. In the paragraphs below, we summarize current knowledge about the classification, predicted structures, regulation by substrates, and signaling capabilities of amino acid transporters in yeast and other fungi.

## 4.2 Classification of Fungal Amino Acid Transporters

The fungal amino acid transporters (fAATs) functionally characterized to date belong to several transporter families (Table 4.1). These families often include fungal proteins transporting compounds other than amino acids, and most of them are conserved in plants and animals. Several of these families are phylogenetically related and grouped into superfamilies, including the highly diverse Aminoacid-Polyamine-organoCation (APC) superfamily and the Major Facilitator Superfamily (MFS) (Saier et al. 2014). The recently increasing number of available transporter atomic structures has made it possible to define a limited number of structural folds shared by transporter families that are often unrelated at sequence level (Shi 2013). These common structural folds now tend to be used as an additional criterion for transporter classification. For instance, the APC superfamily now includes several additional transporter families previously viewed as unrelated to APC proteins (Västermark et al. 2014).

Most characterized fAATs belong to the APC superfamily. The study of these APC proteins began with the pioneering work of Marcelle Grenson and others, who used transport assays coupled to the isolation of mutants to define yeast genes likely to encode plasma membrane amino acid permeases, e.g. mutations in the same gene caused total loss of a permease activity or alteration of its apparent $K_m$ (Wiame et al. 1985). With the emergence of gene cloning methods, the deduced sequences of highly specific amino acid permeases (eg. the *S. cerevisiae* Can1, Hip1, Lyp1, Dip5, Put4 permeases, and the *A. nidulans* PrnB permease), broad-specificity amino acid permeases (eg. the *S. cerevisiae* Agp1, Gnp1, Bap2 permeases), and that of the

**Table 4.1** Fungal amino acid transporters of known function

| Superfamily | Family | Organism | Name | Location | Specificity | References |
|---|---|---|---|---|---|---|
| APC | YAT | *S. cerevisiae* | Agp1 | pm | Broad-specificity, relatively-low-affinity, neutral amino-acid permease | Iraqui et al. (1999), Regenberg et al. (1999), and Schreve et al. (1998) |
| APC | YAT | *S. cerevisiae* | Agp2 | pm | Proposed role in carnitine, polyamine and amino-acid transport | Schreve and Garrett (2003) and van Roermund et al. (1999) |
| APC | YAT | *S. cerevisiae* | Agp3 | pm | Low-affinity uptake of branched-chain and aromatic amino acids | Schreve and Garrett (2003) |
| APC | YAT | *S. cerevisiae* | Alp1 | pm | Arginine, unknown conditions of expression | Regenberg et al. (1999) and Sychrova and Chevallier (1994) |
| APC | YAT | *S. cerevisiae* | Bap2 | pm | Branched chain and aliphatic amino acids | Grauslund et al. (1995) and Regenberg et al. (1999) |
| APC | YAT | *S. cerevisiae* | Bap3 | pm | Broad-specificity, neutral amino-acid permease | Didion et al. (1998), Iraqui et al. (1999), and Regenberg et al. (1999) |
| APC | YAT | *S. cerevisiae* | Can1 | pm | High affinity Arg permease, also transports Lys and His with lower affinity | Grenson et al. (1966) and Hoffmann (1985) |
| APC | YAT | *S. cerevisiae* | Dip5 | pm | Asp, Glu permease, also transports Gln, Asn, Ser, Ala, and Gly | Regenberg et al. (1998, 1999) |
| APC | YAT | *S. cerevisiae* | Gap1 | pm | General amino acid permease | Grenson et al. (1970) Jauniaux and Grenson (1990) |
| APC | YAT | *S. cerevisiae* | Gnp1 | pm | Broad-specificity, relatively-low-affinity, neutral amino-acid permease | Iraqui et al. (1999), Regenberg et al. (1999), and Zhu et al. (1996) |
| APC | YAT | *S. cerevisiae* | Hip1 | pm | His | Tanaka and Fink (1985) |
| APC | YAT | *S. cerevisiae* | Lyp1 | pm | Lys | Sychrova and Chevallier (1993) |
| APC | YAT | *S. cerevisiae* | Mmp1 | pm | s-methylmethionine permease | Rouillon et al. (1999) |
| APC | YAT | *S. cerevisiae* | Put4 | pm | High affinity Pro permease, also transports Gly, Ala, GABA and L-AZC | Vandenbol et al. (1989) |
| APC | YAT | *S. cerevisiae* | Sam3 | pm | S-adenosylmethionine permease | Rouillon et al. (1999) and Uemura et al. (2007) |

(continued)

**Table 4.1** (continue)

| Superfamily | Family | Organism | Name | Location | Specificity | References |
|---|---|---|---|---|---|---|
| APC | YAT | *S. cerevisiae* | Ssy1 | pm | Sensor of extracellular amino acids (mostly neutral amino acids) | Didion et al. (1998) and Iraqui et al. (1999) |
| APC | YAT | *S. cerevisiae* | Tat1 | pm | Transporter of Val, Leu, Ile and Tyr; low-affinity Trp and His transporter | Bajmoczi et al. (1998), Regenberg et al. (1999), and Schmidt et al. (1994) |
| APC | YAT | *S. cerevisiae* | Tat2 | pm | High affinity for Trp and Tyr, also contributes to Phe, Ala, Gly and Cys uptake | Regenberg et al. (1999) and Schmidt et al. (1994) |
| APC | YAT | *S. pastorianus* | Tat3 | pm | Aromatic amino acids, Leu | Omura et al. (2007) |
| APC | YAT | *A. nidulans* | PrnB | pm | Pro | Sophianopoulou and Scazzocchio (1989) |
| APC | YAT | *A. nidulans* | AgtA | pm | Asp, Glu | Apostolaki et al. (2009) |
| APC | LAT | *S. cerevisiae* | Mup1 | pm | High affinity for Met, also involved in Cys transport | Isnard et al. (1996) and Kosugi et al. (2001) |
| APC | LAT | *S. cerevisiae* | Mup3 | pm | Low affinity Met permease | Isnard et al. (1996) |
| APC | ACT | *S. cerevisiae* | Uga4 | pm | GABA | André et al. (1993) |
| APC | ACT | *S. cerevisiae* | Hnm1 | pm | Choline, ethanolamine, and carnitine | Li et al. (1991) |
| APC | ACT | *A. nidulans* | GabA | pm | GABA | Hutchings et al. (1999) |
| APC | AAAP | *S. cerevisiae* | Avt1 | vac | Tyr, Gln, Asn, Ile and Leu transporter | Russnak et al. (2001) |
| APC | AAAP | *S. cerevisiae* | Avt3 | vac | Tyr, Gln, Asn, Ile and Leu exporter | Russnak et al. (2001) |
| APC | AAAP | *S. cerevisiae* | Avt4 | vac | Tyr, Gln, Asn, Ile and Leu exporter | Russnak et al. (2001) |
| APC | AAAP | *S. cerevisiae* | Avt6 | vac | Glu and Asp exporter | Chahomchuen et al. (2009) |
| MFS | DHA1 | *S. cerevisiae* | Aqr1 | ves | Vesicular amino-acid transporter involved in excretion of homoserine, Thr and other aa | Velasco et al. (2004) |
| MFS | V-BAAT | *S. cerevisiae* | Vba1 | vac | Transporter involved in vacuolar sequestration of Lys and His | Shimazu et al. (2005) |

| | | | | | | |
|---|---|---|---|---|---|---|
| MFS | V-BAAT | *S. cerevisiae* | Vba2 | vac | Transporter involved in vacuolar sequestration of Arg, His, Lys and Tyr | Shimazu et al. (2005) |
| MFS | DHA2 | *S. cerevisiae* | Vba3 | vac | Transporter involved in vacuolar sequestration of His and Lys | Shimazu et al. (2005) |
| MFS | UMF1 | *S. cerevisiae* | Atg22 | vac | Transporter catalyzing efflux of amino acids from the vacuole during autophagy | Yang et al. (2006) |
| MFS | ACS | *S. cerevisiae* | Yct1 | pm | Cysteine | Kaur and Bachhawat (2007) |
| TOG | LCT | *S. cerevisiae* | Ypq1 | vac | Basic amino acid exporter of the vacuole | Jézégou et al. (2012) |
| TOG | LCT | *S. cerevisiae* | Ypq2 | vac | Basic amino acid exporter of the vacuole | Jézégou et al. (2012) |
| TOG | LCT | *S. cerevisiae* | Ypq3 | vac | Basic amino acid exporter of the vacuole | Jézégou et al. (2012) |
| TOG | LCT | *S. cerevisiae* | Ers1 | vac | Vacuolar and endosomal cystine exporter | Gao et al. (2005) |
| MC | MC | *S. cerevisiae* | Crc1 | mit | Mitochondrial carnitine transporter | Palmieri et al. (2000) |
| MC | MC | *S. cerevisiae* | Ort1 | mit | Mitochondrial ornithine exporter | Palmieri et al. (1997) |
| MC | MC | *S. cerevisiae* | Agc1 | mit | Mitochondrial glutamate uniporter and Asp-Glu exchanger | Cavero et al. (2003) |

Abbreviations used in the table but not in the text: *UMF1* unidentified major facilitator 1, *pm.* plasma membrane, *vac* vacuole, *mit* mitochondria, *ves* vesicles yeast general amino acid permease (Gap1) defined a large family of highly similar transporters containing 12 predicted transmembrane domains (TMs) (Table 4.1). This family is very well conserved in bacteria as well. The Transporter Classification (TC) system proposes the name "YAT" (Yeast Amino acid Transport; TC: 2.A.3.10) for this fAAT family. Another, more distantly related member of the YAT family is the Ssy1 sensor of external amino acids, which lacks transporter activity but has conserved the ability to bind external amino acids. In response to amino acid detection, Ssy1 activates a pathway leading ultimately to transcriptional induction of several YAT genes, including *AGP1*, encoding broad-specificity neutral amino acid permeases (Didion et al. 1998; Iraqui et al. 1999; Klasson et al. 1999). Further studies have revealed other yeast plasma membrane amino acid permeases that do not belong to the YAT family. The highly similar Mup1 and Mup3 methionine-specific permeases (Isnard et al. 1996), for example, belong to the L-type Amino acid Transporter family (LAT) (TC: 2.A.3.8), and the GABA-specific permease (Uga4) (André et al. 1993) is a member of the Amino Acid/Choline Transporter (ACT) family (TC: 2.A.3.4). Finally, five permeases (Avt1 to −5) mediating amino acid transport at the vacuolar membrane (Russnak et al. 2001) have been classified in the Amino Acid/Auxin Permease (AAAP) family (TC: 2.A.18). Interestingly, this family also includes two proteins characterized in other fungal species (*Neurospora crassa* and *Penicillium chrysogenum*) as plasma membrane permeases mediating uptake of aliphatic and aromatic amino acids (Trip et al. 2004). The YAT, LAT, ACT, and AAAP families are members of the APC superfamily. As discussed below, APC transporters are now defined as proteins with a common structural fold (called the LeuT or 5+5 fold) consisting of two intertwined 5-TM-helix repeats (TMs 1-5 and TMs 6-10) sharing a twofold inverted pseudo symmetry. In YAT proteins, this LeuT fold is formed by TMs 1-10, which tend to be more conserved than TM11 and TM12.

Transporters of the MFS superfamily share another structural fold, consisting of two 6-TM-helix domains sharing a twofold pseudo symmetry. Each 6-TM domain comprises a pair of inverted 3-TM-helix repeats (TMs 1-3 and TMs 4-6 in the N-terminal domain, and TMs 7-9 and TMs 10-12 in the C-terminal domain). In addition to the core MFS fold, MFS transporters often contain extra TMs. The MFS superfamily is subdivided into multiple families of more closely related proteins, several of which include a few characterized fAATs (Table 4.1). For example, the Drug:$H^+$ Antiporter-1 (12 Spanner) family (DHA1) (TC: 2.A.1.2) includes Aqr1, a vesicular membrane amino acid transporter involved in excretion of homoserine, threonine, and other amino acids (Velasco et al. 2004). This family also includes Dtr1, a plasma membrane transporter induced during sporulation and excreting bisformyl dityrosine, the major building block of the spore wall (Felder et al. 2002). The Drug:$H^+$ Antiporter-2 family (14 Spanner) (DHA2) (TC: 2.A.1.3) includes the Vba3, while the Vacuolar Basic Amino Acid Transporter (V-BAAT) Family (TC: 2.A.1.48) includes Vba1 and -2 transporters involved in vacuolar sequestration of basic amino acids (Shimazu et al. 2005). The Anion:Cation Symporter family (ACS) (TC: 2.A.1.14), comprising several yeast transporters for carboxylic acids or vitamins, also includes Yct1, a cysteine-specific plasma-membrane permease (Kaur and Bachhawat 2007). Finally, the Atg22 protein, which belongs to a less well-characterized MFS family, is a vacuolar membrane transporter catalyzing efflux of leucine and other amino acids during autophagy (Yang et al. 2006).

Other characterized fAATs belong to transporter families unrelated to the APC and MFS superfamilies. The Lysosomal Cystine Transporter (LCT) family (TC: 2.A.43) is represented by seven proteins in *S. cerevisiae* including Ypq1, -2 and -3, three transporters catalyzing export of basic amino acids from the vacuole (Jézégou et al. 2012), and Ers1, a probable exporter of cystine from the vacuole (Gao et al. 2005). The LCT family belongs to a transporter superfamily which also includes small transporter monomers containing only 3 TMs, e. g. the bacterial SemiSWEET proteins, whose atomic structure has recently been solved (Xu et al. 2014). The LCT proteins consist of two repeats of this 3-TM-helix block (TMs 1-3 and TMs 5-7), sharing a twofold pseudo symmetry and connected to each other by TM4.

Finally, the large Mitochondrial Carrier (MC) family (TC: 2.A.29) comprises no less than 35 members in *S. cerevisiae*, including an ornithine carrier (Arg11) (Palmieri et al. 1997) and a glutamate carrier (Agc1) (Cavero et al. 2003).

## 4.3 Structure-Function Relationships and Determination of Specificity in the YAT Family

### *4.3.1 The YAT-Family Proteins Adopt the LeuT Fold*

Obtaining high-resolution crystal structures of membrane proteins has proved highly challenging, mainly because of their partially hydrophobic surface, their flexibility, and their lack of stability. In 2003, the first crystal structures of two transporters, namely the lactose permease (lacY) and the glycerol-3-phosphate permease (GlpT) of *Escherichia coli*, both of the MFS superfamily, were solved (Abramson et al. 2003; Huang et al. 2003). Both structures confirmed the predicted highly $\alpha$-helical transmembrane nature of transporters. The LacY 3D structure turned out to be impressively similar to a tentative model based on a plethora of genetic and biochemical data (Sorgen et al. 2002). This shows that alternative approaches used to study structure-function relationships in membrane proteins can produce highly informative data.

The subsequent crystallization of four prokaryotic transporters classified into unrelated families unexpectedly revealed that they adopt a common structural fold. These carriers were the amino acid transporter LeuT (Yamashita et al. 2005) of the Neurotransmitter Sodium Symporter (NSS) family, the benzylhydantoin transporter Mhp1 (Weyand et al. 2008) of Nucleobase Cation Symporter (NCS1) family 1, the galactose transporter vSGLT (Faham et al. 2008) of the Solute Sodium Symporter (SSS) family, and the glycine-betaine permease BetP (Ressl et al. 2009) of the Betaine Choline Carnitine transporter (BCCT) family. As mentioned above, the so-called LeuT or 5+5 structural fold shared by these transporters consists of two V-shaped inverted repeat domains, each containing 5 TMs intertwined to form a helical sub-domain. The first repeat is related to the second by a twofold pseudo symmetry around an axis through the center of the membrane plane. The substrate-binding site of all four proteins was found to involve residues located in the unwound part of the first kinked α-helix of each of the two 5-TMs domains.

A subsequent *in silico* study based on the similarity of hydropathy profiles predicted that this common structural fold would also be adopted by members of the APC superfamily (Lolkema and Slotboom 2008). On the basis of these results, a model structure of a YAT, namely the proline transporter (PrnB) of *Aspergillus nidulans*, was built using the crystal structure of LeuT as a template (Vangelatos et al. 2009). Together with data from mutational analyses and transport assays, this model suggested that the substrate-binding pocket of YATs and other APC members is formed mainly by conserved residues of two unwound TMs, TM1 and TM6.

Crystal structures have then been determined for two APC proteins, namely the AdiC arginine/agmatine antiporters of *E. coli* and *Salmonella*, and for ApcT, a broad-specificity proton-coupled amino acid transporter of *Methanocaldococcus jannaschii* (Fang et al. 2010; Gao et al. 2009, 2010; Kowalczyk et al. 2011; Shaffer et al. 2009). These crystal structures confirmed that the structural core of APC transporters indeed corresponds to the LeuT fold and that Arg recognition by AdiC is similar to leucine recognition by LeuT. Interestingly, the currently available crystal structures of APC proteins, including AdiC, ApcT, and the glutamate/GABA antiporter GadC (Ma et al. 2012), correspond to different conformations and thus reveal a series of snapshots along the transport cycle (Diallinas 2008; Fotiadis et al. 2013). The general scheme deduced from a comparison of these conformations fits well with the alternating-access transport model (Jardetzky 1966), where an outward-facing (OF) conformation of the transporter binds the substrate and subsequently undergoes several conformational changes that successively close access of the binding site to the OF side and open it to the inward-facing (IF) side. The substrate is then released and the transporter returns to its original conformation. The different conformations thus far solved for transporters adopting the LeuT-fold reveal significant movements of several TMs, depicting transitions between successive states of the transport cycle. These broad movements, combined with more local rearrangements of individual side chains, cause either the occlusion or the opening of gates whose role is to hamper or facilitate substrate diffusion (Diallinas 2008).

The solved structures of transporters sharing the LeuT fold now provide a structural framework for deciphering the mechanisms of amino acid binding, specificity, and translocation by APC transporters. As no YAT structure has yet been solved, the most suitable model for YAT proteins appears to be AdiC, given the reasonable sequence similarity it shares with them and the availability of structures for AdiC trapped in different states, in the absence or presence of its substrate arginine. These atomic level structural analyses also open another promising prospect: to better understand the molecular mechanisms by which YATs are regulated by their own substrates (see below). The caveat of these studies is that it has been impossible so far to model the structures of large amino- and carboxy-terminal YAT cytosolic regions, known to play an important role in regulation, as they are not conserved in prokaryotic transporters.

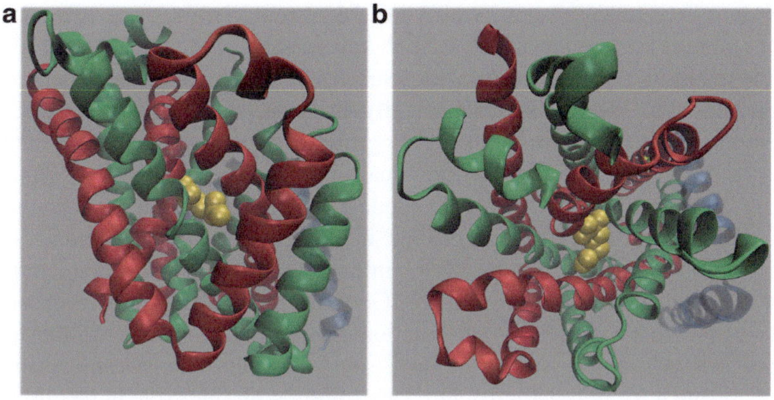


**Fig. 4.1** Side (**a**) and top (**b**) view of the three-dimensional model of the Gap1 permease in the OF occluded state, built using the AdiC crystal structure (PDB code 3L1L) as a template. The protein is depicted as a *ribbon diagram*. The first and second 5-TMs repeats are colored in *red* and *green*, respectively. The last two TMs (TM11 and TM12), which are not part of the "5+5" fold, are depicted in *blue*. The arginine ligand is shown as *yellow* spheres. The loops are not shown for the sake of clarity

## *4.3.2 Using Structural Models to Study Substrate Binding and Translocation in YATs*

The substrate-bound, OF occluded conformation of AdiC (Gao et al. 2010) has been used as a template to build the 3D structure of different YATs in order to investigate their transport mechanism and substrate specificity. The studied YATs include four *S. cerevisiae* permeases, namely the general amino acid permease Gap1 (Ghaddar et al. 2014b) (Fig. 4.1), the arginine permease Can1 (Ghaddar et al. 2014a), the aromatic amino acid transporter Tat2 (Usami et al. 2014), and the branched chain amino acid permease Bap2 (Kanda and Abe 2013), and one *A. nidulans* permease: the proline transporter PrnB (Gournas et al. 2015). In this section, we review the outcomes of these studies and re-evaluate previous reports in the light of these modeled structures.

### 4.3.2.1 Residues Mediating Recognition of the Amino-Acid Backbone

The modeled structures of different YATs, coupled with docking calculations and biochemical and mutational data, now provide a coherent picture of the residues responsible for substrate binding and translocation. The YAT substrate-binding site is formed mainly by residues of TM1, TM3, TM6, TM8, and TM10. In the two unwound segments of TM1 and TM6 lie two highly conserved motifs, which are also present in AdiC (Tables 4.2 and 4.3): the GTG motif in TM1 and the (F/Y)(S/A/T)(F/Y)xGxE motif in TM6. These correspond, respectively, to the

$GSG^{27}$ and $WSFIGVE^{208}$ motifs of AdiC. Most remarkably, these two motifs and nearby residues contain the amino acids which, in AdiC, have been shown to interact with the invariant part of an amino acid substrate, i. e. the α-carboxyl and α-amino groups (Gao et al. 2010). More specifically, $Ile^{23}$, $Trp^{202}$, and $Ile^{205}$ of AdiC accept three hydrogen bonds from the α-amino group of arginine, while $Ser^{26}$ and $Gly^{27}$ form two hydrogen bonds with the α-carboxyl group (Table 4.2). Remarkably, docking of arginine in the Can1 3D model, of arginine and phenylalanine in the Gap1 model (Fig. 4.2), and of proline in the PrnB and Put4 models suggests that the amino acid substrate backbone interacts with residues equivalent to those found in AdiC (Ghaddar et al. 2014a; Gournas et al. 2015). These observations are corroborated by the results of mutational analyses, since substitutions of these residues in PrnB, Gap1, Tat2, and Bap2 severely affect or even abolish transport activity (Table 4.3). Interestingly, and consistently with their role in recognizing the backbone of amino acid substrates, several substitutions of these conserved residues have been shown to affect both the functionality and the kinetic characteristics of the transporters (Table 4.3).

#### 4.3.2.2 Residues in Several TMs Mediate Recognition of Amino Acid Side Chains

In contrast to the α-carboxyl and α-amino groups, the side chain of the arginine substrate in the OF occluded crystal structure of AdiC interacts with residues in TM3, TM6, TM8, and TM10, showing poor conservation in YATs. More specifically, the guanidinium group stacks against $Trp^{293}$ in TM8, probably through a cation–π interaction (Gao et al. 2010). In addition, the nitrogen atoms of the guanidinium group are at approximate hydrogen-bond distances from four oxygen atoms in AdiC: the side chains of $Ser^{357}$ in TM10 and of $Asn^{101}$ in TM3 and the carbonyl groups of $Ala^{96}$ and $Cys^{97}$ in TM3. The aliphatic portion of Arg interacts with the side chains of $Met^{104}$ in TM3 and $Trp^{202}$ and $Ile^{205}$ in TM6 (Gao et al. 2010). The variable nature of the corresponding residues in YATs (Table 4.4) suggests a possible role in determining the substrate specificity of the permeases. This has been proven in the case of Can1, whose substrate binding pocket, as revealed by docking of Arg in the modeled structure of Can1, differs at only two positions from that of the Lys-specific permease Lyp1. These positions are $Ser^{176}$ and $Thr^{456}$, corresponding to $Asn^{193}$ and $Ser^{478}$ in Lyp1 (Ghaddar et al. 2014a). $Ser^{176}$ of Can1 appears essential to efficient arginine transport, since its replacement by Asn or Ala results, respectively, in loss or a strong reduction of transport capacity. Substitution of Ser for $Thr^{456}$ converts Can1 to a transporter capable of translocating both basic amino acids, while replacement of $Thr^{456}$ with Ala only slightly increases the apparent affinity for lysine. Most importantly, combining in Can1 the two Lyp1-mimicking substitutions S176N and T456S results in conversion of Can1 to a lysine-specific permease displaying kinetic characteristics almost identical to those of Lyp1 (Ghaddar et al. 2014b).

**Table 4.2** Summary of AdiC mutants

| | Backbone recognition | | Side-chain recognition | | | | |
|---|---|---|---|---|---|---|---|
| **Transporter**/Allele | TM1 | TM6 | TM3 | TM8 | TM10 | Phenotype | Reference |
| **AdiC** | NIMGSG$^{27}$ | WSFI-GVE$^{208}$ | YWLACWIGNIAM$^{104}$ | GSLGGW$^{293}$ | SV I FTLVP Y$^{365}$ | | |
| N22A | AIMGSG$^{27}$ | WSFI-GVE$^{208}$ | YWLACWIGNIAM$^{104}$ | GSLGGW$^{293}$ | SV I FTLVP Y$^{365}$ | Improved affinity for Arg$^{+}$ | Gao et al., (2010) |
| S26K | NIMGKG$^{27}$ | WSFI-GVE$^{208}$ | YWLACWIGNIAM$^{104}$ | GSLGGW$^{293}$ | SV I FTLVP Y$^{365}$ | No-apparent transport of Agm$^{2+}$ | Gao et al., (2009) |
| S26A | NIMGAG$^{27}$ | WSFI-GVE$^{208}$ | YWLACWIGNIAM$^{104}$ | GSLGGW$^{293}$ | SV I FTLVP Y$^{365}$ | No change in the selectivity of Arg$^{+}$/Arg$^{2+}$ | Tsai and Miller, (2013) |
| Y93L | NIMGSG$^{27}$ | WSFI-GVE$^{208}$ | LWLACWIGNIAM$^{104}$ | GSLGGW$^{293}$ | SV I FTLVP Y$^{365}$ | Significantly reduced Arg$^{+}$ transport | Fang et al., (2009) |
| Y93A | NIMGSG$^{27}$ | WSFI-GVE$^{208}$ | AWLACWIGNIAM$^{104}$ | GSLGGW$^{293}$ | SV I FTLVP Y$^{365}$ | No Agm$^{2+}$ binding, reduced Arg$^{-}$ binding reduced Agm$^{2+}$ transport | Gao et al., (2009) |
| Y93K | NIMGSG$^{27}$ | WSFI-GVE$^{208}$ | KWLACWIGNIAM$^{104}$ | GSLGGW$^{293}$ | SV I FTLVP Y$^{365}$ | No Agm$^{2+}$ binding, nearly no Arg$^{-}$ binding and Agm$^{2+}$ transport | Gao et al., (2009) |
| N101A | NIMGSG$^{27}$ | WSFI-GVE$^{208}$ | YWLACWIGAIAM$^{104}$ | GSLGGW$^{293}$ | SV I FTLVP Y$^{365}$ | Much reduced $K_d$ and $V_{m,}$ stabilized in an OF open conformation | Kowalczyk et al., (2010) |
| N101D | NIMGSG$^{27}$ | WSFI-GVE$^{208}$ | YWLACWIGDIAM$^{104}$ | GSLGGW$^{293}$ | SV I FTLVP Y$^{365}$ | 2-fold increased $V_m$ and $K_m$ | Kowalczyk et al., (2010) |
| W202L | NIMGSG$^{27}$ | LSFI-GVE$^{208}$ | YWLACWIGNIAM$^{104}$ | GSLGGW$^{293}$ | SV I FTLVP Y$^{365}$ | Reduced Arg$^{+}$ uptake | Fang et al., (2009) |
| E208D | NIMGSG$^{27}$ | WSFI-GVD$^{208}$ | YWLACWIGNIAM$^{104}$ | GSLGGW$^{293}$ | SV I FTLVP Y$^{365}$ | Nearly loss of Agm$^{2+}$ transport | Gao et al., (2009) |
| E208A | NIMGSG$^{27}$ | WSFI-GVA$^{208}$ | YWLACWIGNIAM$^{104}$ | GSLGGW$^{293}$ | SV I FTLVP Y$^{365}$ | Nearly loss of Agm$^{2+}$ transport | Gao et al., (2009) |
| W293Y | NIMGSG$^{27}$ | WSFI-GVE$^{208}$ | YWLACWIGNIAM$^{104}$ | GSLGGY$^{293}$ | SV I FTLVP Y$^{365}$ | 3 fold increased $K_m$, 8-fold decreased $V_{m,}$ no apparent $K_d$ | Kowalczyk et al., (2010) |
| W293L | NIMGSG$^{27}$ | WSFI-GVE$^{208}$ | YWLACWIGNIAM$^{104}$ | GSLGGL$^{293}$ | SV I FTLVP Y$^{365}$ | Loss of Agm$^{2+}$ transport | Fang et al., (2009) |
| Y365A | NIMGSG$^{27}$ | WSFI-GVE$^{208}$ | YWLACWIGNIAM$^{104}$ | GSLGGW$^{293}$ | SV I FTLVP A$^{365}$ | Binds but does not transport Agm$^{2+}$, nearly loss of Arg$^{+}$ binding and transport | Gao et al., (2009) |

The AdiC residues involved in substrate binding are colored in green. In blue are the residues forming the distal gate, and substitutions are indicated in red

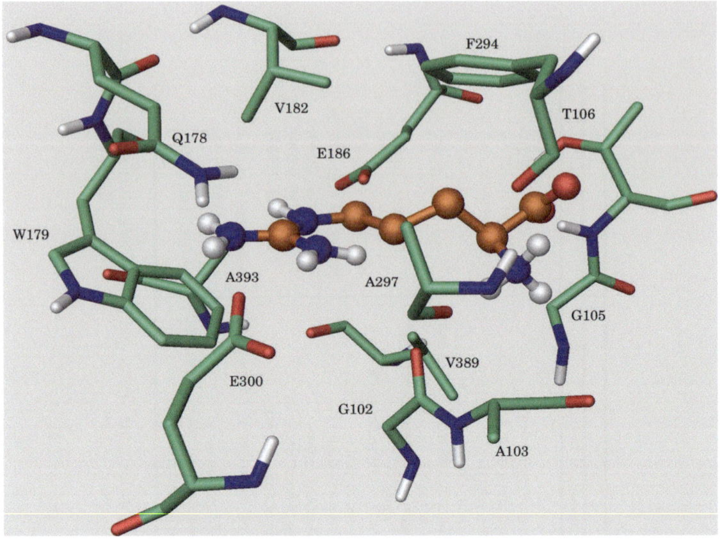


**Fig. 4.2** Close view of the binding pocket of Gap1 bound to arginine. The substrate Arg and Gap1 residues within 5 Å of the substrate are shown, respectively, as ball-and-stick and stick models. The arginine substrate is colored according to the following scheme: *brown*, carbon; *red*, oxygen; *blue*, nitrogen

The same approach applied to PrnB has yielded impressively similar results (Gournas et al. 2015). In this case, the substrate-interacting residues of the PrnB permease, which is highly specific for proline (Gournas et al. 2015; Tavoularis et al. 2003; Tazebay et al. 1995), were replaced with those of Put4, its *S. cerevisiae* ortholog capable of transporting, in addition to proline, the amino acids alanine, glycine, and γ-aminobutyric acid (GABA) and the toxic analog L-azetidine-2-carboxylic acid (AZC) (Andréasson et al. 2004; Horák and Ríhová 1982; Moretti et al. 1998; Omura et al. 2005; Regenberg et al. 1999). PrnB variants containing the Put4-mimicking substitution S130C, W351F, or T414S recognize Ala and AZC more efficiently than the wild-type permease, while showing no significant change in apparent $K_m$ for proline (Table 4.4, Gournas et al. 2015). Interestingly, residues $Ser^{130}$ and $Thr^{414}$ of PrnB are equivalent, respectively, to the specificity-determining residues $Ser^{176}$ and $Thr^{456}$ of Can1. Remarkably, furthermore, combining S130C with T414S suppresses the thermosensitivity caused by S130C and the low functionality due to T414S and leads to an improved $K_i$ for AZC, Ala, and Gly. The synergistic interaction observed for these residues in PrnB (as for the corresponding residues in Can1) suggests that these positions, in addition to interacting directly with substrates, could control substrate specificity by affecting structural transitions of the permeases during their transport cycle (Ghaddar et al. 2014a; Gournas et al. 2015). Two other Put4-mimicking substitutions introduced separately into PrnB, F252L and S253G in TM6, also lead to improved recognition of Ala and AZC. Most importantly, a quintuple mutant of PrnB (5 M in Table 4.4) featuring Put4 residues

**Table 4.3** YAT transporter mutants in residues involved in the recognition of the backbone of amino acids

| Transporter/Allele | TM1 | TM6 | Phenotype | Reference |
|---|---|---|---|---|
| **AdiC** | IMGSG$^{27}$ | VTLWSFI-GVE$^{208}$ | | |
| **Can1** | TIGTG$^{127}$ | NAAFTFQ-GTE$^{301}$ | | |
| **Lyp1** | TIGTG$^{105}$ | NAAFTYQ GTE$^{323}$ | | |
| **Gap1** | AIGTG$^{107}$ | TAAFSFA-GSE$^{300}$ | | |
| G107N | AIGTN$^{107}$ | TAAFSFA-GSE$^{300}$ | Loss of function | Ghaddar et al. (2014b) |
| **Put4** | AIGTG$^{127}$ | KGAFAFILGPE$^{328}$ | | |
| **PrnB** | CIGTG$^{58}$ | KSGFSFIFSPE$^{255}$ | | |
| C54A | AIGTG$^{58}$ | KSGFSFIFSPE$^{255}$ | $V_m$ is ~ 50% reduced | Kafasla et al., (2007) |
| G56A | CIATG$^{58}$ | KSGFSFIFSPE$^{255}$ | $V_m$ is ~ 70% of the wt | Vangelatos et al., (2009) |
| G56P | CIPTG$^{58}$ | KSGFSFIFSPE$^{255}$ | Loss of function | Gournas et al., (2015) |
| G56S | CISTG$^{58}$ | KSGFSFIFSPE$^{255}$ | Loss of function | Vangelatos et al., (2009) |
| T57A | CIGAG$^{58}$ | KSGFSFIFSPE$^{255}$ | Loss of function | Vangelatos et al., (2009) |
| T57S | CIGSG$^{58}$ | KSGFSFIFSPE$^{255}$ | Nearly wt | Vangelatos et al., (2009) |
| T57N | CIGNG$^{58}$ | KSGFSFIFSPE$^{255}$ | 6 times decreased $K_m$, significantly reduced $V_i$ | Gournas et al., (2015) |
| G58Q | CIGTQ$^{58}$ | KSGFSFIFSPE$^{255}$ | Loss of function | Gournas et al., (2015) |
| F248L | CIGTG$^{58}$ | KSGLSFIFSPE$^{255}$ | 4 times increased Km, $V_m$ is ~ 50% reduced | Tavoularis et al., (2003) |
| F248Y | CIGTG$^{58}$ | KSGYSFIFSPE$^{255}$ | Nearly wt | Gournas et al., (2015) |
| F248W | CIGTG$^{58}$ | KSGWSFIFSPE$^{255}$ | 2 times decreased $K_m$, significantly reduced $V_i$ | Gournas et al., (2015) |
| F250Y | CIGTG$^{58}$ | KSGFSYIFSPE$^{255}$ | 3 times decreased $K_m$, significantly reduced $V_i$ | Gournas et al., (2015) |
| F248W/F250W | CIGTG$^{58}$ | KSGWSWIFSPE$^{255}$ | 20 times decreased $K_m$, significantly reduced $V_i$ | Gournas et al., (2015) |
| E255D | CIGTG$^{58}$ | KSGFSFIFSPD$^{255}$ | 19 times decreased $K_m$, significantly reduced $V_i$ | Gournas et al., (2015) |
| E255Q | CIGTG$^{58}$ | KSGFSFIFSPQ$^{255}$ | Loss of function | Gournas et al., (2015) |
| **Bap2** | GIGTG$^{112}$ | TAYFSFG-GME$^{305}$ | | |
| I109T | GTGTG$^{112}$ | TAYFSFG-GME$^{305}$ | Loss of function | Usami et al., (2014) |
| G110V | GIVTG$^{112}$ | TAYFSFG-GME$^{305}$ | Loss of function | Usami et al., (2014) |
| T111A | GIGAG$^{112}$ | TAYFSFG-GME$^{305}$ | Loss of function | Usami et al., (2014) |
| E305D | GIGTG$^{112}$ | TAYFSFG-GMD$^{305}$ | Loss of function | Usami et al., (2014) |
| F299S | GIGTG$^{112}$ | TAYSSFG-GME$^{305}$ | Reduced transport capacity, transport of $Rb^+$ | Wright et al., (1997) |
| **Tat2** | SIGTG$^{99}$ | VASYSLG -GI E$^{286}$ | | |
| I96T | STGTG$^{99}$ | VASYSLG -GI E$^{286}$ | Loss of function | Kanda et al., (2013) |
| G97V | SIVTG$^{99}$ | VASYSLG -GI E$^{286}$ | Loss of function | Kanda et al., (2013) |
| T98A | SIGAG$^{99}$ | VASYSLG -GI E$^{286}$ | Loss of function | Kanda et al., (2013) |
| E286D | SIGTG$^{99}$ | VASYSLG -GI D$^{286}$ | Loss of function | Kanda et al., (2013) |
| **Hip1** | AIGTG$^{107}$ | TAAFSYS -GI E$^{301}$ | | |
| F295S | AIGTG$^{107}$ | TAASSYS -GI E$^{301}$ | Reduced transport capacity, transport of $Rb^+$ | Wright et al., (1997) |

Color coding is as in Table 4.2

**Table 4.4** YAT transporter mutants in residues involved in specificity

| Transporter/Allele | TM3 | TM6 | TM8 | TM10 | Phenotype | Reference |
|---|---|---|---|---|---|---|
| **AdiC** | YWLACWIGNIAM$^{104}$ | VTLWSFI-GVE$^{208}$ | GSLGGW$^{293}$ | LVSSVSV I FTLVP Y$^{365}$ | | |
| **Can1** | YWFSWAITFALE$^{184}$ | NAAFTFQ-GTE$^{301}$ | SAANSN$^{393}$ | WLLN ITGVAGFFAW$^{464}$ | | |
| Y173(D/H) | XWFSWAITFALE$^{184}$ | NAAFTFQ-GTE$^{301}$ | SAANSN$^{393}$ | WLLN ITGVAGFFAW$^{464}$ | Gained citrulline transport | Regenberg et al., (2001) |
| S176N | YWFNWAITFALE$^{184}$ | NAAFTFQ-GTE$^{301}$ | SAANSN$^{393}$ | WLLN ITGVAGFFAW$^{464}$ | Loss of function | Ghaddar et al., (2014a) |
| T180R | YWFSWAIRFALE$^{184}$ | NAAFTFQ-GTE$^{301}$ | SAANSN$^{393}$ | WLLN ITGVAGFFAW$^{464}$ | Loss of function | Ghaddar et al., (2014b) |
| E184D | YWFSWAITFALD$^{184}$ | NAAFTFQ-GTE$^{301}$ | SAANSN$^{393}$ | WLLN ITGVAGFFAW$^{464}$ | 10-fold improved Km, reduced $V_i$ | Ghaddar et al., (2014b) |
| E184Q | YWFSWAITFALQ$^{184}$ | NAAFTFQ-GTE$^{301}$ | SAANSN$^{393}$ | WLLN ITGVAGFFAW$^{464}$ | Loss of function | Ghaddar et al., (2014a) |
| W451(L/S/C) | YWFSWAITFALE$^{184}$ | NAAFTFQ-GTE$^{301}$ | SAANSN$^{393}$ | XLLN ITGVAGFFAW$^{464}$ | Gained citrulline transport | Regenberg et al., (2001) |
| F461S | YWFSWAITFALE$^{184}$ | NAAFTFQ-GTE$^{301}$ | SAANSN$^{393}$ | WLLN ITGVAGSFAW$^{464}$ | Gained citrulline transport | Regenberg et al., (2001) |
| T456S | YWFSWAITFALE$^{184}$ | NAAFTFQ-GTE$^{301}$ | SAANSN$^{393}$ | WLLN ISGVAGFFAW$^{464}$ | High affinity and capacity Arg and Lys transport | Ghaddar et al., (2014a) |
| T456A | YWFSWAITFALE$^{184}$ | NAAFTFQ-GTE$^{301}$ | SAANSN$^{393}$ | WLLN IAGVAGFFAW$^{464}$ | Slight impovement in the $K_m$ and $V_i$ of Lys | Ghaddar et al., (2014a) |
| S176N/T456S | YWFNWAITFALE$^{184}$ | NAAFTFQ-GTE$^{301}$ | SAANSN$^{393}$ | WLLN ISGVAGFFAW$^{464}$ | High affinity and capacity, specific Lys transporter | Ghaddar et al., (2014a) |
| **Lyp1** | YWFNWAITYAVE$^{201}$ | NAAFTYQ GTE$^{323}$ | SAANSN$^{415}$ | WLI N I STLAGLCAW$^{486}$ | | |
| **Gap1** | YMLQWLVVLPLE$^{186}$ | TAAFSFA-GSE$^{300}$ | SVGNSA$^{393}$ | WLLALSGLSSLFTW$^{464}$ | | |
| W179L | YMLQLLVVLPLE$^{186}$ | TAAFSFA-GSE$^{300}$ | SVGNSA$^{393}$ | WLLALSGLSSLFTW$^{464}$ | 4-fold decrease in the $K_m$ of Phe, impaired transport | Ghaddar et al., (2014b) |
| L185V | YMLQWLVVLPVE$^{186}$ | TAAFSFA-GSE$^{300}$ | SVGNSA$^{393}$ | WLLALSGLSSLFTW$^{464}$ | Citrulline non-transporting, glycine transporting mutant | Risinger et al., (2006) |
| A297V | YMLQWLVVLPLE$^{186}$ | TAAFSFV-GSE$^{300}$ | SVGNSA$^{393}$ | WLLALSGLSSLFTW$^{464}$ | Citrulline non-transporting, glycine transporting mutant | Risinger et al., (2006) |
| S388C | YMLQWLVVLPLE$^{186}$ | TAAFSFA-GSE$^{300}$ | CVGNSA$^{393}$ | WLLALSGLSSLFTW$^{464}$ | Impaired activity and signaling upon MTSEA treatment | Van Zeebroeck et al., (2009) |
| V389C | YMLQWLVVLPLE$^{186}$ | TAAFSFA-GSE$^{300}$ | SCGNSA$^{393}$ | WLLALSGLSSLFTW$^{464}$ | Impaired activity and signaling upon MTSEA treatment | Van Zeebroeck et al., (2009) |
| **Gnp1** | FCLQWLCVCPLE$^{246}$ | TAAFAFG-MSE$^{359}$ | SVANGA$^{452}$ | WLLAI SGLSQLFTW$^{523}$ | | |
| W239C | FCLQCLCVCPLE$^{246}$ | TAAFAFG-MSE$^{359}$ | SVANGA$^{452}$ | WLLAI SGLSQLFTW$^{523}$ | Gained citrulline transport | Regenberg et al., (2001) |
| **PrnB** | YWYSFAMLLASE$^{138}$ | KSGFSFIFSPE$^{255}$ | SSGNAW$^{351}$ | WFTNI TTVGGFI NW$^{422}$ | | |
| S130C | YWYCFAMLLASE$^{138}$ | KSGFSFIFSPE$^{255}$ | SSGNAW$^{351}$ | WFTNI TTVGGFI NW$^{422}$ | Thermosensitive, improved recognition of Ala and AZC | Gournas et al., (2015) |
| E138(D/Q) | YWYSFAMLLASX$^{138}$ | KSGFSFIFSPE$^{255}$ | SSGNAW$^{351}$ | WFTNI TTVGGFI NW$^{422}$ | Loss of function | Gournas et al., (2015) |
| K245L | YWYSFAMLLASE$^{138}$ | LSGFSFIFSPE$^{255}$ | SSGNAW$^{351}$ | WFTNI TTVGGFI NW$^{422}$ | 9-fold increase in $K_m$of proline | Tavoularis et al., (2003) |
| K245E | YWYSFAMLLASE$^{138}$ | ESGFSFIFSPE$^{255}$ | SSGNAW$^{351}$ | WFTNI TTVGGFI NW$^{422}$ | 2-fold increase in $K_m$of proline, impaired $V_i$ | Tavoularis et al., (2003) |
| K245C | YWYSFAMLLASE$^{138}$ | CSGFSFIFSPE$^{255}$ | SSGNAW$^{351}$ | WFTNI TTVGGFI NW$^{422}$ | Loss of function | Kafasla et al., (2007) |
| F252L | YWYSFAMLLASE$^{138}$ | KSGFSFILSPE$^{255}$ | SSGNAW$^{351}$ | WFTNI TTVGGFI NW$^{422}$ | Improved recognition of AZC | Gournas et al., (2015) |
| S253G | YWYSFAMLLASE$^{138}$ | KSGFSFIFGPE$^{255}$ | SSGNAW$^{351}$ | WFTNI TTVGGFI NW$^{422}$ | Improved recognition of Ala and AZC | Gournas et al., (2015) |
| W351F | YWYSFAMLLASE$^{138}$ | KSGFSFIFSPE$^{255}$ | SSGNAF$^{351}$ | WFTNI TTVGGFI NW$^{422}$ | Improved $K_i$ of AZC and Ala, impaired $V_i$, Ala transport | Gournas et al., (2015) |
| W351E | YWYSFAMLLASE$^{138}$ | KSGFSFIFSPE$^{255}$ | SSGNAE$^{351}$ | WFTNI TTVGGFI NW$^{422}$ | Impaired activity, Ala transport | Gournas et al., (2015) |
| W251A | YWYSFAMLLASE$^{138}$ | KSGFSFIFSPE$^{255}$ | SSGNAA$^{351}$ | WFTNI TTVGGFI NW$^{422}$ | Loss of function | Gournas et al., (2015) |
| T414A | YWYSFAMLLASE$^{138}$ | KSGFSFIFSPE$^{255}$ | SSGNAW$^{351}$ | WFTNI ATVGGFI NW$^{422}$ | Significantly impaired activity | Gournas et al., (2015) |
| T414S | YWYSFAMLLASE$^{138}$ | KSGFSFIFSPE$^{255}$ | SSGNAW$^{351}$ | WFTNI STVGGFI NW$^{422}$ | Improved $K_i$ of AZC, Ala and Gly, impaired $V_i$, Ala and Gly transport | Gournas et al., (2015) |
| S130C/T414S | YWYCFAMLLASE$^{138}$ | KSGFSFIFSPE$^{255}$ | SSGNAW$^{351}$ | WFTNI STVGGFI NW$^{422}$ | Improved recognition of Ala, Gly and AZC | Gournas et al., (2015) |
| 5M | YWYCFAMLLASE$^{138}$ | KSGFSFILGPE$^{255}$ | SSGNAF$^{351}$ | WFTNI STVGGFI NW$^{422}$ | Impoved $K_i$ of AZC, Ala, Gly and GABA, Ala and Gly transport | Gournas et al., (2015) |
| **Put4** | YFYCYV I LVAAE$^{211}$ | KGAFAFILGPE$^{328}$ | SAANAF$^{424}$ | WFSNI STI SGFLGW$^{495}$ | | |
| **Bap2** | YCFQWLTVLPLE$^{192}$ | TAYFSFG-GME$^{305}$ | SVANSS$^{398}$ | WLAAI AGLSELFTW$^{469}$ | | |
| Y181L | LCFQWLTVLPLE$^{192}$ | TAYFSFG-GME$^{305}$ | SVANSS$^{398}$ | WLAAI AGLSELFTW$^{469}$ | Loss of function | Usami et al., (2014) |
| V394A | YCFQWLTVLPLE$^{192}$ | TAYFSFG-GME$^{305}$ | SAANSS$^{398}$ | WLAAI AGLSELFTW$^{469}$ | Loss of function | Usami et al., (2014) |
| W469L | YCFQWLTVLPLE$^{192}$ | TAYFSFG-GME$^{305}$ | SVANSS$^{398}$ | WLAAI AGLSELFTL$^{469}$ | Functional | Usami et al., (2014) |
| **Tat2** | YVLQWFFVLPLE$^{178}$ | VASYSLG -GI E$^{286}$ | SVGNSC$^{375}$ | WLMAI AGLATCIVW$^{446}$ | | |
| Y167L | LVLQWFFVLPLE$^{178}$ | VASYSLG -GI E$^{286}$ | SVGNSC$^{375}$ | WLMAI AGLATCIVW$^{446}$ | Loss of function | Kanda et al., (2013) |
| S370G | YVLQWFFVLPLE$^{178}$ | VASYSLG -GI E$^{286}$ | GVGNSC$^{375}$ | WLMAI AGLATCIVW$^{446}$ | Loss of function | Kanda et al., (2013) |
| V371A | YVLQWFFVLPLE$^{178}$ | VASYSLG -GI E$^{286}$ | SAGNSC$^{375}$ | WLMAI AGLATCIVW$^{446}$ | Loss of function | Kanda et al., (2013) |
| W446L | YVLQWFFVLPLE$^{178}$ | VASYSLG -GI E$^{286}$ | SVGNSC$^{375}$ | WLMAI AGLATCIVL$^{446}$ | Partial loss of function | Kanda et al., (2013) |

Color coding is as in Table 4.2

surrounding the substrate-binding pocket shows a specificity profile remarkably similar to that of Put4 (Gournas et al. 2015).

It is also noteworthy that only proline-specific YATs have a lysine at the position corresponding to $Lys^{245}$ in TM6 of PrnB (Table 4.4). In induced-fit docking calculations, the corresponding residue in Put4 was also found to interact with the substrate proline. The residues at equivalent positions in other YATs of known specificity show interesting patterns, suggesting that this position could be another specificity determinant of YATs (Table 4.4) (Gournas et al. 2015; Vangelatos et al. 2009).

An interesting difference between the substrate-binding pockets of YATs and AdiC is the presence of a highly conserved Glu residue in TM3 of the former, corresponding to $Met^{104}$ in AdiC. This acidic residue is absolutely necessary for PrnB transport activity, since even the most conservative substitutions, i. e. its replacement with Asp or Gln, cause loss of function. Docking calculations suggest that its side chain interacts with the α-imino group of proline (Gournas et al. 2015). Substitution of Gln for the corresponding $Glu^{184}$ in Can1 produces a nonfunctional transporter, while the E184D Can1 variant displays a tenfold-lower apparent $K_m$ for Arg and significantly reduced transport activity (Ghaddar et al. 2014a, b). The results of protonation probability calculations suggest that the protonation state of $Glu^{184}$ could influence the orientation of arginine and lysine substrate side chains in the binding pocket of Can1. As AdiC functions as an antiporter while YATs are proton symporters, it is tempting to speculate that this residue, which is highly conserved in YATs, is involved in coupling proton transport to substrate transport.

#### 4.3.2.3 Role of Proximal and Middle Gate Residues

In the OF occluded crystal structure of AdiC, the arginine substrate is sandwiched between two bulky aromatic residues, $Trp^{202}$ of TM6 and $Trp^{293}$ of TM8 (Gao et al. 2010). It is proposed that these residues form two gates, the proximal and the middle gate, respectively. $Trp^{202}$ would act to control access to the binding pocket from the periplasm, while $Trp^{293}$ would move to allow the Arg substrate to "sink" down along the translocation pathway. Induced-fit dockings in the Can1 and Gap1 binding sites have revealed a similar sandwich configuration for arginine in Can1 and for both arginine and phenylalanine in Gap1 (Ghaddar et al. 2014a, b), but the middle-gate aromatic residue is provided by TM3 and not by TM8 as in AdiC. This residue is thus conserved in the 3D structures of AdiC and YATs but not in the corresponding aligned sequences. In Gap1, replacement of this aromatic residue ($Trp^{179}$) with Leu causes a strong reduction in intrinsic phenylalanine transport activity ($V_m$) and a significant decrease in apparent $K_m$ (Table 4.4). Furthermore, a genetic screen for Gnp1 mutants capable of transporting citrulline yielded a variant where the equivalent $Trp^{239}$ was replaced by Cys (Regenberg and Kielland-Brandt 2001).

#### 4.3.2.4 Role of Distal Gate Residues

In the OF crystal structures of AdiC, a third layer of residues located below the middle gate blocks access of the substrate to the cytoplasm. This distal gate is composed of Tyr$^{93}$ of TM3, Glu$^{208}$ of TM6, and Tyr$^{365}$ of TM10. These three residues interact with each other through H-bonds. Exit of the substrate from the binding pocket requires a movement of Trp$^{293}$ that is proposed to break the interactions between distal-gate residues and to promote the conformational changes enabling AdiC to adopt an IF open conformation (Gao et al. 2010). This model is supported by the crystal structure of GadC in the IF conformation (Ma et al. 2012).

Interestingly, induced-fit docking of proline in PrnB and of alanine in Put4 has revealed a potential second binding site, conserved in all YATs, just upstream from the distal gate (Gournas et al. 2015). In this pocket, substrates interact with the side chains of two of the residues of the PrnB distal gate, Tyr$^{127}$ and Glu$^{255}$ (corresponding to Tyr$^{93}$ and Glu$^{208}$ of AdiC). A negative charge at position 255 appears to be required for transport, since the E255Q variant is totally inactive whereas the E255D mutant retains some activity, displaying up to a 19-fold decreased $K_{m/i}$ for several tested ligands. In line with this, a Glu-to-Asp substitution at the corresponding position in Tat2 or Bap2 has been shown to cause loss of function. Similarly, in these two permeases, replacements of the residues corresponding to Tyr$^{93}$ and Tyr$^{365}$ of AdiC significantly affect or even abolish transport (Kanda and Abe 2013; Usami et al. 2014), suggesting that the putative distal-gate residues of YATs are indeed important for their function. In further support of this view, replacement of Tyr$^{173}$ of Can1 (corresponding to Tyr$^{93}$ of AdiC) by Glu or His was observed among mutants having gained the ability to transport citrulline, albeit with low affinity (Regenberg and Kielland-Brandt 2001).

### *4.3.3 Genetically Isolated Specificity-Affecting YAT Mutants Altered in TM Domains*

The properties of permease mutants isolated in genetic screens also highlight the importance of certain non-conserved residues of TM3, TM6, TM8, and TM10 in YAT substrate binding and specificity. For example, several substitutions in TM3, TM6, and TM10 have been shown to affect the substrate selectivity of Gap1 and Can1 (Table 4.4) (Regenberg and Kielland-Brandt 2001; Risinger et al. 2006). Notably, the Gap1 variants resulting from the L185V substitution in TM3 or the A297V substitution in TM6 can still transport glycine but not citrulline, while substitutions in TM3 (Tyr$^{173}$) or TM10 (Trp$^{451}$, Phe$^{461}$) of Can1 confer the ability to transport citrulline. In Ssy1, substitution of Ile for Thr$^{639}$ in TM8 results in a hyporesponsive variant displaying defective recognition of specific amino acids (Bernard and André 2001; Poulsen et al. 2008), while in contrast, replacements of residue Thr$^{382}$ in TM3 confer constitutive signaling (in the absence of amino acids) and hyperresponsiveness to low concentrations of amino acids (Gaber et al. 2003; Poulsen et al. 2008).

### 4.3.4 Role of Intracellular and Extracellular Loops in Substrate Selectivity

Examination of the crystal structures of several APC-superfamily transporters suggests that intracellular and extracellular TM-connecting loops could control access to the substrate binding site. Specifically, the intracellular loop connecting TM2 and TM3 occludes the intracellular surface of LeuT, while the extracellular TM7-TM8 loop forms an α-helix making numerous contacts with TM1 (Krishnamurthy and Gouaux 2012; Yamashita et al. 2005). Interestingly, Can1 variants featuring mutations in the TM2-3 loop (P148L, V149F, S152F) or the TM7-8 loop (Δ354–355, Y356N, Y356H) have been isolated as citrulline-transporting mutants (Regenberg and Kielland-Brandt 2001), while a Gap1 mutant having Ala substitutions at the equivalent TM2-3 loop positions is not functional, though it is normally targeted to the plasma membrane (Merhi et al. 2011). Ssy1 variants altered in the TM2-3 loop or in TM3 ($Ser^{351}$, $Val^{354}$, $Phe^{358}$) and showing hyperresponsiveness to low concentrations of amino acids have also been isolated. Since it is the OF conformation of Ssy1 that is believed to activate signaling (see Sect. 4.5.1), perhaps these mutant permeases preferably adopt this conformation, either because the IF conformation is destabilized or because of impaired binding of cytosolic amino acids to the substrate binding pocket. Interestingly, these substitutions are close to the residue corresponding to $Tyr^{74}$ in the TM2-3 loop of AdiC, shown to act as a pH sensor occluding the substrate binding site from the cytoplasm at high pH (Wang et al. 2014).

Substitutions in the external TM1-2 loop (P113L) or the internal TM6-7 loop (G308A, P313S) also confer to Can1 the ability to transport citrulline (Regenberg and Kielland-Brandt 2001). These substitutions might impact substrate specificity indirectly, by affecting the positioning of TM1 and TM6 (important for substrate binding and conformational transitions) (Gao et al. 2010). Such a role is compatible with the non-functionality of a Gap1 mutant ($SESV^{310}$/AAAA) carrying four Ala substitutions at equivalent positions in the intracellular TM6-7 loop (Merhi et al. 2011). Apart from the above-mentioned mutants altered in TM3 and TM6 (Table 4.4), Gap1 mutants having specifically lost citrulline uptake activity have been found to contain substitutions in the internal TM10-11 loop (A479V) or the external TM7-8 loop (V363G, A365V/T590A) (Risinger et al. 2006). This further supports the idea that loops play an important role in determining the YAT substrate specificity.

### 4.3.5 Summary

The above-summarized studies support the view that YATs adopt the LeuT fold, with formation of a substrate binding pocket by residues of TM1, TM3, TM6, TM8, and TM10. Residues in the unwound segments of TM1 and TM6 (Table 4.3), present in highly conserved motifs, promote recognition of the invariant α-carboxyl and α-amino groups of amino acids. On the other hand, permease selectivity is determined by residues in TM3, TM6, TM8, and TM10 (Table 4.4), which are variable among YATs, and by residues in loops connecting specific TMs. The identified specificity-determining residues seem to play this role directly or indirectly and in many ways. First, residues in TM3, TM6, and TM10 (corresponding to $Ala^{96}$, $Cys^{97}$, $Asn^{101}$, $Trp^{202}$, $Ile^{205}$, and $Ser^{357}$ in AdiC and $Lys^{245}$ in PrnB) likely interact with the side chain of the amino acid substrate and/or affect structural transitions in the transport cycle. Second, residues of the unwound segment of TM6 (equivalent to $Phe^{252}$ and $Ser^{253}$ of PrnB) might interact with the substrate and/or affect the flexibility of TM6 (Gournas et al. 2015). Third, residues of TM3 and TM8 (the equivalent of $Trp^{293}$ and $Tyr^{93}$ of AdiC), shaping the middle and distal gates, might interact directly with substrates and/or impose steric constraints. Fourth, residues in the internal TM2-3 loop and the external TM7-8 loop could affect selectivity by either participating in occlusion of the substrate-binding site or affecting the positioning of nearby TMs. These conclusions largely converge with those of studies focusing on structure-function relationships in several bacterial transporters closely related to the YAT proteins. Mutations affecting the activity or specificity of the PheP, AroP, GabP, and CadB permeases have indeed been located in TM1, TM3, TM6, TM8, TM9, and TM10 and in loops TM2-3 and TM8-9 (for a summary, see Vangelatos et al. 2009). The regions altered by these mutations overlap impressively with those found to be important in substrate transport by YATs.

Finally, it is worth mentioning that hydrophobic/steric interactions, which are more difficult to identify in modeling and docking studies, are likely important for the specificity of YATs towards amino acids with hydrophobic side chains. Two studies have approached this question. Measurements of L-[$^3$H]-proline uptake via PrnB in the presence of various competing proline analogs suggest that the size of the hydrophobic side chain ring is important in recognition (Gournas et al. 2015). The second study, based on the calculated inhibition of Bap2-mediated uptake of radiolabelled leucine in the presence of several known Bap2 substrates, came up with an interesting finding: that for all substrates except tryptophan, the degree of inhibition correlated very well with the *logP* value (octanol–water partition coefficient) (Usami et al. 2014). The authors thus suggest that the primary determinant of Bap2 selectivity, rather than side-chain size or interactions, is the efficiency of substrate partition to the buried Bap2 binding pocket.

## 4.4 Substrate-Mediated Endocytosis

### *4.4.1 Inhibition of Transporters by Selective Endocytosis: General Principles*

Selective sorting into the endocytic pathway is a widespread means of inhibiting the function of specific transporters in fungi (Haguenauer-Tsapis and André 2004). The subsequent fate of the protein resembles the diverse situations encountered for cell surface receptors in more complex cells. One possible ultimate fate is delivery of the transporter into the lumen of the vacuole (the lysosome in fungi), where it is degraded. Alternatively, the internalized transporter may be recycled to the plasma membrane. Situations have also been described where the transporter is stored in internal compartments and delivered to the cell surface in response to specific stimuli (Kim et al. 2007; Roberg et al. 1997). In a recent study, the Jen1 lactate transporter internalized after glucose addition was found to traffic from early endosomes to the Golgi, whence it can be recycled to the plasma membrane (if the glucose signal is only transient) or be sorted to late endosomes and targeted to the vacuole for degradation (when the glucose signal is sustained) (Becuwe and Léon 2014). These diverse situations illustrate the complexity of transporter regulation at intracellular traffic level.

It is well established that initial sorting of most fungal transporters to the endocytic pathway is induced by their ubiquitylation (Duprè et al. 2004; Lauwers et al. 2010). This modification is catalyzed by a unique HECT-family ubiquitin (Ub) ligase, Rsp5 in *S. cerevisiae* and HulA in *A. nidulans*. Both of these are orthologous to the Nedd4-type Ub ligases of mammalian cells. Yeast Rsp5 interacts with its target proteins via its PPXY-sequence-binding WW domains (Shcherbik et al. 2004). As fungal transporters have no PPXY sequences, their recognition by Rsp5 relies on adaptors possessing such motifs. The main class of Rsp5 adaptors is represented by arrestin-like proteins named ART (Arrestin-related trafficking) or alpha-arrestins (Lin et al. 2008; Nikko et al. 2008). Initial studies on this class of proteins focused on *A. nidulans* and led to characterization of CreD and PalF, involved, respectively, in carbon catabolite repression and ambient pH signaling (Boase and Kelly 2004; Herranz et al. 2005). Subsequent work on yeast, which possesses 13 alpha-arrestins (Art1 to −11, Bul1 and Bul2), has highlighted the central role of ARTs in Rsp5-mediated ubiquitylation of many transporters and receptors (Alvaro et al. 2014; Becuwe et al. 2012; MacGurn et al. 2012). A central question addressed in several recent studies is: what are the molecular events inducing transporter ubiquitylation by the ART-Rsp5 machinery? The general picture emerging from these studies is that transporter ubiquitylation is stimulated by molecular events affecting either the adaptors (which would acquire the ability to interact with transporters) or the transporters themselves (which would gain the ability to recruit adaptors). As detailed below, these principles are particularly well illustrated by studies focusing on several yeast amino acid transporters.

### *4.4.2 Substrate-Induced Endocytosis of Fungal Amino Acid Transporters*

Many fungal transporters undergo Ub-dependent endocytosis in response to excess substrates. This was initially observed in the case of the inositol permease Itr1 (Lai et al. 1995) and later in studies on several metal transporters (Alr1, Zrt1, Smf1,

Smf2, Ctr1), the uracil permease Fur4, the phosphate permease Pho84, and the high-affinity hexose transporter Hxt6/7 (Haguenauer-Tsapis and André 2004; Krampe et al. 1998; Nikko and Pelham 2009). Endocytosis induced by excess substrates has also been reported for several fAATs, such as Gap1 (Rubio-Texeira et al. 2012; Stanbrough and Magasanik 1995), Can1 (Opekarová et al. 1998), Lyp1, Mup1 (Lin et al. 2008), Dip5 (Hatakeyama et al. 2010), and Tat2 (Nikko and Pelham 2009). Although the precise mechanisms mediating substrate-induced downregulation of most of these transporters remain elusive, investigators have dissected this process in two recent studies, focused on Gap1 and Can1, and have unraveled mechanisms that might be valid for other transporters as well (Ghaddar et al. 2014b; Merhi and André 2012).

#### 4.4.2.1 Endocytosis in Response to Increased Intracellular Levels of Amino Acids

The Gap1 permease is active and stable at the plasma membrane in cells growing under poor nitrogen supply conditions. The role of Gap1 (a high-affinity permease) under such conditions is to scavenge any amino acids that might be available in the medium even at low concentration. When a high concentration of amino acids is provided to the cells, Gap1 is rapidly ubiquitylated on one of two possible lysines ($Lys^9$ or $Lys^{16}$) present in its N-terminal tail, and this induces its downregulation. Gap1 ubiquitylation depends on the Rsp5 Ub ligase and on the redundant Bul1 and Bul2 arrestin-like adaptor proteins (Hein et al. 1995; Helliwell et al. 2001; Soetens et al. 2001; Springael and André 1998). At least two pathways promote Gap1 ubiquitylation under these conditions. The first pathway responds to an increase in the internal concentration of amino acids. It can also be stimulated by providing cells with $NH_4^+$, provided that $NH_4^+$ can be assimilated into amino acids. The increase in internal amino acids results in relief from inhibition exerted on the Bul proteins by the Npr1 kinase. Under poor nitrogen supply conditions, the Bul adaptors are indeed phosphorylated in an Npr1-dependent manner. This phosphorylation promotes association of the Bul adaptors with the 14-3-3 proteins, which inhibit them. When the internal pool of amino acids increases, the TORC1 kinase complex is activated, in turn causing phosphoinhibition of the Npr1 kinase. Furthermore, the PP2A-type Sit4 phosphatase, also under TORC1 control, promotes dephosphorylation of the Bul proteins. Once dephosphorylated, the Bul adaptors dissociate from the 14-3-3 proteins and induce Gap1 ubiquitylation (Merhi and André 2012). This ubiquitylation requires an N-terminal region of Gap1 very close to the Ub acceptor lysines and likely corresponding to an interaction site for the dephosphorylated Buls (Ghaddar et al. 2014b; Merhi et al. 2011). This first pathway, inducing Gap1 ubiquitylation in response to internal amino acids, is thus due to stimulation of the Bul adaptors. It is also fully operational in the case of Gap1 mutants having lost their transport activity.

#### 4.4.2.2 Activity-Dependent Endocytosis

In contrast, the second pathway promoting Gap1 ubiquitylation in response to amino acids occurs only upon substrate transport (Ghaddar et al. 2014b). This pathway was unraveled thanks to isolation of a Gap1 mutant resistant to ubiquitylation in response to internal amino acids (namely Gap1-112, altered in the putative N-terminal binding site for dephosphorylated Buls). This Gap1 mutant still undergoes ubiquitylation and endocytosis in response to transport of substrates, e. g. phenylalanine (Phe). The results obtained in the study of this endocytosis (see below) support the model that Gap1, when catalyzing substrate transport, adopts a transient conformation promoting its ubiquitylation. This substrate-transport-induced ubiquitylation again depends on Rsp5 and on the redundant adaptors Bul1 and Bul2, but several observations support the view that the Buls are not stimulated under these conditions. In other words, the ubiquitylation-inducing signal seems to come from the Gap1 permease itself. For instance, specific stimulation of this pathway does not coincide with any detectable change in Bul-protein post-translational modification patterns. Namely, two active Gap1 mutants undergo Bul-dependent constitutive endocytosis, suggesting that they adopt a conformation mimicking that induced by substrate transport. Even though the arrestin-like Bul proteins are still phosphorylated and thus likely associated with the 14-3-3 proteins, they promote ubiquitylation of these Gap1 mutants. It thus seems that the Bul adaptors, even when bound to the 14-3-3 proteins, are able to associate with Gap1 in a particular conformation adopted during substrate transport (Ghaddar et al. 2014b). The cytosolic regions of Gap1 recognized by the Bul adaptors under these conditions are thus likely unveiled when the permease is catalytically active. Control of Gap1 permease ubiquitylation by amino acids is thus complex, being stimulated by both internal and external amino acids. Each pathway can, however, be studied separately thanks to Gap1 mutants insensitive to only one pathway.

Regarding Can1, the arginine-specific permease, a recent study reported that this protein, unlike Gap1, does not undergo endocytosis upon TORC1 activation or loss of Npr1 function, but it does undergo endocytosis upon substrate transport (Ghaddar et al. 2014b). Furthermore, a Can1 mutant converted to a lysine-specific permease (as a result of the S176N and T456S substitutions) has gained the property of undergoing Lys-induced endocytosis. This endocytosis depends on the Rsp5 Ub ligase and on the arrestin-like protein Art1. Interestingly, the post-translational modification state of the Art1 adaptor does not change detectably upon Can1-mediated arginine uptake. This supports a model in which the Can1 permease, upon substrate transport, adopts a conformation capable of recruiting the Art1 adaptor, thereby causing its own Rsp5-dependent ubiquitylation.

Regarding the inability of Can1 to undergo endocytosis in response to TORC1 activation, other authors in fact report other conclusions: that the Npr1 kinase inhibits by phosphorylation the ability of Art1 to promote Can1 endocytosis, and that this inhibition is relieved in response to TORC1 stimulation by internal amino acids (MacGurn et al. 2011). The causes of this apparent discrepancy between studies on Can1 regulation remain unclear, and could be linked to the use of different conditions to trigger an increase in internal amino acids: addition of $NH_4^+$ to cells growing on a poor nitrogen source (Ghaddar et al. 2014b) or addition of cycloheximide (MacGurn et al. 2011).

### 4.4.3 Structural Determinants of Activity-Dependent Substrate-Induced Endocytosis

It seems likely that the Ub-dependent endocytosis of many other fungal transporters is stimulated by conformational changes coupled to transport catalysis. For example, such a process was initially reported for two fungal permeases, the uric acid-xanthine transporter (UapA) of *A. nidulans* (Gournas et al. 2010) and the uracil permease (Fur4) of *S. cerevisiae* (Keener and Babst 2013; Séron et al. 1999). Interestingly, it now seems possible to identify from studies on UapA, Fur4, Can1, and Gap1 endocytosis some features that might be common to transporters undergoing this type of control by their own substrates.

First, intracellular accumulation of the substrates does not seem to be the trigger of endocytosis upon substrate transport. For instance, the Gap1-112 mutant (resistant to endocytosis stimulated by TORC1 activation) undergoes endocytosis in response to external amino acids, but is stable at the plasma membrane of cells growing on a high concentration of $NH_4^+$, conditions under which the internal pool of amino acids is high (Merhi and André 2012). Similarly, the UapA permease, downregulated by external uric acid, appears stable at the plasma membrane in two situations where the internal uric acid concentration is high: in a mutant lacking uricase and after excessive external supply of the uric acid precursor hypoxanthine. Further study of wild-type UapA and of mutants with an altered $K_m$ has demonstrated a correlation between the lowest concentration of substrate required to trigger permease endocytosis and the $K_m$ (Gournas et al. 2010). As for the Can1 arginine permease, it appears insensitive to Arg accumulation in an arginase mutant (Gournas and André, unpublished observations), and inactive Can1 variants (S176N, T180R, S176N/T456S) do not undergo endocytosis in response to external Arg, even if the amino acid actively enters the cell via functional Can1 and Gap1 permeases coproduced with the mutant Can1 (Ghaddar et al. 2014b). These situations suggest that activity-dependent substrate-induced endocytosis of permeases does not result from binding of the substrate to a cytosolic region of the permease. Yet this does not rule out the possibility that specific permeases might undergo endocytosis via such a mechanism. For instance, an inactive Fur4 variant (K272A) unable to bind external uracil also fails to undergo downregulation in response to an increase in internal uracil, in contrast to wild-type Fur4 (Keener and Babst 2013). Whether a YAT prone to endocytosis upon uptake of external substrate can have a similar fate in response to intracellular accumulation of this substrate might depend on the intrinsic stability of the OF occluded state, in which the substrate-binding site is not accessible to internal amino acids, and on the relative affinity of the cytosol-facing, substrate-binding site for internal amino acids, accessible when the protein switches to more IF states.

Second, external ligand binding is necessary but not sufficient to elicit endocytosis. A shift of the permease to a particular conformation is most likely also required. It is true that transport-abolishing substitutions predicted to prevent substrate binding confer full protection against endocytosis in response to substrate transport, as illustrated, for instance, by observations on the Can1(T180R), Gap1(G107N), and UapA(N409D), and (E356Q) mutants (Ghaddar et al. 2014b; Gournas et al. 2010). Yet other observations indicate that substrate binding is not sufficient to elicit endocytosis. For example, the Gap1(W179L) mutant, which seems to bind Phe normally (its apparent $K_m$ for Phe transport is even lower than that of the wild type) but displays a strongly reduced intrinsic Phe uptake activity ($V_m$), is largely resistant to Phe-transport-triggered endocytosis (Ghaddar et al. 2014b). Furthermore, UapA resists endocytosis in the presence of 3-methyl-xanthine (Gournas et al. 2010), a non-transportable high-affinity competitive inhibitor of xanthine uptake (Papageorgiou et al. 2008).

Third, completion of the transport cycle is not needed to elicit endocytosis. This conclusion is based on observed situations where permeases undergo endocytosis in the presence of non-transportable ligands. For instance, the inactive Can1(E184Q) mutant is still efficiently internalized in the presence of Arg; the Can1(T456S/S176N) mutant converted to a lysine-specific permease undergoes endocytosis in the presence of thialysine, a toxic analog of Lys, without being able to catalyze its uptake (Ghaddar et al. 2014b). Similarly, the inactive Fur4(K272A) mutant has lost the ability to catalyze uptake of the immunosuppressant drug leflunomide, but it can still undergo endocytosis in the presence of this compound (Keener and Babst 2013). Finally, Gap1 is reported to undergo ubiquitylation in response to the L-Asp-γ-Phe peptide, without being able to catalyze its uptake (Van Zeebroeck et al. 2014).

Fourth, highly active Can1 and UapA mutants with a broadened selectivity have been found to resist endocytosis. A prominent example is the T456S mutant of Can1, a high-capacity transporter of both Arg and Lys (Ghaddar et al. 2014a). This mutant unexpectedly fails to undergo endocytosis in the presence of external Arg or Lys (Ghaddar et al. 2014b). Similarly, certain UapA selectivity mutants, also able to transport hypoxanthine (F528A, T526M and T526L), have been found to be partially or totally resistant to endocytosis stimulated by uric acid or hypoxanthine (Gournas et al. 2010). Although these observations may seem to contradict the above conclusions, they could mean that the endocytosis-inducing conformation adopted during the transport cycle must be stable enough to promote efficient downregulation. In the case of active variants that resist endocytosis, this transient conformation might not be sufficiently stable and populated to promote efficient endocytosis. In support of this model, several of the mutants just mentioned display an increased intrinsic transport capacity ($V_m$), suggesting that some of the conformational transitions involved in transport occur at a faster rate. Interestingly, the T456S substitution in Can1 creates a void in TM10 (Ghaddar et al. 2014a), while the above-mentioned substitutions in UapA lie within the gating domain of the transporter, which controls access to the substrate-binding site (Diallinas 2014). These features might indeed facilitate structural transitions during the transport cycle.

Finally, the substrate-elicited conformation that induces permease endocytosis most likely differs from the basal state by a remodeling of cytosolic regions (tails and/or loops), allowing recruitment of arrestin-like adaptors and thus favoring the action of the Ub ligase. As comparison of the OF open and OF occluded conformations of AdiC revealed only minor changes in the cytoplasmic part of the transporter (Gao et al. 2010), it seems logical to hypothesize that substrate-induced endocytosis is promoted by a shift to a subsequent conformation, i. e. to an IF state. In support of this view, structural analyses of the dopamine transporter and LeuT have shown that interactions formed in the OF open conformation between intracellular loops and membrane-proximal residues of the N-terminal tail are disrupted when the permease shifts to the IF open conformation. In particular, TM1 undergoes an ample movement that could induce a significant conformational change in the directly connected N-terminal tail (Kniazeff et al. 2008; Krishnamurthy and Gouaux 2012). Such conformational changes might be the cause of substrate-transport-induced permease ubiquitylation. For instance, a study of uracil-induced downregulation of the uracil permease, which as a member of the NCS1 family adopting the LeuT fold (Krypotou et al. 2015; Krypotou et al. 2012), has yielded a model where a Loop Interaction Domain (LID) in the N-terminal tail interacts with internal loops only in the OF conformation (Keener and Babst 2013). The conformational changes coupled to transport catalysis or due to misfolding under various stress conditions could disrupt these LID-loop interactions, thereby exposing N-terminal Ub-acceptor lysines. A similar mechanism might account for substrate-induced endocytosis of other proteins adopting the LeuT fold. Consistently, Gap1 mutants altered in the N-terminal tail or the second intracellular loop undergo constitutive Ub-dependent endocytosis (Merhi et al. 2011). These mutant permeases do not seem to be misfolded, as they traffic normally to the cell surface, and when stabilized at the plasma membrane, they are active (albeit less than the wild type) and display near-normal apparent $K_m$ values. Thus, these cytosolic regions of Gap1 might interact and undergo conformational changes during the transport cycle, thereby resulting in Gap1 ubiquitylation via the Bul arrestins and the Rsp5 Ub ligase (Ghaddar et al. 2014b).

In support of the view that substrate-transport-induced permease ubiquitylation is due solely to a particular conformation adopted by the permease during transport catalysis, the arrestin-like adaptors interacting with the Ub ligase do not seem to undergo, in this pathway, any detectable change in their post-translational modification state. An illustration is the case, discussed above, of the Bul adaptors in the context of Phe-transport induced endocytosis of Gap1 (Ghaddar et al. 2014b). Similarly, the phosphorylation and ubiquitylation states of the Art1 adaptor do not undergo any detectable change upon Arg-induced Can1 endocytosis (Ghaddar et al. 2014b), in contrast to cycloheximide-induced Can1 endocytosis (Lin et al. 2008; MacGurn et al. 2011). In the case of the UapA permease, two situations have been shown to promote ubiquitylation of a single, carboxy-terminal Ub acceptor lysine residue: substrate transport and addition of $NH_4^+$ (Gournas et al.

2010). In both situations the same arrestin-like adaptor, ArtA, is essential to UapA ubiquitylation (Karachaliou et al. 2013). Furthermore, ubiquitylation of ArtA appears important in both pathways. Yet ArtA ubiquitylation increases in response to $NH_4^+$ addition but not in response to substrate transport. Lastly, substrate-induced endocytosis of the yeast glutamate/aspartate permease Dip5 requires the Aly1/2 adaptors (Hatakeyama et al. 2010; O'Donnell et al. 2013). In this case, calcineurin-dependent dephosphorylation of Aly1 is necessary for Aly1–mediated Dip5 endocytosis but neither causes Aly1 to dissociate from the 14-3-3 proteins nor affects its ubiquitylation (O'Donnell et al. 2013). It has not been clarified in this case whether Dip5 endocytosis is due to transport catalysis, intracellular accumulation of aspartate/glutamate, or both.

## 4.5 Signaling Mediated by YAT Proteins

As mentioned above, the successive conformational changes typically undergone by an amino acid permease during transport catalysis also affect parts of the protein exposed to the cytosol. These structural changes can thus potentially be translated into intracellular responses, such as recruitment of arrestin-like adaptors causing Rsp5-mediated ubiquitylation and endocytosis (see above) or, perhaps, activation of signaling pathways. Accordingly, the general amino acid permease Gap1 is reported to activate protein kinase A (PKA) in response to transport of at least some of its substrates (Donaton et al. 2003; Van Zeebroeck et al. 2009, 2014). In cases where transporter-mediated signaling is elicited by a conformation preceding substrate release into the cytosol, the transporter might secondarily loose its transport activity while conserving its signaling function. Such an evolutionary scenario might have given rise to the Ssy1 sensor of external amino acids, very similar in sequence to YAT-family transporters.

### *4.5.1 Amino-Acid Signaling Mediated by the Ssy1 Sensor*

The Ssy1 sensor is essential to transcriptional induction of several amino acid permease genes in response to a wide variety of external amino acids (Didion et al. 1998; Iraqui et al. 1999; Klasson et al. 1999). Although Ssy1 itself has not been associated with any detectable amino-acid uptake activity, it very likely undergoes some conformational changes in response to amino acid binding. This leads ultimately to stimulation of the Ssy5 endoprotease, which in turn activates by endoproteolytic processing the cytoplasmic precursor forms of the Stp1 and Stp2 transcription factors (Abdel-Sater et al. 2004; Andréasson et al. 2006; Andréasson and Ljungdahl 2002; Wielemans et al. 2010). The Ptr3 protein is essential to signal transduction from Ssy1 to Ssy5, and as it interacts with both proteins, this transduction is likely direct (Bernard and André 2001; Klasson et al. 1999; Liu et al.

2008). The Ssy1, Ptr3, and Ssy5 proteins are thus thought to assemble into a plasma-membrane-associated complex (named SPS), although the cell-surface localization of Ptr3 and Ssy5 has yet to be visualized by imaging techniques. Casein kinase 1 (CK1) and the $SCF^{Grr1}$ ubiquitin-ligase complex, two factors involved in several other signaling pathways, are also essential to transduction of the amino-acid signal from Ssy1 to Ssy5 (Abdel-Sater et al. 2004; Iraqui et al. 1999). In the absence of external amino acids, the N-terminal inhibitory subunit of Ssy5, generated by autocleavage, is bound to the endoprotease's catalytic subunit, which is then inactive (Abdel-Sater et al. 2004; Andréasson et al. 2006). In response to amino-acid binding to Ssy1, the inhibitory subunit is phosphorylated by CK1, and this modification elicits its ubiquitylation by the $SCF^{Grr1}$ complex (Abdel-Sater et al. 2011; Omnus et al. 2011). The exact mechanism by which this ubiquitylation activates the Ssy5 catalytic subunit remains to be clarified. In one model, ubiquitylation of the Ssy5 prodomain promotes its degradation by the proteasome, thereby relieving the inhibition exerted on the Ssy5 catalytic domain (Omnus et al. 2011; Pfirrmann et al. 2010). In an alternative model, ubiquitylation of the prodomain is sufficient to relieve inhibition of the catalytic subunit, and both subunits remain bound to each other; subsequent degradation of the prodomain is a secondary, slower process inessential to activating the catalytic subunit (Abdel-Sater et al. 2011).

Although Ssy1 is related in sequence to classical amino acid permeases, it has four particular features. First, its N-terminal cytosolic tail, essential to signaling, is much longer, and is reported to interact with Ptr3 (Bernard and André 2001; Didion et al. 1998; Iraqui et al. 1999; Klasson et al. 1999; Liu et al. 2008). Second, the GTG motif present in TM1 of all YATs (Table 4.2) and crucial to their transport activity (Table 4.3) is not conserved in Ssy1. This difference could at least partially explain why this non-conventional YAT lacks transport activity. Third, two external loops (the TM5-6 loop and the TM7-8 loop) are also much larger than those of amino acid permeases. Remarkably, the TM7-8 loop of many LeuT-fold-adopting transporters, which participates in occlusion of the OF conformation (Sect. 3.4), contains a conserved α-helical region which interacts with residues of TM1 (Krishnamurthy and Gouaux 2012). This raises the question of whether the particularly large size of this loop in Ssy1 might be related to its signaling function and/or to its lack of transport activity. Finally, the glutamate residue of TM3, which is highly conserved in YATs and thought to contribute to proton symport (see above), corresponds to a glutamine in Ssy1. This difference could contribute to the lack of transport activity of Ssy1.

The signaling-eliciting conformation adopted by Ssy1 in response to amino acid binding remains unknown. In the model proposed by Kielland-Brandt and collaborators (Wu et al. 2006), Ssy1 would mainly adopt an IF open state in the absence of external amino acids. Upon amino acid binding, the sensor would be stabilized in an OF conformation, which would promote signaling. This model is supported by the interesting observation that intracellular accumulation of leucine competitively inhibits sensing of external amino acids. A high internal leucine concentration would thus stabilize the IF, non-signaling conformation (Wu et al. 2006). A prediction of this model is that Ssy1 should sense the ratio of amino acid concentrations across the plasma membrane, rather than the absolute extracellular concentration of amino acids. This type of sensing should be more feasible for transporter-like proteins than for classical receptors unable to bind ligands on the cytosolic side (Wu et al. 2006).

### 4.5.2 Amino-Acid Signaling Mediated by the Gap1 Transporter

The group of J. Thevelein has reported that the Gap1 permease activates protein kinase A (PKA) upon detection of its amino acid substrates (Donaton et al. 2003). According to the proposed model, this activation of PKA does not result from intracellular accumulation of the transported amino acid but from a signaling effect mediated by Gap1 having switched to a specific substrate-elicited conformation (Kriel et al. 2011; Schothorst et al. 2013). The term "transceptor" was therefore coined to designate proteins such as Gap1, acting both as transporters and as receptors. Other permeases reported to act as transceptors include the Pho84 phosphate permease (Popova et al. 2010), the Mep2 ammonium permease (Van Nuland et al. 2006), and the Sul1 and Sul2 sulfate permeases (Kankipati et al. 2015).

The role of Gap1 in signaling has typically been studied in yeast cells grown first on a rich nitrogen medium and then transferred for 24 h to a glucose-rich medium devoid of any nitrogen source. In these nitrogen-starved cells, the Gap1 permease is present at the plasma membrane, although a large fraction is also present in the vacuolar lumen (Rubio-Texeira et al. 2012), probably because of stress-elicited Gap1 endocytosis (Crapeau et al. 2014). PKA activity is low in these nitrogen-starved cells, as reflected, e. g., by decreased activity of trelahase, well known to be under PKA control. Upon addition of citrulline, which can enter cells only via Gap1 (Grenson et al. 1970), the trehalase activity rapidly increases in a Gap1-dependent manner (Donaton et al. 2003). That this activation in response to amino acids is mediated by PKA is supported by the fact that it correlates with phosphorylation of trehalase residues Ser21 and Ser83, whose amino-acid sequence contexts meet the criteria for consensus sites of PKA phosphorylation (Schepers et al. 2012). Yeast cells synthesize three catalytic PKA subunits, Tpk1, Tpk2, and Tpk3. As trehalase is still activated by amino acids in each *tpk* double mutant, it has been deduced that each Tpk subunit, including the less active Tpk3, is sufficient to promote trehalase activation after Gap1-mediated uptake of amino acids. Furthermore, as this activation does not correlate with any increase in cAMP, activation of each Tpk subunit is proposed to occur through an unusual mechanism that does not involve cAMP (Donaton et al. 2003; Durnez et al. 1994; Hirimburegama et al. 1992). Trehalase activation by amino acids is also reported to be defective in a *sch9* mutant lacking the S6 kinase (Crauwels et al. 1997), known to be positively controlled by the TORC1 kinase complex. As Sch9 is related to PKA (both belong to the AGC kinase family) and as Sch9 and PKA can phosphorylate the same residues of some target proteins (Cocklin and Goebl 2011), it seems that Sch9 might also play a role in Gap1-mediated trehalase activation.

That Gap1-dependent trehalase activation in response to citrulline is due to signaling is supported by several observations. First, it occurs normally in an *arg1* mutant lacking the first enzyme of citrulline catabolism. Second, uptake of an equal amount of citrulline into *gap1* mutant cells (as measured after addition of 50 mM citrulline in a 10-sec uptake assay) fails to activate trehalase (in the typical 30-min interval of the enzyme assay) (Donaton et al. 2003). Third, beta-alanine and D-histidine, two Gap1 substrates that cannot be used as nitrogen sources, stimulate trehalase in a Gap1-dependent manner (Donaton et al. 2003). Fourth, two mutant Gap1 permeases have been shown to cause constitutive activation of trehalase (Donaton et al. 2003), although further analysis revealed that this phenotype is observed only in cells harboring an as yet uncharacterized genomic mutation. This mutation, proposed to result in stabilization of the mutant Gap1 permeases at the plasma membrane, has been shown to confer Gap1-mediated increased resistance to rapamycin, an inhibitor of the TORC1 kinase complex (Lagatie 2007). Another observation further supports the view that trehalase activation following Gap1-mediated transport is actually due to signaling: at high concentration, the L-Leu-Gly dipeptide, which competitively inhibits Gap1-mediated citrulline uptake but is not significantly transported into cells via Gap1, activates trehalase in a partially Gap1-dependent manner (this activation is reduced 2-fold in a *gap1* mutant) (Van Zeebroeck et al. 2009). It has thus been proposed that the Gap1-mediated signaling (hereafter called Gap1-m-S) causing trehalase activation is elicited by a specific conformation of Gap1, adopted after substrate binding and preceding (or not necessarily coupled to) substrate release into the cytosol (Kriel et al. 2011; Schothorst et al. 2013). A prediction of this model is that it should, in principle, be possible to isolate a Gap1 mutant lacking transport activity but still competent for Gap1-m-S.

Further analysis of Gap1-m-S has shown it to depend on two Gap1 residues, Ser$^{388}$ and Val$^{389}$, located in TM8. The S388C and V389C Gap1 variants show normal transport activity and Gap1-m-S, but unlike wild-type Gap1, they undergo inhibition of both activities in response to addition of 2-aminoethylmethane-thiosulfonate, a compound reacting covalently with exposed sulfhydryl groups. Furthermore, citrulline protects both Gap1 mutants against these inhibitory effects. The conclusion has been drawn that the side chains of Ser$^{388}$ and Val$^{389}$ are exposed in the substrate-binding pocket of the permease, and that binding of amino acids or peptides to this pocket is crucial to initiating Gap1-m-S (Van Zeebroeck et al. 2009). Accordingly, analysis of the AdiC-based structural model of Gap1 (Ghaddar et al. 2014b) shows that the side chain of Val$^{389}$ is indeed oriented towards the substrate-recognizing pocket of Gap1. Furthermore, replacement of Ala$^{393}$, located just one helix turn downstream, inhibits Gap1 activity (our unpublished result). On the other hand, as Ser$^{388}$ precedes Val$^{389}$ immediately in a helix, its side chain is unlikely to be oriented also towards the substrate binding site, at least in the occluded state.

Although amino acids or peptides must interact with the substrate-binding pocket of Gap1 to elicit Gap1-m-S, this binding is not sufficient, since certain compounds such as L-Asp-γ-Phe can competitively inhibit Gap1-mediated transport without promoting Gap1-m-S (Van Zeebroeck et al. 2009). Furthermore, the amino acids

Trp, His, and Lys are reported to be efficiently transported by Gap1 and to compete with citrulline uptake but not to promote efficient Gap1-m-S (Van Zeebroeck et al. 2014). This suggests that the Gap1 conformation that mediates signaling must be stable enough and sufficiently populated to promote efficient trehalase activation, and that whether these conditions are met depends on the compound interacting with the substrate-binding pocket of the permease (Kriel et al. 2011; Schothorst et al. 2013).

The specific Gap1 conformation that elicits signaling remains unknown. It seems to be more populated under high saturation, since cintrulline does not trigger Gap1-m-S at concentrations below 0.5 mM, a concentration well above the apparent $K_m$ of Gap1 for citrulline transport (0.07 mM) (Van Zeebroeck et al. 2014). The signaling conformation also seems to differ from that which Gap1 adopts in response to substrate binding and which favors recruitment of the Bul arrestins and ubiquitylation (Ghaddar et al. 2014b), since some substrates (e. g. His) promote efficient Gap1 endocytosis but not Gap1-m-S (Van Zeebroeck et al. 2014). Further work is thus needed to determine the exact Gap1 conformation which elicits signaling, and to elucidate the molecular mechanisms through which this particular Gap1 conformation is sensed and the signal transduced to downstream targets.

## 4.6 Prospects

Current studies focusing on the structural modeling and mutational analysis of fAATs will certainly shed further light on the mechanisms of amino acid recognition and translocation by these transporters. To gain a detailed view of these mechanisms, however, it is necessary to solve the atomic structures of several of these proteins, ideally bound to their substrates and adopting different conformations of the transport cycle. Although this still seems highly challenging, recent efforts to determine the crystal structures of specific eukaryotic secondary transporters have been successful (Penmatsa et al. 2013). Furthermore, there is hope that new technologies such as single-particle electron cryomicroscopy (Cheng et al. 2015) will lift the current limitations to determining membrane-protein tertiary structure. Another highly challenging but crucial goal for the future is to learn how substrate transport influences the structures of fAAT cytosolic loops and tails. This should help to unravel the molecular details of fAAT recognition by factors involved in transporter ubiquitylation and signaling. Another prospect is to extend these studies to amino acid transporters of more complex species, including *H. sapiens* (Hinz et al. 2015; Peura et al. 2013). This line of research should continue to benefit from the invaluable knowledge gained through the study of fAATs.

**Acknowledgments** We thank all lab members for regular and fruitful discussions. Work in the laboratories of B.A. and M.P. was funded by the F.R.S-FRNS (Fonds National de Recherche Scientifique, Belgium, 3.4559.12, T.1107.15) and an ARC grant (AUWB 2010-15-2) of the Fédération Wallonie-Bruxelles. M.P. is a Senior Research Associate and C.G. and E.V.K. are postdoctoral Researchers at the FRS-FNRS (Belgium).

## References

Abdel-Sater F, El-Bakkoury M, Urrestarazu A, Vissers S, André B (2004) Amino acid signaling in yeast: casein kinase I and the Ssy5 endoprotease are key determinants of endoproteolytic activation of the membrane-bound Stp1 transcription factor. Mol Cell Biol 24:9771–9785. doi:10.1128/MCB.24.22.9771-9785.2004

Abdel-Sater F, Jean C, Merhi A, Vissers S, André B (2011) Amino acid signaling in yeast: activation of Ssy5 protease is associated with its phosphorylation-induced ubiquitylation. J Biol Chem 286:12006–12015. doi:10.1074/jbc.M110.200592

Abramson J, Smirnova I, Kasho V, Verner G, Kaback HR, Iwata S (2003) Structure and mechanism of the lactose permease of Escherichia coli. Science 301:610–615. doi:10.1126/science.1088196

Alvaro CG, O'Donnell AF, Prosser DC, Augustine AA, Goldman A, Brodsky JL, Cyert MS, Wendland B, Thorner J (2014) Specific α-arrestins negatively regulate Saccharomyces cerevisiae pheromone response by down-modulating the G-protein coupled receptor Ste2. Mol Cell Biol. doi:10.1128/MCB.00230-14

André B, Hein C, Grenson M, Jauniaux JC (1993) Cloning and expression of the UGA4 gene coding for the inducible GABA-specific transport protein of Saccharomyces cerevisiae. Mol Gen Genet 237:17–25

Andréasson C, Ljungdahl PO (2002) Receptor-mediated endoproteolytic activation of two transcription factors in yeast. Genes Dev 16:3158–3172. doi:10.1101/gad.239202

Andréasson C, Neve EPA, Ljungdahl PO (2004) Four permeases import proline and the toxic proline analogue azetidine-2-carboxylate into yeast. Yeast 21:193–199. doi:10.1002/yea.1052

Andréasson C, Heessen S, Ljungdahl PO (2006) Regulation of transcription factor latency by receptor-activated proteolysis. Genes Dev 20:1563–1568. doi:10.1101/gad.374206

Apostolaki A, Erpapazoglou Z, Harispe L, Billini M, Kafasla P, Kizis D, Peñalva MA, Scazzocchio C, Sophianopoulou V (2009) AgtA, the dicarboxylic amino acid transporter of Aspergillus nidulans, is concertedly down-regulated by exquisite sensitivity to nitrogen metabolite repression and ammonium-elicited endocytosis. Eukaryot Cell 8:339–352. doi:10.1128/EC.00270-08

Bajmoczi M, Sneve M, Eide DJ, Drewes LR (1998) TAT1 encodes a low-affinity histidine transporter in Saccharomyces cerevisiae. Biochem Biophys Res Commun 243:205–209. doi:10.1006/bbrc.1998.8082

Becuwe M, Léon S (2014) Integrated control of transporter endocytosis and recycling by the arrestin-related protein Rod1 and the ubiquitin ligase Rsp5. Elife. doi:10.7554/eLife.03307

Becuwe M, Herrador A, Haguenauer-Tsapis R, Vincent O, Léon S (2012) Ubiquitin-mediated regulation of endocytosis by proteins of the arrestin family. Biochem Res Int 2012:242764. doi:10.1155/2012/242764

Bernard F, André B (2001) Genetic analysis of the signalling pathway activated by external amino acids in *Saccharomyces cerevisiae*. Mol Microbiol 41:489–502

Boase NA, Kelly JM (2004) A role for creD, a carbon catabolite repression gene from Aspergillus nidulans, in ubiquitination. Mol Microbiol 53:929–940. doi:10.1111/j.1365-2958.2004.04172.x

Boles E, André B (2004) Role of transporter-like sensors in glucose and amino acid signalling in yeast. Top Curr Genet 9:121–153

Cavero S, Vozza A, del Arco A, Palmieri L, Villa A, Blanco E, Runswick MJ, Walker JE, Cerdán S, Palmieri F, Satrústegui J (2003) Identification and metabolic role of the mitochondrial aspartate-glutamate transporter in Saccharomyces cerevisiae. Mol Microbiol 50:1257–1269. doi:10.1046/j.1365-2958.2003.03742.x

Chahomchuen T, Hondo K, Ohsaki M, Sekito T, Kakinuma Y (2009) Evidence for Avt6 as a vacuolar exporter of acidic amino acids in Saccharomyces cerevisiae cells. J Gen Appl Microbiol 55:409–417

Cheng Y, Grigorieff N, Penczek PA, Walz T (2015) A primer to single-particle cryo-electron microscopy. Cell 161:438–449. doi:10.1016/j.cell.2015.03.050

Cocklin R, Goebl M (2011) Nutrient sensing kinases PKA and Sch9 phosphorylate the catalytic domain of the ubiquitin-conjugating enzyme Cdc34. PLoS ONE 6:e27099. doi:10.1371/journal.pone.0027099
Conrad M, Schothorst J, Kankipati HN, Van Zeebroeck G, Rubio-Texeira M, Thevelein JM (2014) Nutrient sensing and signaling in the yeast Saccharomyces cerevisiae. FEMS Microbiol Rev 38:254–299. doi:10.1111/1574-6976.12065
Crapeau M, Merhi A, André B (2014) Stress conditions promote yeast Gap1 permease ubiquitylation and down-regulation via the arrestin-like Bul and Aly proteins. J Biol Chem 289:22103–22116. doi:10.1074/jbc.M114.582320
Crauwels M, Donaton MC, Pernambuco MB, Winderickx J, de Winde JH, Thevelein JM (1997) The Sch9 protein kinase in the yeast Saccharomyces cerevisiae controls cAPK activity and is required for nitrogen activation of the fermentable-growth-medium-induced (FGM) pathway. Microbiology 143(Pt 8):2627–2637
Diallinas G (2008) Biochemistry. An almost-complete movie. Science 322:1644–1645. doi:10.1126/science.1168107
Diallinas G (2014) Understanding transporter specificity and the discrete appearance of channel-like gating domains in transporters. Front Pharmacol 5:207. doi:10.3389/fphar.2014.00207
Didion T, Regenberg B, Jørgensen MU, Kielland-Brandt MC, Andersen HA (1998) The permease homologue Ssy1p controls the expression of amino acid and peptide transporter genes in Saccharomyces cerevisiae. Mol Microbiol 27:643–650
Donaton MCV, Holsbeeks I, Lagatie O, Van Zeebroeck G, Crauwels M, Winderickx J, Thevelein JM (2003) The Gap1 general amino acid permease acts as an amino acid sensor for activation of protein kinase A targets in the yeast *Saccharomyces cerevisiae*. Mol Microbiol 50:911–929
Duprè S, Urban-Grimal D, Haguenauer-Tsapis R (2004) Ubiquitin and endocytic internalization in yeast and animal cells. Biochim Biophys Acta 1695:89–111. doi:10.1016/j.bbamcr.2004.09.024
Durnez P, Pernambuco MB, Oris E, Argüelles JC, Mergelsberg H, Thevelein JM (1994) Activation of trehalase during growth induction by nitrogen sources in the yeast Saccharomyces cerevisiae depends on the free catalytic subunits of cAMP-dependent protein kinase, but not on functional Ras proteins. Yeast 10:1049–1064. doi:10.1002/yea.320100807
Faham S, Watanabe A, Besserer GM, Cascio D, Specht A, Hirayama BA, Wright EM, Abramson J (2008) The crystal structure of a sodium galactose transporter reveals mechanistic insights into Na+/sugar symport. Science 321:810–814. doi:10.1126/science.1160406
Fang P, Li X, Wang J, Niu L, Teng M (2010) Structural basis for the specificity of the GAE domain of yGGA2 for its accessory proteins Ent3 and Ent5. Biochemistry 49:7949–7955. doi:10.1021/bi1010255
Felder T, Bogengruber E, Tenreiro S, Ellinger A, Sá-Correia I, Briza P (2002) Dtr1p, a multidrug resistance transporter of the major facilitator superfamily, plays an essential role in spore wall maturation in Saccharomyces cerevisiae. Eukaryot Cell 1:799–810. doi:10.1128/EC.1.5.799-810.2002
Fotiadis D, Kanai Y, Palacín M (2013) The SLC3 and SLC7 families of amino acid transporters. Mol Aspects Med 34:139–158. doi:10.1016/j.mam.2012.10.007
Gaber RF, Ottow K, Andersen HA, Kielland-Brandt MC (2003) Constitutive and hyperresponsive signaling by mutant forms of Saccharomyces cerevisiae amino acid sensor Ssy1. Eukaryot Cell 2:922–929
Gao X, Wang J, Keppler-Ross S, Dean N (2005) ERS1 encodes a functional homologue of the human lysosomal cystine transporter. FEBS J 272:2497–2511. doi:10.1111/j.1742-4658.2005.04670.x
Gao X, Lu F, Zhou L, Dang S, Sun L, Li X, Wang J, Shi Y (2009) Structure and mechanism of an amino acid antiporter. Science 324:1565–1568. doi:10.1126/science.1173654
Gao X, Zhou L, Jiao X, Lu F, Yan C, Zeng X, Wang J, Shi Y (2010) Mechanism of substrate recognition and transport by an amino acid antiporter. Nature 463:828–832. doi:10.1038/nature08741

Ghaddar K, Krammer E, Mihajlovic N, Brohée S, André B, Prévost M (2014a) Converting the yeast arginine Can1 permease ito a lysine permease. J Biol Chem 289:7232–7246
Ghaddar K, Merhi A, Saliba E, Krammer E, Prévost M, André B (2014b) Substrate-induced ubiquitylation and endocytosis of yeast amino Acid permeases. Mol Cell Biol 34:4447–4463. doi:10.1128/MCB.00699-14
Gournas C, Amillis S, Vlanti A, Diallinas G (2010) Transport-dependent endocytosis and turnover of a uric acid-xanthine permease. Mol Microbiol 75:246–260. doi:10.1111/j.1365-2958.2009.06997.x
Gournas C, Evangelidis T, Athanasopoulos A, Mikros E, Sophianopoulou V (2015) The Aspergillus nidulans proline permease as a model for understanding the factors determining substrate binding and specificity of fungal amino acid transporters. J Biol Chem. doi:10.1074/jbc.M114.612069
Grauslund M, Didion T, Kielland-Brandt MC, Andersen HA (1995) BAP2, a gene encoding a permease for branched-chain amino acids in Saccharomyces cerevisiae. Biochim Biophys Acta 1269:275–280
Grenson M, Mousset M, Wiame JM, Béchet J (1966) Multiplicity of the amino acid permeases in *Saccharomyces cerevisiae*. I. Evidence for a specific arginine-transporting system. Biochim Biophys Acta 127:325–338
Grenson M, Hou C, Crabeel M (1970) Multiplicity of the amino acid permeases in Saccharomyces cerevisiae. IV. Evidence for a general amino acid permease. J Bacteriol 103:770–777
Haguenauer-Tsapis R, André B (2004) Membrane trafficking of yeast transporters: mechanisms and physiological control of downregulation. Springer, Berlin/Heidelberg, pp 273–323
Hatakeyama R, Kamiya M, Takahara T, Maeda T (2010) Endocytosis of the aspartic acid/glutamic acid transporter Dip5 is triggered by substrate-dependent recruitment of the Rsp5 ubiquitin ligase via the arrestin-like protein Aly2. Mol Cell Biol 30:5598–5607. doi:10.1128/MCB.00464-10
Hein C, Springael JY, Volland C, Haguenauer-Tsapis R, André B (1995) *NPI1*, an essential yeast gene involved in induced degradation of Gap1 and Fur4 permeases, encodes the Rsp5 ubiquitin-protein ligase. Mol Microbiol 18:77–87
Helliwell SB, Losko S, Kaiser CA (2001) Components of a ubiquitin ligase complex specify polyubiquitination and intracellular trafficking of the general amino acid permease. J Cell Biol 153:649–662. doi:10.1083/jcb.153.4.649
Herranz S, Rodríguez JM, Bussink H, Sánchez-Ferrero JC, Arst HN, Peñalva MA, Vincent O (2005) Arrestin-related proteins mediate pH signaling in fungi. Proc Natl Acad Sci U S A 102:12141–12146. doi:10.1073/pnas.0504776102
Hinz KM, Meyer K, Kinne A, Schülein R, Köhrle J, Krause G (2015) Structural insights into thyroid hormone transport mechanisms of the L-Type Amino Acid Transporter 2. Mol Endocrinol. doi:10.1210/me.2015-1044
Hirimburegama K, Durnez P, Keleman J, Oris E, Vergauwen R, Mergelsberg H, Thevelein JM (1992) Nutrient-induced activation of trehalase in nutrient-starved cells of the yeast Saccharomyces cerevisiae: cAMP is not involved as second messenger. J Gen Microbiol 138:2035–2043
Hoffmann W (1985) Molecular characterization of the CAN1 locus in Saccharomyces cerevisiae. A transmembrane protein without N-terminal hydrophobic signal sequence. J Biol Chem 260:11831–11837
Horák J, Ríhová L (1982) L-Proline transport in Saccharomyces cerevisiae. Biochim Biophys Acta 691:144–150
Huang Y, Lemieux MJ, Song J, Auer M, Wang D (2003) Structure and mechanism of the glycerol-3-phosphate transporter from Escherichia coli. Science 301:616–620. doi:10.1126/science.1087619
Hutchings H, Stahmann KP, Roels S, Espeso EA, Timberlake WE, Arst HN, Tilburn J (1999) The multiply-regulated gabA gene encoding the GABA permease of Aspergillus nidulans: a score of exons. Mol Microbiol 32:557–568. doi:10.1046/j.1365-2958.1999.01371.x

Iraqui I, Vissers S, Bernard F, De Craene JO, Boles E, Urrestarazu A, André B (1999) Amino acid signaling in *Saccharomyces cerevisiae*: a permease-like sensor of external amino acids and F-Box protein Grr1p are required for transcriptional induction of the AGP1 gene, which encodes a broad-specificity amino acid permease. Mol Cell Biol 19:989–1001

Isnard AD, Thomas D, Surdin-Kerjan Y (1996) The study of methionine uptake in Saccharomyces cerevisiae reveals a new family of amino acid permeases. J Mol Biol 262:473–484. doi:10.1006/jmbi.1996.0529

Jardetzky O (1966) Simple allosteric model for membrane pumps. Nature 211:969–970

Jauniaux JC, Grenson M (1990) GAP1, the general amino acid permease gene of Saccharomyces cerevisiae. Nucleotide sequence, protein similarity with the other bakers yeast amino acid permeases, and nitrogen catabolite repression. Eur J Biochem 190:39–44. doi:10.1111/j.1432-1033.1990.tb15542.x

Jézégou A, Llinares E, Anne C, Kieffer-Jaquinod S, O'Regan S, Aupetit J, Chabli A, Sagné C, Debacker C, Chadefaux-Vekemans B, Journet A, André B, Gasnier B (2012) Heptahelical protein PQLC2 is a lysosomal cationic amino acid exporter underlying the action of cysteamine in cystinosis therapy. Proc Natl Acad Sci U S A 109:E3434–E3443. doi:10.1073/pnas.1211198109

Kafasla P, Bouzarelou D, Frillingos S, Sophianopoulou V (2007) The proline permease of Aspergillus nidulans: functional replacement of the native cysteine residues and properties of a cysteine-less transporter. Fungal Genet Biol 44:615–626. doi:10.1016/j.fgb.2007.01.011

Kanda N, Abe F (2013) Structural and functional implications of the yeast high-affinity tryptophan permease Tat2. Biochemistry 52:4296–4307. doi:10.1021/bi4004638

Kankipati HN, Rubio-Texeira M, Castermans D, Diallinas G, Thevelein JM (2015) Sul1 and Sul2 sulfate transceptors signal to protein kinase a upon exit of sulfur starvation. J Biol Chem. doi:10.1074/jbc.M114.629022

Karachaliou M, Amillis S, Evangelinos M, Kokotos AC, Yalelis V, Diallinas G (2013) The arrestin-like protein ArtA is essential for ubiquitination and endocytosis of the UapA transporter in response to both broad-range and specific signals. Mol Microbiol 88:301–317. doi:10.1111/mmi.12184

Kaur J, Bachhawat AK (2007) Yct1p, a novel, high-affinity, cysteine-specific transporter from the yeast Saccharomyces cerevisiae. Genetics 176:877–890. doi:10.1534/genetics.107.070342

Keener JM, Babst M (2013) Quality control and substrate-dependent downregulation of the nutrient transporter Fur4. Traffic 14:412–427. doi:10.1111/tra.12039

Kim Y, Deng Y, Philpott CC (2007) GGA2- and ubiquitin-dependent trafficking of Arn1, the ferrichrome transporter of Saccharomyces cerevisiae. Mol Biol Cell 18:1790–1802. doi:10.1091/mbc.E06-09-0861

Klasson H, Fink GR, Ljungdahl PO (1999) Ssy1p and Ptr3p are plasma membrane components of a yeast system that senses extracellular amino acids. Mol Cell Biol 19:5405–5416

Kniazeff J, Shi L, Loland CJ, Javitch JA, Weinstein H, Gether U (2008) An intracellular interaction network regulates conformational transitions in the dopamine transporter. J Biol Chem 283:17691–17701. doi:10.1074/jbc.M800475200

Kosugi A, Koizumi Y, Yanagida F, Udaka S (2001) MUP1, high affinity methionine permease, is involved in cysteine uptake by Saccharomyces cerevisiae. Biosci Biotechnol Biochem 65:728–731

Kowalczyk L, Ratera M, Paladino A, Bartoccioni P, Errasti-Murugarren E, Valencia E, Portella G, Bial S, Zorzano A, Fita I, Orozco M, Carpena X, Vázquez-Ibar JL, Palacín M (2011) Molecular basis of substrate-induced permeation by an amino acid antiporter. Proc Natl Acad Sci U S A 108:3935–3940. doi:10.1073/pnas.1018081108

Krampe S, Stamm O, Hollenberg CP, Boles E (1998) Catabolite inactivation of the high-affinity hexose transporters Hxt6 and Hxt7 of *Saccharomyces cerevisiae* occurs in the vacuole after internalization by endocytosis. FEBS Lett 441:343–347

Kriel J, Haesendonckx S, Rubio-Texeira M, Van Zeebroeck G, Thevelein JM (2011) From transporter to transceptor: signaling from transporters provokes re-evaluation of complex trafficking and regulatory controls: endocytic internalization and intracellular trafficking of nutrient transceptors may, at least in part, be governed by their signaling function. Bioessays 33:870–879. doi:10.1002/bies.201100100

Krishnamurthy H, Gouaux E (2012) X-ray structures of LeuT in substrate-free outward-open and apo inward-open states. Nature 481:469–474. doi:10.1038/nature10737

Krypotou E, Kosti V, Amillis S, Myrianthopoulos V, Mikros E, Diallinas G (2012) Modeling, substrate docking, and mutational analysis identify residues essential for the function and specificity of a eukaryotic purine-cytosine NCS1 transporter. J Biol Chem 287:36792–36803. doi:10.1074/jbc.M112.400382

Krypotou E, Evangelidis T, Bobonis J, Pittis AA, Gabaldón T, Scazzocchio C, Mikros E, Diallinas G (2015) Origin, diversification and substrate specificity in the family of NCS1/FUR transporters. Mol Microbiol. doi:10.1111/mmi.12982

Lagatie O (2007) Amino acid sensing and signalling by Gap1 in the yeast Saccharomyces cerevisiae. PhD thesis, KU Leuven, Leuven, Belgium

Lai K, Bolognese CP, Swift S, McGraw P (1995) Regulation of inositol transport in *Saccharomyces cerevisiae* involves inositol-induced changes in permease stability and endocytic degradation in the vacuole. J Biol Chem 270:2525–2534

Lauwers E, Erpapazoglou Z, Haguenauer-Tsapis R, André B (2010) The ubiquitin code of yeast permease trafficking. Trends Cell Biol 20:196–204. doi:10.1016/j.tcb.2010.01.004

Li ZY, Haase E, Brendel M (1991) Hyper-resistance to nitrogen mustard in Saccharomyces cerevisiae is caused by defective choline transport. Curr Genet 19:423–427

Lin CH, MacGurn JA, Chu T, Stefan CJ, Emr SD (2008) Arrestin-related ubiquitin-ligase adaptors regulate endocytosis and protein turnover at the cell surface. Cell 135:714–725. doi:10.1016/j.cell.2008.09.025

Liu Z, Thornton J, Spírek M, Butow RA (2008) Activation of the SPS amino acid-sensing pathway in Saccharomyces cerevisiae correlates with the phosphorylation state of a sensor component, Ptr3. Mol Cell Biol 28:551–563. doi:10.1128/MCB.00929-07

Lolkema JS, Slotboom DJ (2008) The major amino acid transporter superfamily has a similar core structure as Na+-galactose and Na+-leucine transporters. Mol Membr Biol 25:567–570. doi:10.1080/09687680802541177

Ma D, Lu P, Yan C, Fan C, Yin P, Wang J, Shi Y (2012) Structure and mechanism of a glutamate-GABA antiporter. Nature 483:632–636. doi:10.1038/nature10917

MacGurn JA, Hsu P, Smolka MB, Emr SD (2011) TORC1 regulates endocytosis via Npr1-mediated phosphoinhibition of a ubiquitin ligase adaptor. Cell 147:1104–1117. doi:10.1016/j.cell.2011.09.054

MacGurn JA, Hsu P, Emr SD (2012) Ubiquitin and membrane protein turnover: from cradle to grave. Annu Rev Biochem 81:231–259. doi:10.1146/annurev-biochem-060210-093619

Merhi A, André B (2012) Internal amino acids promote Gap1 permease ubiquitylation via TORC1/Npr1/14-3-3-dependent control of the Bul arrestin-like adaptors. Mol Cell Biol 32:4510–4522. doi:10.1128/MCB.00463-12

Merhi A, Gérard N, Lauwers E, Prévost M, André B (2011) Systematic mutational analysis of the intracellular regions of yeast Gap1 permease. PLoS ONE 6:e18457. doi:10.1371/journal.pone.0018457

Moretti MB, García SC, Batlle A (1998) UGA4 gene expression in Saccharomyces cerevisiae depends on cell growth conditions. Cell Mol Biol (Noisy-le-Grand) 44:585–590

Nikko E, Pelham HRB (2009) Arrestin-mediated endocytosis of yeast plasma membrane transporters. Traffic 10:1856–1867. doi:10.1111/j.1600-0854.2009.00990.x

Nikko E, Sullivan JA, Pelham HRB (2008) Arrestin-like proteins mediate ubiquitination and endocytosis of the yeast metal transporter Smf1. EMBO Rep 9:1216–1221. doi:10.1038/embor.2008.199

O'Donnell AF, Huang L, Thorner J, Cyert MS (2013) A calcineurin-dependent switch controls the trafficking function of α-arrestin Aly1/Art6. J Biol Chem 288:24063–24080. doi:10.1074/jbc.M113.478511

Omnus DJ, Pfirrmann T, Andréasson C, Ljungdahl PO (2011) A phosphodegron controls nutrient-induced proteasomal activation of the signaling protease Ssy5. Mol Biol Cell 22:2754–2765. doi:10.1091/mbc.E11-04-0282

Omura F, Fujita A, Miyajima K, Fukui N (2005) Engineering of yeast Put4 permease and its application to lager yeast for efficient proline assimilation. Biosci Biotechnol Biochem 69:1162–1171

Omura F, Hatanaka H, Nakao Y (2007) Characterization of a novel tyrosine permease of lager brewing yeast shared by Saccharomyces cerevisiae strain RM11-1a. FEMS Yeast Res 7:1350–1361. doi:10.1111/j.1567-1364.2007.00310.x

Opekarová M, Caspari T, Pinson B, Bréthes D, Tanner W (1998) Post-translational fate of CAN1 permease of *Saccharomyces cerevisiae.* Yeast 14:215–224. doi:10.1002/(SICI)1097-0061(199802)14:33.0.CO;2-3

Palmieri L, De Marco V, Iacobazzi V, Palmieri F, Runswick MJ, Walker JE (1997) Identification of the yeast ARG-11 gene as a mitochondrial ornithine carrier involved in arginine biosynthesis. FEBS Lett 410:447–451

Palmieri L, Lasorsa FM, Iacobazzi V, Runswick MJ, Palmieri F, Walker JE (2000) Identification of the mitochondrial carnitine carrier in Saccharomyces cerevisiae. FEBS Lett 462:472–476

Papageorgiou I, Gournas C, Vlanti A, Amillis S, Pantazopoulou A, Diallinas G (2008) Specific interdomain synergy in the UapA transporter determines its unique specificity for uric acid among NAT carriers. J Mol Biol 382:1121–1135. doi:10.1016/j.jmb.2008.08.005

Penmatsa A, Wang KH, Gouaux E (2013) X-ray structure of dopamine transporter elucidates antidepressant mechanism. Nature 503:85–90. doi:10.1038/nature12533

Peura L, Malmioja K, Huttunen K, Leppänen J, Hämäläinen M, Forsberg MM, Gynther M, Rautio J, Laine K (2013) Design, synthesis and brain uptake of LAT1-targeted amino acid prodrugs of dopamine. Pharm Res 30:2523–2537

Pfirrmann T, Heessen S, Omnus DJ, Andréasson C, Ljungdahl PO (2010) The prodomain of Ssy5 protease controls receptor-activated proteolysis of transcription factor Stp1. Mol Cell Biol 30:3299–3309. doi:10.1128/MCB.00323-10

Popova Y, Thayumanavan P, Lonati E, Agrochão M, Thevelein JM (2010) Transport and signaling through the phosphate-binding site of the yeast Pho84 phosphate transceptor. Proc Natl Acad Sci U S A 107:2890–2895. doi:10.1073/pnas.0906546107

Poulsen P, Gaber RF, Kielland-Brandt MC (2008) Hyper- and hyporesponsive mutant forms of the Saccharomyces cerevisiae Ssy1 amino acid sensor. Mol Membr Biol 25:164–176. doi:10.1080/09687680701771917

Regenberg B, Kielland-Brandt MC (2001) Amino acid residues important for substrate specificity of the amino acid permeases Can1p and Gnp1p in Saccharomyces cerevisiae. Yeast 18:1429–1440. doi:10.1002/yea.792

Regenberg B, Holmberg S, Olsen LD, Kielland-Brandt MC (1998) Dip5p mediates high-affinity and high-capacity transport of L-glutamate and L-aspartate in Saccharomyces cerevisiae. Curr Genet 33:171–177

Regenberg B, Düring-Olsen L, Kielland-Brandt MC, Holmberg S (1999) Substrate specificity and gene expression of the amino-acid permeases in Saccharomyces cerevisiae. Curr Genet 36:317–328

Ressl S, van Scheltinga ACT, Vonrhein C, Ott V, Ziegler C (2009) Molecular basis of transport and regulation in the Na(+)/betaine symporter BetP. Nature 458:47–52. doi:10.1038/nature07819

Risinger AL, Cain NE, Chen EJ, Kaiser CA (2006) Activity-dependent reversible inactivation of the general amino acid permease. Mol Biol Cell 17:4411–4419. doi:10.1091/mbc.E06-06-0506

Roberg KJ, Rowley N, Kaiser CA (1997) Physiological regulation of membrane protein sorting late in the secretory pathway of Saccharomyces cerevisiae. J Cell Biol 137:1469–1482. doi:10.1083/jcb.137.7.1469

Rouillon A, Surdin-Kerjan Y, Thomas D (1999) Transport of sulfonium compounds. Characterization of the s-adenosylmethionine and s-methylmethionine permeases from the yeast Saccharomyces cerevisiae. J Biol Chem 274:28096–28105. doi:10.1074/jbc.274.40.28096

Rubio-Texeira M, Van Zeebroeck G, Thevelein JM (2012) Peptides induce persistent signaling from endosomes by a nutrient transceptor. Nat Chem Biol 8:400–408. doi:10.1038/nchembio.910
Russnak R, Konczal D, McIntire SL (2001) A family of yeast proteins mediating bidirectional vacuolar amino acid transport. J Biol Chem 276:23849–23857. doi:10.1074/jbc.M008028200
Saier MH, Reddy VS, Tamang DG, Västermark A (2014) The transporter classification database. Nucleic Acids Res 42:D251–D258. doi:10.1093/nar/gkt1097
Schepers W, Van Zeebroeck G, Pinkse M, Verhaert P, Thevelein JM (2012) In vivo phosphorylation of Ser21 and Ser83 during nutrient-induced activation of the yeast protein kinase A (PKA) target trehalase. J Biol Chem 287:44130–44142. doi:10.1074/jbc.M112.421503
Schmidt A, Hall MN, Koller A (1994) Two FK506 resistance-conferring genes in Saccharomyces cerevisiae, TAT1 and TAT2, encode amino acid permeases mediating tyrosine and tryptophan uptake. Mol Cell Biol 14:6597–6606
Schothorst J, Kankipati HN, Conrad M, Samyn DR, Van Zeebroeck G, Popova Y, Rubio-Texeira M, Persson BL, Thevelein JM (2013) Yeast nutrient transceptors provide novel insight in the functionality of membrane transporters. Curr Genet 59:197–206. doi:10.1007/s00294-013-0413-y
Schreve JL, Garrett JM (2003) Yeast Agp2p and Agp3p function as amino acid permeases in poor nutrient conditions. Biochem Biophys Res Commun 313:745–751
Schreve JL, Sin JK, Garrett JM (1998) The Saccharomyces cerevisiae YCC5 (YCL025c) gene encodes an amino acid permease, Agp1, which transports asparagine and glutamine. J Bacteriol 180:2556–2559
Séron K, Blondel MO, Haguenauer-Tsapis R, Volland C (1999) Uracil-induced down-regulation of the yeast uracil permease. J Bacteriol 181:1793–1800
Shaffer PL, Goehring A, Shankaranarayanan A, Gouaux E (2009) Structure and mechanism of a Na+-independent amino acid transporter. Science 325:1010–1014. doi:10.1126/science.1176088
Shcherbik N, Kee Y, Lyon N, Huibregtse JM, Haines DS (2004) A single PXY motif located within the carboxyl terminus of Spt23p and Mga2p mediates a physical and functional interaction with ubiquitin ligase Rsp5p. J Biol Chem 279:53892–53898. doi:10.1074/jbc.M410325200
Shi Y (2013) Common folds and transport mechanisms of secondary active transporters. Annu Rev Biophys 42:51–72. doi:10.1146/annurev-biophys-083012-130429
Shimazu M, Sekito T, Akiyama K, Ohsumi Y, Kakinuma Y (2005) A family of basic amino acid transporters of the vacuolar membrane from Saccharomyces cerevisiae. J Biol Chem 280:4851–4857. doi:10.1074/jbc.M412617200
Soetens O, De Craene JO, André B (2001) Ubiquitin is required for sorting to the vacuole of the yeast general amino acid permease, Gap1. J Biol Chem 276:43949–43957. doi:10.1074/jbc.M102945200
Sophianopoulou V, Scazzocchio C (1989) The proline transport protein of Aspergillus nidulans is very similar to amino acid transporters of Saccharomyces cerevisiae. Mol Microbiol 3:705–714. doi:10.1111/j.1365-2958.1989.tb00219.x
Sorgen PL, Hu Y, Guan L, Kaback HR, Girvin ME (2002) An approach to membrane protein structure without crystals. Proc Natl Acad Sci U S A 99:14037–14040. doi:10.1073/pnas.182552199
Springael JY, André B (1998) Nitrogen-regulated ubiquitination of the Gap1 permease of *Saccharomyces cerevisiae*. Mol Biol Cell 9:1253–1263
Stanbrough M, Magasanik B (1995) Transcriptional and posttranslational regulation of the general amino acid permease of *Saccharomyces cerevisiae*. J Bacteriol 177:94–102
Sychrova H, Chevallier MR (1993) Cloning and sequencing of the Saccharomyces cerevisiae gene LYP1 coding for a lysine-specific permease. Yeast 9:771–782. doi:10.1002/yea.320090711
Sychrova H, Chevallier MR (1994) APL1, a yeast gene encoding a putative permease for basic amino acids. Yeast 10:653–657. doi:10.1002/yea.320100509
Tanaka J, Fink GR (1985) The histidine permease gene (HIP1) of Saccharomyces cerevisiae. Gene 38:205–214

Tavoularis SN, Tazebay UH, Diallinas G, Sideridou M, Rosa A, Scazzocchio C, Sophianopoulou V (2003) Mutational analysis of the major proline transporter (PrnB) of Aspergillus nidulans. Mol Membr Biol 20:285–297. doi:10.1080/0968768031000106339

Tazebay UH, Sophianopoulou V, Cubero B, Scazzocchio C, Diallinas G (1995) Post-transcriptional control and kinetic characterization of proline transport in germinating conidiospores of Aspergillus nidulans. FEMS Microbiol Lett 132:27–37

Trip H, Evers ME, Driessen AJM (2004) PcMtr, an aromatic and neutral aliphatic amino acid permease of Penicillium chrysogenum. Biochim Biophys Acta 1667:167–173. doi:10.1016/j.bbamem.2004.09.014

Tsai M, Miller C (2013) Substrate selectivity in arginine-dependent acid resistance in enteric bacteria. Proc Natl Acad Sci U S A 110:5893–5897. doi:10.1073/pnas.1301442110

Uemura T, Kashiwagi K, Igarashi K (2007) Polyamine uptake by DUR3 and SAM3 in Saccharomyces cerevisiae. J Biol Chem 282:7733–7741. doi:10.1074/jbc.M611105200

Usami Y, Uemura S, Mochizuki T, Morita A, Shishido F, Inokuchi J, Abe F (2014) Functional mapping and implications of substrate specificity of the yeast high-affinity leucine permease Bap2. Biochim Biophys Acta 1838:1719–1729. doi:10.1016/j.bbamem.2014.03.018

Van Belle D, André B (2001) A genomic view of yeast membrane transporters. Curr Opin Cell Biol 13:389–398

Van Nuland A, Vandormael P, Donaton M, Alenquer M, Lourenço A, Quintino E, Versele M, Thevelein JM (2006) Ammonium permease-based sensing mechanism for rapid ammonium activation of the protein kinase A pathway in yeast. Mol Microbiol 59:1485–1505. doi:10.1111/j.1365-2958.2005.05043.x

van Roermund CW, Hettema EH, van den Berg M, Tabak HF, Wanders RJ (1999) Molecular characterization of carnitine-dependent transport of acetyl-CoA from peroxisomes to mitochondria in Saccharomyces cerevisiae and identification of a plasma membrane carnitine transporter, Agp2p. EMBO J 18:5843–5852. doi:10.1093/emboj/18.21.5843

Van Zeebroeck G, Bonini BM, Versele M, Thevelein JM (2009) Transport and signaling via the amino acid binding site of the yeast Gap1 amino acid transceptor. Nat Chem Biol 5:45–52. doi:10.1038/nchembio.132

Van Zeebroeck G, Rubio-Texeira M, Schothorst J, Thevelein JM (2014) Specific analogues uncouple transport, signalling, oligo-ubiquitination and endocytosis in the yeast Gap1 amino acid transceptor. Mol Microbiol 93:213–233. doi:10.1111/mmi.12654

Vandenbol M, Jauniaux JC, Grenson M (1989) Nucleotide sequence of the Saccharomyces cerevisiae PUT4 proline-permease-encoding gene: similarities between CAN1, HIP1 and PUT4 permeases. Gene 83:153–159

Vangelatos I, Vlachakis D, Sophianopoulou V, Diallinas G (2009) Modelling and mutational evidence identify the substrate binding site and functional elements in APC amino acid transporters. Mol Membr Biol 26:356–370. doi:10.1080/09687680903170546

Västermark A, Wollwage S, Houle ME, Rio R, Saier MH (2014) Expansion of the APC superfamily of secondary carriers. Proteins. doi:10.1002/prot.24643

Velasco I, Tenreiro S, Calderon IL, André B (2004) Saccharomyces cerevisiae Aqr1 is an internal-membrane transporter involved in excretion of amino acids. Eukaryot Cell 3:1492–1503. doi:10.1128/EC.3.6.1492-1503.2004

Wang S, Yan R, Zhang X, Chu Q, Shi Y (2014) Molecular mechanism of pH-dependent substrate transport by an arginine-agmatine antiporter. Proc Natl Acad Sci U S A 111:12734–12739. doi:10.1073/pnas.1414093111

Weyand S, Shimamura T, Yajima S, Suzuki S, Mirza O, Krusong K, Carpenter EP, Rutherford NG, Hadden JM, O'Reilly J, Ma P, Saidijam M, Patching SG, Hope RJ, Norbertczak HT, Roach PCJ, Iwata S, Henderson PJF, Cameron AD (2008) Structure and molecular mechanism of a nucleobase-cation-symport-1 family transporter. Science 322:709–713. doi:10.1126/science.1164440

Wiame JM, Grenson M, Arst HN (1985) Nitrogen catabolite repression in yeasts and filamentous fungi. Adv Microb Physiol 26:1–88

Wielemans K, Jean C, Vissers S, André B (2010) Amino acid signaling in yeast: post-genome duplication divergence of the Stp1 and Stp2 transcription factors. J Biol Chem 285:855–865. doi:10.1074/jbc.M109.015263

Wright MB, Ramos J, Gomez MJ, Moulder K, Scherrer M, Munson G, Gaber RF (1997) Potassium transport by amino acid permeases in Saccharomyces cerevisiae. J Biol Chem 272:13647–13652. doi:10.1074/jbc.272.21.13647

Wu B, Ottow K, Poulsen P, Gaber RF, Albers E, Kielland-Brandt MC (2006) Competitive intra- and extracellular nutrient sensing by the transporter homologue Ssy1p. J Cell Biol 173:327–331. doi:10.1083/jcb.200602089

Xu Y, Tao Y, Cheung LS, Fan C, Chen L, Xu S, Perry K, Frommer WB, Feng L (2014) Structures of bacterial homologues of SWEET transporters in two distinct conformations. Nature. doi:10.1038/nature13670

Yamashita A, Singh SK, Kawate T, Jin Y, Gouaux E (2005) Crystal structure of a bacterial homologue of Na+/Cl–dependent neurotransmitter transporters. Nature 437:215–223. doi:10.1038/nature03978

Yang Z, Huang J, Geng J, Nair U, Klionsky DJ (2006) Atg22 recycles amino acids to link the degradative and recycling functions of autophagy. Mol Biol Cell 17:5094–5104. doi:10.1091/mbc.E06-06-0479

Zhu X, Garrett J, Schreve J, Michaeli T (1996) GNP1, the high-affinity glutamine permease of S. cerevisiae. Curr Genet 30:107–114

# Chapter 5
# Water Transport in Yeasts

**Farzana Sabir, Catarina Prista, Ana Madeira, Teresa Moura, Maria C. Loureiro-Dias, and Graça Soveral**

**Abstract** Water moves across membranes through the lipid bilayer and through aquaporins, in this case in a regulated manner. Aquaporins belong to the MIP superfamily and two subfamilies are represented in yeasts: orthodox aquaporins considered to be specific water channels and aquaglyceroporins (heterodox aquaporins). In *Saccharomyces cerevisiae* genome, four aquaporin isoforms were identified, two of which are genetically close to orthodox aquaporins (ScAqy1 and ScAqy2) and the other two are more closely related to the aquaglyceroporins (ScFps1 and ScAqy3). Advances in the establishment of water channels structure are reviewed in this chapter in relation with the mechanisms of selectivity, conductance and gating. Aquaporins are important for key aspects of yeast physiology. They have been shown to be involved in sporulation, rapid freeze-thaw tolerance, osmo-sensitivity, and modulation of cell surface properties and colony morphology, although the underlying exact mechanisms are still unknown.

**Keywords** Aquaporins • Aquaglyceroporins • Genome sequence • Water transport

---

F. Sabir (✉)
LEAF, Instituto Superior de Agronomia, Universidade de Lisboa, Tapada da Ajuda, 1349-017 Lisbon, Portugal

Research Institute for Medicines (iMed.ULisboa), Faculty of Pharmacy, Universidade de Lisboa, 1649-003 Lisbon, Portugal
e-mail: fsabir@isa.ulisboa.pt

C. Prista • M.C. Loureiro-Dias
LEAF, Instituto Superior de Agronomia, Universidade de Lisboa, Tapada da Ajuda, 1349-017 Lisbon, Portugal
e-mail: cprista@isa.ulisboa.pt; mcdias@isa.ulisboa.pt

A. Madeira • T. Moura • G. Soveral
Research Institute for Medicines (iMed.ULisboa), Faculty of Pharmacy, Universidade de Lisboa, 1649-003 Lisbon, Portugal
e-mail: capsmadeira@ff.ulisboa.pt; teresa@fct.unl.pt; gsoveral@ff.ulisboa.pt

 
J. Ramos et al. (eds.), *Yeast Membrane Transport*, Advances in Experimental Medicine and Biology 892, DOI 10.1007/978-3-319-25304-6_5

## 5.1 Introduction

Water homeostasis is a fundamental requirement for survival and adaptation of all living beings. Although initially it was assumed that water diffuses only through the lipid bilayer, the first studies in the late 1950s on water transport in mammalian red blood cells (Paganelli and Solomon 1957; Vieira et al. 1970; Macey et al. 1972) demonstrated that water permeability in these cells was much higher than expected by diffusion across the bilayer, which provided clues for the existence of specialized water channels. Also, water transport exhibited low activation energy and was inhibited by mercurials (Macey et al. 1972; Farmer and Macey 1970; Sidel and Solomon 1957). Based on the ratio of osmotic to diffusion water permeabilities, a single file diffusion mechanism of water transport within the putative channel was proposed even before the molecular identification of water channels (Moura et al. 1984; Finkelstein 1984).

Only in the nineties, Agre and co-workers identified the first water channel protein, now known as aquaporin 1 (AQP1), in red blood cells (Preston et al. 1992). Meanwhile, numerous aquaporin and aquaporin-like sequences have been identified in nearly all living organisms (Nehls and Dietz 2014), including 2 in bacteria, 4 in yeasts (Kruse et al. 2006), 35 in plants and 13 in mammals (reviewed by Kaldenhoff and Fischer 2006).

Now it is generally accepted that water moves through membranes by two different pathways: through the lipid bilayer, in a high activation energy process, and through aquaporins in a low activation energy process, leading to a determinant role of temperature in the relative importance of both mechanisms in nature. Particularly in *Saccharomyces cerevisiae*, it was shown that water channels are important for cells osmotic adjustments at lower temperature (Soveral et al. 2006). In either case of lipid path or channel, the driving force for water movement is the gradient of chemical potential of water (osmotic and/or hydrostatic pressure) between both sides of the membrane and predicted to occur in either direction (Finkelstein 1984). Aquaporins provide regulation of the water status in the cells by different mechanisms. For example, in yeast and plants the activity of water channels is affected by phosphorylation, pH, pressure, tension, solute gradients and temperature (Chaumont et al. 2005; Maurel 2007; Tornroth-Horsefield et al. 2010; Soveral et al. 2008; Leitao et al. 2014).

Aquaporins belong to a highly conserved group of membrane proteins called the major intrinsic proteins (MIPs) that form a large family comprising more than 1700 integral membrane proteins (Abascal et al. 2014) and they are involved in the transport of water and small solutes such as glycerol (Maurel et al. 1994; Carbrey et al. 2003; Wang et al. 2005), $H_2O_2$ (Hooijmaijers et al. 2012), ammonia, boron (Dynowski et al. 2008), nitrate (Ikeda et al. 2002) and urea (Mitani-Ueno et al. 2011), and also gases like $CO_2$ (Navarro-Ródenas et al. 2012). The MIP superfamily (MIP, 1.A.8) includes several subfamilies of proteins: (i) aquaporins 1.A.8.6. (called orthodox, ordinary, conventional or classical aquaporins) which are considered to be specific water channels (Takata et al. 2004), (ii) aquaglyceroporins 1.A.8.2. (called unconventional or heterodox aquaporins) and iii. aquaporins with unusual NPA (Asn-Pro-Ala) boxes (called unorthodox superaquaporins or subcellular aquaporins), a recently described third subfamily that is only present in animals but not in plants, fungi and bacteria (reviewed by Ishibashi et al. 2011; Ishibashi et al. 2014).

In *S. cerevisiae* genome, four aquaporin isoforms were identified, two of which are genetically close to orthodox aquaporins (ScAqy1 and ScAqy2) and the other more closely related to the aquaglyceroporins (YFL054Cp, recently reclassified as ScAqy3 (Patel 2013), and ScFps1) (Bonhivers et al. 1998; Carbrey et al. 2001). While Aqy- and Fps1-like aquaporins have been mainly localized in the plasma membrane, Aqy3-like aquaporins have been mainly localized in the vacuoles. Occurrence of water channels showed diverse patterns in yeasts. In some yeasts (*S. cerevisiae* strain Σ1278B and *Candida glabrata*) both functional orthodox aquaporins were found, while in some species (*Schizosaccharomyces pombe*, *Zygosaccharomyces rouxii* and *Z. bailli*), these genes were not present at all. In such cases, presence of the aquaglyceroporin (*FPS1*) was always observed. In several strains of *S. cerevisiae*, one of the genes (especially, *AQY2*) was truncated during evolution (Carbrey et al. 2001; Laizé et al. 2000). Whereas, strain S288C (the first strain used for sequencing of *S. cerevisiae* genome) carries inactivating mutations in both aquaporin genes and their deletion did not cause distinct growth phenotype (Laizé et al. 2000; Karpel and Bisson 2006).

Aquaporins are important for key aspects of yeast physiology. They are involved in sporulation (Sidoux-Walter et al. 2004; Will et al. 2010), rapid freeze-thaw tolerance (Tanghe et al. 2002; Fischer et al. 2009), osmo-sensitivity, and modulation of cell surface properties and colony morphology (Furukawa et al. 2009). These depicted roles of aquaporins are exclusively based on observed phenotypes caused by their deletion/overexpression or differential gene expression pattern, although the underlying exact mechanisms are still unknown.

## 5.2 Genes Coding for Water Channels in Yeasts

Protein sequences of orthodox aquaporins (Aqy1 and Aqy2) and aquaglyceroporin (Fps1) within the yeasts genome were searched by BLASTp and tBLASTn tool from the available genome database at Saccharomyces Genome Database (SGD, http://www.yeastgenome.org), Génolevures (http://www.genolevures.org) and NCBI (http://www.ncbi.nlm.nih.gov). These sequences were compared with sequences of functional versions of both Aqy1 and Aqy2 proteins of *S. cerevisiae* Σ1278B strain. In Table 5.1, we present a detailed compilation of obtained MIP sequences within the genome of 14 yeasts representing 8 species and 7 different strains of *S. cerevisiae* collected from different niches. Sequences annotated in this study were designated according to their highest percentage identity with ScAqy1, ScAqy2 (not shown in Table 5.1) and ScFps1. Among the different species, Aqy1 of *S. paradoxus* (SpAqy1) showed the highest identity (98 %), while *Debaryomyces hansenii* (DhAqy1) showed the lowest identity (34 %) with ScAqy1. An obvious high identity

**Table 5.1** **Yeast aquaporins.** Yeast aquaporin sequences were obtained by BLASTp, tBLASTn search with ScAqy1, ScAqy2 and ScFps1 against available yeasts genome databases at NCBI and SGD. Sequences were compared with ScAqy1, ScAqy2 and ScFps1 by clustalW analysis and were designated according to their highest percentage identity

| Species | Strain | Sequence designation | Length (aa) | Accession number | % identity with Σ1278B_ScAqy1 | % identity with Σ1278B _ScFps1 | References |
|---|---|---|---|---|---|---|---|
| *Candida albicans* | SC5314 | CaAqy1 | 273 | XP_715831[a] | 53 | 21 | Carbrey et al. (2001) and Tanghe et al. (2005) |
| | | CaFps1 | Not found | – | – | – | |
| *Candida glabrata* | CBS138 | CgAqy2.1 | 293 | XP_444824 | 72 | 24 | Dujon et al. (2004) |
| | | CgAqy2.2 | 290 | XP_445420 | 71 | 22 | Dujon et al. (2004) |
| | | CgFps1 | 652 | XP_445318.1 | 21 | 51 | Dujon et al. (2004) |
| *Dekkera bruxellensis* | AWRI1499 | DbAqy2 | 273 | EIF49326.1 | 56 | 22 | Curtin et al. (2012) |
| | | DbFps1 | Not found | – | – | – | |
| *Debaryomyces hansenii* | CBS767 | DhAqy1 | 310 | XP_461517 | 34 | 21 | Dujon et al. (2004) |
| | | DhFps1 | Not found | – | – | – | |
| *Kluyveromyces lactis* | NRRL Y-1140 | KlAqy1 | 273 | CAH02367 | 52 | 27 | Dujon et al. (2004) |
| | | KlFps1 | 563 | XP_453974.1 | 19 | 47 | Dujon et al. (2004) |
| *Kluyveromyces marxianus* | DMKU3-1042 | KmAqy2 | 309 | BAO39643.1 | 51 | 23 | Lertwattanasakul N et al. 2011-direct submission |
| | | KmFps1 | 572 | BAO42722.1 | 19 | 44 | Lertwattanasakul N et al. 2011-direct submission |
| *Pichia pastoris* | GS115 | PpAqy1 | 279 | XP_002492992[a] | 54 | 21 | Karlsson et al. (2003) and Daniels et al. (2006) |
| | | PpFps1 | 326 | XP_002494229.1 | 18 | 36 | De Schutter et al. (2009) |

| | | | | | | | |
|---|---|---|---|---|---|---|---|
| *Saccharomyces bayanus* | 623-6C YM4911 | SbAqy1 | 274 | AL401051.1 | 91 | 24 | Reviewed by Soveral et al. (2011) |
| | | SbFps1 | 661 | gb\|AACG02000085.1 | 25 | 89 | Sabir et al. unpublished results |
| *Saccharomyces cerevisiae* strains | | | | | | | |
| Bakery isolates, Vietnam | CLIB324 | CLIB324_ScAqy1 | 327 | gb\|AEWM01001183.1 | 99 | 20 | Sabir et al. unpublished results |
| | | CLIB324_ScAqy2 | 149 | gb\|AEWM01001569.1 | 69 | 24 | Sabir et al. unpublished results |
| | | CLIB324_ScFps1 | Not assembled | gb\|AEWM01002257.1 | nd | nd | Sabir et al. unpublished results |
| | | | | gb\|AEWM01002256.1 | | | |
| Commercial wine strain | EC1118 | EC1118_ScAqy1 | 327 | C8ZJM1.1 | 99 | 20 | Novo et al. (2009) |
| | | EC1118_ScAqy2 | 149 | C8ZCS3.1 | 69 | 24 | Novo et al. (2009) |
| | | EC1118_ScFps1 | 669 | CAY81197.1 | 20 | 100 | Novo et al. (2009) |
| Sake, Japan | K7 | K7_ScAqy1-1 | 294 | GAA27146.1 | 96 | 21 | Akao et al. (2011) |
| | | K7_ScAqy1-2 | 240 | GAA22751.1 | 98 | 22 | Akao et al. (2011) |
| | | K7_ScFps1 | 646 | GAA23030.1 | 23 | 22 | Akao et al. (2011) |
| Vinyard, California | RM11-1a | RM11-1a_ScAqy1 | 327 | gb\|AAEG01000033.1 | 99 | 20 | Sabir et al. unpublished results |
| | | RM11-1a_ScAqy2.2 | 149 | B3LTC5.1 | 69 | 25 | Birren et al. 2005-direct submission |
| | | RM11-1a_ScFps1 | 646 | EDV12566.1 | 23 | 22 | Birren et al. 2005-direct submission |
| Laboratory strain | S288c | S288c_ScAqy1 | 305 | NP_015518.1 | 96 | 21 | Bussey et al. (1997) |
| | | S288c_ScAqy2 | 152 | NP_013047.1 | 85 | 23 | Johnston et al. (1997) |
| | | S288c_ScFps1 | 669 | NP_013057.1 | 20 | 99 | Johnston et al. (1997) |

(continued)

**Table 5.1** (continued)

| Species | Strain | Sequence designation | Length (aa) | Accession number | % identity with Σ1278B_ScAqy1 | % identity with Σ1278B _ScFps1 | References |
|---|---|---|---|---|---|---|---|
| Laboratory strain | Σ1278B | Σ1278B_ScAqy1 | 327 | P0CD92[a] | 100 | 20 | Bonhivers et al. (1998) and Laizé et al. (2000) |
| | | Σ1278B_ScAqy2 | 289 | P0CD89[a] | 87 | 25 | Soveral et al. (2006), Bonhivers et al. (1998) and Laizé et al. (2000) |
| | | Σ1278B_ScFps1 | 669 | gb\|ACVY01000031.1 | 20 | 100 | Sabir et al. unpublished results |
| Cocconut, Philippines | Y10 | Y10_ScAqy1 | 305 | ADC55549.1 | 96 | 21 | Will et al. (2010) |
| | | Y10_ScAqy2 | 43 | ADC55257.1 | 16 | 21 | Will et al. (2010) |
| | | Y10_ScFps1 | Not assembled | gb\|AEWK01002276.1 | nd | nd | Sabir et al. unpublished results |
| | | | | gb\|AEWK01002277.1 | | | |
| *Saccharomyces paradoxus* | NRRL Y-17217 | SpAqy1 | 278 | gb\|AABY01000011 | 98 | 21 | Reviewed by Soveral et al. (2011) |
| | | SpFps1 | 673 | gb\|AABY01000032.1 | 20 | 98 | Sabir et al. unpublished results |
| *Saccharomyces pastorianus* | Weihenstephan 34/70 | SpaAqy2 | 289 | gb\|ABPO01000065 | 82 | 24 | Reviewed by Soveral et al. (2011) |
| | | SpaFPs1 | 669 | gb\|AZAA01000411.1 | 20 | 99 | Sabir et al. unpublished results |

| | | | | | | | |
|---|---|---|---|---|---|---|---|
| *Schizosaccharomyces pombe* | 972 h- | SchAqy | Not found | – | – | – | |
| | | SchpFps1 | 598 | NP_592788.1 | 22 | 22 | Wood et al. (2002) |
| *Zygosaccharomyces rouxii* | CBS 732 | ZrAqy | Not found | – | – | – | |
| | | ZrFps1 | 692 | XP_002498330.1 | 21 | 52 | Genoscope |
| *Zygosaccharomyces* bailii | CLIB 213 | ZbAqy | Not found | – | – | – | |
| | | ZbFps1 | 686 | CDF89479.1 | 19 | 52 | Galeote et al. 2013-direct submission |

Not assembled: protein is not annotated in database and is present in more than one contig
*nd* not done
[a]Protein characterized for water transport activity

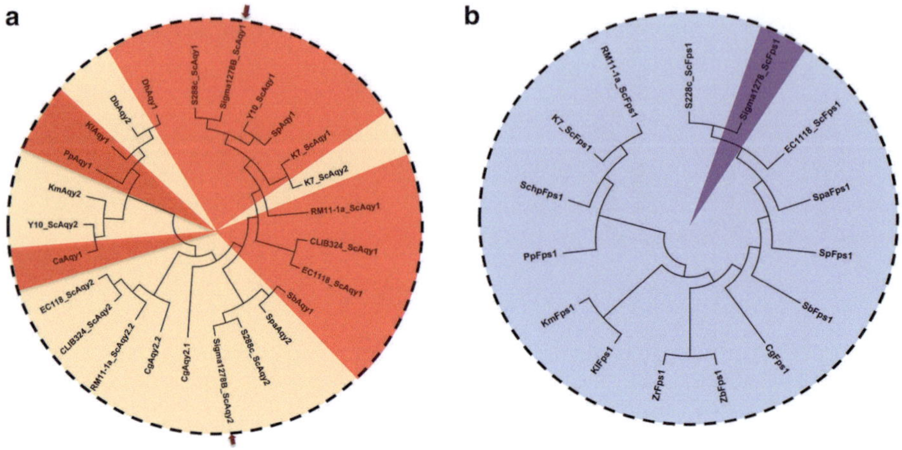


**Fig. 5.1 Phylogenetic tree based on the protein sequences of yeast aquaporins.** Dendrograms depicting the phylogenetic relationship of *S. cerevisiae* orthodox aquaporins (Aqy1 and Aqy2) (**a**), and Fps1 aquaglyceroporin (**b**) with the sequences detected in other yeasts. Dendograms were generated by neighbor-joining method (applied to 1000 bootstrap data sets) using the MEGA5.1 programme. Details of terminology and accession numbers are listed in Table 5.1

of sequences (up to 99 %) was observed within the strains of *S. cerevisiae*. ScAqy1 and ScAqy2 are highly similar (87 % identical) proteins, indicating the probable recent gene duplication during evolution (reviewed by Pettersson et al. 2005). Contrarily, ScFps1 showed very low similarity with ScAqy1 (20 %), suggesting that it belongs to a distinct subfamily (aquaglyceroporin) of MIPs (Abascal et al. 2014). It showed high variability (22–98 %) even within the alignment of Fps1 proteins of yeasts. Except for SpaFps1 (99 %) and SpFps1 (98 % identical), lower similarities of ScFps1 with other yeasts as well as within the strains of *S. cerevisiae* were found. The phylogenetic tree for sequences listed in Table 5.1 was constructed by MEGA5.1 software using neighbor-joining method (Tamura et al. 2011). It shows the distance among all obtained orthodox aquaporins (Fig. 5.1a) and among aquaglyceroporins (Fig. 5.1b). Most of the Aqy1 were grouped together, although CaAqy1, PpAqy1, KlAqy1 and DhAqy1, respectively, were more divergent from ScAqy1 (Fig. 5.1a). In the case of Fps1, the aquaglyceroporin of *Sch. pombe* was the most distant among different species, although Fps1 of *S. cerevisiae* RM11-1a strain was even more divergent than Fps1 of *Sch. pombe* (Fig. 5.1b).

The number of genes of aquaporins/aquaglyceroporins family present in yeasts exhibits an interesting pattern. Almost all the selected yeasts possess at least one water/solute channel protein (either Aqy1, Aqy2 or Fps1). Yeasts (*Sch. pombe*, *Z. rouxii* and *Z. bailii*), which do not have any orthodox aquaporins, present at least one aquaglyceroporin, suggesting its putative versatile role as water/solute channel. In addition, even the orthodox aquaporins are also reported to be involved in transport of solutes other than water (reviewed by Pettersson et al. 2005).

## 5.3 PpAqy1 Structure and Gating

In the last decade several high resolution aquaporin structures have been solved, and currently structures of 14 aquaporins from 9 different organisms have been published. The overall structures of all aquaporins are essentially identical, regardless of the subfamily or the host organism (Klein et al. 2014). The reported structures revealed a tetrameric assembly of four identical monomers each behaving as an independent channel and sharing a conserved overall typical hourglass fold (Murata et al. 2000). *Pichia pastoris* Aqy1 is the only fungal AQP structure solved so far (Fischer et al. 2009). PpAqy1 along with ScAqy1 were used as reference sequences to compare the topology and important amino acid residues in the sequences of yeasts. All the yeast aquaporins in this study exhibited the typical topology of the MIP family. Figure 5.2a shows different views of PpAqy1 tetrameric structure (side, top and bottom). Similarly to all other AQP members, PpAqy1 monomers interact with two of their neighbours and form the tetramer with a central pore that excludes water molecules.

Figure 5.2b shows the typical topology of aquaporin monomers for PpAqy, where the pore region is indicated by the red mesh. Each monomer is comprised by six membrane-spanning helices (H1-6) and two elongated loops that form short half helices (loops B and E). These two half helices contain highly conserved NPA motifs (asparagine-proline-alanine) that are critical for water or other substrate permeation

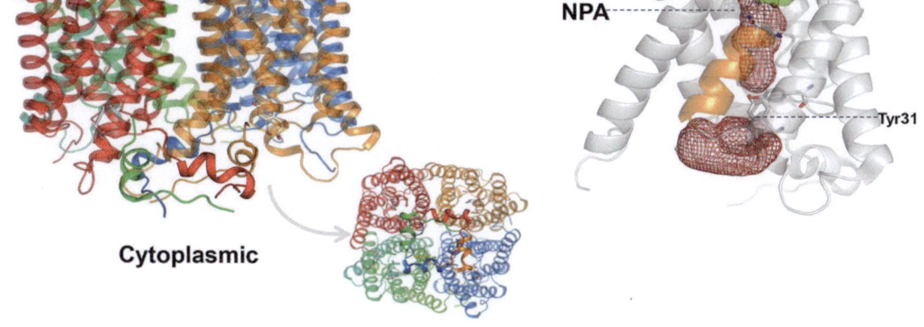


**Fig. 5.2 Structure of PpAqy1.** (**a**) Tetrameric structure showing side, top and bottom views. (**b**) Monomeric structure highlighting the two half-helices entering from loops B and E to form a pseudo seventh transmembrane helix, and the selectivity filters regions ar/R and NPA. The *red mesh* represents the residues lining the channel, which is in the closed configuration, with Tyr31 protruding into the pore (Protein Data Bank code: 2W2E)

(Jung et al. 1994). The NPA repeats are located at the end of loops B and E that fold back into the membrane and form the seventh transmembrane pseudo-helix. The NPA signatures are oriented 180° to each other in the centre of the channel and form part of the surface of the aqueous pathway.

Both amino and carboxy termini are located in the cytoplasm, and similarly to other yeasts, the N-terminal of all selected AQPs is unusually long when compared with isoforms of other organisms (Tornroth-Horsefield et al. 2010), as depicted in Fig. 5.2a for PpAqy1. From the obtained crystal structure of PpAqy1 a functional role for the yeast aquaporin N terminus is inferred, which has been associated with mechanisms controlling the transport activity of the channel (gating) (Fischer et al. 2009).

## 5.3.1 Selectivity and Mechanism of Conductance

Classical aquaporins are still considered mostly specific for water. In the pore region, the water specificity is achieved by the presence of particular residues conferring size constrictions and/or charge characteristics that enable water molecules to pass through, while preventing permeation to protons or any solutes above 2.8 Å. The cytoplasmic and periplasmatic entry of the pore in the aquaporin monomer offers several water-wall residue interactions, mainly with carbonyl groups. After passing these first interactions, wall regions with different hydrophobic/hydrophilic characteristics determine selectivity, conduction rate and open/closed state of the pore. Two main constrictions sites that act as selectivity filters were identified in AQP channels: one is positioned on the extracellular face of the channel in the aromatic/arginine (ar/R) constriction region, whereas the other is located in the central part of the channel at the NPA region (Hub et al. 2009; Soveral et al. 2011; de Groot and Grubmuller 2001). The location of the two half helices and some of the residues lining the pore wall relevant for selectivity (ar/R and NPA filters) and gating are depicted in Fig 5.2b.

Beginning on the exoplasmic side and running along the pore, after passing the first interactions with the carbonyl groups (not shown), the narrowest region of the pore is found to be the aromatic/R constriction site (ar/R) containing highly conserved aromatic and arginine residues. In PpAqy1 the ar/R constriction is formed by phenylalanine/arginine with Arg227 and His212 creating the hydrophilic surface where the polar interactions water-protein occur, whereas Phe92 pushes away water molecules. Sequence alignment showed that these residues are highly conserved in all the yeasts selected in Table 5.1. Further down in the middle of the pore, water molecules find the second selective filter as they bypass the dual NPA motifs. In this region, water repulsion by a hydrophobic residue that intrudes into the pore, forces water molecules to interact strongly with Asn224 and Asn112 of the NPA loops, which are also conserved in all yeasts of Table 5.1. These residues act as hydrogen donors to the oxygen atoms of passing water molecules, compelling them to acquire a specific orientation (Murata et al. 2000; Sui et al. 2001; de Groot et al. 2003; de Groot and Grubmuller 2005). In addition, water molecules that enter this region are reoriented by the electrostatic field produced by the half helices HB and HE, resulting in hydrogen bonds disruption between neighboring water molecules (de Groot et al. 2003; de Groot and Grubmuller 2005). Additionally, hydrophobic residues lining the pore near the NPA filter provide a hydrophobic surface that prevents water-protein interaction (Soveral et al. 2011).

### *5.3.2 PpAqy1 Gating Mechanism*

The resolution of multiple AQP structures has disclosed common patterns of gating among different isoforms. These mechanisms directly affect the protein conformation, which in turn impacts its transport activity.

Besides the described selectivity size filters, an additional and significantly narrower constriction site at the cytoplasmic entry and an insertion of a residue in a specific 3D position of the protein structure is a common feature to all gated AQPs (reviewed by Tornroth-Horsefield et al. 2010). This residue was found to be Tyr31 for PpAqy1 (Fig 5.2b).

Of notice, the existing PpAqy1 structure captures the channel in the closed configuration, depicting the N-terminal Tyr31 residue intruding and occluding the pore. In this structure, the kink introduced by Pro29 allows Tyr31 to be inserted into the channel and to participate in a network of hydrogen bonds with two water molecules and the backbone carbonyl oxygen atoms of Gly108 and Gly109. These interactions narrow the cytoplasmic side of the pore to around 0.8 Å diameter, hindering water permeation.

Fischer et al. (2009) reported that conformational changes able to dislodge Tyr31 from its blocking position could be attained by mechanical forces and/or by phosphorylation of Ser107. Ser107 locates at the cytoplasm end of helix 2 in loop B and connects to Tyr31 via a sequence of H-bonds, including Gly108 and a bridging water molecule thus helping the pore blockage. Through molecular dynamic simulation studies on PpAqy1, it was possible to resolve the question of how mechanical forces (bending the membrane towards the cytoplasmic side) are transmitted from membrane to the gate of the channel. The results pointed to Leu189, Ala190 and Val191 as responsible for this transmission via coupled movements of helices 4, 5 and 6 but this same opening mechanism can be triggered by Ser107 phosphorylation (Fischer et al. 2009). In agreement, functional studies have suggested the closure of yeast Aqy1 by membrane surface tension (stretching the membrane) (Soveral et al. 2008), an experimental approach in which the mechanical forces work in the inverse direction of the above-described mechanism thus producing the opposite effect.

Sequence alignment of PpAqy1 with other yeasts aquaporins showed that Tyr31 is not highly conserved in all selected yeasts, since it is only present in *C. albicans* and *D. hansenii*, whereas Pro29 was found to be conserved only in *C. albicans*, supporting the possibility of different gating mechanism in some yeasts (Cui and Bastien 2012). On the other hand, Gly108, Gly109, Ala190 and Val191 along with putative phosphorylation site Ser107 are conserved in all yeast species.

## 5.4 Assessment of Water Transport in Yeasts

A direct correlation between phylogeny and function is not obvious for fungal MIPs (Nehls and Dietz 2014). Therefore, prediction of MIP relevance is not possible without expression studies and further functional analysis.

Functional studies to access aquaporin activity are centred on following cell and/or vesicle volume changes resulting directly from water fluxes driven by osmotic and/or pressure gradients. The observed volume changes (swelling or shrinking) are directly proportional to water fluxes and consequently to the driving forces (the pressure gradients), with the osmotic permeability coefficient $P_f$ as the proportionality constant. The rate at which the volume changes occur depends on the fraction of water that follows the channel (aqueous pathway) versus the lipid pathway. In addition, compared with water flow across a hydrophobic lipid bilayer, water fluxes through a hydrophilic channel pore need lower activation energy $E_a$. Thus, high $P_f$ and low $E_a$ values are an indication of an aqueous path (active aquaporins present), whereas low $P_f$ and high $E_a$ point mainly to lipid pathway (aquaporin absence or inactivation). For a comprehensive overview on the equations and parameters used to evaluate water permeability, see Soveral et al. (2011).

The techniques used to measure volume changes take advantage of volume dependent physical parameters based on optical properties, e.g. light transmission or absorbance, light scattering and fluorescence of volume sensitive dyes.

Different preparations involving different methodologies have been used for the functional characterization of yeast aquaporins. These include the heterologous expression of yeast aquaporins in *Xenopus laevis* oocytes (systems with low intrinsic water permeability) (Bonhivers et al. 1998; Carbrey et al. 2001), yeast protoplasts (Soveral et al. 2006; Laize et al. 1999), yeast secretory vesicles (Coury et al. 1999; Meyrial et al. 2001) and yeast intact cells (Soveral et al. 2007). In some laboratories permeability has been assessed by monitoring the kinetics of bursting of osmotically challenged protoplasts as a decrease in optical density (Pettersson et al. 2006), while in others the changes in cell volume by an image analysis system connected to an inverted light microscope have been followed (Prudent et al. 2005).

A frequently used method to follow rapid volume changes in a cell/vesicle suspension is the stopped flow spectroscopy. Here cell/vesicle suspensions are subjected to osmotic challenges by rapid mixing with an equal volume of hypo- or hyperosmotic solution. For vesicle or protoplast suspensions, light of a chosen wavelength is directed to the observation chamber through an optical fibre and the change in scattered light is followed until a stable signal is attained. An alternative approach uses cell/vesicles loaded with volume sensitive fluorescent dyes. Yeast cells are loaded with the membrane permeable nonfluorescent precursor carboxyfluorescein diacetate, which is cleaved intracellularly by non-specific esterases to form the fluorescent free form. Changes in fluorescence intensity resulting from osmotically induced volume changes can be monitored by stopped-flow fluorescence. Signals are then calibrated into volumes and analysed for permeability evaluation.

For protoplasts and vesicles, the signals obtained with small osmotic perturbations can be described by an exponential function whose rate constant allows the direct evaluation of the osmotic permeability (Soveral et al. 2006; Bonhivers et al. 1998). For walled cells however, evaluation of osmotic permeability must take into account possible gradients of hydrostatic pressure that arise when cells are in low osmotic buffers; this experimental situation was further used to disclose aquaporin gating by membrane surface tension (Soveral et al. 2008).

Gating of aquaporin function by pH, phosphorylation or specific inhibitors, can be screened through simple measurements of $P_f$ and $E_a$. However, aquaporin gating by physical parameters such as membrane surface tension implies the design of specific protocols that can only be applied to vesicle systems (Soveral et al. 1997) or walled cells (Soveral et al. 2008) that can sustain membrane tensions without rupture. Using protocols that induce an increase in membrane surface tension just before the osmotic shock, Aqy1 from *S. cerevisiae* was found to be regulated by tension and to behave as a pressure regulated water channel (Soveral et al. 2008), supporting its combined involvement with the aquaglyceroporin Fps1 in yeast osmoregulation (Hohmann et al. 2007).

## 5.5 Aquaporins in *Saccharomyces* from Different Ecological Niches

In nature, wild *S. cerevisiae* strains exist in a wide range of environments, from cold (like oak soil) to hot climates, and are commonly found in sugar rich conditions (like in flowers, fruits, sap, grape must). The high sugar environment provides ideal conditions for their fermentation, reproduction and growth as single cell as well as pseudohyphae of attached cells. *S. cerevisiae* strains are also the most domesticated yeasts by human activity throughout history, due to their role in the production of important food and beverages such as bread, beer, wine and sake among several others. This role led to yeast improvement and selection for more efficient production and higher quality products, much before Pasteur reported its role in fermentation.

Evolutionary studies suggest that the whole genome of *S. cerevisiae* was duplicated around 100 million years ago, followed by partial loss of duplicated genes. The remaining 15 % of duplicated genes have evolved as functionally divergent from their parental genes, and have been maintained, providing selective benefits to the cells (Botstein and Fink 2011; Dujon 2010). The analysis of recently available genomes from wild and domesticated *S. cerevisiae* strains revealed the existence of two orthodox aquaporins, Aqy1 and Aqy2. Most of the laboratory and domesticated strains harbor at least one non-functional aquaporin. Exceptionally, *S. cerevisiae* Σ1278B laboratory strain contains both functional Aqy1 and Aqy2 (Carbrey et al. 2001).

The reason of the existence of aquaporins is still speculative, due to the lack of clear and direct correlation between yeast growth and overexpression/deletion of functional orthodox aquaporins (reviewed by Ahmadpour et al. 2014). Since microbial cells are of small size resulting in large surface-to-volume ratio, enhanced water permeability appeared irrelevant for water movement during osmotic adjustment (reviewed by Tanghe et al. 2006). So, what is the physiological role of microbial aquaporins? How does their presence correlate with adaptive evolution in different ecological niches?

Recently, important progress occurred in the answer to this issue and may provide the insight on physiological importance of aquaporins in yeasts (Will et al. 2010). Up to now, several purposes for aquaporins in yeasts have been pointed, namely: (1) adaptation to environmental conditions under which water flux through the lipid bilayer becomes more restrictive, in particular under low temperature (Soveral et al. 2006) and freeze-thaw cycles (Tanghe et al. 2002), (2) involvement in developmental stages where water transport becomes critical (Sidoux-Walter et al. 2004; Will et al. 2010), (3) control of water fluxes through the plasma membrane (as well as other small polar molecules) to adjust their internal osmotic pressure (Nehls and Dietz 2014; Soveral et al. 2008) and (4) modulation of cell surface properties for substrate adhesion and formation of cell biofilms (Furukawa et al. 2009). These findings are mostly supported by phenotypes observed in deletion mutants or aquaporin overexpressing strains as well as by the gene expression pattern during cell cycle and under various stress conditions (low temperature, osmotic shock and nutrients depletion), suggesting that their expression is differentially regulated and they appear to perform similar as well as different functions. The expression of *AQY1* was up-regulated under starvation, contrarily, *AQY2* was up-regulated in exponential phase of cells growing in rich medium (Meyrial et al. 2001). During cell cycle, only *AQY1* expression appeared to be tightly linked with sporulation in *S. cerevisiae* SK1 (Sidoux-Walter et al. 2004) and YPS163 (Will et al. 2010) strains. Moreover, the level of aquaporins expression is highly correlated with rapid freeze-thaw tolerance. Higher expression provided improved tolerance (Tanghe et al. 2002), while deletion caused susceptibility to freeze/thaw cycles (Tanghe et al. 2002). Interestingly, Will et al (Will et al. 2010) reported that although the presence of a functional allele of aquaporins provides freeze-thaw tolerance (useful in oak soil), their absence offers them fitness during growth on high-sugar substrates. Additionally, together with the loss of functional aquaporins during evolution, their ancestral need of aquaporins for spore formation was also lost.

*AQY2* .appeared to behave as osmosensor in yeasts, since its expression was down regulated under hyperosmotic shock in a Hog1 dependent manner, and was recovered under lower osmotic conditions (reviewed by Pettersson et al. 2005). Moreover, overexpression of *AQY1* and *AQY2* also affected the cell surface properties and colony morphology of the yeasts. Their deletion enhanced the hydrophobicity of cell surface and cell flocculence (Carbrey et al. 2001), while their overexpression increased the plastic adhesion of cell surface, agar invasion and colony fluffiness (Furukawa et al. 2009; Šťovíček et al. 2010).

In order to test the role of aquaporins during wine fermentation, Karpel and Bisson (Karpel and Bisson 2006) investigated five native wine yeast strains and found only a functional Aqy1. They observed that yeast adaptation to stress during wine fermentation was not dependent on aquaporins. Our recent study on the effect of ethanol on water fluxes on yeast demonstrated that a low concentration of ethanol (4 %) had a remarkable inhibitory effect on aquaporin activity (Madeira et al. 2010), supporting the idea that aquaporins play a poor role in wine fermentation.

Strains isolated from oak soil, harbouring functional aquaporins, probably represent an ancestral state of evolution. In association with humans, *S. cerevisiae* strains migrated worldwide. Human-facilitated migration may have significantly increased exposure of *S. cerevisiae* to various environments, imposing new selective pressures when strains occupied new ecological niches, where activity of functional aquaporins was deleterious and knock-out mutations in *AQY* genes brought benefits for their progeny.

## References


Abascal F, Irisarri I, Zardoya R (2014) Diversity and evolution of membrane intrinsic proteins. Biochim Biophys Acta 1840:1468–1481

Ahmadpour D, Geijer C, Tamas MJ, Lindkvist-Petersson K, Hohmann S (2014) Yeast reveals unexpected roles and regulatory features of aquaporins and aquaglyceroporins. Biochim Biophys Acta 1840:1482–1491

Akao T, Yashiro I, Hosoyama A, Kitagaki H, Horikawa H et al (2011) Whole-genome sequencing of sake yeast saccharomyces cerevisiae Kyokai no. 7. DNA Res 18:423–434

Bonhivers M, Carbrey JM, Gould SJ, Agre P (1998) Aquaporins in *Saccharomyces*: genetic and functional distinctions between laboratory and wild-type strains. J Biol Chem 273:27565–27572

Botstein D, Fink GR (2011) Yeast: an experimental organism for 21st century biology. Genetics 189:695–704

Bussey H, Storms RK, Ahmed A, Albermann K, Allen E et al (1997) The nucleotide sequence of *Saccharomyces cerevisiae* chromosome XVI. Nature 387:103–105

Carbrey JM, Bonhivers M, Boeke JD, Agre P (2001) Aquaporins in *Saccharomyces*: characterization of a second functional water channel protein. Proc Natl Acad Sci U S A 98:1000–1005

Carbrey JM, Gorelick-Feldman DA, Kozono D, Praetorius J, Nielsen S et al (2003) Aquaglyceroporin AQP9: solute permeation and metabolic control of expression in liver. Proc Natl Acad Sci 100:2945–2950

Chaumont F, Moshelion M, Daniels MJ (2005) Regulation of plant aquaporin activity. Biol Cell 97:749–764

Coury LA, Hiller M, Mathai JC, Jones EW, Zeidel ML et al (1999) Water transport across yeast vacuolar and plasma membrane-targeted secretory vesicles occurs by passive diffusion. J Bacteriol 181:4437–4440

Cui Y, Bastien DA (2012) Molecular dynamics simulation and bioinformatics study on yeast aquaporin Aqy1 from Pichia pastoris. Int J Biol Sci 8:1026–1035

Curtin CD, Borneman AR, Chambers PJ, Pretorius IS (2012) De-novo assembly and analysis of the heterozygous triploid genome of the wine spoilage yeast *Dekkera bruxellensis* AWRI1499. PLoS One 7:e33840
Daniels MJ, Wood MR, Yeager M (2006) In vivo functional assay of a recombinant aquaporin in Pichia pastoris. Appl Environ Microbiol 72:1507–1514
de Groot BL, Grubmuller H (2001) Water permeation across biological membranes: mechanism and dynamics of aquaporin-1 and GlpF. Science 294:2353–2357
de Groot BL, Grubmuller H (2005) The dynamics and energetics of water permeation and proton exclusion in aquaporins. Curr Opin Struct Biol 15:176–183
de Groot BL, Frigato T, Helms V, Grubmuller H (2003) The mechanism of proton exclusion in the aquaporin-1 water channel. J Mol Biol 333:279–293
De Schutter K, Lin Y-C, Tiels P, Van Hecke A, Glinka S et al (2009) Genome sequence of the recombinant protein production host Pichia pastoris. Nat Biotechnol 27:561–566
Dujon B (2010) Yeast evolutionary genomics. Nat Rev Genet 11:512–524
Dujon B, Sherman D, Fischer G, Durrens P, Casaregola S et al (2004) Genome evolution in yeasts. Nature 430:35–44
Dynowski M, Mayer M, Moran O, Ludewig U (2008) Molecular determinants of ammonia and urea conductance in plant aquaporin homologs. FEBS Lett 582:2458–2462
Farmer REL, Macey RI (1970) Perturbation of red cell volume: rectification of osmotic flow. Biochim Biophys Acta Biomembr 196:53–65
Finkelstein A (1984) Water movement through membrane channels. In: Felix B (ed) Current topics in membranes and transport. Academic Press, New York, pp 295–308
Fischer G, Kosinska-Eriksson U, Aponte-Santamaria C, Palmgren M, Geijer C et al (2009) Crystal structure of a yeast aquaporin at 1.15 angstrom reveals a novel gating mechanism. PLoS Biol 7:e1000130
Furukawa K, Sidoux-Walter F, Hohmann S (2009) Expression of the yeast aquaporin Aqy2 affects cell surface properties under the control of osmoregulatory and morphogenic signalling pathways. Mol Microbiol 74:1272–1286
Hohmann S, Krantz M, Nordlander B (2007) Yeast osmoregulation. Methods Enzymol 428:29–45
Hooijmaijers C, Rhee JY, Kwak KJ, Chung GC, Horie T et al (2012) Hydrogen peroxide permeability of plasma membrane aquaporins of *Arabidopsis thaliana*. J Plant Res 125:147–153
Hub JS, Grubmuller H, de Groot BL (2009) Dynamics and energetics of permeation through aquaporins. What do we learn from molecular dynamics simulations? Handb Exp Pharmacol 190:57–76
Ikeda M, Beitz E, Kozono D, Guggino WB, Agre P et al (2002) Characterization of aquaporin-6 as a nitrate channel in mammalian cells requirement of pore-lining residue threonine 63. J Biol Chem 277:39873–39879
Ishibashi K, Kondo S, Hara S, Morishita Y (2011) The evolutionary aspects of aquaporin family. Am J Physiol Regul Integr Comp Physiol 300:R566–R576
Ishibashi K, Tanaka Y, Morishita Y (2014) The role of mammalian superaquaporins inside the cell. Biochim Biophys Acta 1840:1507–1512
Johnston M, Hillier L, Riles L, Albermann K, André B et al (1997) The nucleotide sequence of *Saccharomyces cerevisiae* chromosome XII. Nature 387:87–90
Jung JS, Preston GM, Smith BL, Guggino WB, Agre P (1994) Molecular structure of the water channel through aquaporin CHIP the hourglass model. J Biol Chem 269:14648–14654
Kaldenhoff R, Fischer M (2006) Aquaporins in plants. Acta Physiol 187:169–176
Karlsson M, Fotiadis D, Sjövall S, Johansson I, Hedfalk K et al (2003) Reconstitution of water channel function of an aquaporin overexpressed and purified from Pichia pastoris. FEBS Lett 537:68–72
Karpel JE, Bisson LF (2006) Aquaporins in *Saccharomyces cerevisiae* wine yeast. FEMS Microbiol Lett 257:117–123
Klein N, Neumann J, O'Neil JD, Schneider D (2014) Folding and stability of the aquaglyceroporin GlpF: implications for human aqua(glycero)porin diseases. Biochim Biophys Acta 1848:622–633

Kruse E, Uehlein N, Kaldenhoff R (2006) The aquaporins. Genome Biol 7:206

Laize V, Gobin R, Rousselet G, Badier C, Hohmann S et al (1999) Molecular and functional study of AQY1 from *Saccharomyces cerevisiae*: role of the C-terminal domain. Biochem Biophys Res Commun 257:139–144

Laizé V, Tacnet F, Ripoche P, Hohmann S (2000) Polymorphism of *Saccharomyces cerevisiae* aquaporins. Yeast 16:897–903

Leitao L, Prista C, Loureiro-Dias MC, Moura TF, Soveral G (2014) The grapevine tonoplast aquaporin TIP2;1 is a pressure gated water channel. Biochem Biophys Res Commun 450:289–294

Macey R, Karan D, Farmer RL (1972) Properties of water channels in human red cells. In: Kreuzer F, Slegers JFG (eds) Biomembranes: passive permeability of cell membranes. Springer US, Boston, pp 331–340

Madeira A, Leitão L, Soveral G, Dias P, Prista C et al (2010) Effect of ethanol on fluxes of water and protons across the plasma membrane of *Saccharomyces cerevisiae*. FEMS Yeast Res 10:252–258

Maurel C (2007) Plant aquaporins: novel functions and regulation properties. FEBS Lett 581:2227–2236

Maurel C, Reizer J, Schroeder JI, Chrispeels MJ, Saier MH (1994) Functional characterization of the *Escherichia coli* glycerol facilitator, GlpF, in *Xenopus oocytes*. J Biol Chem 269:11869–11872

Meyrial V, Laize V, Gobin R, Ripoche P, Hohmann S et al (2001) Existence of a tightly regulated water channel in *Saccharomyces cerevisiae*. Eur J Biochem 268:334–343

Mitani-Ueno N, Yamaji N, Zhao F-J, Ma JF (2011) The aromatic/arginine selectivity filter of NIP aquaporins plays a critical role in substrate selectivity for silicon, boron, and arsenic. J Exp Bot 62:4391–4398

Moura T, Macey R, Chien D, Karan D, Santos H (1984) Thermodynamics of all-or-none water channel closure in red cells. J Membr Biol 81:105–111

Murata K, Mitsuoka K, Hirai T, Walz T, Agre P et al (2000) Structural determinants of water permeation through aquaporin-1. Nature 407:599–605

Navarro-Ródenas A, Ruíz-Lozano JM, Kaldenhoff R, Morte A (2012) The aquaporin TcAQP1 of the desert truffle *Terfezia claveryi* is a membrane pore for water and $CO_2$ transport. Mol Plant-Microbe Interact 25:259–266

Nehls U, Dietz S (2014) Fungal aquaporins: cellular functions and ecophysiological perspectives. Appl Microbiol Biotechnol 98:8835–8851

Novo M, Bigey F, Beyne E, Galeote V, Gavory F et al (2009) Eukaryote-to-eukaryote gene transfer events revealed by the genome sequence of the wine yeast Saccharomyces cerevisiae EC1118. Proc Natl Acad Sci U S A 106:16333–16338

Paganelli CV, Solomon AK (1957) The rate of exchange of tritiated water across the human red cell membrane. J Gen Physiol 41:259–277

Patel D (2013) Characterization of vacuole aquaporin function and its implication in membrane fission. Masters thesis Concordia University, Montreal, Quebec, Canada http://spectrum.library.concordia.ca/977797/1/Patel_MSc_F2013.pdf

Pettersson N, Filipsson C, Becit E, Brive L, Hohmann S (2005) Aquaporins in yeasts and filamentous fungi. Biol Cell 97:487–500

Pettersson N, Hagstrom J, Bill RM, Hohmann S (2006) Expression of heterologous aquaporins for functional analysis in *Saccharomyces cerevisiae*. Curr Genet 50:247–255

Preston GM, Carroll TP, Guggino WB, Agre P (1992) Appearance of water channels in *Xenopus oocytes* expressing red cell CHIP28 protein. Science 256:385–387

Prudent S, Marty F, Charbonnier M (2005) The yeast osmosensitive mutant fps1Delta transformed by the cauliflower BobTIP1;1 aquaporin withstand a hypo-osmotic shock. FEBS Lett 579:3872–3880

Sidel VW, Solomon AK (1957) Entrance of water into human red cells under an osmotic pressure gradient. J Gen Physiol 41:243–257

Sidoux-Walter F, Pettersson N, Hohmann S (2004) The *Saccharomyces cerevisiae* aquaporin Aqy1 is involved in sporulation. Proc Natl Acad Sci U S A 101:17422–17427

Soveral G, Macey RI, Moura TF (1997) Membrane stress causes inhibition of water channels in brush border membrane vesicles from kidney proximal tubule. Biol Cell 89:275–282
Soveral G, Veiga A, Loureiro-Dias MC, Tanghe A, Van Dijck P et al (2006) Water channels are important for osmotic adjustments of yeast cells at low temperature. Microbiology 152:1515–1521
Soveral G, Madeira A, Loureiro-Dias MC, Moura TF (2007) Water transport in intact yeast cells as assessed by fluorescence self-quenching. Appl Environ Microbiol 73:2341–2343
Soveral G, Madeira A, Loureiro-Dias MC, Moura TF (2008) Membrane tension regulates water transport in yeast. Biochim Biophys Acta 1778:2573–2579
Soveral G, Prista C, Moura TF, Loureiro-Dias MC (2011) Yeast water channels: an overview of orthodox aquaporins. Biol Cell 103:35–54
Št'ovíček V, Váchová L, Kuthan M, Palková Z (2010) General factors important for the formation of structured biofilm-like yeast colonies. Fungal Genet Biol 47:1012–1022
Sui H, Han BG, Lee JK, Walian P, Jap BK (2001) Structural basis of water-specific transport through the AQP1 water channel. Nature 414:872–878
Takata K, Matsuzaki T, Tajika Y (2004) Aquaporins: water channel proteins of the cell membrane. Prog Histochem Cytochem 39:1–83
Tamura K, Peterson D, Peterson N, Stecher G, Nei M et al (2011) MEGA5: molecular evolutionary genetics analysis using maximum likelihood, evolutionary distance, and maximum parsimony methods. Mol Biol Evol 28:2731–2739
Tanghe A, Van Dijck P, Dumortier F, Teunissen A, Hohmann S et al (2002) Aquaporin expression correlates with freeze tolerance in baker's yeast, and overexpression improves freeze tolerance in industrial strains. Appl Environ Microbiol 68:5981–5989
Tanghe A, Carbrey JM, Agre P, Thevelein JM, Van Dijck P (2005) Aquaporin expression and freeze tolerance in Candida albicans. Appl Environ Microbiol 71:6434–6437
Tanghe A, Van Dijck P, Thevelein JM (2006) Why do microorganisms have aquaporins? Trends Microbiol 14:78–85
Tornroth-Horsefield S, Hedfalk K, Fischer G, Lindkvist-Petersson K, Neutze R (2010) Structural insights into eukaryotic aquaporin regulation. FEBS Lett 584:2580–2588
Vieira FL, Sha'afi RI, Solomon AK (1970) The state of water in human and dog red cell membranes. J Gen Physiol 55:451–466
Wang Y, Schulten K, Tajkhorshid E (2005) What makes an aquaporin a glycerol channel? A comparative study of AqpZ and GlpF. Structure 13:1107–1118
Will JL, Kim HS, Clarke J, Painter JC, Fay JC et al (2010) Incipient balancing selection through adaptive loss of aquaporins in natural *Saccharomyces cerevisiae* populations. PLoS Genet 6:e1000893
Wood V, Gwilliam R, Rajandream MA, Lyne M, Lyne R et al (2002) The genome sequence of *Schizosaccharomyces pombe*. Nature 415:871–880

# Chapter 6
# Sugar and Glycerol Transport in *Saccharomyces cerevisiae*

**Linda F. Bisson, Qingwen Fan, and Gordon A. Walker**

**Abstract** In *Saccharomyces cerevisiae* the process of transport of sugar substrates into the cell comprises a complex network of transporters and interacting regulatory mechanisms. Members of the large family of hexose (*HXT*) transporters display uptake efficiencies consistent with their environmental expression and play physiological roles in addition to feeding the glycolytic pathway. Multiple glucose-inducing and glucose-independent mechanisms serve to regulate expression of the sugar transporters in yeast assuring that expression levels and transporter activity are coordinated with cellular metabolism and energy needs. The expression of sugar transport activity is modulated by other nutritional and environmental factors that may override glucose-generated signals. Transporter expression and activity is regulated transcriptionally, post-transcriptionally and post-translationally. Recent studies have expanded upon this suite of regulatory mechanisms to include transcriptional expression fine tuning mediated by antisense RNA and prion-based regulation of transcription. Much remains to be learned about cell biology from the continued analysis of this dynamic process of substrate acquisition.

**Keywords** HXT • Prion • PMA1 • Post-translational modification • Stress

## 6.1 Introduction

Substrate transport and end-product excretion enable living cells to reproduce, obtain energy, interact efficiently with and adapt to their environment and maintain cytoplasmic homeostasis of a variety of critical cellular parameters including redox potential, pH and solute concentrations. Efficient nutrient depletion can enhance competitiveness with other species of organisms in the natural environment and release of metabolites upon senescence may serve to feed members of the same species or genotype. Regulated transport across the cell surface as well as within the

---

L.F. Bisson (✉) • Q. Fan • G.A. Walker
Department of Viticulture and Enology, University of California, Davis, CA 95616, USA
e-mail: lfbisson@ucdavis.edu; qfan@ucdavis.edu; gawalker@ucdavis.edu



J. Ramos et al. (eds.), *Yeast Membrane Transport*, Advances in Experimental Medicine and Biology 892, DOI 10.1007/978-3-319-25304-6_6

cellular compartments is essential to maintaining cellular function and functionality and underlies population dynamics in complex ecosystems.

The yeast *Saccharomyces cerevisiae* is found naturally in sugar-rich environments and not surprisingly has evolved mechanisms assuring efficient utilization of sugar substrates. In the native environment of grape juice, the principle sugars available to this yeast are the hexoses glucose and fructose (Boulton et al. 1996). These sugars are present in roughly equimolar ratios and are found initially at concentrations in excess of 200 g/L. Lower concentrations of the disaccharide sucrose can also be found in grape but this substrate is present in higher concentrations in other natural yeast habitats. *S. cerevisiae* is capable of efficient sugar utilization spanning the molar to millimolar concentration range.

*S. cerevisiae* displays a phenomenon termed the Crabtree effect meaning that in the presence of sufficiently high concentrations of utilizable hexoses a fermentative mode of metabolism will be operational regardless of the presence or absence of molecular oxygen if other nutrients, particularly nitrogen, are also available (De Deken 1966; Diaz-Ruiz et al. 2011). In conditions of nutrient and energy source excess fermentation allows for a more rapid production of energy at a lower metabolic cost to the cell. The principle site of control of glycolytic flux occurs at the level of substrate transport. Early studies of glucose uptake and metabolism in *S. cerevisiae* noted a striking correlation of initial uptake rates and subsequent metabolic rates through the pathway and concluded that transport was the rate limiting step of glycolysis (Becker and Betz 1972; Gancedo and Serrano 1989). This tight linkage occurred over a range of substrate concentrations and implied a very exquisite control of sugar transport. However, an inherent risk of glycolysis at these high rates of metabolism is the potential for uncoupling of the energy consumption of the early steps of substrate level phosphorylation from the net energy production that occurs at later steps in the pathway. If the rate of substrate phosphorylation is not tightly tied to subsequent metabolism and energy generation ATP levels will become depleted and cells will be unable to sustain viability or metabolic activities (Teusink et al. 1998). Thus the process of transport might not in itself be a pacemaker of metabolism per se but instead is exquisitely regulated to guarantee the rate of transport is in balance with the rate of cellular metabolism and that ATP is not depleted. Indeed, subsequent research supports this view of the complexity and plasticity of hexose transporter regulation in *Saccharomyces*. At high substrate levels glycolytic flux controls hexose uptake while at lower concentrations the reverse seems to be the case and transport controls glycolytic flux rates (Elbing et al. 2004).

This review will cover the status of our understanding of hexose and sugar transport and the regulation of transport, both transcriptional and post-translational. Transcriptional regulation of transporter genes is complex displaying synergistic and antagonistic features. In addition to control at the level of transcription, transporter activity is tightly controlled post-translationally and these mechanisms of regulation are actively being investigated. Recent research indicates a role of heritable protein complexes or prions in the control of metabolic rates and transport which will also be discussed. The connection between glycerol and the many roles of glycerol within the cell will be considered especially in relation to osmotic tolerance, transport, and metabolism. Several excellent earlier reviews of this topic have appeared (Bisson et al. 1993; Boles and Hollenberg 1997; Horák 2013; Kruckeberg 1996; Leandro et al. 2009; Peeters and Thevelein 2014; Rolland et al. 2002; Santangelo 2006).

## 6.2 The Process of Sugar Transport

### *6.2.1 General Types of Transport Processes*

There are three main types of processes that enable translocation of substrates across lipid bilayers (Fig. 6.1). Transport across membranes may occur via simple diffusion, facilitated diffusion or active transport (reviewed in Eddy and Barnett 2007). Some neutral substrates such as ethanol or protonated acids appear to transit across cellular membranes via simple diffusion. In these cases diffusion can be impacted by plasma membrane composition and the cellular proton motive force across the plasma membrane but specific transporter proteins are not likely involved. In other cases a pore or channel may mediate simple diffusion across a membrane in a more controlled or gated fashion.

Facilitated diffusion systems involve proteinaceous transport systems and serve to move substrates along a concentration gradient. Such systems are saturable and subject to regulation of activity but are not concentrative, that is, the depletion of the substrate within the cell provides the energetic drive for continued uptake. If cellular concentrations of the metabolite exceed those of the medium then the substrate will be excreted from the cell. Hexoses are transported via facilitated diffusion. The high

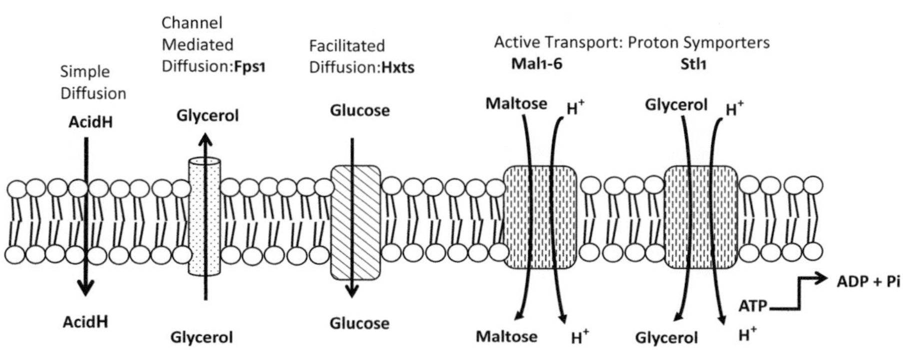


**Fig. 6.1** The types of plasma membrane transport processes of cells. Neutral substrates may enter cells via simple diffusion or via channels. Substrates moving along a concentration gradient utilize protein-mediated facilitated diffusion. Substrates moving against a concentration gradient are transported at the expense of energy. In proton symport the plasma membrane ATPase is required for removal of co-transported protons and maintenance of cytoplasmic proton homeostasis

rates of phosphorylation of hexoses coupled to subsequent metabolism provides the driving force for continued translocation into the cell. In contrast active transport systems serve to move a substrate against a concentration gradient at the expense of cellular energy and from an area of potential lower concentration to one of higher concentration.

Amino acid transport is a primary example of active transport as amino acids are generally present in the natural environment at concentrations far lower than those found in the cytoplasm. Amino acids are taken up by symport meaning the translocation of the component moving against its concentration gradient is thermodynamically coupled to movement of a component along its concentration gradient. In the case of amino acids amino acid movement is coupled to proton uptake or to potassium. The low pH values of the native grape juice environment (pH values between 3 and 4) compared to the near neutrality pH values of the cytoplasm generates a strong gradient of protons across the membrane. The protons transported into the cell are then pumped from the cell by the plasma membrane ATPase that couples proton excretion to ATP hydrolysis thus providing the needed thermodynamic energy to export protons from the cell. With respect to sugars, the majority of sugar transporters are facilitated diffusion systems as the high rates of metabolism accompanying uptake keep cytoplasmic levels of the substrate low and transport unidirectional. The primary exception to this rule is maltose transport. Maltose transport is coupled to proton movements and therefore requires the expenditure of energy and activation of the plasma membrane ATPase (Serrano 1977).

### *6.2.2 The Central Role of the Plasma Membrane ATPase Pma1*

Given the relatively low pH of the yeast native environment of grape juice it is not surprising that many active transport systems take advantage of the proton gradient and couple movements of protons to movements of substrates. The plasma membrane proton pump is encoded by a single gene the *PMA1* gene (Serrano et al. 1986). The Pma1 protein is the major protein species found in the plasma membrane and is essential for maintenance of pH homeostasis of the cytoplasm. The Pma1 protein has a finite concentration within the membrane and therefore a finite ability to excrete protons from the cell. Protons arise in the cytoplasm from several sources. Hexose catabolism produces protons that are extruded by the pump and protons may also gain entry into the cytoplasm via passive proton flux. Substrate uptake via proton symport systems brings protons into the cells and protons may also enter via simple diffusion of neutral or protonated acid species across the membrane.

There is an intrinsic relationship between transport and the ability of the Pma1 protein to maintain pH homeostasis within the cell (Fig. 6.2). As fermentation of high sugar substrates occurs and ethanol is produced ethanol leads to enhanced rates of passive proton flux directly and via the accompanying changes to the plasma

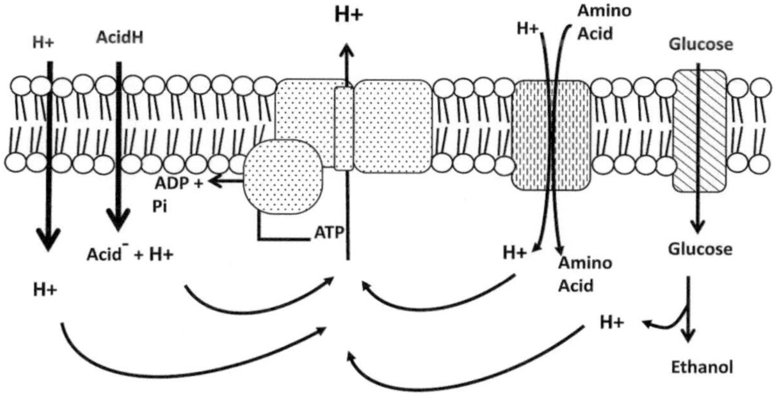


**Fig. 6.2** The Central role of the plasma membrane ATPase in cellular homeostasis. Protons are generated within the cytoplasm via multiple routes. Passive proton flux leads to acidification of the cytoplasm if protons are not extruded. Undissociated acid species may diffuse across the membrane or enter via the glycerol transporter and then release a proton within the cytoplasm. Active proton symporters require a functional ATPAse in order to remove the co-transported proton and sustain the proton gradient across the membrane. Protons are also released during catabolism

membrane that allow adaptation to loss of water of hydration and the presence of ethanol. Nitrogen compound transport is decreased as ethanol increases again in order for the cell to maintain homeostasis. Rates of glycolysis and sugar transport may also be reduced as needed to maintain the balance between energy generation, energy consumption in removal of excess cytoplasmic acidity, and transport of proton-coupled substrates.

## 6.3 The Hexose Transporter Gene Family

Hexose transport in *Saccharomyces* is mediated by a gene family of 17 putative membrane proteins that display high amino acid sequence similarity (Bisson et al. 1993: Boles and Hollenberg 1997; Horák 2013; Kruckeberg 1996; Leandro et al. 2009; Peeters and Thevelein 2014; Rolland et al. 2002; Santangelo 2006). These proteins consist of cytoplasmic C and N terminal domains of varying lengths with 12 membrane spanning domains. The region between membrane spanning domains 6 and 7 defines a cytoplasmic loop. It has been postulated that the membrane spanning domains form a barrel like structure with an inner pore and that the cytoplasmic domains can associate with that inner core of the barrel structure thereby impacting transporter function. The hexose transporters catalyze the uptake of glucose, fructose and mannose via facilitated diffusion (Bisson et al. 1993).

### *6.3.1 Discovery of the Family of Hexose Transporters*

The first genes impacting sugar transport in *Saccharomyces* were identified from the analysis of mutations impacting utilization of the trisaccharide raffinose (Neigeborn and Carlson 1984). Raffinose contains one moiety each of glucose, fructose and galactose and is cleaved to the monomer hexoses outside of the cell. The enzyme invertase is responsible for cleaving sucrose externally as well as the glucose-fructose bond of raffinose. The efficiency of cleavage of raffinose leads to the production of low levels of both glucose and fructose. Thus use of this substrate requires efficient transport into the cell of the low levels of these monomer compounds produced externally. The original mutations impacting raffinose use as carbon and energy source were designated "snf" for sucrose non-fermenting (Neigeborn and Carlson 1984).

Several lines of evidence suggested that one of these genes, *SNF3*, encoded a hexose transporter. The *snf3* mutants showed a defect in growth on low monomer substrate concentrations in general, which was especially pronounced in the presence of a respiratory inhibitor. Direct influx uptake assays showed a decrease in efficiency of glucose and fructose transport at low substrate concentration in *snf3* mutations (Bisson et al. 1987; Celenza et al. 1988). Sequence analysis of *SNF3* showed a relatedness of amino acid composition and proposed structure to previously described transporter proteins of bacteria and mammalian systems (Celenza et al. 1988). Therefore the initial phenotypic characterization of loss of function mutations of the *SNF3* gene were consistent with a role in hexose uptake at low substrate concentrations and Snf3 was thought to be a high affinity hexose carrier (Bisson et al. 1987). However, *SNF3* was subsequently shown to be a sensor of low glucose concentrations that functions in the regulation of expression genes encoding the bonafide hexose transporters (*HXT* genes) rather than playing a direct role in sugar uptake (Coons et al. 1997; Liang and Gaber 1996; Ozcan et al. 1996a). Thus the Snf3 protein functions not as a transporter but as a nutrient transceptor – signaling the presence of substrate to the cell while retaining similarity of form and function to catabolic transporters (Holsbeeks et al. 2004; Kriel et al. 2011; Van Zeebroeck et al. 2009).

Suppressors of the *snf3* mutation appeared readily on selective media and several have been identified using a variety of strategies (Marshall-Carlson et al. 1991). Spontaneous suppressors and those induced by mutagens were obtained as were multicopy suppressors identified in the attempt to clone the wild type version of the *SNF3* gene. There were two major classes of multicopy suppressors. The first class of multicopy suppressor identified genes that when overexpressed restored glucose uptake and growth on low glucose concentrations to *snf3* null strains. DNA sequence analysis of these genes showed that they too were similar in amino acid content and induced structure to the superfamily of transporter proteins (Kruckeberg and Bisson 1990; Lewis and Bisson 1991). Several genes with similar yet different map locations and sequences were identified and named "*HXT*" for hexose transporter, and designated the *HXT1* through *HXT7* genes (Kruckeberg 1996). These genes also showed features in common with the galactose transporter encoded by the *GAL2* gene (Kruckeberg 1996). The genome sequence of *Saccharomyces* revealed the existence of 10 additional *HXT* genes, *HXT8* through *HXT17* (Leandro et al. 2009).

Several wine strains of *S. cerevisiae* of the EC118 lineage where found to contain a specific fructophilic proton symporter encoded by the fungal *FSY1* gene that arose from a horizontal gene transfer event (Anjos et al. 2013; Galeote et al. 2010; Novo et al. 2009). Subsequent research has determined that multiple independent horizontal gene transfer events of the *FSY1* gene have occurred throughout the Ascomycota suggesting that horizontal gene transfer is common and that this particular transporter confers a growth advantage leading to its maintenance in the genome (Coelho et al. 2012). Other wine strains efficiently utilizing fructose carry a mutation of the *HXT3* gene altering substrate specificity (Guillaume et al. 2007).

### 6.3.2 Characterization of the HXT Gene Family

Comparative analyses of the biochemical characteristics of these transporters demonstrated that they possessed differing substrate affinities for glucose (Kruckeberg and Bisson 1990). Transporters could be classified into three broad groups with respect to glucose substrate affinity: low affinity ($K_m$ values of 40–100 mM), moderate affinity ($K_m$ values of 10–15 mM) and high affinity ($K_m$ values of 1–3 mM) (Reifenberger et al. 1997). These carriers in general display lower affinities for fructose as compared to glucose (Boles and Hollenberg 1997). The Hxt2 transporter appears to be able to exist in more than one conformation and alternately displays high ($K_m$ of 2 mM) and moderate ($K_m$ of 10 mM) affinity (Reifenberger et al. 1997). The affinity for substrate of Hxt2 appears to be correlated with external substrate concentrations. When Hxt2 was present as the sole hexose transporter, biphasic transport kinetics were observed under high substrate concentrations suggesting that this carrier can exist in multiple conformations (Reifenberger et al. 1997). Deletion analysis indicated that six *HXT* genes, *HXT1, HXT2, HXT3, HXT4, HXT6* and *HXT7,* comprise the major transporters functional during growth on glucose and fructose under laboratory conditions (Liang and Gaber 1996; Reifenberger et al. 1995). Similarly deletion of these carriers in a wine strain likewise prevented growth and fermentation of grape juice (Karpel et al. 2008; Luyten et al. 2002; Perez et al. 2005). Hxt1 and Hxt3 were transporters responsible for the bulk of sugar fermentation during wine production. The *HXT2* gene was expressed early in fermentation when sugar concentrations were high and *HXT6* and *HXT7* were found expressed at the end of fermentation (Luyten et al. 2002). Of the major Hxt carriers, *HXT1* and *HXT3* are low affinity transporters and *HXT6* and *HXT7* are high affinity transporters with $K_m$ values of 1–2 mM. *HXT6* and *HXT7* are located adjacent to each other on Chromosome IV and show high sequence identity with only two differing amino acids, 293 (Val/Ile) and 556 (Thr/Ala). Some laboratory strains show chimeras of these genes suggesting recombination events within the genome

(Boles and Hollenberg 1997). The Hxt4 transporter displays a moderate affinity with a $K_m$ value of 9 mM. Analysis of gene expression of these transporters in relation to external substrate concentrations showed a high level of correlation between $K_m$ values and the external inducing concentrations of substrates (Buziol et al. 2008) suggesting that the variation in affinity is needed for optimal sugar depletion across a range of substrate concentrations. The *HXT* genes can also function in the excretion of glucose. Jansen et al. (2002) showed that efflux of glucose from excessive maltose uptake required the function of members of the *HXT* transporter family. The $K_m$ values of countertransport or efflux for each of the major Hxt transporters matched in magnitude the substrate uptake values (Maier et al. 2002).

Over-expression of *HXT5* and *HXT8* was found to partially restore growth on low glucose levels to snf3 null strains suggesting that these two transporters could function in catabolism but suggesting that perhaps that is not their major metabolic role. Conditions of stress also lead to increased expression of *HXT5* (Diderich et al. 1999; Gasch et al. 2000; Verwaal et al. 2002). Low levels of cAMP are required to maintain expression of *HXT5* in the absence of glucose (Bermejo et al. 2013). The presence of glucose leads to a higher spike of cAMP levels and a loss of HXT5 expression (Bermejo et al. 2013). Thus, Hxt5 is thought to play an important role during starvation in enabling the cells to rapidly detect glucose and trigger a signaling cascade. This role of transporters in readiness to adapt to substrates has been termed an "ajar" pathway (Bermejo et al. 2013). Two other HXT transporters (Hxt6 and Hxt7) have also been found to be expressed mainly under starvation conditions similarly to Hxt5 (Klockow et al. 2008) raising the intriguing possibility that these transporters are perhaps directly linked to distinct internal signaling pathways or mechanisms enabling rapid re-tooling of cellular metabolism to take optimal advantage of available substrate.

*HXT9*, *HXT11* and *HXT12* form a subset of transporters with highly sequence identity (greater than 97 %). In some genetic backgrounds the *HXT12* gene carries a disruption likely inactivating the transporter function. The *HXT9* and *HXT11* genes have been shown to play a role in pleiotropic drug resistance although one of them, *HXT11*, appears to function as a sugar carrier (Boles and Hollenberg 1997). These two proteins appear to play an important role in drug resistance, perhaps enabling localized sugar uptake and energy generation in response to the presence of specific inhibitors in the environment.

Mutation of *HXT13, HXT14, HXT15, HXT16* or *HXT17* does not appear to impact glucose transport in any obvious way although it appears that these genes are expressed under certain conditions and encode functional transporters (Diderich et al. 1999; Boles and Hollenberg 1997). *HXT5, HXT13* and *HXT15* expression has been shown to increase in the presence of non-fermentable carbon sources (Greatrix and van Vuuren 2006). This same study found that *HXT17* gene expression was elevated at higher pH in the presence of galactose or raffinose. The *GAL2* gene encodes a transporter used specifically for galactose uptake and shows over 70 % identity to members of the *HXT* family. Gal2 is fully able to transport glucose and will do so with a relatively high affinity (Boles and Hollenberg 1997).

Since facilitated diffusion systems are optimally active around their substrate affinity value or $K_m$ the spread in substrate affinities of this group of proteins was believed to enable assimilation of available glucose and other hexose sugars across the spectrum of hexose concentrations (molar to millimolar) found in the yeast environment. Yeast strains that have deletions in seven *HXT* genes, *HXT1-HXT7,* are unable to grow on glucose, suggesting that the remaining Hxt proteins are not physiologically relevant for laboratory growth conditions or may have cellular functions other than as a catabolic transporter (Reifenberger et al. 1995; Wieczorke et al. 1999). The Snf3 protein was found to regulate expression of key members of this superfamily of hexose transporters and was required for the expression of high affinity transporters under low glucose conditions (Bisson et al. 1987; Kruckeberg and Bisson 1990; Bisson et al. 1993; Kruckeberg 1996). The *HXT* genes were therefore identified as multicopy suppressors because in high gene copy numbers transporter expression was elevated to levels enabling growth (Theodoris et al. 1994; Theodoris and Bisson 2001).

### 6.3.3 The Physiological Roles of the HXT Genes

Physiological analyses indicated that while *HXT1, 2, 3, 4, 6* and *7* were the principle hexose carriers associated with growth over a wide range of glucose and fructose concentrations, other *HXT* genes when expressed in multicopy in a multiple *HXT* delete strain could restore sugar uptake and permit growth (Wieczorke et al. 1999). Comparative analysis of the *HXT1* (low affinity) and *HXT2* (high affinity) genes and the generation of chimeric proteins of the two coding regions (Kasahara et al. 2007) revealed that transmembrane domains 1, 5, 7 and 8 appear to specify substrate affinity and specific residues within those domains were identified. These two proteins show a difference of 75 residues across the 12 membrane spanning domains. These four domains are thought to face the inner central pore of the transporter a finding confirmed by the location of the amino acid residues defining transporter affinity (Kasahara et al. 2007).

*HXT2* is induced by medium alkalinization which requires Snf1 and the calcineurin pathway, specifically the Crz1 gene (Ruiz et al. 2008). Under alkaline conditions glucose utilization is impaired likely due to the lack of a proton motive force across the cellular plasma membrane. This proton motive force is correlated with efficient oxygen uptake so it is not surprising that minimization of the differential across the plasma membrane would lead to inhibition of glucose utilization. It is possible that the activation mediated by external calcium serves to increase expression of key proteins involved in glucose uptake thereby restoring metabolism and energy production under these growth-limiting conditions (Ruiz et al. 2008). These findings are consistent with the timing of expression of *HXT2* observed during wine fermentation. This gene was found to be expressed early in lag phase when substrate is present in excess (Karpel et al. 2008; Perez et al. 2005) inconsistent with a simple role as a high affinity transporter but consistent with activity in an altered membrane state or in growth initiation.

A hybrid of the *HXT1* and *HXT7* genes called TM6* was found to override the Crabtree effect and enable high rates of respiration in the presence of high external concentrations of glucose (Bosch et al. 2007; Otterstedt et al. 2004). This property appeared to be associated with the *HXT7* gene and yields of ethanol when *HXT7* is the sole transporter present are lower than that for other *HXT* transporters suggesting a redirection of carbon to respiration. This observation implies a role for the transport process in the control of the shift between fermentation and respiration. The occurrence of natural recombinants between the *HXT6* and *HXT7* gene may be driven by selection for or against maintenance of respiration under fermentative conditions. Alternately, selection for enhanced utilization of glucose after continued growth under glucose limitation yielded mutants with amplifications and alterations of the *HXT6/HXT7* region (Brown et al. 1998).

## 6.4 Other Mechanisms of Multicopy Suppression of *snf3* Mutations

The screen for multicopy suppressors of *snf3* null mutations yielded identified other modes of suppression. In some cases the suppressor plasmid did not contain an intact *HXT* gene but only the regulatory region of that gene. Subsequent analysis identified the domains within promoter regions of the *HXT* genes that resulted in suppression presumably via the titration of negative regulatory factors called DDSE (*D*NA sequence *d*ependent *s*uppressing *e*lement) (Theodoris et al. 1994). DDSE sequences were found in the promoter regions of several *HXT* genes as well as in the promoters of other genes such as *SKS1* protein kinase gene (Yang and Bisson 1996). Subsequent work on the low affinity *HXT1* gene revealed that overexpression of Hxt1 from another promoter was insufficient to complement the defect in transport caused by loss of the Snf3 sensor suggesting that the DDSE binding domains of the native *HXT1* promoter and not *HXT1* itself was responsible for multicopy suppression (Boles and Hollenberg 1997). Other genes that were identified also seemed to suppress *snf3* null mutations by virtue of their promoter regions and similar titration of inhibitors of transcription. Interestingly overexpression of the *SKS1* gene independently of its promoter was also shown to suppress loss of Snf3 function and defines a glucose-independent pathway for adaptation to low substrate (Yang and Bisson 1996; Vagnoli and Bisson 1998).

## 6.5 Other Classes of Suppressors of the Loss of the *SNF3* Gene

In addition to multicopy suppressors, genomic recessive and dominant mutations suppressing the *snf3Δ* growth defect have also been identified (Marshall-Carlson et al. 1991). Mutations of *RGT, SSN6* and *TUP1* (Marshall-Carlson et al. 1991) or of the *IRA2* gene (Ramakrishnan et al. 2006) have been shown to suppress loss of *snf3*. The *IRA2* gene encodes a regulator of the RAS-cAMP pathway in yeast (Tanaka et al. 1990) that is involved in the control of glycogen synthesis and is a GTPAse activating protein (Tanaka et al. 1990). The role of the Ira2 protein in suppression of growth on low glucose was shown to be unrelated to its role in reduction of carbon flow to glycogen (Ramakrishnan et al. 2006). Since the cAMP pathway also regulates expression of the glycolytic pathway with secondary effects on expression of *HXT* genes the *IRA2* gene may function as a suppressor by virtue of its impact on cAMP signaling.

As described in detail below, the *RGT1* gene encodes a transcriptional repressor that functions to block expression of a subset of the *HXT* genes (Ozcan et al. 1996b; Ozcan and Johnston 1996). Activation of the Snf3 protein leads to proteolytic degradation of the Rgt1 protein and expression of the target *HXT* genes. Loss of the Rgt1 protein due to mutation therefore results in expression of the *HXT* genes similarly to what occurs after activation of Snf3. The *SSN6* and *TUP1* genes both encode general repressor proteins that negatively regulate expression of some members of the *HXT* gene family. Loss of one or both of these proteins has a global impact on glucose repression and results in a bypass of the requirement of the Snf3 glucose signal for gene expression.

All of these suppressors are recessive. Diploid strains homozygous for loss of *snf3* but carrying a single wild type copy of either *RGT1, SSN6, TUP1* or *IRA2* do not display the *snf3* phenotype and are fully able to grow on low substrate concentrations. These suppressor studies led to a model in which the *SNF3* protein plays a role in substrate detection and the signal that glucose is present is subsequently transduced via a series of regulatory proteins to ultimately impact the level of expression of the hexose transporter genes. Loss of *SNF3* can be bypassed directly by mutation of downstream repressor proteins or indirectly by impacting the functioning of other pathways that modulate cellular responses to glucose.

Suppressor analyses also identified a dominant suppressor, the *RGT2* gene (Marshall-Carlson et al. 1991). When present in the diploid strain in single copy the mutated version of *RGT2* led to active suppression of *snf3* and restored growth on low glucose to a diploid homozygous for loss of *snf3*. Deletion of *RGT2* had no impact on suppression of *snf3*, indicating it was a gain of function, not loss of function, suppressor mutation. *RGT2* was later shown to encode a low-affinity glucose sensor with high sequence relatedness to the *SNF3* gene (Ozcan and Johnston 1996; Moriya and Johnston 2004). The dominant *RGT2* mutations allow for constitutive expression of several *HXT*s (Ozcan et al. 1996b; Ozcan and Johnston 1999).

The *SNF3* gene was originally discovered in a screen of a laboratory strain derived from S288C (Neigeborn and Carlson 1984). The starting strain was described as congenic to S288C having undergone both outcrossing and back-crossing (Neigeborn and Carlson 1984). This parental strain was mutagenized with subsequent selection for strains impaired in raffinose utilization. The *snf3* phenotype was extensively characterized in the original strain background. However loss of snf3 in other strain backgrounds including all wine strains examined did not lead to a defect in hexose uptake or growth on low glucose concentrations. The original strain used to isolate the *snf3* mutation has been found to carry a mutation of the *HXT7* gene, inactivating this gene (Place and Bisson 2013). The restoration of *HXT7* to the original strain used to isolate snf3 restores glucose transport in this strain as well thus confirming the role of *HXT7* in suppression of the *SNF3* phenotype.

## 6.6 Transport of Other Sugars

The transport of α-glucosides such as maltose in *Saccharomyces* is mediated by an active proton symport mechanism (Serrano 1977). The *MAL* loci consist of three genes, the maltose transporter, maltase, and a transcriptional regulator responsible for maltose induction of the other two genes (Charron et al. 1986). Five highly homologous *MAL* loci differing in chromosomal location have been identified (*MAL1-MAL4* and *MAL6*) (Charron et al. 1989; Chow et al. 1989). The three genes within each locus are designated by a second number that defines function. The designation "1" is used for the transporter proteins, "2" is used for the maltose α-glucosidase and "3" is used for the transcriptional inducer. Thus there are five known maltose transporter genes: *MAL11*, *MAL21. MAL31, MAL41* and *MAL61* that are largely functionally interchangeable (Horák 2013). However, detailed sequencing analysis of the MAL transporter family has shown that although there is a high degree of functional and sequence similarity there are differences in primary sequence that may impact substrate specificity (Alves et al. 2008). Expression of the transporters may also vary (Alves et al. 2014). The Mal21 transporter catalyzes the uptake of trehalose and sucrose in addition to maltose with a low affinity (90–100 mM Km) (Stambuk and de Araujo 2001). The affinity for maltose is much higher (Km = 5 mM).

The *AGT1/MAL11* gene of *S. cerevisiae* encodes an α-glucoside proton symporter with a broader substrate specificity than the other MAL transporter genes (Alves et al. 2008). The Agt1/Mal11 transporter has a higher affinity for trehalose and sucrose (Km of 7 mM) and a lower affinity for maltose (Km of 18 mM). This transporter is also responsible for high affinity uptake of maltotriose with an affinity similar to that of maltose (Alves et al. 2008). Lager brewing strains of *S. pastorianus* contain an additional gene, MTY1 that displays homology to AGT1 and also functions as an α-glucoside transporter (Cousseu et al. 2013; Salema-Oom et al. 2005). Differences in α-glucoside transport efficiencies across brewing strains are due to both the numbers of loci of the *MAL* genes, presence of the *AGT1* allele and *MTY1*, and the specific alleles encoded by these genes (Duval et al. 2010). Two other genes YDL247w and YJR160c have also been shown to encode disaccharide transporters that along with *AGT1* can also confer glucose uptake to a strain deleted of all of the known *HXT* transporters (Wieczorke et al. 1999).

## 6.7 The Regulation of Hexose Transport

The process of hexose transport is controlled at many levels within the cell. Several transcriptional regulators have been identified that are either specific to regulation of the *HXT* family or are general regulators that also impact expression of these genes. Some of these regulatory systems respond to the levels of glucose in the medium and efficiently coordinate the affinity of expressed transporters with the level of substrate present in the medium (Daran-Lapujade et al. 2004; Elbing et al. 2004). Regulation of transporter expression also shows glucose independent elements (Pasula et al. 2007) The RNA helicase Dbp2 that is found associated with actively transcribed chromatin and appears to regulate *HXT* gene expression (Beck et al. 2014). Dbp2 is linked to long non-coding RNA (lncRNA) or antisense RNA. Antisense RNAs for the *HXT* genes have been identified shown to either stimulate or enhance repression of *HXT* gene transcription (Beck et al. 2014). Hxt proteins are also regulated post-translationally and protein turnover can be rapid and occurs via the normal endocytosis pathway (Krampe et al. 1998; Kresnowati et al. 2006; Roy et al. 2014b). The yeast glucose sensors Snf3 and Rgt2 are also degraded via endocytosis (Roy and Kim 2014).

### *6.7.1 Transcriptional Regulators of the HXT Family*

Since glucose and fructose are present in nature at a wide range of concentrations, yeast cells have evolved exquisite systems to enable maximal efficiency of use of different levels of glucose (Broach 2012; Compagno et al. 2014). Glucose uptake involves the crosstalk between three main signaling pathways in yeast cells: the Rgt2/Snf3 glucose sensing pathway, Snf1-Mig1 glucose repression pathway and the cAMP-PKA signaling cascade (Broach 2012; Conrad et al. 2014; Gancedo 2008; Kim et al. 2013; Rødkær and Færgeman 2014). Several genes are involved in this regulatory pathway and are listed in Fig. 6.3. As a result, glucose represses expression of genes that are redundant for its metabolism and induces expression of other genes required for its utilization (Geladé et al. 2003; Johnston and Kim 2005; Weinhandl et al. 2014). Among the genes induced by the presence of glucose, the *HXT* genes are of great importance. These genes encode an extensive set of glucose transporters that function as facilitated diffusion carriers with various affinities and capacities. The glucose induction of these *HXT* genes is mainly based on a repression mechanism mediated by a transcriptional factor—Rgt1. By recruiting another two transcriptional factors Std1 and Mth1, Rgt1 interacts with a general corepressor Cyc8-Tup1 and forms a repressor complex that inhibits the expression of *HXT* genes in the absence of glucose (Fig. 6.4).

| Protein | Function Description |
|---|---|
| Rgt2 | High glucose sensor |
| Snf3 | Low glucose sensor |
| Yck1/2 | Yeast casein kinase complex |
| Cyc8-Tup1 | General co-repressor complex that interacts with Rgt1 and regulates expression of *HXT* genes |
| Rgt1 | Transcriptional activator and repressor of *HXT* genes in response to glucose |
| Mth1 | Regulator of the glucose-sensing signal transduction pathway |
| Std1 | Regulator of the glucose-sensing signal transduction pathway |

**Fig. 6.3** Descriptive list of the transcriptional and non-transcriptional regulatory factors involved in control of hexose transporter gene expression

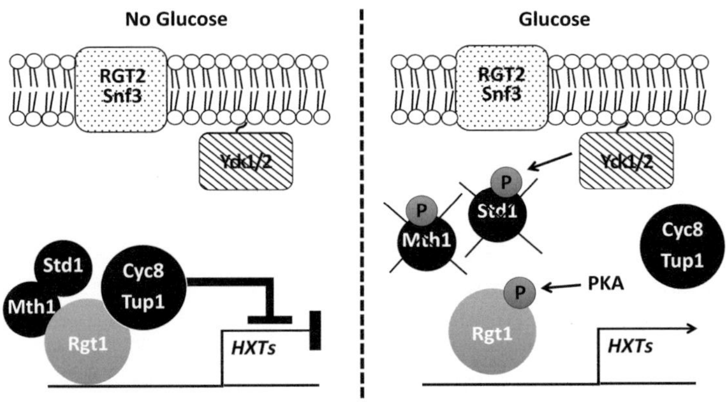


**Fig. 6.4** Current model of glucose induction of *HXT* gene expression. In the absence of glucose, transcription factors Mth1 and/or Std1 bind to Rgt1, enabling the interaction of Rgt1 with the general co-repressor Cyc8-Tup1. Expression of *HXT* genes is inhibited. When glucose becomes available, it binds to the transmembrane domain of Rgt2 or Snf3, causing conformational switch of the sensors which then activates membrane-bound casein kinase I (Yck1/2). This activation further leads to the recruitment of transcriptional factors Mth1 and Std1 from nucleus to the plasma membrane. Yck1/2 subsequently phosphorylates these proteins marking them for degradation through the ubiquitin-proteasome pathway. The decrease in protein levels of Mth1 and Std1 prevents the association between Rgt1 and Cyc8-Tup1 thereby freeing Rgt1 from Cyc8-Tup1 corepressor. As a result repression is relieved and glucose transporters are expressed

### *6.7.2 General Co-repressor Complex Cyc8-Tup1*

Cyc8-Tup1 is a general transcriptional corepressor complex composed of one Cyc8 and four Tup1 subunits (Varanasi et al. 1996). It is required for transcriptional repression of distinct sets of genes. Among them are: glucose-repressible genes, oxygen-regulated genes, cell-type regulatory genes and DNA damage-reducible genes (Keleher et al. 1992; Tzamarias and Struhl 1994). Through interaction with diverse pathway-specific DNA-binding transcriptional factors, the Ssn6-Tup1 complex is recruited to promoters of the above genes, mediating their expression (Keleher et al. 1992; Papamichos-Chronakis et al. 2000; Wahi et al. 1998). The C-terminal 7 WD40 motifs in Tup1 subunits and the N-terminal 10 tetratrico peptide repeat (TPR) motifs in the Cyc8 (SSn6) subunit are the functional domains responsible for protein-protein interactions with specific repressors such as cell-type regulator alpha 2 (Malave and Dent 2006; Sprague et al. 2000; Schultz et al. 1990; Wu et al. 2001). Different combinations of TPRs in Cyc8 are required for repression of different genes (Tzamarias and Struhl 1995). The N-terminal portion of Tup1 folds into a helical structure that is important for Tup1 tetramerization and for interactions with Cyc8 (Smith et al. 1995; Watson et al. 2000). Primarily, the Cyc8-Tup1 complex inhibits transcription by masking activation domains of recruiting proteins (Wong and Struhl 2011). However, specific metabolic signals, such as mitochondrial dysfunction, may convert the Cyc8-Tup1 transcriptional corepressor into a transcriptional coactivator (Conlan et al. 1999).

Rgt1-mediated repression of the *HXT* genes occurs by a mechanism that requires the association with Cyc8-Tup1 complex (Ozcan et al. 1996b). There is evidence showing that Rgt1 binds to its target promoter in the absence of Cyc8-Tup1, but this binding does not in itself repress transcription. Cyc8-Tup1 negatively regulates the DNA-binding ability of Rgt1 by repressing expression of the *MTH1* gene (Roy et al. 2013). Dissociation of Cyc8-Tup1 from Rgt1 is the most essential step for glucose induction of *HXT* gene expression. It is necessary and sufficient to relive the repression brought by Rgt1.

### 6.7.3 *Snf3/Rgt2 Glucose Sensing Pathway*

The Snf3/Rgt2 pathway is a receptor-mediated signaling pathway that couples the expression of multiple *HXT* transporters to glucose concentration in the environment (Westergaard et al. 2007). Glucose availability is detected via two glucose sensing membrane proteins: Snf3 and Rgt2. In a current model for transduction of the glucose signal from Snf3/Rgt2 to *HXT* genes, glucose binds to the transmembrane domain of Rgt2 or Snf3, causing conformational switch of the sensors which then activates membrane-bound casein kinase I (Yck1/2) (Moriya and Johnston 2004; Pasula et al. 2010). This activation further leads to the recruitment of transcriptional factors Mth1 and Std1 from nucleus to the plasma membrane. Yck1/2 subsequently phosphorylates these proteins marking them for degradation through the ubiquitin-proteasome pathway (Flick et al. 2003; Kim et al. 2006; Lafuente et al. 2000; Schmidt et al. 1999). The decrease in protein levels of Mth1 and Std1 prevents the association between Rgt1 and Cyc8-Tup1 thereby freeing Rgt1 from Cyc8-Tup1 corepressor (Lakshmanan et al. 2003; Polish et al. 2005; Roy et al. 2013). As a result repression is relieved and glucose transporters are expressed.

Snf3 and Rgt2: Snf3 and Rgt2 are membrane proteins that are structurally similar to each other. They both consist of two functionally distinct domains, a predicted 12-membrane-spanning domains and an unusually long C-terminal tail facing the cytoplasmic side of the yeast plasma membrane. Both of these proteins show similarity in structure to the glucose transporter family but appear to have lost the ability to transport glucose into the cell; instead, they act as glucose sensors monitoring extracellular glucose levels (Dlugai et al. 2001). Snf3 functions a high-affinity receptor sensing low glucose concentrations (Dlugai et al. 2001). It is required for the expression of *HXT* genes encoding high- and moderate-affinity glucose transporters. Rgt2, however, serves as a low-affinity receptor sensing high glucose concentration and is essential for the maximum activation of *HXT1* expression. However, it remains unclear whether the abundance of these two glucose sensors present on plasma membrane is associated with their affinity for glucose. Previous studies have shown that unique C-terminal tails of Snf3 and Rgt2 are essential for signaling the presence of hexose. First, deletion of the entire C-terminal tail results in the loss of the ability to induce *HXT* expression in the presence of glucose. Second, tethering the overexpressed C-terminal tail of Snf3 and Rgt2 by themselves to the cell membrane causes constitutive expression of *HXT* genes. Thus the current model suggests that the association of the sensors with glucose or fructose leads to a change in conformation and an exposure of the C-terminal regions to the cytoplasm. Factors interacting or responding to the presence of those regions then launch a regulatory cascade leading to expression of the target *HXT* genes.

Mth1/Std1: Mth1 and Std1 are key transcriptional factors known to transduce the glucose signal from the membrane glucose receptors to nuclear repressor Rgt1, thus modulating glucose-related *HXT* gene expression (Horák 2013). When glucose is absent from the environment, both Mth1 and Std1 are required for the repression of *HXT* gene transcription by Rgt1 and its repressor function is controlled by the abundance of these two proteins (Lafuente et al. 2000). The association of Mth1 and Std1 with Rgt1 facilitates the recruitment of the general corepressor complex Cyc8-Tup1 to Rgt1, resulting in the inhibition of *HXT* gene expression (Fig. 6.5a).

However, when glucose becomes available in the environment, Mth1 and Std1 are recruited from nucleus to plasma membrane (Fig. 6.5b). Upon binding to the long C-terminal cytoplasmic tails of the glucose sensors Snf3 and Rgt2, Mth1 and Std1 are subsequently phosphorylated by the yeast casein kinase I complex Yck1/2 (Schmidt et al. 1999). Although reports are conflicting, Mth1 has been reported to interact with the cytoplasmic domains of either of the membrane glucose sensors, Snf3 and Rgt2. Phosphorylated Mth1 and Std1 are further taken by the $SCF^{Grr1}$ ubiquitin-protein-ligase complex to the proteasome for degradation, causing a decreased abundance. Depletion of Mth1 and Std1 in the cytoplasm exposes Rgt1 to phosphorylation, thus triggering an intramolecular interaction between a short sequence adjacent to its zinc cluster DNA-binding domain and the central region (Kim et al. 2003). This further inhibits its ability to bind to *HXT* promoters hence relieves the repression mediated by this protein.

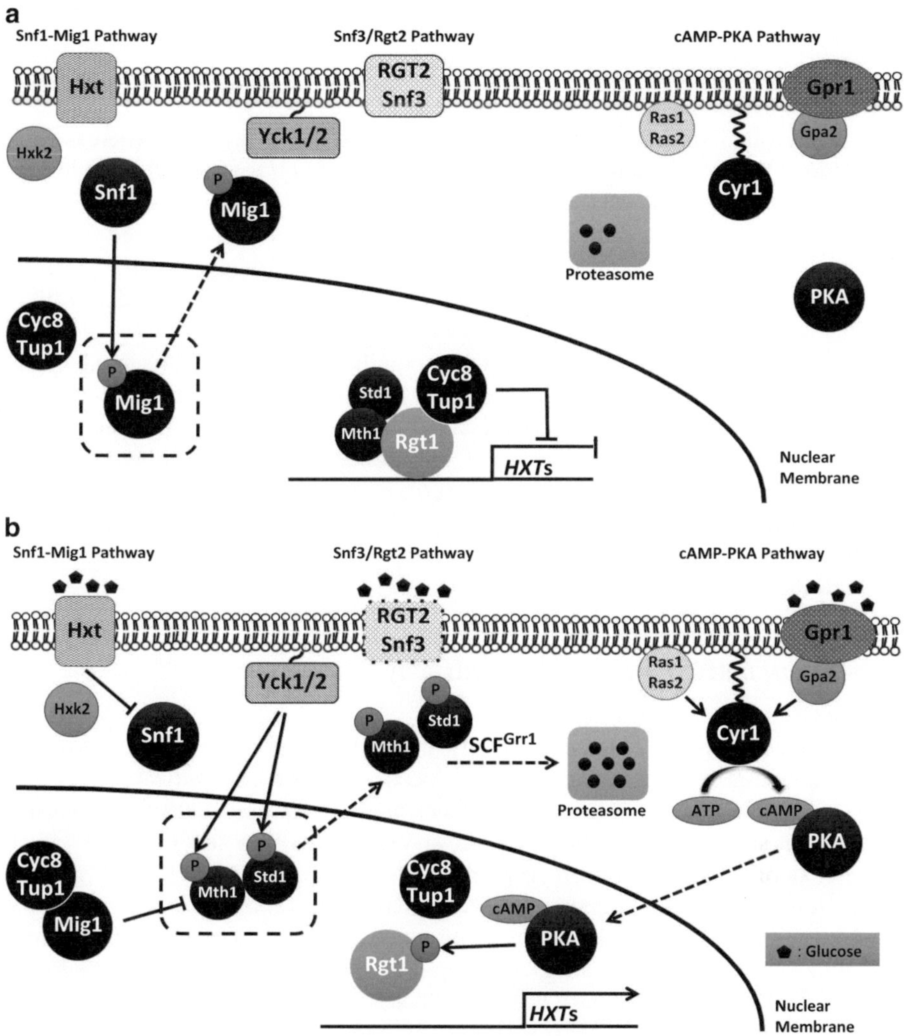


**Fig. 6.5** **Panel a** Crosstalk between three main glucose signaling pathways in yeast in absence of glucose. *Solid arrows* indicate activation or phosphorylation, *solid bars* refer to repression, while *dotted arrows* indicate direction of translocalization. When glucose is not available to yeast cells, Mig1 is phosphorylated by Snf1 and transported into cytoplasm, thus dissociated from the general repressor Cyc8-Tup1 leading to the expression of glucose repressed genes. Therefore alternative carbon sources can be taken up and utilized by the cell. Meanwhile, Rgt1 associated with Mth1 or Std1 is capable of interacting with Cyc8-Tup1, inhibiting the expression of hexose transporter genes (HXTs). **Panel b** Crosstalk between three main glucose signaling pathways in yeast in presence of glucose. *Solid arrows* indicate activation or phosphorylation, *solid bars* refer to repression, while *dotted arrows* indicate direction of translocalization. Transmembrane glucose sensors Snf3 and Rgt2 send out a signal to the yeast casein kinase complex Yck1/2 when glucose becomes available in the environment. Transcription factors Mth1 and Std1 then get phosphorylated by Yck1/2 and are recognized by the SCFGrr1 ubiquitin ligase, therefore taken to proteasome to degradation. Expression of Mth1 gene is also repressed by Mig1 coupled with Cyc8-Tup1 in the presence of glucose, leading to faster depletion of Mth1 in the nucleus. Lack of Mth1 or Std1 inside nucleus exposes the phosphorylation sites of Rgt1 by cAMP coupled PKA. In this way, Cyc8-Tup1 is no longer binding to phosphorylated Rgt1, liberating Rgt1 from the promoter region of hexose transporter genes

Although Mth1 and Std1 are primarily dedicated to regulate Rgt1 function, their individual contributions appear to differ. Mutant analysis shows that in a *snf3 rgt2* background strain, Mth1 plays a role in invertase repression whereas Std1 is required for its activation. Further, mutations in *STD1* specifically affect low-glucose signaling, while mutations in *MTH1* affect *HXT* expression even in the absence of glucose (Schmidt et al. 1999). Mth1, but not Std1, is required for DNA binding of Rgt1 and inhibits phosphorylation of Rgt1. Moreover, Mth1 and Std1 levels behave differently in response to glucose. Repression of *MTH1* transcription by glucose via the Snf1-Mig1 repression pathway accelerates the glucose-stimulated degradation of Mth1, leading to a rapid and significant decline in the abundance of Mth1 (Kaniak et al. 2004; Schmidt et al. 1999). The levels of Std1, however, change little upon glucose addition due to the fact that glucose induction of its transcription offsets degradation of the protein (Flick et al. 2003; Moriya and Johnston 2004). Thus, Mth1 is proposed to serve as a primary regulator of Rgt1 function and therefore of *HXT* gene expression, whereas Std1 may only provide a buffer that modulates the speed and extent of inhibition and activation of *HXT* gene expression (Sabina and Johnston 2009).

### *6.7.4 Snf1-Mig1 Repression Pathway*

The preferential use of glucose as energy and carbon source by yeast cells results in the induction of gene expression for glucose transport and utilization but also leads to the inhibition of expression of genes required for use of alternative carbon sources. If sugar concentration is high enough to engage the Crabtree effect, repression of genes involve in respiration also occurs. The Snf1-Mig1 glucose repression pathway is the central regulator that implements this inhibition in the presence of glucose (Horák 2013). In this pathway, three key players, the Snf1 protein kinase complex, the Mig1 transcription repressor complex, and the Reg1-Glc7 protein phosphatase I have been identified.

<u>Snf1:</u> The *SNF1* gene encodes a protein that serves as the catalytic alpha subunit of a trimeric protein complex, the Snf1 protein kinase complex. In addition to Snf1, it consists of one of the scaffolding and targeting alternative beta subunits Sip1, Sip2, or Gal83, as well as the regulatory gamma subunit Snf4 (Nayak et al. 2006; Rudolph et al. 2005). It is required primarily for the adaptation of yeast cells to glucose limitation, environmental stresses and for growth on alternative carbon sources (Hedbacker and Carlson 2008; Young et al. 2003). It is also involved in various cellular developmental processes, such as meiosis and sporulation, aging, haploid invasive growth, and diploid pseudohyphal growth (Ashrafi et al. 2000; Honigberg and Lee 1998; Kuchin et al. 2002).

The N-terminal of Snf1 is the kinase domain that is responsible for the phosphorylation of serine/threonine within its target, while the C-terminal region contains a regulatory domain that is involved in the interaction with other subunits of the complex. Snf1 is phosphorylated and activated in response to glucose limitation by phosphorylation at conserved residue T210 in a reaction catalyzed by one of the three upstream protein kinases, Sak1, Elm1 or Tos3 (Jiang and Carlson 1996; McCartney and Schmidt 2001; Nayak et al. 2006; Rubenstein et al. 2008; Wilson et al. 1996). Although the upstream glucose signal that triggers the phosphorylation of Snf1 is still unclear, it has been reported that Hxk2, the hexokinase 2 isoenzyme encoded by the *HXK2* gene, is required for the signal transduction (Ahuatzi et al. 2007; Westergaard et al. 2007). On the other hand, Snf1 is inactivated by dephosphorylation of T210 mediated by the type I protein phosphatase Glc7, which is targeted to Snf1 by the Reg1 and Reg2 subunits (Sanz et al. 2000).

The ultimate target of the Snf1 protein kinase complex in the glucose signaling network is Mig1, which forms a repressor complex in conjunction with the general corepressor Cyc8-Tup1 and inhibits expression of genes required for utilization of alternative carbon sources in the unphosphorylated form. Snf1 phosphorylates Mig1 at least at four serine residues thus inactivating its function (Papamichos-Chronakis et al. 2004; Treitel et al. 1998). Previous study reveals that the subcellular localization of Mig1 is regulated by glucose because Mig1 is observed to be imported into the nucleus soon after the addition of glucose and is just as rapidly transported back to cytoplasm when glucose is removed. The fact that this localization of Mig1 from nucleus to cytoplasm happens to be coincident with the phosphorylation of Mig1 further leads to the proposal that the ability of Mig1 to act as a repressor depends on its localization, which, in turn, depends on its phosphorylation by Snf1 (De Vit et al. 1997). However, a more recent study argues that though Snf1-dependent phosphorylation is still the trigger of Mig1 dysfunction, it is the release of phosphorylated Mig1 from Cyc8-Tup1 that leads to the de-repression of Mig1 on its target genes in the presence of glucose, but not Mig1 localization (Bendrioua et al. 2014).

Mig1: Mig1 is a sequence specific binding protein that contains two $Cys_2His_2$ zinc finger motifs. Upon binding to GC box within the promoter region of target genes, it recruits the general co-repressor Cyc8-Tup1 and negatively regulates glucose-related genes such as *SUC*, *GAL* and *MAL* genes (Ostling et al. 1996; Treitel and Carlson 1995). Repression by Mig1 can be both direct and indirect, through repression of genes encoding transcriptional activators. Its interaction with Cyc8-Tup1 appears to be disrupted by the conformational change of Mig1 via phosphorylation by Snf1, leading to de-repression of its target genes mainly (Wu and Trumbly 1998). Yeast has two other zinc finger proteins that are closely related to Mig1: Mig2 and Mig3. Mig2 appears to be a minor player in glucose repression as Mig1 itself is sufficient to repress many glucose-related genes in addition to those that can only be repressed in the presence of both wild type Mig1 and Mig2, whereas no genes have been reported to be repressed only by Mig2 (Westholm et al. 2008).

### 6.7.5 cAMP-PKA Signaling Cascade

The third pathway that's known to affect glucose transport is the cAMP-PKA signaling cascade. Protein kinase A (PKA) is a family of enzymes whose activity is dependent on cytoplasmic levels of cyclic AMP (cAMP) (Wilson and Roach 2002).

Protein kinase A has many functions in the cell, including regulation of several metabolic pathways, such as glycolysis and gluconeogenesis (Hunter and Plowman 1997; Thevelein and De Winde 1999). cAMP-PKA signaling cascade controls many yeast behaviors besides metabolism: cell growth, cell aging and stress resistance are all mediated by this signaling cascade (Ghillebert et al. 2011; Peeters and Thevelein 2014).

PKA: Yeast PKA consists of two catalytic subunits Tpk1, Tpk2, Tpk3, encoded by one of the three redundant *TPK* genes: *TPK1*, *TPK2*, *and TPK3* and two regulatory subunits Bcy1 encoded by *BCY1* gene (Cannon and Tatchell 1987; Kim and Johnston 2006). PKA exert its function based on the association of cAMP with its regulatory subunit Bcy1 (Toda et al. 1987a; Toda et al. 1987b). This further liberates Bcy1 from the catalytic subunits Tpk1, Tpk2 and Tpk3, resulting in an active PKA capable of phosphorylating downstream targets like Rgt1. Upon addition of glucose to cells previously grown on non-fermentable carbon sources, trans-membrane receptor Gpr1 coupled to a G-protein alpha unit Gpa2 activates the adenylate cyclase, Cyr1. Activated Cyr1 further catalyzes the synthesis of cAMP from ATP, which acts as an intracellular secondary messenger and stimulate PKA function by binding to the regulatory subunit of PKA (Bendrioua et al. 2014; Kraakman et al. 1999; Peeters et al. 2006). At the same time, as discussed above, Mth1 and Std1 are taken to degradation via the Snf3/Rgt2 signaling pathway, leaving the PKA phosphorylation sites in the N-terminal region of Rgt1 exposed. Thus Rgt1 is phosphorylated by PKA, leading to the dissociation of Cyc8-Tup1 from Rgt1 (Kim and Johnston 2006). As a result, Rgt1 can no longer repress *HXT* gene expression; thereby more glucose transporters are produced.

There is another way PKA can be activated in response to glucose, which is through the interaction of Ras1/Ras2 with Cyr1. Ras1 and Ras2 are both monomeric GTPase that exert function based on GDP/GTP exchange and GTP hydrolysis (Colombo et al. 2004). The actual mechanism of how glucose metabolism triggers Ras1/Ras2 activation is still unclear, yet evidence has indicated that addition of glucose induces a fast increase in the GTP loading state of Ras2 and is in consistency with the increase in cellular cAMP level.

## 6.8 Rgt1-Interacting Proteins Grr1, Mth1, Std1 on Low and High Glucose

Rgt1 is the ultimate target of both Snf3/Rgt2 and cAMP/PKA pathways in glucose transport. It negatively regulates the expression of *HXT* genes and the $SCF^{Grr1}$ ubiquitin-protein-ligase complex (Jouandot et al. 2011; Kim and Johnston 2006; Ozcan et al. 1996b; Polish et al. 2005; Roy et al. 2014a). Rgt1 is a constitutively expressed protein localized to the nucleus, which contains a $C_6$-$Zn_2$ “zinc cluster” DNA-binding domain that recognizes the sequence 5′-CGGANNA-3′ presents in the promoter region of all the *HXT* genes (Kim et al. 2003; Kim 2009). Rgt1 exerts a repressor role on glucose-induced genes by recruiting the transcriptional co-repressor complex Cyc8-Tup1 and another two transcriptional factors Mth1 and Rgt1 in the absence of glucose (Roy et al. 2014a; Flick et al. 2003). It can also act as an activator that is required for full induction of *HXT1* expression under high glucose concentration (Mosley et al. 2003). The function of Rgt1 is regulated by an intra-molecular interaction between the middle part of Rgt1 and its N terminus, which includes the DNA binding domain (Kim 2009).

## 6.9 Transcriptional Regulatory Mechanisms Specific to Individual *HXT* Genes

There are in total of 17 genes encoding the hexose transporter family, all with differing affinity and capacity. Among them, Hxt1-4, Hxt6-7 are the main glucose transporters as expression of any single one of these transporter is sufficient for substantial glucose utilization (Greatrix and van Vuuren 2006; Kuttykrishnan et al. 2010; Maier et al. 2002; Ozcan and Johnston 1999) Previous studies revealed that varying numbers of Rgt1-binding sites (5'-CGGANNA-3') exist in *HXT* promoter regions: the *HXT3* promoter contains 15 sites, the *HXT2* promoter contains only 3 sites as well as *HXT4* promoter, while 13 Rgt1-binding sites were identified in the *HXT1* promoter region (Kim et al. 2003). Multiple Rgt1-binding sites present in individual *HXT* promoters potentially allow various architectures of clusters to form. Work done by Kim (2009) demonstrates that Rgt1 interacts with three clusters, in total of 8 Rgt1-binding sites, present in *HXT1* promoter region. Association of Rgt1 to different combinations of these clusters results in differing levels of repression suggesting that Rgt1 binding on multiple sites synergistically mediates *HXT1* expression. However, beside the indigenous difference between *HXT* promoters' structure, other factors also contribute to different expression regulation profiles.

To utilize glucose at maximum efficiency, yeast has developed both transcriptional and post-transcriptional mechanisms to control the amount of glucose transporters present on cell membrane according to extracellular glucose level. Earlier studies have revealed three types of glucose mediated *HXT* gene expression patterns: 1. Induction only at high glucose concentration (*HXT1*); 2. Induction at low glucose level and repression at high glucose level (*HXT2*, *HXT4*); 3. Induction by glucose independent of concentration (*HXT3*) (Diderich et al. 1999; Ozcan and Johnston 1995). The implementation of these expression profiles is mainly based on a regulatory network formed by all three discussed (Snf3/Rgt2, Snf1/Mig1 and cAMP/PKA) glucose-induced pathways intertwined together. The general de-repression of Rgt1 on *HXT* genes induced by glucose happens through the Snf3/Rgt2 pathway and is reinforced via the cAMP-PKA signaling cascade: glucose sensors Snf2 and Rgt2 transduces the signal to Yck1/2 complex, which leads to transcriptional factors Mth1 and Std1 being phosphorylated and taken by $SCF^{Grr1}$ ubiquitin-protein-ligase complex to degradation. Mth1 mRNA level is further down-regulated upon presence of glucose via Snf1-Mig1 pathway. Quickly depletion of Mth1 impairs the interaction between Rgt1 and Cyc8-Tup1, thus liberating Rgt1 from the promoter region. At the same time, dissociation of Mth1 and Std1 from Rgt1 exposes the binding region for PKA, which in turn, results in the hyperphosphorylated dysfunctional Rgt1 that can no longer maintain repression of *HXT* gene expression.

### 6.9.1 Induction Only at High Glucose Level (HXT1)

Hxt1 is a low affinity, high capacity glucose transporter; its expression reaches maximum level only under high glucose condition (4 % glucose). Repression of *HXT1* in the absence of glucose is via the Snf3/Rgt2 signaling pathway, yet the regulatory mechanism responsible for induction at high glucose concentration is more complex. Peak expression level of *HXT1* is achieved not only by the general de-repression of Rgt1 via Snf2/Rgt2 pathway and cAMP-PKA signaling cascade, but also the transformation of Rgt1 into an activator activating *HXT1* expression. Previous study also found out that Hxk2 glucose kinase and Reg1 are necessary in the complete induction as well. Moreover, even in the absence of Rgt1, *HXT1* expression is still inducible by high glucose concentration. All the above evidence indicates that more than two pathways are involved in the full activation of *HXT1* expression at high glucose level. Besides the fact that PKA plays a role in the general de-repression of Rgt1 on *HXT* genes, it also regulates the stability of Hxt1 protein. When glucose depleted, *HXT1* is internalized and degraded via endocytosis (Roy et al. 2014b).

The *HXT1* gene is also induced by hyperosmotic stress (Boles and Hollenberg 1997) mediated by the HOG-MAP kinase cascade (Tomas-Cobos et al. 2004). This cascade also induces glycerol synthesis as an osmoprotectant and perhaps the overexpression of *HXT1* under these circumstances serves to maintain glucose transport rates under these conditions of membrane stabilization (Belinchon and Gancedo 2007).

### 6.9.2 Induction at Low Glucose Level and Repression at High Glucose Level (HXT2/HXT4)

Hxt2 and Hxt4 are high affinity, low capacity glucose transporters. Two independent transcription repression pathways apply in this mechanism—Mig1 prevents expression at high levels of glucose, and Rgt1 inhibits expression in the absence of glucose (Ozcan and Johnston 1996). This conclusion is based on the following observations: 1. Deletion of *RGT1* causes constitutive expression of *HXT2* and

*HXT4* in the absence of glucose but has no effect on expression of these genes at high concentrations of glucose. 2. Deletion of *MIG1* causes expression of *HXT2* and *HXT4* to become inducible by high levels of glucose yet has no effect on Rgt1-mediated general repression in the absence of glucose. Only when glucose is scarce, both of Mig1 and Rgt1 are inactive, resulting in the complete activation of *HXT2* and *HXT4* gene expression.

### *6.9.3 Induction by Glucose Independent of Concentration (HXT3)*

Hxt3 is a low affinity high capacity glucose transporter. It is expressed only on glucose medium but the induction is independent of glucose concentration. Its expression is induced about tenfold by both low and high concentration of glucose. Previous report shows that Snf3 activates expression of *HXT1-4* via the Snf3/Rgt2 signaling pathway at a low glucose level (Ozcan and Johnston 1995). However, complete induction of *HXT3* is independent of Hxt2 and Reg1, indicating a different mechanism employed under high glucose condition as also seen for *HXT1* expression.

## 6.10 Post-translational Regulation of Hexose Transport

Although transcriptional regulation of the *HXT* gene family has been explored in detail, several lines of evidence suggest that post-translational regulation of this family of proteins is equally important for the tight control of substrate entry and coupling of metabolic and transport rates at the levels seen during fermentative growth. Dynamic turnover of Hxts is achieved through post-transcriptional control and is essential for adaptation to changing conditions.

High affinity hexose transporters are rapidly degraded upon addition of high concentrations of glucose to the medium (Krampe et al. 1998). This degradation occurs within the vacuole and requires genes involved in endocytosis although the mechanism of targeting of these proteins for degradation is unknown (Krampe et al. 1998). Moreover, Hxt2 is endocytosed and degraded in the vacuole when cells previously grown on glucose are transferred to glucose free medium.

Recent evidence suggests that Hxt3 is endocytosed and degraded in vacuole when cells are starved for glucose, Hxt7 is processed similarly under nitrogen limiting conditions (Snowdon and van der Merwe 2012). The signal for turnover is mediated by the TORC1, and Ras/cAMP/PKA pathways discussed previously. Degradation of Hxt3/7 is dependent on components of the Vid30 complex, an E3 ubiqutin ligase (Snowdon and van der Merwe 2012).

## 6.11 Post-translational Regulation and Adaptive Stress Response Networks and Maintenance of Cellular Homeostasis

Maintaining cellular homeostasis is a prime concern for yeast. The ability to quickly and efficiently adapt to changing extra and intracellular environmental conditions is crucial for survival. Yeast has a multitude of cell signaling cascades and sensor networks that allow cells to exert exquisite dynamic control, linking cell physiology and metabolism at the transcriptional, translational, and post-translation level (Chen and Thorner 2007). Yeast are non-motile organisms that can successfully endure a wide variety of environmental insults and stresses like: osmotic stress, nutrient starvation, oxidative stress, temperature shock, and a myriad of toxic compounds. Regulation at the post-translational level in yeast is especially crucial for withstanding these environmental insults, as well as adaptation to future stresses.

Post-translational modifications (PTMs) help to regulate the activity, stability, expression, and interactions of many of the proteins involved in growth, transport, and metabolism. An interconnected network of proteins like kinases/phosphatases, ubiquitin ligases, and acetyltransferases/deacetylases regulate PTMs as dictated by intracellular signals. These PTMs not only modify target proteins, but sequentially amplify entire cell signaling cascades that titer cellular adaptation to environmental conditions (Broach 2012). There are a variety of important PTMs that allow a cell to quickly and dynamically respond to environmental conditions. These include, but are not limited to: phosphorylation, ubiquination, acetylation, sumoylation, glycosylation, allosteric regulation, and chaperone-mediated refolding (Oliveira and Sauer 2012; Tripodi et al. 2015; Xiong et al. 2011).

As described earlier, yeast is extremely responsive to the presence of glucose. The availability of a carbon source is one the most important environmental considerations for yeast. The ability to recognize a carbon source and quickly respond to its presence or depletion can dictate the fate of a whole population. Glucose induces global changes in gene expression, mRNA transcript stability, and a host of key regulatory PTMs (Brion et al. 2013; Broach 2012; Cakir et al. 2007; Horák 2013).

The activity of the primary plasma membrane $H^+$ ATPase Pma1 is strongly up-regulated in response to glucose (Eraso et al. 2006; Serrano 1983). Pma1 is essential for generating proton motive force and maintaining cytosolic pH, thus it is a key regulatory target for coupling rate of transport to metabolic flux. The activity of Pma1 is regulated through the C-terminal end, which under glucose starvation conditions is tightly folded, inhibiting ATPase activity (Mason et al. 2014). Two key residues in the C-terminal tail regulate this process. Phosphorylation of Threonine-912 shifts the C-terminal tail into a low activity, low affinity form under carbon-starvation (Lecchi et al. 2007). A sequential phosphorylation of Serine-911 fully opens the C-terminal tail, fully activating Pma1 ATPase function when glucose is present (Lecchi et al. 2007). Pma1 activity can also be negatively regulated in the presence of glucose via phosphorylation of Glu-504 by the Yck1/2 kinases; providing antagonistic nuanced control of Pma1 activity and metabolic flux within the cell (Estrada et al. 1996). The presence of misfolded Pma1 activates the osmotic stress response, further underscoring the importance of Pma1 as a key regulatory node (Eraso et al. 2011).

As discussed, yeast are highly adapted to take advantage of fermentable sugars. An essential part of this adaptation is the ability to withstand extreme osmotic stress in a medium like grape juice, which can contain upwards of 200 g/L of glucose:fructose and other solutes (Boulton et al. 1996). When presented with osmotic stress, yeast will activate the high osmolarity glycerol (HOG) signaling network. Activation of the HOG pathway results in expression of Hog1, a mitogen-activated protein (MAP) kinase that affects gene regulation, signaling cascades, cell cycle arrest, and glycerol accumulation in response to osmotic stress (Abu Irqeba et al. 2014; Hohmann et al. 2007; Klipp et al. 2005). Hog1p promotes uptake and synthesis of glycerol as an intracellular solute (osmolyte) to maintain cell turgor and water homeostasis under osmotic stress conditions (Brewster et al. 1993; Hohmann 2002). Under normal conditions, glycerol and other small solutes are free to diffuse out of the cell via the aqua-glyceroporin Fps1 (Ahmadpour et al. 2014). Within a minute of exposure to hyperosmotic stress, the Fps1 channel is inactivated via Hog1, thus trapping glycerol inside the cell (Duskova et al. 2015; Tamas et al. 1999). Through the HOG network and the activity of Hog1p, expression of *STL1*, a glycerol/$H^+$ symporter is strongly up-regulated (Ferreira et al. 2005). The glycerol uptake activity of Stl1 is strongly but transiently induced upon exposure to osmotic stress, and inactivated in the presence of glucose. Stl1 activity is dependent on proton motive force, tying the import of glycerol to pH homeostasis and the activity of Pma1. The HOG network also stimulates production of glycerol via the *GPD1/2* and *GPP1/2* enzymes, which produce glycerol from dihydroxyacetone-phosphate (DHAT) in a two-step reaction (Albertyn et al. 1994; Ferreira et al. 2005; Norbeck et al. 1996). The effects of the HOG network are attenuated by the intracellular concentration of glycerol, adapting the genetic and physiological response as a function of the osmotic stress being experienced by the cell (Klipp et al. 2005).

In addition to its role in osmotolerence, can serve as a non-fermentable respiratory carbon source. Yeast can aerobically metabolize glycerol, although it is a poor energy source, providing a net gain of only 1 ATP molecule per molecule of glycerol. Glycerol is first converted into glycerol-3-phosphate (G3P) by Gut1p before being transported into mitochondria where Gut2p converts G3P into DHAT, which returns to the cytoplasm and is utilized in glycolysis or gluconeogenesis (Pavlik et al. 1993; Ronnow and Kielland-Brandt 1993). Expression of *GUT1/GUT2* is down-regulated as part of glucose-repression, but is induced on non-fermentable carbon sources via the Snf1p kinase and a co-transcription activating complex (Grauslund and Ronnow 2000). Snf1 is an AMP-activated serine/threonine kinase that as is active in a heterotrimeric complex, where the subunits determine the target specificity (Celenza and Carlson 1984; Celenza et al. 1989; Jiang and Carlson 1996). The active Snf1p kinase complex has a wide range of targets, playing a regulatory role in chromatin remodeling, transcription, translation, glycogen and lipid biosynthesis (Ashe et al. 2000; Hardy et al. 1994; Kuchin et al. 2000; Lo et al. 2001). The Snf1p complex also plays general regulatory role in several types of stress response (Alepuz et al. 1997; Hahn and Thiele 2004; Mayordomo et al. 2002).

Aerobic respiration of non-fermentable carbon source like glycerol in the mitochondria leads to the generation of reactive oxygen species (ROS) which activate the oxidative stress response (OSR). Recognizing and coping with ROS is important for maintaining cellular homeostasis and continued growth. In yeast the main ROS species are hydrogen peroxide and superoxide, which in addition to their disruptive effect on cell processes can act as signaling molecules; activating a multitude of stress response pathways (Temple et al. 2005). ROS are produced as a normal part of catabolic activities, but are especially prevalent during the diauxic shift from glucose to non-fermentable carbon sources, and during quiescence (Boveris 1984; Gray et al. 2004). Mtl1 is one of several transmembrane ROS sensors that are part of the cell wall integrity (CWI) pathway (Vilella et al. 2005). Mtl1 is N-glycosylated and highly O-mannosylated, forming a functional PTM domain that has been hypothesized to act a mechanical ROS sensor (Dupres et al. 2009; Petkova et al. 2012). Mtl1 regulates several stress response pathways during quiescence, linking the physical state of the cell to the physiological and metabolic processes occurring within (de la Torre-Ruiz et al. 2015). Downstream targets of Mtl1, Rho1 and Rom2 facilitate the response to oxidative stress by inactivating TORC1 and Ras/cAMP, indirectly leading to expression of general stress transcription factors (TF) Msn2/4 (Lee et al. 2011; Park et al. 2005; Serrano et al. 2006).

Msn2/4 are semi-redundant TFs that regulate ∼200 genes by binding Stress Response Elements (STREs) in response to environmental stresses (Causton et al. 2001; Martinez-Pastor et al. 1996). Msn2/4 are activated as part of several stress response pathways, and have been shown to be important regulators of the cellular adaptation to stress (Gasch et al. 2000). Other than the response to osmotic and oxidative stress, yeast has several other interconnected important environmental response networks that are crucial for tolerance, and adaptation to stress. The environmental stress response (ESR) is a broad response encompassing all kinds of environmental insults, it is reviewed here (Gasch 2003). Through the action of Msn2/4 and plethora of downstream target the ESR ameliorates the current stress, but is geared for future adaptation to stress. It has also been hypothesized that the ESR can moderate mutagenesis caused by proteotoxic stress, implying a role for the ESR in long-term genetic adaptation (Shor et al. 2013).

The goal of these stress response networks is to maintain cellular homeostasis and normal cellular functions (Fig. 6.6). Various stresses can cause loss of protein homeostasis, or proteostasis, severely interfering with normal cellular functions and potentially leading to proteotoxicity. One of the best characterized stress response pathways is the Heat Shock Response (HSR); heat shock transcription factors (HSFs) mediate a highly conserved series of changes in gene regulation, leading to expression and activation of a family of cytoprotectant proteins known as Heat Shock Proteins (HSPs) (Pirkkala et al. 2001; Westerheide et al. 2012; Wu 1995). Due to molecular crowding caused by extremely high concentrations of protein in

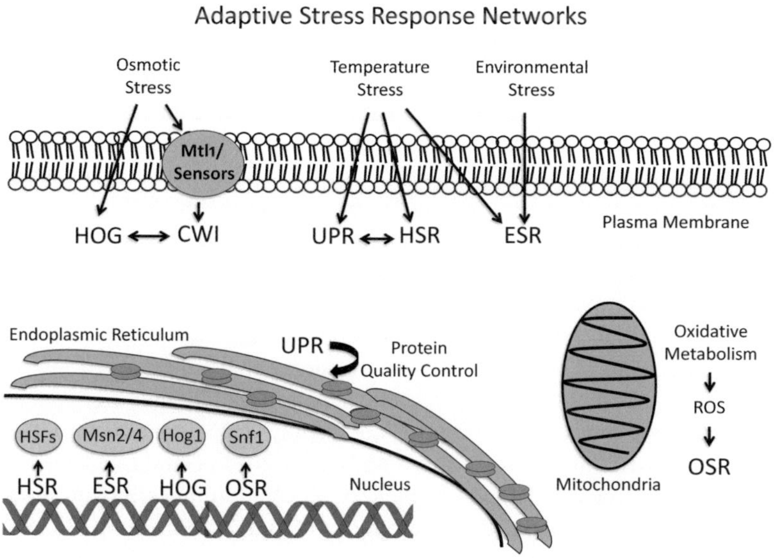


**Fig. 6.6** The cumulative action of adaptive stress response networks (ASRNs) is to mitigate the immediate impact of stress and help the cell adapt to future insults. Many of these ASRNs are linked by interactions of overlapping proteins, and sensing of extra and intracellular conditions. The complexity of these ASRNs allow yeast to dynamically titer their responses to stress, and adjust their metabolic and physiological programming to environmental conditionsTypes of Internal Stress: ROS, cAMP, mis-folded proteins, protein aggregates, proton imbalanceTypes of Extracellular Stress: Osmotic, temperature, nutrient limitation, toxins, competition

the cytoplasm 300 mg/mL, it is essential for the cell maintain proteostasis (Frydman 2001). Complex networks of molecular chaperones, including some HSPs help to maintain proteostasis within the cell. Some HSPs are constitutively expressed and help to mediate proper protein folding, transport across membranes, and protein turnover (Ellis 1987; Gong et al. 2009). However, under conditions that cause proteotoxic stress, HSPs are especially active in refolding proteins, regulating protein turnover, disaggregating amyloids, and restoring growth by rescuing cell cycle arrest (Verghese et al. 2012). The Unfolded Protein Response (UPR) involves many of the same HSPs and is triggered by similar conditions as the HSR, but its effects are more concentrated with rescuing misfolded peptides in the endoplasmic reticulum (ER) (Shamu et al. 1994). The UPR is linked to the HSR by the histone deacetylase Sir2, which mediates regulation of HSFs (Weindling and Bar-Nun 2015). Cross-talk between these stress response networks is achieved with proteins like Snf1 and Msn2/4, which serve to coordinate cell signaling cascades and the adaptive cellular response to stress (Ferrer-Dalmau et al. 2015; Gasch 2003).

The cumulative action of all these adaptive stress response networks (ASRNs) is to help populations not only survive an initial insult, but also prepare for further insults. In addition to the multitude of transcriptional, post-translational, and physiological responses yeast can spontaneously induce heritable epigenetic "switches" known as prions when proteostasis is sufficiently disrupted (Halfmann and Lindquist 2010).

Yeast prions are proteins capable of at least two structurally and functionally distinct states, one of which is self-templating (Garcia and Jarosz 2014). A population of yeast exposed to stress will activate the appropriate ASRNs, undergoing global changes in PTMs, metabolite flux, and relative energy ratios. A portion of this population will experience sufficient disruption of cellular homeostasis for these proto-prion proteins to spontaneously refold. Many yeast prions contain domains rich in asparagine (N) and glutamine (Q) that in conjunction with HSPs, generate the self-templating [*PRION*$^+$] form (Alberti et al. 2010). The brackets denote non-Mendelian inheritance pattern, while the capitals explain the dominant genetic characteristic of these prion states. General characteristics of yeast prions are that they are reversible/can be cured, they arise at higher frequencies than random mutation, they are infectious, and display 4:0 segregation of the prion phenotype (Alberti et al. 2010; DiSalvo and Serio 2011; Garcia and Jarosz 2014; Halfmann and Lindquist 2010). Prion phenotypes often mimic a mutant phenotype, either by sequestration of the target protein within the cell, or by a structurally driven change in the activity and/or association (Li and Kowal 2012). Proto-prion proteins are frequently TFs or other proteins that sit at important regulatory nodes (Alberti et al. 2009). The adaptive advantage of inducing a prion is allows the cell to quickly and reversibly change its phenotypic output and modulate metabolism and/or gene expression without the permanence of a mutation or energetic cost of reprogramming a series of metabolic pathways.

There has been some debate as to whether yeast prions are truly advantageous bet-hedging mechanisms, or instead deleterious selfish genetic elements (McGlinchey et al. 2011; Wickner et al. 2011). However, recent work has demonstrated that yeast prions are definitively advantageous given the appropriate environmental conditions (Halfmann et al. 2012; Holmes et al. 2013). It has been observed that yeast from different environmental niches exhibit different frequencies for particular prion states, reinforcing the concept of prions as adaptive cellular mechanisms (Halfmann et al. 2012; Jarosz et al. 2014). The majority of yeast prions that have been characterized form amyloid fibers and are dependent on Hsp104 for propagation. Recently it has been hypothesized that Hsp70 and Hsp40 are also required for faithful propagation of prions (Sharma and Masison 2009). Furthermore a recently described prion known as [*GAR*$^+$] is non-amyloid in character and is solely dependent on yeast Hsp70 *SSA1* for propagation (Brown and Lindquist 2009).

## 6.12 Prion-Based Regulation of Cellular Homeostasis

The [*GAR*$^+$] prion represents a novel prion state that in addition to being non-amyloid in character and Hsp70 dependent, is formed through the association of plasma membrane ATPase Pma1, and co-transcriptional regulator Std1. [*GAR*$^+$]

stands for “resistant to glucose-associated repression”. Cells that harbor the [*GAR*$^+$] prion can circumvent glucose-associated repression of alternative carbon sources (Brown and Lindquist 2009). This phenotype allows yeast to grow on alternative and non-fermentable carbon substrates even in the presence of glucose. The heritable epigenetic ability to bypass glucose repression represents a major paradigm shift for how yeast interacts with its environment and regulates its metabolism.

Yeast prions are usually induced by a particular type of stress, however the exact reasons for [*GAR*$^+$] induction remain to be elucidated. Under laboratory conditions, [*GAR*$^+$] can be induced on media containing glucosamine, a non-hydrolyzable glucose mimetic, along with glycerol as an alternative carbon source. Depending on strain background, and environmental niche, yeast will auto-induce [*GAR*$^+$] allowing them to grow on this selective media. In a more “natural” context, it has been clearly demonstrated that in a density dependent manner, growth of specific species and strains bacteria induce the [*GAR*$^+$] prion in yeast (Jarosz et al. 2014). These bacteria export a heat stable, low molecular weight, diffusible factor that is a signal to the yeast to induce [*GAR*$^+$] (Jarosz et al. 2014). Bacterial induction of [*GAR*$^+$] prion represents an example of cross-kingdom communication, and hints at an intriguing ecological role for this prion in mixed microbial communities.

Relatively few significant fold changes occur in gene expression upon induction of [*GAR*$^+$], except for a ~40-fold decrease in *HXT3* transcripts (BL09). This is significant because *HXT3* is a ubiquitous high capacity, low affinity sugar transporter that has been shown to be important for the fermentative capacity of cells (Karpel et al. 2008). Given the reduction in *HXT3* expression, it was hypothesized that [*GAR*$^+$] cells would struggle in fermentation. Fermentations performed in synthetic juice showed that [*GAR*$^+$] displayed slower rates of fermentation and made less ethanol (in prep). When [*GAR*$^+$] fermentations were performed in Chardonnay juice, the effects if [*GAR*$^+$] were even more evident. In grape juice [*GAR*$^+$] appeared to severely affect the fermentative capacity of yeast, causing sluggish/stuck fermentations with high levels of bacterial activity. This indicated that not only were [*GAR*$^+$] cells unable to ferment efficiently, but were also less dominant of microbial competition.

This lack of competitiveness in fermentation is particularly significant given the ability of certain wine-spoilage associated bacteria to induce [*GAR*$^+$]. Induction of the [*GAR*$^+$] prion benefits bacteria, allowing them to remain active throughout fermentation. But, what do the yeast gain from inducing [*GAR*$^+$]? While the exact mechanism of induction remains to be elucidated, bacteria are indicating to the yeast that a change in metabolic and physiological programming could be advantageous. In addition to the ability to bypass glucose repression, [*GAR*$^+$] cells display a number of other physiological changes that explain their behavior in fermentation. By high-performance liquid chromatography analysis [*GAR*$^+$] cells in fermentation displayed a starkly different phospholipid profile. [*GAR*$^+$] cells have a low ratio of phosphatidylcholine (PC), to a high ratio of phosphatidylinositol (PI); this phospholipid ratio is more typical of stationary phase cells has been shown to be associated with problem fermentations (Henderson et al. 2013). Quick depletion of oxygen (within 24 h) is one of strategies *S. cerevisiae* employs to dominate microbial competition within fermentation. However, indicative of reduced membrane potential, [*GAR*$^+$] cells uptake oxygen more slowly. The ability to deplete of amino acids is also affected in cells harboring [*GAR*$^+$], again implying that transport across the membrane has been affected in these cells. It has also been observed that [*GAR*$^+$] cells are more resistant to certain inhibitors. Similar to *can1Δ* mutants, [*GAR*$^+$] cells are resistant to the toxic amino acid analog, canavanine. Similar to certain *pma1Δ* mutants, [*GAR*$^+$] cells display resistance to vanadate, which exerts proton stress on the cytosol (Walker unpublished observations).

These cumulative observations strongly imply that there has been modulation of the plasma membrane and the structure/function of Pma1 in [*GAR*$^+$] cells (Fig. 6.7).

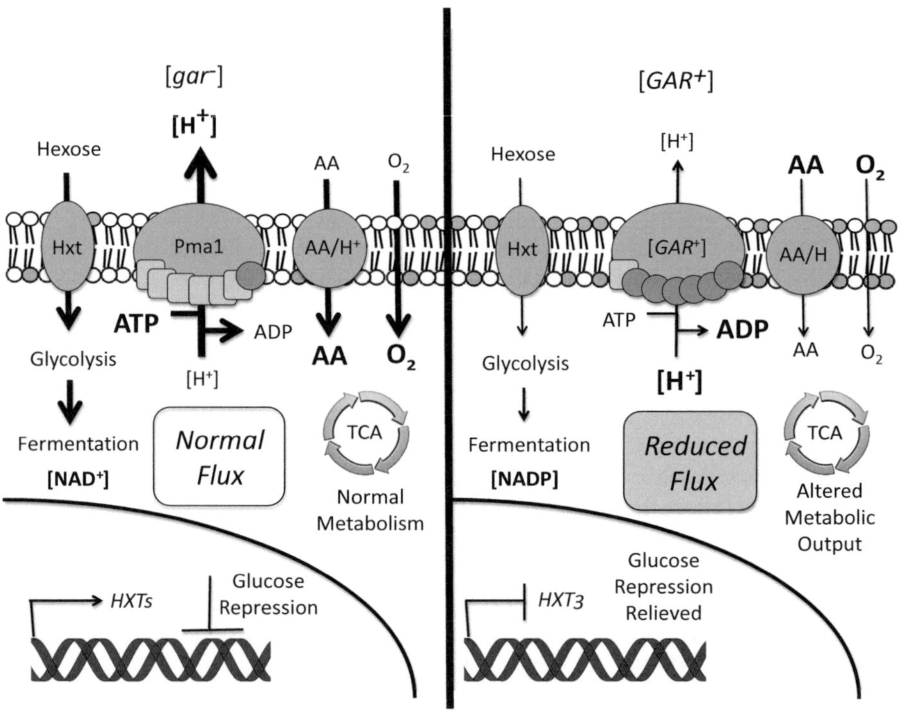


**Fig. 6.7** Model of the impact of the [*GAR*$^+$] prion on cellular homeostasis. The observed behaviors of [*GAR*$^+$] cells are explained by the physiological and metabolic differences between the [*gar*$^-$] and [*GAR*$^+$] states. Full expression of the *HXT* genes allows *S. cerevisiae* to uptake sugar and run glycolysis to produce ATP and maintain pH homeostasis. Alcoholic fermentation occurs regenerate NAD$^+$, facilitating continued high rates of glycolysis and metabolic flux. [*GAR*$^+$] cells behave more like quiescent cells displaying reduced rates of flux, reduction in expression of *HXT3*, higher intracellular pools of protons and reduced NADP. [*GAR*$^+$] cells also display a change in association of Pma1 being bound by Mth1 (*squares*) vs Std1 (*circles*), as well as a shift in the ratio of phospholipids PC (*white head groups*) to PI (*grey head groups*). Overall [*GAR*$^+$] cells display reduced metabolic flux and altered metabolic output, more closely attenuating their intracellular environment with external conditions

Regulation of Pma1 activity is essential to maintaining cellular homeostasis. We propose that post-translational modification of Pma1 serves as a central regulatory node for linking physiological processes with metabolic flux in *S. cerevisiae*. Furthermore the bacterial induction of [$GAR^+$] implies a role for regulation of Pma1 as an environmental sensor. It has been shown in fission yeast that a mutation in *pma1* that affects activity also extends chronological lifespan (Naito et al. 2014). We hypothesize that the change in structure/activity of Pma1 in response to [$GAR^+$] induction also extends the lifespan of *S. cerevisiae*. The full effects of [$GAR^+$] on yeast metabolism are still unknown, however ongoing experiments to assess the metabolome of [$GAR^+$] will further our understanding of this phenomenon.

Yeast prions have been proposed to be evolutionarily conserved bet-hedging mechanisms that promote dynamic adaptation to stressful environmental conditions at frequencies higher than that of random mutation (lots of refs). In the context of a wine fermentation, the induction of [$GAR^+$] appears to heavily favor bacteria by slowing the rate of fermentation and reducing the competitiveness of the yeast. However, the [$GAR^+$] prion would not be so highly conserved if it did not provide some advantages. Microbial competition provides the signal to induce [$GAR^+$], which in turn modulates yeast behavior. Inducing [$GAR^+$] turns *S. cerevisiae* from the dominant organism in fermentation, to a microbial community member, and allowing them to diversify their metabolism and increase long-term survivability. This is especially relevant when you consider the frequency of [$GAR^+$] observed in yeasts from different environmental niches. Laboratory, clinical, and beer yeast kept as pure cultures have very low frequencies of [$GAR^+$], while wine, insect, and fruit yeast have much higher frequencies of [$GAR^+$]. We hypothesize that the [$GAR^+$] prion gives yeast the capacity to recognize and modulate behavior based on their microbial environment.

## 6.13 Glycerol Transport and Link to Cellular Homeostasis

As discussed earlier, glycerol is an extremely important metabolite for regulating a variety of cellular processes. Other than its role in osmotolerance, the transport and metabolism of glycerol are involved in resistance to temperature stress, redox balance, lipid synthesis, and cycling of inorganic phosphates (Ferreira and Lucas 2007).

Glycerol uptake from the environment occurs through Stl1, a proton symporter similar in structure to members of the *HXT* family (Ferreira et al. 2005). *STL1* expression is repressed by glucose, but this can be circumvented by the HOG pathway, growth on non-fermentable carbon substrates, and at the end of fermentation. *STL1* expression during the diauxic shift is induced by Cat8p, which mediates expression of a set of gluconeogneic genes during growth on non-fermentable carbon sources (Haurie et al. 2001). Expression of *STL1* can also be induced at high and low temperatures regardless of the presence of glucose, indicative of the importance of glycerol in thermotolerance (Ferreira and Lucas 2007; Tulha et al. 2010). Due to the diverse roles of glycerol within the cell, it has also been shown that the activity of Stl1 is important for maintaining redox balance, intracellular pH homeostasis, and intracellular $P_i$ availability (Larsson et al. 1998; Sanders and Slayman 1982; Luyten et al. 1995).

The Glycerol Uptake 1 & 2 (*GUP1/2*) genes were originally hypothesized to be responsible for the proton symport of glycerol into the cell (Holst et al. 2000). It has since been shown that Gup1/2 are instead O-Acyltransferases that play vital roles in plasma membrane maintenance, lipid composition, bud-site selection, the secretory/endocytic pathway, and remodeling of glycophosphatidyl anchors (Bleve et al. 2005; Bosson et al. 2006; Ferreira et al. 2010). While it has been clearly demonstrated that Gup1/2 do not directly transport glycerol, the authors hypothesize that Gup1/2 can regulate transport of glycerol through modification or targeting of Stl1 and possibly Fps1.

*FPS1* and *YFL054C* encode the two non-paralogous aquaglyceroporins in *S. cerevisiae* (Ahmadpour et al. 2014). YFL054C mediates the passive diffusion of glycerol in the presence of ethanol, and is present in a wider range of fungi than Fps1 (Oliveira et al. 2003) Same as aquaporins, Fps1 functions as a homotetramer but has a more flexible pore structure and has longer N- and C- terminal extensions. Fps1 as discussed earlier is an essential part of the HOG pathway. In hyperosmotic condition Fps1 closes, trapping glycerol inside the cell; in hypo-osmotic conditions Fps1 allows glycerol to diffuse out of the cell. Fps1 can also mediate the diffusion of arsenite, antimonite, boric acid, and acetic acid across the plasma membrane (Nozawa et al. 2006; Tamas et al. 1999; Wysocki et al. 2001). As a channel protein, Fps1 mediates diffusion both ways across the membrane, but the only observed phenotypes of Fps1 relate to glycerol export (Ahmadpour et al. 2014). Osmotic balance is mating yeast cells, thus release of glycerol via Fps1 is required for cell fusion to occur (Philips and Herskowitz 1997). Regulation of Fps1 occurs through phosphorylation of the N- and C-terminal extensions by components of the HOG and CWI pathways, tying together cell wall sensing and turgor sensing (Ahmadpour et al. 2014).

In addition to transport, the anabolism and catabolism of glycerol are also tightly regulated in response to extra and intracellular conditions. *GPD1* expression is specifically up-regulated in conditions of osmotic stress, while *GPD2* is expressed in conditions of oxidative stress (Babazadeh et al. 2014; Lee et al. 2012). Gut2 is required for maintaining redox balance in aerobic conditions, by oxidizing excess cytoplasmic NADPH in the mitochrondria (Grandier-Vazeille et al. 2001). Glycerol metabolism is also tightly linked to lipid synthesis since G3P is converted into phosphatidic acid which serves as a precursor for all yeast phospholipids (Sorger and Daum 2003). Thus the transport and metabolism of glycerol responds dynamically to promote adaptation stress and maintain cellular homeostasis.

## 6.14 Conclusions

The *HXT* gene family in *S. cerevisiae* contains both transporters that have been shown to play a key role in catabolic uptake and those for which the precise physiological roles remain to be elucidated. The differing affinities of the Hxt transporters are expressed when they will be most useful, that is, when the external sugar concentration matches the efficacy of transporter activity. The process of sugar transport in *Saccharomyces* is intricately regulated at multiple cellular levels in order to maintain this complexity of expression. Although a primary regulator of expression is hexose level, other regulatory circuits modify, fine-tune or eliminate components of glucose-mediated regulation assuring optimal integration of the critical sugar uptake step of catabolism with cellular physiology. The investigation of transport in this yeast has provided findings directly applicable to eukaryotes in general and many of the processes uncovered are widely present across the fungi. Additional experiments in other eukaryotic systems are elucidating possible roles for prion like states as part of normal physiological adaptation. Post-translational modifications and prion-mediated regulation give yeast populations the capability to rapidly respond and adapt to conditions of stress in a heritable manner and integrate metabolic activities within a microbial community. The impact of glucose on activation of the Pma1 plasma membrane ATPase has been known for decades. The discovery of the [$GAR^{+}$] prion links Pma1 functionality to the regulation of transport via an intricate yet elegantly simple mechanism that senses membrane changes and integrates membrane integrity with catabolic activity. *S. cerevisiae* continues to be a key model organism for the delineation of the mechanics of sugar transport, maintenance of cellular homeostasis, and now integration of metabolic activities of microbially diverse ecologies.

## References

Abu Irqeba A, Li Y, Panahi M et al (2014) Regulating global sumoylation by a MAP kinase Hog1 and its potential role in osmo-tolerance in yeast. PLoS One 9(2):e87306. doi:10.1371/journal.pone.0087306

Ahmadpour D, Geijer C, Tamas MJ et al (2014) Yeast reveals unexpected roles and regulatory features of aquaporins and aquaglyceroporins. Biochim Biophys Acta 1840:1482–1491

Ahuatzi D, Riera A, Pelaez R et al (2007) Hxk2 regulates the phosphorylation state of Mig1 and therefore its nucleocytoplasmic distribution. J Biol Chem 282:4485–4493

Alberti S, Halfmann R, King O et al (2009) A systematic survey identifies prions and illuminates sequence features of prionogenic proteins. Cell 137:146–158

Alberti S, Halfmann R, Lindquist S (2010) Biochemical, cell biological, and genetic assays to analyze amyloid and prion aggregation in yeast. Methods Enzymol 470:709–734

Albertyn J, Hohmann S, Thevelein JM et al (1994) GPD1, which encodes glycerol-3-phosphate dehydrogenase, is essential for growth under osmotic stress in Saccharomyces cerevisiae, and its expression is regulated by the high-osmolarity glycerol response pathway. Mol Cell Biol 14:4135–4144

Alepuz PM, Cunningham KW, Estruch F (1997) Glucose repression affects ion homeostasis in yeast through the regulation of the stress-activated *ENA1* gene. Mol Microbiol 26:91–98

Alves SL, Herberts RA, Hollatz C et al (2008) Molecular analysis of maltotriose active transport and fermentation by *Saccharomyces cerevisiae* reveals a determinant role for the *AGT1* permease. Appl Environ Microbiol 74:1494–1501

Alves SL, Thevelein JM, Stambuk BU (2014) Expression of *Saccharomyces cerevisiae* α-glucosidase transporters under different growth conditions. Braz J Chem Eng 31:1–8

Anjos J, Rodrigues de Sousa H, Roca C et al (2013) Fsy1, the sole hexose proton transporter characterized in *Saccharomyces* yeasts exhibits a variable fructose:$H^+$ stoichiometry. Biochim Biophys Acta 1828:201–207

Ashe MP, De Long SK, Sachs AB (2000) Glucose depletion rapidly inhibits translation initiation in yeast. Mol Biol Cell 11:833–848

Ashrafi K, Lin SS, Manchester JK et al (2000) Sip2p and its partner snf1p kinase affect aging in *S. cerevisiae*. Genes Dev 14:1872–1885

Babazadeh R, Furukawa T, Hohmann S et al (2014) Rewiring yeast osmostress signalling through the MAPK network reveals essential and non-essential roles of Hog1 in osmoadaptation. Sci Rep 4:4697. doi:10.1038/srep04697

Beck ZT, Cloutier SC, Schipma MJ et al (2014) Regulation of glucose-dependent gene expression by the RNA helicase Dbp2 in *Saccharomyces cerevisiae*. Genetics 198:1001–1014

Becker JU, Betz A (1972) Membrane transport as controlling pacemaker of glycolysis in *Saccharomyces carlsbergensis*. Biochim Biophys Acta 274:584–597

Belinchon MM, Gancedo JM (2007) Different signalling pathways mediate glucose induction of *SUC2*, *HXT1* and pyruvate decarboxylase in yeast. FEMS Yeast Res 7:40–47

Bendrioua L, Smedh M, Almquist J et al (2014) Yeast AMP-activated protein kinase monitors glucose concentration changes and absolute glucose levels. J Biol Chem 289:12863–12875

Bermejo C, Haerizadeh F, Sadoine MSC et al (2013) Differential regulation of glucose transport activity in yeast by specific cAMP signatures. Biochem J 452:489–497

Bisson LF, Neigeborn L, Carlson M et al (1987) The *SNF3* gene is required for high-affinity glucose transport in *Saccharomyces cerevisiae*. J Bacteriol 169:1656–1662

Bisson LF, Coons DM, Kruckeberg AL et al (1993) Yeast sugar transporters. Crit Rev Biochem Mol Biol 28:259–308

Bleve G, Zacheo G, Cappello MS, Dellaglio F, Grieco F (2005) Subcellular localization and functional expression of the glycerol uptake protein 1 (GUP1) of *Saccharomyces cerevisiae* tagged with green fluorescent protein. Biochem J 390:145–155

Boles E, Hollenberg CP (1997) The molecular genetics of hexose transport in yeasts. FEMS Microbiol Rev 21:85–111

Bosch D, Johansson M, Ferndahl C et al (2007) Characterization of glucose transport mutants of *Saccharomyces cerevisiae* during a nutritional upshift reveals a correlation between metabolite levels and glycolytic flux. FEMS Yeast Res 8:10–25

Bosson R, Jaquenoud M, Conzelmann A (2006) GUP1 of *Saccharomyces cerevisiae* encodes an O-acyltransferase involved in remodeling of the GPI anchor. Mol Biol Cell 17:2636–2645

Boulton RB, Singleton VS, Bisson LF et al (1996) Principles and practices of winemaking. Chapman Hall, New York

Boveris A (1984) Determination of the production of superoxide radicals and hydrogen peroxide in mitochondria. Methods Enzymol 105:429–435

Brewster JL, de Valoir T, Dwyer ND et al (1993) An osmosensing signal transduction pathway in yeast. Science 259:1760–1763

Brion C, Ambroset C, Sanchez I et al (2013) Differential adaptation to multi-stressed conditions of wine fermentation revealed by variations in yeast regulatory networks. BMC Genomics 14:681. doi:10.1186/1471-2164-14-681

Broach JR (2012) Nutritional control of growth and development in yeast. Genetics 192:73–105

Brown JC, Lindquist S (2009) A heritable switch in carbon source utilization driven by an unusual yeast prion. Genes Dev 23:2320–2332

Brown CJ, Todd KM, Rosenzweig RF (1998) Multiple duplications of yeast hexose transport genes in response to selection in a glucose –limited environment. Mol Biol Evol 15:931–942

Buziol S, Warth L, Magario I et al (2008) Dynamic response of the expression of hxt1, hxt5 and hxt7 transport proteins in *Saccharomyces cerevisiae* to perturbations in the extracellular glucose concentration. J Biotechnol 134:203–210

Cakir T, Kirdar B, Onsan ZI et al (2007) Effect of carbon source perturbations on transcriptional regulation of metabolic fluxes in *Saccharomyces cerevisiae*. BMC Syst Biol 1:18. doi:10.1186/1752-0509-1-18

Cannon JF, Tatchell K (1987) Characterization of *Saccharomyces cerevisiae* genes encoding subunits of cyclic AMP-dependent protein kinase. Mol Cell Biol 7:2653–2663

Causton HC, Ren B, Koh SS et al (2001) Remodeling of yeast genome expression in response to environmental changes. Mol Biol Cell 12:323–337

Celenza JL, Carlson M (1984) Structure and expression of the *SNF1* gene of *Saccharomyces cerevisiae*. Mol Cell Biol 4:54–60

Celenza JL, Marshall-Carlson L, Carlson M (1988) The yeast SNF3 gene encodes a glucose transporter homologous to the mammalian protein. Proc Natl Acad Sci U S A 85:2130–2134

Celenza JL, Eng FJ, Carlson M (1989) Molecular analysis of the *SNF4* gene of *Saccharomyces cerevisiae*: evidence for physical association of the SNF4 protein with the SNF1 protein kinase. Mol Cell Biol 9:5045–5054

Charron MJ, Dubin RA, Michels CA (1986) Structural and functional analysis of the *MAL1* locus of *Saccharomyces cerevisiae*. Mol Cell Biol 6:3891–3899

Charron MJ, Read E, Hunt SR et al (1989) Molecular evolution of the telomere-associated *MAL* loci of *Saccharomyces*. Genetics 122:307–316

Chen RE, Thorner J (2007) Function and regulation in MAPK signaling pathways: lessons learned from the yeast *Saccharomyces cerevisiae*. Biochim Biophys Acta 1773:1311–1340

Chow TH, Sollitti P, Marmur J (1989) Structure of the multigene family of *MAL* loci in *Saccharomyces*. Mol Gen Genet 217:60–69

Coelho MA, Gonçalves C, Sampaio JP et al (2012) Extensive intra-kingdom horizontal gene transfer converging on a fungal fructose transporter gene. PLOS Genet 9(6):e100358. doi:10.1371/journal.pgen.1003587

Colombo S, Ronchetti D, Thevelein JM et al (2004) Activation state of the Ras2 protein and glucose-induced signaling in *Saccharomyces cerevisiae*. J Biol Chem 279:46715–46722

Compagno C, Dashko S, Piškur J (2014) Introduction to carbon metabolism in yeast. In: Piškur J, Compagno C (eds) Molecular mechanisms in yeast carbon metabolism. Springer, Heidelberg, pp 1–19

Conlan RS, Gounalaki N, Hatzis P et al (1999) The Tup1-Cyc8 protein complex can shift from a transcriptional co-repressor to a transcriptional co-activator. J Biol Chem 274:205–210

Conrad M, Schothorst J, Kankipati HN et al (2014) Nutrient sensing and signaling in the yeast *Saccharomyces cerevisiae*. FEMS Microbiol Rev 38:254–299

Coons DM, Vagnoli P, Bisson LF (1997) The C-terminal domain of Snf3p is sufficient to complement the growth defect of snf3 null mutations in *Saccharomyces cerevisiae*: *SNF3* functions in glucose recognition. Yeast 13:9–20

Cousseu FEM, Alves SL, Trichez D et al (2013) Characterization of maltotriose transport from the *Saccharomyces eubayanus* subgenome of the hybrid *Saccharomyces pastorianus* lager brewing yeast strain Weihenstephan 34170. Lett Appl Microbiol 56:21–29

Daran-Lapujade P, Jansen MLA, Darant JM et al (2004) Role of transcriptional regulation in controlling fluxes in central carbon metabolism of *Saccharomyces cerevisiae*. J Biol Chem 279:9125–9138

De Deken RH (1966) The Crabtree effect: a regulatory system in yeast. J Gen Microbiol 44:149–156

De la Torre-Ruiz MA, Pujol N, Sundaran V (2015) Coping with oxidative stress. The yeast model. Curr Drug Targets 16(1):2–12

De Vit MJ, Waddle J, Johnston M (1997) Regulated nuclear translocation of the Mig1 glucose repressor. Mol Biol Cell 8:1603–1618

Diaz-Ruiz R, Rigoulet M, Devin A (2011) Warburg and Crabtree effects: on the origin of cancer cell energy metabolism and of yeast glucose repression. Biochim Biophys Acta 1807:568–576

Diderich JA, Schepper M, van Hoek P et al (1999) Glucose uptake kinetics and transcription of *HXT* genes in chemostat cultures of *Saccharomyces cerevisiae*. J Biol Chem 274:15350–15359

DiSalvo S, Serio TR (2011) Insights into prion biology: integrating a protein misfolding pathway with its cellular environment. Prion 5:76–83

Dlugai S, Hippler S, Wieczorke R et al (2001) Glucose-dependent and-independent signalling functions of the yeast glucose sensor Snf3. FEBS Lett 505:389–392

Dupres V, Alsteens D, Wilk S et al (2009) The yeast Wsc1 cell surface sensor behaves like a nanospring in vivo. Nat Chem Biol 5:857–862

Duskova M, Borovikova D, Herynkova P et al (2015) The role of glycerol transporters in yeast cells in various physiological and stress conditions. FEMS Microbiol Lett 362:1–8

Duval EH, Alves SL Jr, Dunn B et al (2010) Microarray karyotyping of maltose-fermenting yeasts with differing maltotriose utilization profiles reveals copy number variation in genes involved in maltose and maltotriose utilization. J Appl Microbiol 109:248–259

Eddy AA, Barnett JA (2007) A history of research on yeasts II. The study of solute transport: the first 90 years, simple and facilitated diffusion. Yeast 24:1023–1059

Elbing K, Larsson C, Bill RM et al (2004) Role of hexose transport in control of glycolytic flux in *Saccharomyces cerevisiae*. Appl Environ Microbiol 70:5323–5330

Ellis J (1987) Proteins as molecular chaperones. Nature 328:378–379

Eraso P, Mazon MJ, Portillo F (2006) Yeast protein kinase Ptk2 localizes at the plasma membrane and phosphorylates in vitro the C-terminal peptide of the H+-ATPase. Biochim Biophys Acta 1758:164–170

Eraso P, Mazon MJ, Posas F et al (2011) Gene expression profiling of yeasts overexpressing wild type or misfolded Pma1 variants reveals activation of the Hog1 MAPK pathway. Mol Microbiol 79:1339–1352

Estrada E, Agostinis P, Vandenheede JR et al (1996) Phosphorylation of yeast plasma membrane H+-ATPase by casein kinase I. J Biol Chem 271:32064–32072

Ferreira C, Lucas C (2007) Glucose repression over *Saccharomyces cerevisiae* glycerol/H+ symporter gene STL1 is overcome by high temperature. FEBS Lett 581:1923–1927

Ferreira C, van Voorst F, Martins A et al (2005) A member of the sugar transporter family, Stl1p is the glycerol/H+ symporter in *Saccharomyces cerevisiae*. Mol Biol Cell 16:2068–2076

Ferreira C, Silva S, Faria-Oliveira F et al (2010) *Candida albicans* virulence and drug-resistance requires the O-acyltransferase Gup1p. BMC Microbiol 10:238. doi:10.1186/1471-2180-10-238

Ferrer-Dalmau J, Randez-Gil F, Marquina M et al (2015) Protein kinase Snf1 is involved in the proper regulation of the unfolded protein response in *Saccharomyces cerevisiae*. Biochem J 468(1):33–47. doi:10.1042/BJ20140734

Flick KM, Spielewoy N, Kalashnikova TI et al (2003) Grr1-dependent inactivation of Mth1 mediates glucose-induced dissociation of Rgt1 from *HXT* gene promoters. Mol Biol Cell 14:3230–3241

Frydman J (2001) Folding of newly translated proteins in vivo: the role of molecular chaperones. Ann Rev Biochem 70:603–647

Galeote V, Novo M, Salema-Oom M et al (2010) FSY1 a horizontally transferred gene in the *Saccharomyces cerevisiae* EC1118 wine yeast strain, encodes a high-affinity fructose/$H^+$ symporter. Microbiology 156:3754–3761

Gancedo JM (2008) The early steps of glucose signaling in yeast. FEMS Microbiol Rev 32:673–704

Gancedo C, Serrano R (1989) Energy-yielding metabolism in yeasts. In: Rose AH, Harrison JS (eds) The yeasts. Academic, London, pp 205–259

Garcia DM, Jarosz DF (2014) Rebels with a cause: molecular features and physiological consequences of yeast prions. FEMS Yeast Res 14:136–147

Gasch A (2003) The environmental stress response: a common yeast response to diverse environmental stresses. In: Hohmann S, Mager W (eds) Yeast stress responses, vol 1. Topics in current genetics. Springer, Berlin/Heidelberg, pp 11–70

Gasch AP, Spellman PT, Kao CM et al (2000) Genomic expression programs in the response of yeast cells to environmental changes. Mol Biol Cell 11:4241–4257
Geladé R, Van de Velde S, Van Dijck P et al (2003) Multi-level response of the yeast genome to glucose. Genome Biol 4:233–233
Ghillebert R, Swinnen E, Wen J et al (2011) The AMPK/SNF1/SnRK1 fuel gauge and energy regulator: structure, function and regulation. FEBS J 278:3978–3990
Gong Y, Kakihara Y, Krogan N et al (2009) An atlas of chaperone-protein interactions in Saccharomyces cerevisiae: implications to protein folding pathways in the cell. Mol Syst Biol 5:275. doi:10.1038/msb.2009.26
Grandier-Vazeille X, Bathany K, Chaignepain S et al (2001) Yeast mitochondrial dehydrogenases are associated in a supramolecular complex. Biochemistry 40:9758–9769
Grauslund M, Ronnow B (2000) Carbon source-dependent transcriptional regulation of the mitochondrial glycerol-3-phosphate dehydrogenase gene, *GUT2*, from *Saccharomyces cerevisiae*. Can J Microbiol 46:1096–1100
Gray JV, Petsko GA, Johnston GC et al (2004) "Sleeping beauty": quiescence in *Saccharomyces cerevisiae*. Microbiol Mol Biol Rev 68:187–206
Greatrix BW, van Vuuren HJJ (2006) Expression of the *HXT13*, *HXT15* and *HXT17* genes in *Saccharomyces cerevisiae* and stabilization of the *HXT1* gene transcript by sugar-induced osmotic stress. Curr Genet 49:205–217
Guillaume C, Delobel P, Sablayrolles JM et al (2007) Molecular basis of fructose utilization by the wine yeast *Saccharomyces cerevisiae*: a mutated *HXT3* allele enhances fructose fermentation. Appl Environ Microbiol 73:2432–2439
Hahn JS, Thiele DJ (2004) Activation of the *Saccharomyces cerevisiae* heat shock transcription factor under glucose starvation conditions by Snf1 protein kinase. J Biol Chem 279:5169–5176
Halfmann R, Lindquist S (2010) Epigenetics in the extreme: prions and the inheritance of environmentally acquired traits. Science 330:629–632
Halfmann R, Jarosz DF, Jones SK et al (2012) Prions are a common mechanism for phenotypic inheritance in wild yeasts. Nature 482:363–368
Hardy TA, Huang D, Roach PJ (1994) Interactions between cAMP-dependent and SNF1 protein kinases in the control of glycogen accumulation in *Saccharomyces cerevisiae*. J Biol Chem 269:27907–27913
Haurie V, Perrot M, Mini T et al (2001) The transcriptional activator Cat8p provides a major contribution to the reprogramming of carbon metabolism during the diauxic shift in *Saccharomyces cerevisiae*. J Biol Chem 276:76–85
Hedbacker K, Carlson M (2008) SNF1/AMPK pathways in yeast. Front Biosci 13:2408–2420
Henderson CM, Lozada-Contreras M, Jiranek V et al (2013) Ethanol production and maximum cell growth are highly correlated with membrane lipid composition during fermentation as determined by lipidomic analysis of 22 *Saccharomyces cerevisiae* strains. Appl Environ Microbiol 79:91–104
Hohmann S (2002) Osmotic stress signaling and osmoadaptation in yeasts. Microbiol Mol Biol Rev 66:300–372
Hohmann S, Krantz M, Nordlander B (2007) Yeast osmoregulation. Methods Enzymol 428:29–45
Holmes DL, Lancaster AK, Lindquist S et al (2013) Heritable remodeling of yeast multicellularity by an environmentally responsive prion. Cell 153:153–165
Holsbeeks I, Lagatie O, Van Nuland A et al (2004) The eukaryotic plasma membrane as a nutrient-sensing device. Trends Biochem Sci 29:556–564
Holst B, Lunde C, Lages F et al (2000) *GUP1* and its close homologue *GUP2*, encoding multimembrane-spanning proteins involved in active glycerol uptake in *Saccharomyces cerevisiae*. Mol Microbiol 37:108–124
Honigberg SM, Lee RH (1998) Snf1 kinase connects nutritional pathways controlling meiosis in *Saccharomyces cerevisiae*. Mol Cell Biol 18:4548–4555
Horák J (2013) Regulations of sugar transporters: insights from yeast. Curr Genet 59:1–31
Hunter T, Plowman GD (1997) The protein kinases of budding yeast: six score and more. Trends Biochem Sci 22:18–22

Jansen MLA, De Winde JH, Pronk JT (2002) Hxt-carrier-mediated glucose efflux upon exposure of Saccharomyces cerevisiae to excess maltose. Appl Environ Microbiol 68:4259–4265
Jarosz DF, Brown JC, Walker GA et al (2014) Cross-kingdom chemical communication drives a heritable, mutually beneficial prion-based transformation of metabolism. Cell 158:1083–1093
Jiang R, Carlson M (1996) Glucose regulates protein interactions within the yeast SNF1 protein kinase complex. Genes & Dev 10:3105–3115
Johnston M, Kim J (2005) Glucose as a hormone: receptor-mediated glucose sensing in the yeast *Saccharomyces cerevisiae*. Biochem Soc Trans 33:247–252
Jouandot D 2nd, Roy A, Kim JH (2011) Functional dissection of the glucose signaling pathways that regulate the yeast glucose transporter gene (*HXT*) repressor Rgt1. J Cell Biochem 112:3268–3275
Kaniak A, Xue Z, Macool D, Kim JH et al (2004) Regulatory network connecting two glucose signal transduction pathways in *Saccharomyces cerevisiae*. Eukaryot cell 3:221–231
Karpel JE, Place WR, Bisson LF (2008) Analysis of the major hexose transporter genes in wine strains of *Saccharomyces cerevisiae*. Am J Enol Vitic 59:265–275
Kasahara T, Maeda M, Ishiguro M et al (2007) Identification by comprehensive chimeric analysis of a key residue responsible for high affinity glucose transport by yeast *HXT2*. J Biol Chem 282:13146–13150
Keleher CA, Redd MJ, Schultz J, Carlson M et al (1992) Ssn6-Tup1 is a general repressor of transcription in yeast. Cell 68:709–719
Kim JH (2009) DNA-binding properties of the yeast Rgt1 repressor. Biochimie 91:300–303
Kim JH, Johnston M (2006) Two glucose-sensing pathways converg on Rgt1 to regulate expression of glucose transporter genes in *Saccharomyces cerevisiae*. J Biol Chem 281:26144–26149
Kim JH, Polish J, Johnston M (2003) Specificity and regulation of DNA binding by the yeast glucose transporter gene repressor Rgt1. Mol Cell Biol 23:5208–5216
Kim JH, Brachet V, Moriya H, Johnston M (2006) Integration of transcriptional and posttranslational regulation in a glucose signal transduction pathway in *Saccharomyces cerevisiae*. Eukaryot Cell 5:167–173
Kim JH, Roy A, Jouandot D 2nd, Cho KH (2013) The glucose signaling network in yeast. Biochim Biophys Acta 1830:5204–5210
Klipp E, Nordlander B, Kruger R et al (2005) Integrative model of the response of yeast to osmotic shock. Nat Biotechnol 23:975–982
Klockow C, Stahl F, Scheper T et al (2008) In vivo regulation of glucose transporter genes at glucose concentrations between 0 and 500 mg/L in a wild type of *Saccharomyces cerevisiae*. J Biotechnol 135:161–167
Kraakman L, Lemaire K, Ma P, Teunissen AWRH et al (1999) A *Saccharomyces cerevisiae* G-protein coupled receptor, Gpr1, is specifically required for glucose activation of the cAMP pathway during the transition to growth on glucose. Mol Microbiol 32:1002–1012
Krampe S, Stamm O, Hollenberg CP et al (1998) Catabolite inactivation of the high-affinity hexose transporters Hxt6 and Hxt7 of *Saccharomyces cerevisiae* occurs in the vacuole after internalization by endocytosis. FEBS Lett 441:343–347
Kresnowati MT, van Winden WA, Almering MJH et al (2006) When transcriptome meets metabolome: fast cellular responses of yeast to sudden relief of glucose limitation. Mol Syst Biol 2: n/a. doi:10.1038/msb4100083
Kriel J, Haesendonckx S, Rubio-Texeira M et al (2011) From transporter to transceptor: signaling from transporters provokes re-evaluation of complex trafficking and regulatory controls: endocytic internalization and intracellular trafficking of nutrient transceptors may, at least in part, be governed by their signaling function. BioEssays 33:870–879
Kruckeberg AL (1996) The hexose transporter family of *Saccharomyces cerevisiae*. Arch Microbiol 166:283–292
Kruckeberg AL, Bisson LF (1990) The *HXT2* gene of *Saccharomyces cerevisiae* is required for high-affinity glucose transport. Mol Cell Biol 10:5903–5913
Kuchin S, Treich I, Carlson M (2000) A regulatory shortcut between the Snf1 protein kinase and RNA polymerase II holoenzyme. Proc Natl Acad Sci U S A 97:7916–7920

Kuchin S, Vyas VK, Carlson M (2002) Snf1 protein kinase and the repressors Nrg1 and Nrg2 regulate *FLO11*, haploid invasive growth, and diploid pseudohyphal differentiation. Mol Cell Biol 22:3994–4000

Kuttykrishnan S, Sabina J, Langton LL et al (2010) A quantitative model of glucose signaling in yeast reveals an incoherent feed forward loop leading to a specific, transient pulse of transcription. Proc Natl Acad Sci U S A 107:16743–16748

Lafuente MJ, Gancedo C, Jauniaux JC et al (2000) Mth1 receives the signal given by the glucose sensors Snf3 and Rgt2 in *Saccharomyces cerevisiae*. Mol Microbiol 35:161–172

Lakshmanan J, Mosley AL, Ozcan S (2003) Repression of transcription by Rgt1 in the absence of glucose requires Std1 and Mth1. Curr Genet 44:19–25

Larsson C, Pahlman IL, Ansell R et al (1998) The importance of the glycerol 3-phosphate shuttle during aerobic growth of *Saccharomyces cerevisiae*. Yeast 14:347–357

Leandro MJ, Fonseca C, Gonçalves P (2009) Hexose and pentose transport in ascomycetous yeasts: an overview. FEMS Yeast Res 9:511–525

Lecchi S, Nelson CJ, Allen KE et al (2007) Tandem phosphorylation of Ser-911 and Thr-912 at the C terminus of yeast plasma membrane H+-ATPase leads to glucose-dependent activation. J Biol Chem 282:35471–35481

Lee ME, Singh K, Snider J et al (2011) The Rho1 GTPase acts together with a vacuolar glutathione S-conjugate transporter to protect yeast cells from oxidative stress. Genetics 188:859–870

Lee YJ, Jeschke GR, Roelants FM et al (2012) Reciprocal phosphorylation of yeast glycerol-3-phosphate dehydrogenases in adaptation to distinct types of stress. Mol Cell Biol 32:4705–4717

Lewis DA, Bisson LF (1991) The *HXT1* gene product of *Saccharomyces cerevisiae* is a new member of the family of hexose transporters. Mol Cell Biol 11:3804–3813

Li L, Kowal AS (2012) Environmental regulation of prions in yeast. PLoS Pathog 8(11):e1002973. doi:10.1371/journal.ppat.1002973

Liang H, Gaber RF (1996) A novel signal transduction pathway in *Saccharomyces cerevisiae* defined by Snf3-regulated expression of *HXT6*. Mol Biol Cell 7:1953–1966

Lo WS, Duggan L, Emre NC et al (2001) Snf1–a histone kinase that works in concert with the histone acetyltransferase Gcn5 to regulate transcription. Science 293:1142–1146

Luyten K, Albertyn J, Skibbe WF et al (1995) Fps1, a yeast member of the MIP family of channel proteins, is a facilitator for glycerol uptake and efflux and is inactive under osmotic stress. EMBO J 14:1360–1371

Luyten K, Riou D, Blondin B (2002) The hexose transporters of *Saccharomyces cerevisiae* play different roles during enological fermentation. Yeast 19:713–726

Maier A, Völker B, Boles E et al (2002) Characterization of glucose transport in *Saccharomyces cerevisiae* with plasma membrane vesicles (countertransport) and intact cells (initial uptake) with single Hxt1, Hxt2, Hxt3, Hxt4, Hxt6 Hxt7 or Gal2 transporters. FEMS Yeast Res 3:539–550

Malave TM, Dent SY (2006) Transcriptional repression by Tup1-Ssn6. Biochem Cell Biol 84(4):437–443

Marshall-Carlson L, Neigeborn L, Coons D et al (1991) Dominant and recessive suppressors that restore glucose transport in a yeast *snf3* mutant. Genetics 128:505–512

Martinez-Pastor MT, Marchler G, Schuller C et al (1996) The *Saccharomyces cerevisiae* zinc finger proteins Msn2p and Msn4p are required for transcriptional induction through the stress response element (STRE). EMBO J 15:2227–2235

Mason AB, Allen KE, Slayman CW (2014) C-terminal truncations of the Saccharomyces cerevisiae PMA1 H+-ATPase have major impacts on protein conformation, trafficking, quality control, and function. Eukaryot Cell 13:43–52

Mayordomo I, Estruch F, Sanz P (2002) Convergence of the target of rapamycin and the Snf1 protein kinase pathways in the regulation of the subcellular localization of Msn2, a transcriptional activator of STRE (Stress Response Element)-regulated genes. J Biol Chem 277:35650–35656

McCartney RR, Schmidt MC (2001) Regulation of Snf1 kinase. Activation requires phosphorylation of threonine 210 by an upstream kinase as well as a distinct step mediated by the Snf4 subunit. J Biol Chem 276:36460–36466
McGlinchey RP, Kryndushkin D, Wickner RB (2011) Suicidal [*PSI+*] is a lethal yeast prion. Proc Natl Acad Sci U S A 108:5337–5341
Moriya H, Johnston M (2004) Glucose sensing and signaling in *Saccharomyces cerevisiae* through the Rgt2 glucose sensor and casein kinas I. Proc Natl Acad Sci U S A 101:1572–1577
Mosley AL, Lakshmanan J, Aryal BK et al (2003) Glucose-mediated phosphorylation converts the transcription factor Rgt1 from a repressor to an activator. J Biol Chem 278:10322–10327
Naito C, Ito H, Oshiro T et al (2014) A new pma1 mutation identified in a chronologically long-lived fission yeast mutant. FEBS Open Bio 4:829–833. doi:10.1016/j.fob.2014.09.006
Nayak V, Zhao K, Wyce A et al (2006) Structure and dimerization of the kinase domain from yeast Snf1, a member of the Snf1/AMPK protein family. Structure 14:477–485
Neigeborn L, Carlson M (1984) Genes affecting the regulation of *SUC2* gene expression by glucose repression in *Saccharomyces cerevisiae*. Genetics 108:845–858
Norbeck J, Pahlman AK, Akhtar N et al (1996) Purification and characterization of two isoenzymes of DL-glycerol-3-phosphatase from *Saccharomyces cerevisiae*. Identification of the corresponding *GPP1* and *GPP2* genes and evidence for osmotic regulation of Gpp2p expression by the osmosensing mitogen-activated protein kinase signal transduction pathway. J Biol Chem 271:13875–13881
Novo M, Bigey F, Beyne E et al (2009) Eukaryote-to-eukaryote gene transfer events revealed by the genome sequence of wine yeast *Saccharomyces cerevisiae* EC1118. Proc Natl Acad Sci U S A 106:16333–16338
Nozawa A, Takano J, Kobayashi M et al (2006) Roles of *BOR1*, *DUR3*, and *FPS1* in boron transport and tolerance in *Saccharomyces cerevisiae*. FEMS Microbiol Lett 262:216–222
Oliveira AP, Sauer U (2012) The importance of post-translational modifications in regulating *Saccharomyces cerevisiae* metabolism. FEMS Yeast Res 12:104–117
Oliveira R, Lages F, Silva-Graca M et al (2003) Fps1p channel is the mediator of the major part of glycerol passive diffusion in *Saccharomyces cerevisiae*: artefacts and re-definitions. Biochim Biophys Acta 1613:57–71
Ostling J, Carlberg M, Ronne H (1996) Functional domains in the Mig1 repressor. Mol Cell Biol 16:753–761
Otterstedt K, Larsson C, Bill RM et al (2004) Switching the mode of metabolism in the yeast *Saccharomyces cerevisiae*. EMBO Rep 5:431–443
Ozcan S, Johnston M (1995) Three different regulatory mechanisms enable yeast hexose transporter (*HXT*) genes to be induced by different levels of glucose. Mol Cell Biol 15:1564–1572
Ozcan S, Johnston M (1996) Two different repressors collaborate to restrict expression of the yeast glucose transporter genes *HXT2* and *HXT4* to low levels of glucose. Mol Cell Biol 16:5536–5545
Ozcan S, Johnston M (1999) Function and regulation of yeast hexose transporters. Microbiol Mol Biol Rev 63:554–569
Ozcan S, Dover J, Rosenwald AG et al (1996a) Two glucose transporters in *Saccharomyces cerevisiae* are glucose sensors that generate a signal for induction of gene expression. Proc Natl Acad Sci U S A 93:12428–12432
Ozcan S, Leong T, Johnston M (1996b) Rgt1p of *Saccharomyces cerevisiae*, a key regulator of glucose-induced genes, is both an activator and a repressor of transcription. Mol Cell Biol 16:6419–6426
Papamichos-Chronakis M, Conlan RS, Gounalaki N, Copf T, Tzamarias D (2000) Hrs1/Med3 Is a Cyc8-Tup1 Corepressor Target in the RNA Polymerase II Holoenzyme. J Biol Chem 275:8397–8403
Papamichos-Chronakis M, Gligoris T, Tzamarias D (2004) The Snf1 kinase controls glucose repression in yeast by modulating interactions between the Mig1 repressor and the Cyc8-Tup1 co-repressor. EMBO Rep 5:368–372

Park JI, Collinson EJ, Grant CM et al (2005) Rom2p, the Rho1 GTP/GDP exchange factor of Saccharomyces cerevisiae, can mediate stress responses via the Ras-cAMP pathway. J Biol Chem 280:2529–2535

Pasula S, Jouandot D, Kim JH (2007) Biochemical evidence for glucose-independent induction of HXT expression in *Saccharomyces cerevisiae*. FEBS Lett 581:3230–3234

Pasula S, Chakraborty S, Choi JH et al (2010) Role of casein kinase 1 in the glucose sensor-mediated signaling pathway in yeast. BMC Cell Biol 11:17

Pavlik P, Simon M, Schuster T et al (1993) The glycerol kinase (*GUT1*) gene of Saccharomyces cerevisiae: cloning and characterization. Curr Genet 24:21–25

Peeters K, Thevelein JM (2014) Glucose sensing and signal transduction in *Saccharomyces cerevisiae*. In: Piškur J, Compagno C (eds) Molecular mechanisms in yeast carbon metabolism. Springer, Heidelberg, pp 21–56

Peeters T, Louwet W, Gelade R et al (2006) Kelch-repeat proteins interacting with the Galpha protein Gpa2 bypass adenylate cyclase for direct regulation of protein kinase A in yeast. Proc Natl Acad Sci U S A 103:13034–13039

Perez M, Lutent K, Michel R et al (2005) Analysis of *Saccharomyces cerevisiae* hexose carrier expression during wine fermentation: both low- and high-affinity Hxt transporters are expressed. FEMS Yeast Res 5:351–361

Petkova MI, Pujol-Carrion N, de la Torre-Ruiz MA (2012) Mtl1 O-mannosylation mediated by both Pmt1 and Pmt2 is important for cell survival under oxidative conditions and TOR blockade. Fungal Genet Biol 49:903–914

Philips J, Herskowitz I (1997) Osmotic balance regulates cell fusion during mating in *Saccharomyces cerevisiae*. J Cell Biol 138:961–974

Pirkkala L, Nykanen P, Sistonen L (2001) Roles of the heat shock transcription factors in regulation of the heat shock response and beyond. FASEB J 15:1118–1131

Place WR, Bisson LF (2013) Identification of *HXT7* as a suppressor of the snf3 growth defect in wine and wild-type strains of *Saccharomyces cerevisiae*. Am J Enol Vitic 64:251–257

Polish JA, Kim JH, Johnston M (2005) How the Rgt1 transcription factor of *Saccharomyces cerevisiae* is regulated by glucose. Genetics 169:583–594

Ramakrishnan V, Theodoris G, Bisson LF (2006) Loss of IRA2 suppresses the growth defect on low glucose caused by the *snf3* mutation in *Saccharomyces cerevisiae*. FEMS Yeast Res 7:67–77

Reifenberger E, Freidel K, Ciriacy M (1995) Identification of novel *HXT* genes in *Saccharomyces cerevisiae* reveals the impact of hexose transporters on glycolytic flux. Mol Microbiol 16:157–167

Reifenberger E, Boles E, Ciriacy M (1997) Kinetic characterization of individual hexose transporters of *Saccharomyces cerevisiae* and their relation to the triggering mechanisms of glucose repression. Eur J Biochem 245:324–333

Rødkær SV, Færgeman NJ (2014) Glucose- and nitrogen sensing and regulatory mechanisms in *Saccharomyces cerevisiae*. FEMS Yeast Res 14:683–696

Rolland F, Winderickx J, Thevelein JM (2002) Glucose-sensing and –signaling mechanisms in yeast. FEMS Yeast Res 2:183–201

Ronnow B, Kielland-Brandt MC (1993) *GUT2*, a gene for mitochondrial glycerol 3-phosphate dehydrogenase of *Saccharomyces cerevisiae*. Yeast 9:1121–1130

Roy A, Kim JH (2014) Endocytosis and vacuolar degradation of the yeast cell surface glucose sensors Rgt2 and Snf3. J Biol Chem 289:7247–7256

Roy A, Shin YJ, Cho KH et al (2013) Mth1 regulates the interaction between the Rgt1 repressor and the Ssn6-Tup1 corepressor complex by modulating PKA-dependent phosphorylation of Rgt1. Mol Biol Cell 24:1493–1503

Roy A, Jouandot D 2nd, Cho KH et al (2014a) Understanding the mechanism of glucose-induced relief of Rgt1-mediated repression in yeast. FEBS Open Bio 4:105–111

Roy A, Kim YB, Cho KH, Kim JH (2014b) Glucose starvation-induced turnover of the yeast glucose transporter Hxt1. Biochim Biophys Acta 1840:2878–2885

Rubenstein EM, McCartney RR, Zhang C et al (2008) Access denied: Snf1 activation loop phosphorylation is controlled by availability of the phosphorylated threonine 210 to the PP1 phosphatase. J Biol Chem 283:222–230

Rudolph MJ, Amodeo GA, Bai Y et al (2005) Crystal structure of the protein kinase domain of yeast AMP-activated protein kinase Snf1. Biochem Biophys Res Commun 337:1224–1228

Ruiz A, Serrano R, Ariño J (2008) Direct regulation of genes involved in glucose utilization by the calcium/calcineurin pathway. J Biol Chem 283:13923–13933

Sabina J, Johnston M (2009) Asymmetric signal transduction through paralogs that comprise a genetic switch for sugar sensing in *Saccharomyces cerevisiae*. J Biol Chem 284:29635–29643

Salema-Oom M, Pinto VV, Gonçalves P et al (2005) Maltotriose utilization by industrial *Saccharomyces* strains: characterization of a new member of the α-glucosidase transporter family. Appl Environ Microbiol 71:5044–5049

Sanders D, Slayman CL (1982) Control of intracellular pH. Predominant role of oxidative metabolism, not proton transport, in the eukaryotic microorganism *Neurospora*. J Gen Physiol 80:377–402

Santangelo GM (2006) Glucose signaling in *Saccharomyces cerevisiae*. Microbiol Mol Biol Rev 70:253–282

Sanz P, Alms GR, Haystead TA et al (2000) Regulatory interactions between the Reg1-Glc7 protein phosphatase and the Snf1 protein kinase. Mol Cell Biol 20:1321–1328

Schmidt MC, McCartney RR, Zhang X et al (1999) Std1 and Mth1 proteins interact with the glucose sensors to control glucose-regulated gene expression in *Saccharomyces cerevisiae*. Mol Cell Biol 19:4561–4571

Schultz J, Marshall-Carlson L, Carlson M (1990) The N-terminal TPR region is the functional domain of *SSN6*, a nuclear phosphoprotein of *Saccharomyces cerevisiae*. Mol Cell Biol 10:4744–4756

Serrano R (1977) Energy requirements for maltose transport in yeast. Eur J Biochem 80:97–102

Serrano R (1983) In vivo glucose activation of the yeast plasma membrane ATPase. FEBS Lett 156:11–14

Serrano R, Kielland-Brandt MC, Fink GR (1986) Yeast plasma membrane ATPAse is essential for growth and has homology with ($Na^+ + K^+$), $K^+$- and $Ca^{2+}$- ATPases. Nature 319:689–693

Serrano R, Martin H, Casamayor A et al (2006) Signaling alkaline pH stress in the yeast *Saccharomyces cerevisiae* through the Wsc1 cell surface sensor and the Slt2 MAPK pathway. J Biol Chem 281:39785–39795

Shamu CE, Cox JS, Walter P (1994) The unfolded-protein-response pathway in yeast. Trends Cell Biol 4:56–60

Sharma D, Masison DC (2009) Hsp70 structure, function, regulation and influence on yeast prions. Protein Pept Lett 16:571–581

Shor E, Fox CA, Broach JR (2013) The yeast environmental stress response regulates mutagenesis induced by proteotoxic stress. PLoS Genet 9(8):e1003680. doi:10.1371/journal.pgen.1003680

Smith RL, Redd MJ, Johnson AD (1995) The tetratricopeptide repeats of Ssn6 interact with the homeo domain of alpha 2. Genes Dev 9:2903–2910

Snowdon C, van der Merwe G (2012) Regulation of Hxt3 and Hxt7 turnover converges on the Vid30 complex and requires inactivation of the Ras/cAMP/PKA pathway in *Saccharomyces cerevisiae*. PLoS One 7(12):e50458. doi:10.1371/journal.pone.0050458

Sorger D, Daum G (2003) Triacylglycerol biosynthesis in yeast. Appl Microbiol Biotechnol 61:289–299

Sprague ER, Redd MJ, Johnson AD et al (2000) Structure of the C-terminal domain of Tup1, a corepressor of transcription in yeast. EMBO J 19:3016–3027

Stambuk BU, de Araujo PS (2001) Kinetics of active α-glucosidase transport in *Saccharomyces cerevisiae*. FEMS Yeast Res 1:73–78

Tamas MJ, Luyten K, Sutherland FC et al (1999) Fps1p controls the accumulation and release of the compatible solute glycerol in yeast osmoregulation. Mol Microbiol 31:1087–1104

Tanaka K, Nakafuku M, Tamanoi F et al (1990) *IRA2*, a second gene of *Saccharomyces cerevisiae* that encodes a protein with a domain homologous to mammalian Ras Gtpase-activating protein. Mol Cell Biol 10:4303–4313

Temple MD, Perrone GG, Dawes IW (2005) Complex cellular responses to reactive oxygen species. Trends Cell Biol 15:319–326

Teusink B, Walsh MC, van Dam K et al (1998) The danger of metabolic pathways with turbo design. Trends Biochem Sci 23:162–169

Theodoris G, Bisson LF (2001) DDSE: downstream targets of the *SNF3* signal transduction pathway. FEMS Microbiol Lett 197:73–77

Theodoris G, Fong NM, Coons DM et al (1994) High-copy suppression of glucose transport defects by *HXT4* and regulatory elements in the promoters of the *HXT* genes in *Saccharomyces cerevisiae*. Genetics 137:957–966

Thevelein JM, De Winde JH (1999) Novel sensing mechanisms and targets for the cAMP–protein kinase A pathway in the yeast *Saccharomyces cerevisiae*. Mol Microbiol 33:904–918

Toda T, Cameron S, Sass P et al (1987a) Cloning and characterization of BCY1, a locus encoding a regulatory subunit of the cyclic AMP-dependent protein kinase in *Saccharomyces cerevisiae*. Mol Cell Biol 7:1371–1377

Toda T, Cameron S, Sass P et al (1987b) Three different genes in *S. cerevisiae* encode the catalytic subunits of the cAMP-dependent protein kinase. Cell 50:277–287

Tomas-Cobos L, Casadome L, Mas G et al (2004) Expression of the *HXT1* low affinity glucose transporter requires the coordinated activities of the HOG and glucose signalling pathways. J Biol Chem 279:22010–22019

Treitel MA, Carlson M (1995) Repression by *SSN6-TUP1* is directed by *MIG1*, a repressor/activator protein. Proc Natl Acad Sci U S A 92:3132–3136

Treitel MA, Kuchin S, Carlson M (1998) Snf1 protein kinase regulates phosphorylation of the Mig1 repressor in *Saccharomyces cerevisiae*. Mol Cell Biol 18:6273–6280

Tripodi F, Nicastro R, Reghellin V et al (2015) Post-translational modifications on yeast carbon metabolism: Regulatory mechanisms beyond transcriptional control. Biochim Biophys Acta 1850:620–627

Tulha J, Lima A, Lucas C et al (2010) *Saccharomyces cerevisiae* glycerol/H+ symporter Stl1p is essential for cold/near-freeze and freeze stress adaptation. A simple recipe with high biotechnological potential is given. Microb Cell Factor 9:82. doi:10.1186/1475-2859-9-82

Tzamarias D, Struhl K (1994) Functional dissection of the yeast Cyc8–Tupl transcriptional co-repressor complex. Nature 369:758–761

Tzamarias D, Struhl K (1995) Distinct TPR motifs of Cyc8 are involved in recruiting the Cyc8-Tup1 corepressor complex to differentially regulated promoters. Genes Dev 9:821–831

Vagnoli P, Bisson LF (1998) The *SKS1* gene of *Saccharomyces cerevisiae* is required for long-term adaptation of snf3 null strains to low glucose. Yeast 14:359–369

Van Zeebroeck G, Bonini BM, Versele M et al (2009) Transport and signaling via the amino acid binding site of the yeast Gap1 amino acid transceptor. Nat Chem Biol 5:45–52

Varanasi US, Klis M, Mikesell PB et al (1996) The Cyc8 (Ssn6)-Tup1 corepressor complex is composed of one Cyc8 and four Tup1 subunits. Mol Cell Biol 16:6707–6714

Verghese J, Abrams J, Wang Y et al (2012) Biology of the heat shock response and protein chaperones: budding yeast (*Saccharomyces cerevisiae*) as a model system. Microbiol Mol Biol Rev 76:115–158

Verwaal R, Paalman JWG, Hogenkamp A et al (2002) *HXT5* expression is determined by growth rates in *Saccharomyces cerevisiae*. Yeast 19:1029–1038

Vilella F, Herrero E, Torres J et al (2005) Pkc1 and the upstream elements of the cell integrity pathway in Saccharomyces cerevisiae, Rom2 and Mtl1, are required for cellular responses to oxidative stress. J Biol Chem 280:9149–9159

Wahi M, Komachi K, Johnson AD (1998) Gene Regulation by the Yeast Ssn6-Tup1 Corepressor. Cold Spring Harb Symp Quant Biol 63:447–458

Watson AD, Edmondson DG, Bone JR et al (2000) Ssn6-Tup1 interacts with class I histone deacetylases required for repression. Genes Dev 14:2737–2744

Weindling E, Bar-Nun S (2015) Sir2 links the unfolded protein response and the heat shock response in a stress response network. Biochem Biophys Res Commun 457:473–478

Weinhandl K, Winkler M, Glieder A et al (2014) Carbon source dependent promoters in yeasts. Microbiol Cell Fact 13(5):1–17

Westergaard SL, Oliveira AP, Bro C et al (2007) A systems biology approach to study glucose repression in the yeast *Saccharomyces cerevisiae*. Biotechnol Bioeng 96:134–145

Westerheide SD, Raynes R, Powell C et al (2012) HSF transcription factor family, heat shock response, and protein intrinsic disorder. Curr Protein Pept Sci 13:86–103

Westholm JO, Nordberg N, Muren E et al (2008) Combinatorial control of gene expression by the three yeast repressors Mig1, Mig2 and Mig3. BMC Genomics 9:601. doi:10.1186/1471-2164-9-601

Wickner RB, Edskes HK, Kryndushkin D et al (2011) Prion diseases of yeast: amyloid structure and biology. Semin Cell Dev Biol 22:469–475

Wieczorke R, Kampe S, Weierstall T et al (1999) Concurrent knock-out of at least 20 transporter genes is required to block uptake of hexoses in *Saccharomyces cerevisiae*. FEBS Lett 464:123–128

Wilson WA, Roach PJ (2002) Nutrient-regulated protein kinases in budding yeast. Cell 111:155–158

Wilson WA, Hawley SA, Hardie DG (1996) Glucose repression/derepression in budding yeast: *SNF1* protein kinase is activated by phosphorylation under derepressing conditions, and this correlates with a high AMP: ATP ratio. Curr Biol 6:1426–1434

Wong KH, Struhl K (2011) The Cyc8-Tup1 complex inhibits transcription primarily by masking the activation domain of the recruiting protein. Genes Dev 25(23):2525–2539

Wu C (1995) Heat shock transcription factors: structure and regulation. Annu Rev Cell Dev Biol 11:441–469

Wu J, Trumbly RJ (1998) Multiple regulatory proteins mediate repression and activation by interaction with the yeast Mig1 binding site. Yeast 14:985–1000

Wu J, Suka N, Carlson M et al (2001) *TUP1* utilizes histone H3/H2B–specific HDA1 deacetylase to repress gene activity in yeast. Mol Cell 7:117–126

Wysocki R, Chery CC, Wawrzycka D et al (2001) The glycerol channel Fps1p mediates the uptake of arsenite and antimonite in *Saccharomyces cerevisiae*. Mol Microbiol 40:1391–1401

Xiong Y, Lei QY, Zhao S et al (2011) Regulation of glycolysis and gluconeogenesis by acetylation of PKM and PEPCK. Cold Spring Harb Symp Quant Biol 76:285–289

Yang Z, Bisson LF (1996) The *SKS1* protein kinase is a multicopy suppressor of the *snf3* mutation of *Saccharomyces cerevisiae*. Yeast 12:1407–1419

Young ET, Dombek KM, Tachibana C et al (2003) Multiple pathways are co-regulated by the protein kinase Snf1 and the transcription factors Adr1 and Cat8. J Biol Chem 278:26146–26158

# Chapter 7
# The CRaZy Calcium Cycle

**Eduardo A. Espeso**

**Abstract** Calcium is an essential cation for a cell. This cation participates in the regulation of numerous processes in either prokaryotes or eukaryotes, from bacteria to humans. *Saccharomyces cerevisiae* has served as a model organism to understand calcium homeostasis and calcium-dependent signaling in fungi. In this chapter it will be reviewed known and predicted transport mechanisms that mediate calcium homeostasis in the yeast. How and when calcium enters the cell, how and where it is stored, when is reutilized, and finally secreted to the environment to close the cycle. As a second messenger, maintenance of a controlled free intracellular calcium concentration is important for mediating transcriptional regulation. Many environmental stimuli modify the concentration of cytoplasmic free calcium generating the "calcium signal". This is sensed and transduced through the calmodulin/calcineurin pathway to a transcription factor, named calcineurin-responsive zinc finger, CRZ, also known as *"crazy"*, to mediate transcriptional regulation of a large number of genes of diverse pathways including a negative feedback regulation of the calcium homeostasis system.

**Keywords** Crz1 • Calcineurin signaling • Calcium pump • Calcium channel • P-type ATPase • Magnesium homeostasis

## 7.1 A Model of Calcium Regulation in Yeasts

In higher eukaryotes entry of calcium in the cell starts concatenated signaling events some of them are of enormous importance in animals such as initiation of the heartbeat or the synapses between neurons. In the budding yeast calcium mediates adaptation to a variety of stimuli such as the presence of mating pheromones (Iida et al. 1990), a damage to endoplasmic reticulum (Bonilla and Cunningham 2003), and different ambient stresses like salinity, alkaline pH or high osmolarity [reviewed

E.A. Espeso (✉)
Department of Cellular and Molecular Biology, Centro de Investigaciones Biológicas, CSIC, Ramiro de Maeztu, 9, 28040 Madrid, Spain
e-mail: eespeso@cib.csic.es



J. Ramos et al. (eds.), *Yeast Membrane Transport*, Advances in Experimental Medicine and Biology 892, DOI 10.1007/978-3-319-25304-6_7

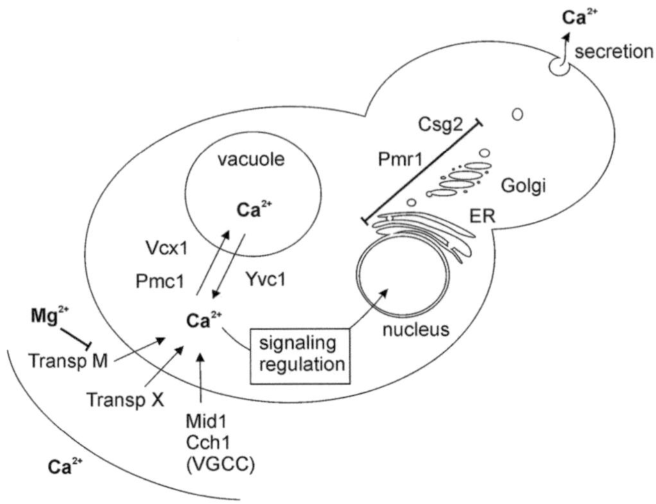


**Fig. 7.1** Main transport elements of calcium in *S. cerevisiae*. The chart depicts the relationships between calcium transporters and their preferential locations in the cell. In the text can be found a detailed description of these components

in Cunningham 2005]. A general model for calcium homeostasis is depicted in Fig. 7.1 and the elements participating in this process will be reviewed in this chapter.

Essentially, calcium enters the cell through different transport mechanisms generating an increase of cytosolic free calcium concentration. A feedback control system enables an stable concentration of calcium in cytoplasm of 50–200 nM (Aiello et al. 2002; Dunn et al. 1994; Miseta et al. 1999). The elevation of calcium levels in the cytoplasm sequentially activates the calcium binding proteins calmodulin (CaM) and the serine/threonine phosphatase calciuneurin, which is composed of the Cna1/Cnb1 or Cmp2/Cnb1 heterodimer. An important final effector for regulation of transcription is the transcription factor Crz1 a target of calcineurin. Dephosphorylation of Crz1 by calcineurin activity causes its immediate entry in the nucleus. Nuclear Crz1 regulates the transcription of a range of genes. Among these Crz1-dependent genes are those encoding for calcium pumps Pmr1 and Pmc1. These two calcium transporters play a key role in regulating cytoplasmic calcium by either pumping calcium to the ER and Golgi, PMR1, or to the vacuole, PMC1. In addition to the previous transport mechanism, calcium is stored in the vacuole through the activity of the $Ca^{2+}/H^{+}$ exchanger Vcx1. In the vacuole, calcium associates with polyphosphates constituting the "non-reusable" stock of calcium. However a low level of vacuolar calcium remains free and it can be transported back to the cytoplasm via the specific channel Yvc1. Release of calcium from the internal stores it is tightly regulated and occurs in response a number of stimuli. Calcium in the ER and Golgi may preferentially follow a secretion process depleting the cell of this cation. In a way, calcium exocytosis closes the cycle of calcium in the yeast.

## 7.2 How Calcium Enters the Cell?

Since calcium is a basic element of cell signaling it is expected that specific transport mechanisms functionally located at the plasma membrane (PM) mediate the influx of this cation. Three different mechanisms of transport have been postulated for $Ca^{2+}$ entry (Fig. 7.1). To date, the only identified transport mechanism is a calcium channel based on two components shown to be responsible of calcium entry in the cell (Courchesne and Ozturk 2003; Locke et al. 2000). Mid1p and Cch1p are proteins of the PM that are thought to assemble into a calcium channel (Paidhungat and Garrett 1997), however evidence also suggests that they may function independently (Locke et al. 2000). Cch1, for calcium channel homologue, was primarily identified as a $Ca^{2+}$ channel homologue due to its sequence similarity to the pore-forming subunit ($\alpha$1) of a plasma membrane, voltage-gated, calcium channel from higher eukaryotes (Paidhungat and Garrett 1997). Cch1p is a large protein of 2039 amino acids with 22–24 predicted transmembrane domains (TMHMM, http://www.yeastgenome.org/locus/S000003449/protein) organized into four hydrophobic repeats (Paidhungat and Garrett 1997).

Mid1, from Mating pheromone-Induced Death, was identified in a screen for mutants deficient in survival after mating differentiation and in calcium uptake (Iida et al. 1994). Later, Mid1 was classified as a stretch-activated channel, with similarities to higher eukaryotes SA-Cat channels (Kanzaki et al. 1999). Mid1 is a 548 amino acid protein with four hydrophobic regions (named H1 to H4) and two cysteine rich regions (C1 and C2) (see PFAM entry PF12929). In addition to a plasma membrane localization, Mid1 was found to be also present in ER as a 200-kDa oligomer by covalent cystein bounding (Yoshimura et al. 2004), probably through the cysteine rich regions. The role of hydrophobic regions in cellular distribution of Mid1 was established, being H1 to H3 required for PM and ER localization and H1 alone for PM localization in response to mating pheromone (Ozeki-Miyawaki et al. 2005).

Importantly, *mid1*$\Delta$ *cch1*$\Delta$ double mutants are indistinguishable of single mutants, this and physiological data early suggested that Mid1 and Cch1 might be actually components of a single yeast $Ca^{2+}$ channel (Paidhungat and Garrett 1997). Both proteins together could act as a voltage-gated $Ca^{2+}$ channel (VGCC) becoming activated in response to depolarization (Catterall 2000; Cui et al. 2009a). The presence of mating pheromone or depletion of manganese from the medium (Paidhungat and Garrett 1997), depletion of calcium from the ER (Bonilla et al. 2002), medium alkalinisation (Viladevall et al. 2004), and cold, osmotic or saline stresses (Matsumoto et al. 2002; Peiter et al. 2005; Viladevall et al. 2004), are among others, signals that trigger calcium entry in the cell through the Mid1/Cch1 VGCC. Of importance to understand how calcium entry is regulated, is the fact that absence of either or both components of yeast VGCC does not completely perturb calcium homeostasis. Based on calcium resistance/tolerance experiments and mathematical models another two transport systems have been postulated (Cui and Kaandorp 2006). These have not been identified so far in *S. cerevisiae*, these transporters or pumps have been termed as transporter X and transporter M. Activities of these $Ca^{2+}$ influx transporters are modulated by extracellular $Mg^{2+}$ and the possible identity of these transporters is speculated below.

Finally, it is interesting to note that as important as those mechanisms present in the yeast are those mechanisms which are absent. Different studies have shown the absence of calcium ATPases of the SERCA family at the plasma membrane (reviewed in Cunningham 2005), as well as the mechanism that directly couples entry of calcium through the PM towards the ER. In animal cells, the store-operated calcium channels (SOCs) allows replenishment of the ER when becomes depleted of calcium by the action of resident calcium ATPases (Zhou et al. 2010a, b). This is a mechanism that ensures the adequate level of exchangeable pool of calcium in animal cells, but lacking these systems in *S. cerevisiae* entry of calcium through the PM is the major bottle neck for appropriate storage of intracellular calcium. Thus it is important to understand how calcium is kept into the yeast cell.

## 7.3 How Calcium is Stored in a Fungal Cell?

Under certain ambient conditions, a massive entry of calcium in the cell occurs and this represents a major stress for the yeast. Excessive free calcium is toxic because it may interact with numerous proteins or oligomolecules (ie. polyphosphate-derived compounds such as NTPs) in the cytoplasm. To prevent deleterious effects, the excess of calcium is rapidly eliminated by the activity of different $Ca^{2+}$ pumps and exchangers and cytosolic $Ca^{2+}$ is maintained at very low concentration, ranging 50–200 nM (see Cui et al. 2009b and references therein). In fact, there are two major mechanisms in this process of calcium sequestration (Fig. 7.1). Calcium can be either stored in the vacuoles or in secretory compartments of ER and Golgi. Most researchers identify the vacuole as the main organelle for storage and sink, and the compartmentalization of $Ca^{2+}$ in the ER/Golgi revealed a pathway for depleting and recycling intracellular calcium (reviewed in Cui et al. 2009a).

For vacuolar storage of $Ca^{2+}$ two principal transporters have been identified; the P-type ATPase Pmc1 pump (Cunningham and Fink 1994a, b) and the calcium/hydrogen exchanger, Vcx1 (Cunningham and Fink 1996; Miseta et al. 1999).

Pmc1 is a 1173 amino acid calcium ATPase. Three highly conserved domains among fungal and higher eukaryote Pmc1 homologues, the E1-E2 ATPase (PF00122), the haloacid dehalogenase-like hydrolase motif (PF00702) and the C-terminal ATPase domain (PF00689). Notably, these motifs are shared with other P-type ATPases, such as the $Na^+/Li^+$ pumps Ena (Ena1, 2 and 5), the plasma membrane proton pump ($H^+$-ATPase), the proton-potassium pump ($H^+,K^+$-ATPase), and the calcium ATPase PMR1.

Since absence of Pmc1 activity is required for tolerance to elevated extracellular calcium levels, it is not surprising that expression of *PMC1* is up-regulated when intracellular calcium levels elevate (Marchi et al. 1999). Loss of Pmc1 function reduces the amount of non-exchangeable calcium in vacuole, however inactivation of calcineurin restores calcium sequestration in the vacuole of a null *pmc1* mutant (Cunningham and Fink 1994b). The explanation for this phenotype is that tolerance of null *pmc1* mutants to calcium is dependent on the activity of Vcx1 which, on the other hand, it is dependent on calcineurin activity (Cunningham and Fink 1996).

Vcx1, named as vacuolar $H^+/Ca^{2+}$ exchanger (Cunningham and Fink 1996), is also known as HUM1, for "high copy number undoes manganese". Mutations in *HUM1* were identified because conferred sensitivity to $Mn^{2+}$ and this phenotype was exacerbated in hypofunctional calcineurin mutants (Pozos et al. 1996). Vcx1 shares similarities to other cation exchangers, among them are those involved in the antiport of $Na^+$, $K^+$ with $H^+$ (Nhx1 and Vnx1) located the vacuole or prevacuolar compartment/vesicles (Nhx1) (Cagnac et al. 2010). Despite the similarity of Vnx1p to other members of the CAX (calcium exchanger) family of transporters, Vnx1p is unable to mediate $Ca^{2+}$ transport but is a low affinity $Na^+/H^+$ and $K^+/H^+$ antiporter (Cagnac et al. 2007).

Biochemical data using purified vacuoles and vacuole membrane vesicles have evidenced that $Ca^{2+}$ transport activity of Pmc1 and Vcx1, more dramatically for the latter, depends on the pH gradient. Calcium uptake is promoted when the interior of these compartments is acid and lost when alkaline. This optimal acidification of vacuoles is maintained by the vacuolar $H^+$ V-ATPase activity (Dunn et al. 1994) and mutations in subunits of this ATPase strongly reduced tolerance to calcium (see below).

In 1994 it was discovered that the main reservoirs of intracellular calcium were vacuoles and ER/Golgi (Fig. 7.1). Pmc1 in the vacuole and a second calcium P-type ATPAse at the ER/Golgi, Pmr1, were essential for viability of *S. cerevisiae* (Cunningham and Fink 1994a). Pmr1 is a high affinity calcium pump. Initially, it was located to the vacuoles but subcellular fractionation studies located to ER and Golgi vesicles (Cunningham and Fink 1994a; Sorin et al. 1997; Strayle et al. 1999). In fact, Pmr1 is a P-type $Ca^{2+}/Mn^{2+}$ ATPase (Antebi and Fink 1992; Rudolph et al. 1989) and the transport of either cation can be specifically modified by affecting distinct amino acids, D778A and Q783A, both locating in transmembrane segment M6 (Mandal et al. 2000). Important for the Pmr1 functionality is the presence of a EF hand like motif at the N-terminal region of this pump (Wei et al. 1999). Mutations in this domain change the affinity of the protein for $Ca^{2+}$, $Mn^{2+}$, or both.

Pmr1 activity is essential for growth of a double *pmc1 vcx1* mutant. The activity of both P-type $Ca^{2+}$ ATPases Pmr1 and Pmc1, and the antiporter Vcx1 are key in maintaining the low cytosolic free calcium concentration which is needed to avoid inappropriate calcineurin activation and other effects due to the presence of this cation (Cunningham and Fink 1994a). Deletion of *PMR1* causes the elevation of cytoplasmic levels of free calcium and a massive accumulation within vacuoles (Halachmi and Eilam 1996). In fact, a deletion of *PMC1* and *PMR1* is a lethal genetic combination, leading to elevation of calcium levels, hyper-activation of calcineurin, and subsequent inactivation of Vcx1 antiporter. Transport activity of Vcx1 is inhibited by calcineurin, possibly in a post-translational mechanism (Cunningham and Fink 1996). Since Rcn2 is a negative regulator of calcineurin (see below), it participates in the regulation of Vcx1 (Kingsbury and Cunningham 2000), which in turn involves the activity of Crz1, that modulates transcription of this calcineurin regulator (Mehta et al. 2009). Expression of Pmc1 is dependent on the transcription factor Crz1 (Matheos et al. 1997; Stathopoulos and Cyert 1997), but also activity of Pmc1 is regulated. In a screen for negative regulators of Pmc1, the Nyv1 protein, a vacuolar v-SNARE, was found to inhibit the calcium transport activity of Pmc1 in the vacuole membrane, without affecting its expression levels (Takita et al. 2001).

In contrast to Vcx1 and Pmc1, Pmr1p activity is not postranscriptionally modulated by calcium binding proteins. Pmr1 apparently lacks the calmodulin binding domain at the C-terminus present in other plasma membrane $Ca^{2+}$ ATPases (PMCAs). However, importance of transport of calcium and manganese to ER extends beyond being an storage for these cations and different studies evidence the role of Pmr1 in the proper functioning of ER, as for the normal secretion of proteins, protein maturation and/or degradation (Durr et al. 1998).

## 7.4 Efflux of Calcium from Internal Stores and Its Recycling

In addition to calcium entry in the cell, a second source for this cation is the return to cytoplasm from the internal stores. Only part of this intracellular calcium, about 10 % of the total, can be released back to the cytoplasm and it is designated as the "exchangeable" pool of calcium (reviewed in Cunningham and Fink 1994a).

### *7.4.1 Release of Calcium from Vacuoles*

In vacuoles calcium is present in two forms: a free and a non-usable pool. The latter is designated as "non-exchangeable" since calcium is associated to polyphosphates. The concentration of free calcium in vacuoles is at the micromolar range (30 μM) meanwhile the total calcium was estimated at the millimolar range (2 mM) (Dunn et al. 1994). However, this reduced free-calcium pool can be returned to the cytoplasm when required. In the yeast vacuole it was identified a ion channel responsible for efflux of vacuolar calcium to the cytoplasm. This activity was designated as the yeast vacuolar conductance and the *YVC1* gene (yeast vacuolar channel 1) was identified (Palmer et al. 2001). Yvc1 is a 675 amino acid protein containing six transmembrane domains and is solely detected in vacuolar membranes. This specific localization in *S. cerevisiae* contrasts with that shown by members of the same family of transporters in higher eukaryotes which locate at the plasma membrane (reviewed in Cunningham 2005). Yvc1, is a calcium-activated cation channel of the transient receptor protein family (Denis and Cyert 2002; Palmer et al. 2001). When yeast cells need the activity of this mechano-sensitive $Ca^{2+}$ channel? In the case of a severe hypertonic shock, the vacuolar free $Ca^{2+}$ can be released into the cytosol through the activity of the Yvc1. This calcium release serves to stimulate the calmodulin/calcineurin network and activates Crz1, thus enabling the transcriptional response to this stress.

### 7.4.2 *Calcium Release from ER and Golgi; the Exocytic Pathway*

The absence genes coding for inositol tri-phosphate (IP3) receptors and ryanodine receptors, RyR, in the yeast genome, and in general all fungal genomes (reviewed in Cunningham 2005), clearly indicates that the role of ER and Golgi compartments in calcium homeostasis is different in fungal cells. Since calcium is stored in the ER and Golgi by the activity of PMR1 ATPases, and release seems not to be immediate because other calcium transporters for this activity are unknown a mayor role of exocytosis is proposed to liberate this compartment of any excess in calcium. In this way, it has been measured that when the calcium concentrations in Golgi and ER exceed their resting levels of 300 μM (Pinton et al. 1998) and 10 μM (Aiello et al. 2002; Strayle et al. 1999), respectively, the calcium in ER and Golgi will be secreted along with the canonical secretory pathways. Most importantly, calcium and manganese, both transported by the Pmr1 ATPase, are necessary for the proper processing and trafficking of peptides and proteins through the secretory pathway. While $Mn^{2+}$ has a role in protein glycosylation, $Ca^{2+}$ is required for normal protein sorting (Durr et al. 1998). As for calcium, the excess of manganese is eliminated via Golgi and the secretory vesicles (Culotta et al. 2005). Hence, exocytosis is a major mechanism to largely reduce the intracellular, compartmentalized, pool of calcium, and other cations, but increasing the extracellular content of these ions, becoming available for a re-start of calcium transport and signaling.

## 7.5 Calcium Signaling

*S. cerevisiae* grows in a wide range of extracellular concentrations of calcium. Yeast cells are able to adapt to large and rapid modifications in environmental calcium, ranging from a low concentration of 1 μM to more than 100 mM (Anraku et al. 1991). Part of this adaptation process relies on the activation of a signaling cascade leading to the modification of the gene expression pattern (Fig. 7.2).

Upon elevation of extracellular levels of calcium a massive influx of this cation occurs in the yeast. For the free calcium now existing in the cytoplasm, one of the targets is the small and essential protein calmodulin, Cmd1 (reviewed in Cyert 2001; see Fig. 7.2a). Calmodulin has 4 EF-hand moieties (Davis et al. 1986). Each EF-hand is able to bind a $Ca^{2+}$ atom, but in the case of *S. cerevisiae* Cmd1 the fourth EF-hand is divergent and most likely is not able to bind calcium (Starovasnik

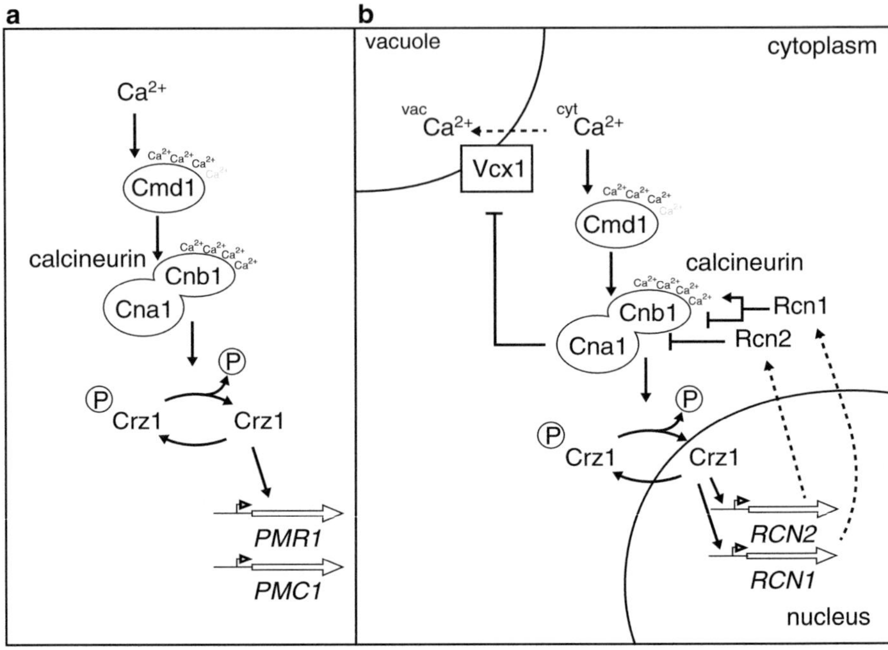


**Fig. 7.2** Calcium signaling cascade in *S. cerevisiae*. (**a**) the calmodulin/calcineurin cascade towards activation of the transcription factor Crz1. The protein phosphatase calcineurin dephosphorylates this transcription factor leading to its transcriptional activation. Among Crz1 targets are the genes *PMR1* and *PMC1,* coding for the P-type ATPases, and (**b**) calcineurin regulators *RCN2* and *RCN1*. Modulation of calcineurin activity by Rcn2 and Rcn1 alters signaling of Crz1 and the negative effect of this phosphatase on Vcx1 activity

et al. 1993). Calmodulin has many roles in the yeast cell which can be classified into calcium independent and dependent functions (Davis et al. 1986). Among the calcium-dependent roles is to activate the protein phosphatase 2B calcineurin (see Fig. 7.2a). In *S. cerevisiae* calcineurin is composed of a catalytic subunit, the A subunit encoded by either *CMP2* or *CNA1* isoforms (Cyert et al. 1991), and the regulatory, B, subunit encoded by *CNB1* (Cyert and Thorner 1992). Activity of Cnb1, and subsequently of the calcineurin, is modulated by the α-arrestin Aly1 (O'Donnell et al. 2013) and through myristoylation (Connolly and Kingsbury 2012) in response to a reduction in calcium signal.

The calcineurin phosphatase may have numerous targets. However, these are directly recognized through a short motif namely the calcineurin docking domain (Rodriguez et al. 2009; Roy et al. 2007). An important target of calcineurin is the Crz1 transcription factor (Stathopoulos and Cyert 1997). Crz1 is a three zinc-finger, two classical Cys2His2 fingers and a non-canonical CysCysHisCys finger, transcription factor that is inactivated and activated in a cyclic, pulsatile dynamic, phosphorylation/dephosphorylation process in the cell (Dalal et al. 2014). Crz1 mediates in yeast tolerance to high concentrations of different cations, alkalinity and other types of stresses (reviewed in Cyert 2003).

Calcineurin modulates the activity of Crz1, at least, regulating its nuclei-cytoplasmic trafficking (Stathopoulos-Gerontides et al. 1999). Crz1 is mainly cytoplasmic when in a phosphorylated state. A nuclear export signal (NES) becomes activated and recognized by the exportin Msn5p (Boustany and Cyert 2002). After calcineurin dependent dephosphorylation, in response to an elevation of the cytoplasmic concentration of calcium, the nuclear import signal is dephosphorylated as is the NES. A nuclear transporter Nmd5 (karyopherin Kap119) recognizes and translocates this dephosphorylated form of Crz1 in nucleus following a general nuclear transport process through the pores located at the nuclear envelope (Polizotto and Cyert 2001). A recent work has shown that Crz1 may enter nucleus in a stochastic mode. This is a response to a brief calcium dependent stimulus that, in a non coordinated mode among cells, promote temporal accumulation of Crz1 and other transcription factors into the nucleus (Dalal et al. 2014). This mechanism is proposed as the basis to provide a fast response to ambient stress signals. However, following our case of an elevation of cytosolic calcium levels, this network of calcium binding proteins and de-phosphorylation process activates the nuclear entry of Crz1 in all cells in a culture.

In the nucleus Crz1 will bind to precise DNA sequences known as CDREs, calcineurin-dependent regulatory elements, present in the promoter of genes under its regulation (Mendizabal et al. 2001; Stathopoulos and Cyert 1997). Among these genes are PMR1 and PMC1 (for review, see Cyert 2001) and Crz1 has a positive role on their expression levels (Fig. 7.2a). Thus, Crz1 is responsible for a rise in the levels of Pmr1 and Pmc1 pumps that would locate at the ER/Golgi and vacuoles, respectively. This effect was early noted by (Beeler et al. 1994; Dunn et al. 1994) when found that a rise in the cytosolic calcium concentration increased the non-exchangeable pool of $Ca^{2+}$. Mutants lacking *PMC1* grow poorly in calcium stress conditions, although growth can be restored by overexpression of *PMR1* or *VCX1*. This reflects the importance of calcium sequestration in tolerance to elevated concentrations of extracellular of calcium (Cunningham 2005; Cunningham and Fink 1994b). Therefore, the activity and an adequate level of expression of these pumps have a direct effect on depleting cytosol of calcium causing the attenuation of calcineurin-dependent signaling.

Crz1 activity also influences calcineurin function. Expression of calcineurin regulators *RCN1* (regulator of calcineurin) and *RCN2* are positively modulated by Crz1 (Fig. 7.2b). Rcn2 is a negative regulator but Rcn1 could act as a positive and negative modulator of calcineurin (Kingsbury and Cunningham 2000; Mehta et al. 2009). Modulation of the phosphorylation levels of Rcn proteins by homologues the glycogen synthase 3 kinase, Gsk3, is the mechanism to activate or inactivate these calcineurin regulators (Hilioti et al. 2004).

A reduction of calcineurin activity has two major consequences in calcium homeostasis. Firstly, the negative effect on Vcx1 activity is reduced and activity of this antiporter is restored, with an immediate consequence in lowering calcium levels in the cytoplasm (Fig. 7.2b). Secondly, a low calcineurin activity will consequently reduce the dephosphorylation process of Crz1, and the kinase activities acting on this TF will restore the pool of inactive Crz1 in the cell. *HRR25* encodes for a casein kinase I with multiple roles in the cell (Hoekstra et al. 1991; Mehlgarten and Schaffrath 2003), among them Hrr25 opposes to the activity of calcineurin on the posttranslational modification of Crz1 (Kafadar et al. 2003). Other kinases (see below) may act on restoring phosphorylated levels of Crz1, rendering a cytoplasmic and, thus, inactive form of this TF.

In summary, calcium homeostasis relies on a meticulous regulatory system that senses, transduces and transcriptionally responds to provide with tools (transporters) to "clean" the excess of calcium, and with regulators to modulate and attenuate signal transduction. Survival of cells to calcium stress not only depends on these specific elements but other players are accessory in some ambient conditions and mutant backgrounds of need to sustain yeast's health.

## 7.6 Other Protein Activities Involved in Calcium Homeostasis

In 1986, Ohya and collaborators published their work on characterizing recessive mutations affecting growth in the presence of high calcium concentration in the media. Eighteen genes, designated as *cls* (calcium sensitive mutants) were identified and many of them showed to be necessary for maintaining the structure of function of the vacuole (Ohya et al. 1986). Among these are genes coding for subunits of the vacuolar proton ATPase. *cls* mutations were isolated in different subunits of the heterocomplex *VMA*, the vacuolar proton ATPase, constituting the type IV *cls* mutants. Among these are Vma3, Vma1, Vma11, Vma13 as parts of the ATPase, and Vma12 the assembly factor (Ohya et al. 1986; Tanida et al. 1996 and references therein). The role of Vma complex is to provide the correct amount of protons at the vacuole to allow exchange with $Ca^{2+}$ via Vcx1 activity. In this screen it was found mutations in other two genes of interest, *csg1* and *csg2*. CSG2 encodes a endoplasmic reticulum membrane protein; required for mannosylation of inositolphosphorylceramide and for growth at high calcium concentrations; protein abundance increases in response to DNA replication stress (Tanida et al. 1996). Csg2 protein is a transmembrane protein, located in the ER. Biochemical and functional studies indicated that Cls2/Csg2 is necessary for mobilization of non-exchangeable pool of calcium distinct from that of the vacuole and plays an important role in calcium tolerance in the yeast, and would cooperate in stimulating the activity of calcineurin (Tanida et al. 1996). It is interesting to note that Csg2 could fulfill the role of ER-calcium efflux transporters from higher eukaryotes (IP3 receptors and RyR), however this might be an specific mechanism of *S. cerevisiae* since orthologues for Csg2 are not found in other fungi. CSG1 (also known as SUR1, suppressor of *rvs161* and *rvs167* mutations, (Beeler et al. 1997)) encodes for a mannosylinositol phosphorylceramide synthase catalytic subunit, and forms a complex with regulatory subunit Csg2 at the ER (Desfarges et al. 1993). Notably, the study of membrane composition, specially the synthesis of sphingolipids and related molecules has discovered a novel role in calcium homeostasis via influx of calcium from the ER to the cytoplasm (Birchwood et al. 2001; Desfarges et al. 1993; Dickson and Lester 2002). Also related with the composition of membranes and specifically in aminophospholipids organisation are the type IV of P-type ATPases, constituted by aminophospholipid translocases APTs or flippases. Of importance in maintaining calcium homeostasis are Drs2 (Ripmaster et al. 1993) and Neo1 (Prezant et al. 1996). These flippases are located in late Golgi and are important for proper constitution of plasma membrane and exocytosis. Also the P-type ATPase, type V, is *SPF1*(YEL031), located to the ER and required for calcium homeostasis. Absence of both Spf1 and Pmr1 function greatly elevate cytoplasmic calcium levels (Cronin et al. 2002). Reinforcing the importance of ER, Golgi and vacuoles in the overall calcium managing in the yeast cell is the participation of proteins located in these compartments, for example, Gdt1, is a transmembrane protein involved in calcium and pH homeostasis in yeast and higher eukaryotes. It localizes to the cis- and medial-Golgi apparatus, but the GFP-fusion protein localizes to the vacuole. The exact role of this protein is unknown, but possibly related to glycosylation since deficiency in its human homologue TMEM165, a human gene which causes congenital disorders of glycosylation (Demaegd et al. 2013). At the vacuole also locate Ccc1 (Cross-Complements $Ca^{2+}$ phenotype of *csg1*), this is a vacuolar $Fe^{2+}/Mn^{2+}$ transporter that also may participate in the respiration process (Fu et al. 1994; Lapinskas et al. 1996), and ECM27 and YDL206W genes coding for two members of the CAX (cation exchangers) family of vacuolar transporters, closely related to Vcx1 and Vnx1,

Finally, it is worth to mention the only mutation found causing the need of large amounts of extracellular calcium for survival, the *cal1-1* mutation (Ohya et al. 1984). This mutation locates in CDC43 gene, that encodes for a β-subunit of geranylgeranyltransferase type I, which catalyzes geranylgeranylation to the cysteine residue in proteins containing a C-terminal CaaX sequence ending in Leu or Phe. In addition to this novel role in calcium homeostasis, the substrates under its regulation are important for morphogenesis (Adams et al. 1990).

As for the latter, the precise role in calcium homeostasis of many of these proteins remains obscure and needs clarification. Some of them open new venues for understanding how calcium is regulated and associated factors and mechanisms that modulate calcium storage and managing. However proteins are not the only new regulatory elements found, but other cations and macromolecules are known to participate in calcium homeostasis.

## 7.7 The Role of Magnesium in Regulating the Calcium Response

Cui and collaborators (2009a) discovered that the calcium sensitivity displayed by a *pmc1* mutant was dependent not only on the concentration of extracellular calcium but on the composition of medium. Further analyses revealed that this suppressing factor was the presence of variable concentrations of magnesium in media. In this way, increasing concentrations of $Mg^{2+}$ elevated the $IC_{50}$ to calcium toxicity of a *pmc1* mutant but also modified sensitivity to calcium caused by the combination of double or triple *pmr1*, *vcx1* and/or calcineurin (*cnb1*) mutations.

Furthermore, this study indicated the existence of a $Mg^{2+}$-inhibited calcium transport system located at the PM. Calcium toxicity displayed by a *yvc1 cch1 pmc1 vcx1* quadruple mutant was suppressed by addition of magnesium to the medium. In fact, a mathematical modeling of calcium transport across the PM predicted the presence of at least two transport systems, in addition to that integrated by Mid1 and Cch1, in the plasma membrane. Only in this way it could be accommodated the experimental data of calcium sensitivities in a variety of single and double mutants involving pumps and channels, and measurements intracellular levels of calcium. Magnesium will have a major role in regulating, at least one of these calcium transport systems. These authors conclude that transporter M is regulated by $Mg^{2+}$ and transporter X respond to a hypertonic calcium shock (see also Fig. 7.1).

One of the most abundant divalent cations in cells is magnesium. $Mg^{2+}$ participates acting a counterion in stabilizing many macromolecules such as RNA and DNA or single nucleotides (i.e. ATP). It also mediates in important catalytic processes and in stabilizing large molecules or membranes (see Wiesenberger et al. 2007 and references there in). Cellular concentrations of $Mg^{2+}$ are in the millimolar range (from 15 to 20 mM), about three orders of magnitude lower than those of $Ca^{2+}$ (100–200 nM). In mammals, entry of $Mg^{2+}$ in the cell is an electrogenic process requiring a negative charge at the inner side of the PM and the activity of two transporters (TRPM6 and TRPM7) (Schlingmann and Gudermann 2005; Schmitz et al. 2003), or by members of a heterogeneous protein family, designated as CorA homologues, and found in lower and high eukaryotes (plants and animals), allowing grow or development even in the presence of very low concentrations of magnesium. In yeast, Mrs2 and Alr1 are orthologues of CorA.

$Mg^{2+}$ enters the cell through the activity of two members of the CorA family of transporters, Alr1 and Alr2 (Graschopf et al. 2001; Macdiarmid and Gardner 1998). These are the metal ion transporter superfamily, MIT. Alr1 and Alr2 may form oligomeric transporter at the plasma membrane constituting a high affinity $Mg^{2+}$ uptake system (Wachek et al. 2006). $Mg^{2+}$ can be stored in two subcellular compartments, the mitochondria, vacuole and ER/Golgi. For the first store the activity of two transporters located in the inner membrane of mitochondria are needed, Mrs2 and Lpe10. $Mg^{2+}$ is stored in the trans-golgi, and possibly in the vacuole by the activity of a $Mg^{2+}/H^{+}$ antiporters (Pisat et al. 2009); (Borrelly et al. 2001). Storage of magnesium in the vacuole requires the activity of the vacuolar proton-ATPase, and for $Mg^{2+}$ efflux it is required the Mrs2 transporter, also belonging to the MIT superfamily (Pisat et al. 2009).

Important for understanding calcium homeostasis were the results of a RNA profile analysis of changes in expression dependent on $Mg^{2+}$ depletion. These showed the intimate relationship between magnesium and calcium homeostasis (Wiesenberger et al. 2007). A reduction in external $Mg^{2+}$ upregulated the

ENA1, encoding the P-type ATPase sodium pump, and PHO89, encoding a sodium/phosphate cotransporter, which are also upregulated under calcium and alkaline pH stress. In that work they demonstrated that $Mg^{2+}$ starvation caused an increase in cytoplasmic calcium. A rise in cytosolic free calcium activated the calmodulin/calcineurin network, which led to the activation of Crz1 TF. ENA1 and PHO89 are among the genes regulated by Crz1 (Hu et al. 2007; Mendizabal et al. 2001).

How $Mg^{2+}$ also influences calcium homeostasis? Immediately from what has been exposed before, a negative role in calcium efflux through the PM is expected, causing a reduction in the intracellular pool of $Ca^{2+}$. To this effect, an elevation of intracellular $Mg^{2+}$ will reduce the release of calcium from the internal stores. But a third effect is also predictable, based on the capacity of EF-hands to bind $Mg^{2+}$ in addition to $Ca^{2+}$ (see review Grabarek 2011). Calmodulin and other EF-hand containing proteins, such as the regulatory subunit of calcineurin, may bind $Mg^{2+}$ rendering an alternative conformation to that originated by calcium, in fact magnesium helps to release calcium from these proteins allowing the pass from the holo-enzyme to the apo-enzyme state (Grabarek 2011). Thus, $Mg^{2+}$ will directly affect the functionality of the calmodulin/calcineurin network, attenuating calcium signaling. Altogether, an excess of $Mg^{2+}$ will cause a reduction on calcium signaling and probably increase tolerance to this cation, meanwhile low extracellular levels of $Mg^{2+}$ increase calcium signaling (Wiesenberger et al. 2007).

## 7.8 Role of Inorganic Phosphate in Calcium Homeostasis

The property of inorganic phosphate or polyphosphates in chelating cations is fundamental to understand calcium and magnesium homeostasis. Actually calcium and phosphate homeostasis have major importance in vertebrates since it is crucial in bone formation, among other cellular and tissue specific processes (Shaker and Deftos 2000).

In the yeast, the presence of polyphosphates in vacuoles is the basis for the immobilization of calcium. In this way, the capacity to maintain the vacuolar $Ca^{2+}$ concentration up to 2 mM is a result of $Ca^{2+}$ binding to vacuolar polyphosphate (Dunn et al. 1994). However, $Ca^{2+}$ can be completely released from isolated vacuoles or from whole cells using the ionophores A23187 or ionomycin, suggesting that the non-exchangeable pool of $Ca^{2+}$ is soluble (Cunningham and Fink 1994a).

The conserved pathway governing phosphate homeostasis in yeast is composed of the PHO genes (reviewed in Tomar and Sinha 2014). Interestingly there is well established interconnection between calcium- and phosphate-dependent regulation, for example at the level of the Crz1 transcription factor. As cited before, Crz1 up-regulates *PHO89* in response to calcium stress or alkalinization, and, in the absence of calcium stress, among the cyclin-dependent kinase Pho85 targets is the transcription factor Crz1p (Sopko et al. 2006).

## 7.9 Concluding Remarks and Future Prospects

Numerous elements have been described in this chapter participating in transport, signaling and storage of a principal messenger in cells, the calcium ion. Along years, *S. cerevisiae* has served as a model to understand how cells deal with a messenger that is essential but lethal at the same time. Most of the principal elements are well known, and many functional analyses have been performed to understand the molecular and biochemical mechanisms involved, e.g. the calmodulin/calcineurin pathway and its fungal effector Crz1. *S. cerevisiae* has served as a model to study calcium homeostasis in other fungi, and in some cases particular variations are found, specially in how Crz homologues are signalized and the transcriptional function of this TF. But a general mechanism underlies in almost all cellular systems under study, from bacteria to human cells, and for that the budding yeast has largely provided with basic and elemental findings to understand this complex homeostatic system. Future research lines will provide with a more detailed view of the exocytic process of calcium and the interrelationships among different ions and ambient stress signals to modulate and generate a coordinated response.

**Acknowledgements** Work at the laboratory of Dr. Espeso at the Centro de Investigaciones Biológicas (CIB), from the Spanish Research Council, CSIC, is supported by the Spanish Ministerio de Economía y Competitividad through grant BFU2012-33142.

## Bibliography

Adams AE, Johnson DI, Longnecker RM, Sloat BF, Pringle JR (1990) CDC42 and CDC43, two additional genes involved in budding and the establishment of cell polarity in the yeast *Saccharomyces cerevisiae*. J Cell Biol 111:131–142

Aiello DP, Fu L, Miseta A, Bedwell DM (2002) Intracellular glucose 1-phosphate and glucose 6-phosphate levels modulate Ca2+ homeostasis in *Saccharomyces cerevisiae*. J Biol Chem 277:45751–45758

Anraku Y, Ohya Y, Iida H (1991) Cell cycle control by calcium and calmodulin in *Saccharomyces cerevisiae*. Biochim Biophys Acta 1093:169–177

Antebi A, Fink GR (1992) The yeast Ca(2+)-ATPase homologue, PMR1, is required for normal Golgi function and localizes in a novel Golgi-like distribution. Mol Biol Cell 3:633–654

Beeler T, Gable K, Zhao C, Dunn T (1994) A novel protein, CSG2p, is required for Ca2+ regulation in *Saccharomyces cerevisiae*. J Biol Chem 269:7279–7284

Beeler TJ, Fu D, Rivera J, Monaghan E, Gable K, Dunn TM (1997) SUR1 (CSG1/BCL21), a gene necessary for growth of *Saccharomyces cerevisiae* in the presence of high Ca2+ concentrations at 37 degrees C, is required for mannosylation of inositolphosphorylceramide. Mol Gen Genet 255:570–579

Birchwood CJ, Saba JD, Dickson RC, Cunningham KW (2001) Calcium influx and signaling in yeast stimulated by intracellular sphingosine 1-phosphate accumulation. J Biol Chem 276:11712–11718

Bonilla M, Cunningham KW (2003) Mitogen-activated protein kinase stimulation of Ca($^{2+}$) signaling is required for survival of endoplasmic reticulum stress in yeast. Mol Biol Cell 14:4296–4305

Bonilla M, Nastase KK, Cunningham KW (2002) Essential role of calcineurin in response to endoplasmic reticulum stress. EMBO J 21:2343–2353

Borrelly G, Boyer JC, Touraine B, Szponarski W, Rambier M, Gibrat R (2001) The yeast mutant vps5Delta affected in the recycling of Golgi membrane proteins displays an enhanced vacuolar Mg2+/H+ exchange activity. Proc Natl Acad Sci U S A 98:9660–9665

Boustany LM, Cyert MS (2002) Calcineurin-dependent regulation of Crz1p nuclear export requires Msn5p and a conserved calcineurin docking site. Genes Dev 16:608–619

Cagnac O, Leterrier M, Yeager M, Blumwald E (2007) Identification and characterization of Vnx1p, a novel type of vacuolar monovalent cation/H+ antiporter of *Saccharomyces cerevisiae*. J Biol Chem 282:24284–24293

Cagnac O, Aranda-Sicilia MN, Leterrier M, Rodriguez-Rosales MP, Venema K (2010) Vacuolar cation/H+ antiporters of *Saccharomyces cerevisiae*. J Biol Chem 285:33914–33922

Catterall WA (2000) Structure and regulation of voltage-gated $Ca^{2+}$ channels. Annu Rev Cell Dev Biol 16:521–555

Connolly S, Kingsbury T (2012) Regulatory subunit myristoylation antagonizes calcineurin phosphatase activation in yeast. J Biol Chem 287:39361–39368

Courchesne WE, Ozturk S (2003) Amiodarone induces a caffeine-inhibited, MID1-dependent rise in free cytoplasmic calcium in *Saccharomyces cerevisiae*. Mol Microbiol 47:223–234

Cronin SR, Rao R, Hampton RY (2002) Cod1p/Spf1p is a P-type ATPase involved in ER function and Ca2+ homeostasis. J Cell Biol 157:1017–1028

Cui J, Kaandorp JA (2006) Mathematical modeling of calcium homeostasis in yeast cells. Cell Calcium 39:337–348

Cui J, Kaandorp JA, Ositelu OO, Beaudry V, Knight A, Nanfack YF, Cunningham KW (2009a) Simulating calcium influx and free calcium concentrations in yeast. Cell Calcium 45:123–132

Cui J, Kaandorp JA, Sloot PM, Lloyd CM, Filatov MV (2009b) Calcium homeostasis and signaling in yeast cells and cardiac myocytes. FEMS Yeast Res 9:1137–1147

Culotta VC, Yang M, Hall MD (2005) Manganese transport and trafficking: lessons learned from *Saccharomyces cerevisiae*. Eukaryot Cell 4:1159–1165

Cunningham KW (2005) Calcium signaling networks in yeast. In: Putney JW (ed) Calcium signaling, 2nd edn. Taylor & Francis Group/CRC Press, Florida, pp 107–201

Cunningham KW, Fink GR (1994a) $Ca^{2+}$ transport in *Saccharomyces cerevisiae*. J Exp Biol 196:157–166

Cunningham KW, Fink GR (1994b) Calcineurin-dependent growth control in *Saccharomyces cerevisiae* mutants lacking *PMC1*, a homolog of plasma membrane Ca2+ ATPases. J Cell Biol 124:351–363

Cunningham KW, Fink GR (1996) Calcineurin inhibits VCX1-dependent H+/Ca2+ exchange and induces Ca2+ ATPases in *Saccharomyces cerevisiae*. Mol Cell Biol 16:2226–2237

Cyert MS (2001) Genetic analysis of calmodulin and its targets in *Saccharomyces cerevisiae*. Annu Rev Genet 35:647–672

Cyert MS (2003) Calcineurin signaling in *Saccharomyces cerevisiae*: how yeast go crazy in response to stress. Biochem Biophys Res Commun 311:1143–1150

Cyert MS, Thorner J (1992) Regulatory subunit (CNB1 gene product) of yeast Ca2+/calmodulin-dependent phosphoprotein phosphatases is required for adaptation to pheromone. Mol Cell Biol 12:3460–3469

Cyert MS, Kunisawa R, Kaim D, Thorner J (1991) Yeast has homologs (*CNA1* and *CNA2* gene products) of mammalian calcineurin, a calmodulin-regulated phosphoprotein phosphatase. Proc Natl Acad Sci U S A 88:7376–7380

Dalal CK, Cai L, Lin Y, Rahbar K, Elowitz MB (2014) Pulsatile dynamics in the yeast proteome. Curr Biol 24:2189–2194

Davis TN, Urdea MS, Masiarz FR, Thorner J (1986) Isolation of the yeast calmodulin gene: calmodulin is an essential protein. Cell 47:423–431

Demaegd D, Foulquier F, Colinet AS, Gremillon L, Legrand D, Mariot P, Peiter E, Van SE, Matthijs G, Morsomme P (2013) Newly characterized Golgi-localized family of proteins is involved in calcium and pH homeostasis in yeast and human cells. Proc Natl Acad Sci U S A 110:6859–6864

Denis V, Cyert MS (2002) Internal Ca(2+) release in yeast is triggered by hypertonic shock and mediated by a TRP channel homologue. J Cell Biol 156:29–34

Desfarges L, Durrens P, Juguelin H, Cassagne C, Bonneu M, Aigle M (1993) Yeast mutants affected in viability upon starvation have a modified phospholipid composition. Yeast 9:267–277

Dickson RC, Lester RL (2002) Sphingolipid functions in *Saccharomyces cerevisiae*. Biochim Biophys Acta 1583:13–25

Dunn T, Gable K, Beeler T (1994) Regulation of cellular Ca2+ by yeast vacuoles. J Biol Chem 269:7273–7278

Durr G, Strayle J, Plemper R, Elbs S, Klee SK, Catty P, Wolf DH, Rudolph HK (1998) The medial-Golgi ion pump Pmr1 supplies the yeast secretory pathway with Ca2+ and Mn2+ required for glycosylation, sorting, and endoplasmic reticulum-associated protein degradation. Mol Biol Cell 9:1149–1162

Fu D, Beeler T, Dunn T (1994) Sequence, mapping and disruption of CCC1, a gene that cross-complements the Ca(2+)-sensitive phenotype of csg1 mutants. Yeast 10:515–521

Grabarek Z (2011) Insights into modulation of calcium signaling by magnesium in calmodulin, troponin C and related EF-hand proteins. Biochim Biophys Acta 1813:913–921

Graschopf A, Stadler JA, Hoellerer MK, Eder S, Sieghardt M, Kohlwein SD, Schweyen RJ (2001) The yeast plasma membrane protein Alr1 controls Mg2+ homeostasis and is subject to Mg2+-dependent control of its synthesis and degradation. J Biol Chem 276:16216–16222

Halachmi D, Eilam Y (1996) Elevated cytosolic free Ca2+ concentrations and massive Ca2+ accumulation within vacuoles, in yeast mutant lacking PMR1, a homolog of Ca2+-ATPase. FEBS Lett 392:194–200

Hilioti Z, Gallagher DA, Low-Nam ST, Ramaswamy P, Gajer P, Kingsbury TJ, Birchwood CJ, Levchenko A, Cunningham KW (2004) GSK-3 kinases enhance calcineurin signaling by phosphorylation of RCNs. Genes Dev 18:35–47

Hoekstra MF, Liskay RM, Ou AC, DeMaggio AJ, Burbee DG, Heffron F (1991) HRR25, a putative protein kinase from budding yeast: association with repair of damaged DNA. Science 253:1031–1034

Hu Z, Killion PJ, Iyer VR (2007) Genetic reconstruction of a functional transcriptional regulatory network. Nat Genet 39:683–687

Iida H, Yagawa Y, Anraku Y (1990) Essential role for induced $Ca^{2+}$ influx followed by $[Ca^{2+}]i$ rise in maintaining viability of yeast cells late in the mating pheromone response pathway. A study of $[Ca^{2+}]i$ in single *Saccharomyces cerevisiae* cells with imaging of fura-2. J Biol Chem 265:13391–13399

Iida H, Nakamura H, Ono T, Okumura MS, Anraku Y (1994) MID1, a novel *Saccharomyces cerevisiae* gene encoding a plasma membrane protein, is required for $Ca^{2+}$ influx and mating. Mol Cell Biol 14:8259–8271

Kafadar KA, Zhu H, Snyder M, Cyert MS (2003) Negative regulation of calcineurin signaling by Hrr25p, a yeast homolog of casein kinase I. Genes Dev 17:2698–2708

Kanzaki M, Nagasawa M, Kojima I, Sato C, Naruse K, Sokabe M, Iida H (1999) Molecular identification of a eukaryotic, stretch-activated nonselective cation channel. Science 285:882–886

Kingsbury TJ, Cunningham KW (2000) A conserved family of calcineurin regulators. Genes Dev 14:1595–1604

Lapinskas PJ, Lin SJ, Culotta VC (1996) The role of the Saccharomyces cerevisiae CCC1 gene in the homeostasis of manganese ions. Mol Microbiol 21:519–528

Locke EG, Bonilla M, Liang L, Takita Y, Cunningham KW (2000) A homolog of voltage-gated Ca($^{2+}$) channels stimulated by depletion of secretory Ca($^{2+}$) in yeast. Mol Cell Biol 20:6686–6694

Macdiarmid CW, Gardner RC (1998) Overexpression of the Saccharomyces cerevisiae magnesium transport system confers resistance to aluminum ion. J Biol Chem 273:1727–1732

Mandal D, Woolf TB, Rao R (2000) Manganese selectivity of pmr1, the yeast secretory pathway ion pump, is defined by residue gln783 in transmembrane segment 6. Residue Asp778 is essential for cation transport. J Biol Chem 275:23933–23938

Marchi V, Sorin A, Wei Y, Rao R (1999) Induction of vacuolar Ca2+-ATPase and H+/Ca2+ exchange activity in yeast mutants lacking Pmr1, the Golgi Ca2+-ATPase. FEBS Lett 454:181–186

Matheos DP, Kingsbury TJ, Ahsan US, Cunningham KW (1997) Tcn1p/Crz1p, a calcineurin-dependent transcription factor that differentially regulates gene expression in Saccharomyces cerevisiae. Genes Dev 11:3445–3458

Matsumoto TK, Ellsmore AJ, Cessna SG, Low PS, Pardo JM, Bressan RA, Hasegawa PM (2002) An osmotically induced cytosolic Ca2+ transient activates calcineurin signaling to mediate ion homeostasis and salt tolerance of Saccharomyces cerevisiae. J Biol Chem 277:33075–33080

Mehlgarten C, Schaffrath R (2003) Mutant casein kinase I (Hrr25p/Kti14p) abrogates the G1 cell cycle arrest induced by *Kluyveromyces lactiszymocin* in budding yeast. Mol Genet Genomics 269:188–196

Mehta S, Li H, Hogan PG, Cunningham KW (2009) Domain architecture of the regulators of calcineurin (RCANs) and identification of a divergent RCAN in yeast. Mol Cell Biol 29:2777–2793

Mendizabal I, Pascual-Ahuir A, Serrano R, de Larrinoa IF (2001) Promoter sequences regulated by the calcineurin-activated transcription factor Crz1 in the yeast *ENA1* gene. Mol Genet Genomics 265:801–811

Miseta A, Kellermayer R, Aiello DP, Fu L, Bedwell DM (1999) The vacuolar Ca2+/H+ exchanger Vcx1p/Hum1p tightly controls cytosolic Ca2+ levels in *S. cerevisiae*. FEBS Lett 451:132–136

O'Donnell AF, Huang L, Thorner J, Cyert MS (2013) A calcineurin-dependent switch controls the trafficking function of alpha-arrestin Aly1/Art6. J Biol Chem 288:24063–24080

Ohya Y, Ohsumi Y, Anraku Y (1984) Genetic study of the role of calcium ions in the cell division cycle of Saccharomyces cerevisiae: a calcium-dependent mutant and its trifluoperazine-dependent pseudorevertants. Mol Gen Genet 193:389–394

Ohya Y, Ohsumi Y, Anraku Y (1986) Isolation and characterization of Ca2+-sensitive mutants of Saccharomyces cerevisiae. J Gen Microbiol 132:979–988

Ozeki-Miyawaki C, Moriya Y, Tatsumi H, Iida H, Sokabe M (2005) Identification of functional domains of Mid1, a stretch-activated channel component, necessary for localization to the plasma membrane and Ca2+ permeation. Exp Cell Res 311:84–95

Paidhungat M, Garrett S (1997) A homolog of mammalian, voltage-gated calcium channels mediates yeast pheromone-stimulated Ca2+ uptake and exacerbates the cdc1(Ts) growth defect. Mol Cell Biol 17:6339–6347

Palmer CP, Zhou XL, Lin J, Loukin SH, Kung C, Saimi Y (2001) A TRP homolog in *Saccharomyces cerevisiae* forms an intracellular Ca(2+)-permeable channel in the yeast vacuolar membrane. Proc Natl Acad Sci U S A 98:7801–7805

Peiter E, Fischer M, Sidaway K, Roberts SK, Sanders D (2005) The *Saccharomyces cerevisiae* Ca2+ channel Cch1pMid1p is essential for tolerance to cold stress and iron toxicity. FEBS Lett 579:5697–5703

Pinton P, Pozzan T, Rizzuto R (1998) The Golgi apparatus is an inositol 1,4,5-trisphosphate-sensitive Ca2+ store, with functional properties distinct from those of the endoplasmic reticulum. EMBO J 17:5298–5308

Pisat NP, Pandey A, Macdiarmid CW (2009) MNR2 regulates intracellular magnesium storage in *Saccharomyces cerevisiae*. Genetics 183:873–884

Polizotto RS, Cyert MS (2001) Calcineurin-dependent nuclear import of the transcription factor Crz1p requires Nmd5p. J Cell Biol 154:951–960

Pozos TC, Sekler I, Cyert MS (1996) The product of HUM1, a novel yeast gene, is required for vacuolar Ca2+/H+ exchange and is related to mammalian Na+/Ca2+ exchangers. Mol Cell Biol 16:3730–3741

Prezant TR, Chaltraw WE Jr, Fischel-Ghodsian N (1996) Identification of an overexpressed yeast gene which prevents aminoglycoside toxicity. Microbiology 142(Pt 12):3407–3414

Ripmaster TL, Vaughn GP, Woolford JL Jr (1993) DRS1 to DRS7, novel genes required for ribosome assembly and function in *Saccharomyces cerevisiae*. Mol Cell Biol 13:7901–7912

Rodriguez A, Roy J, Martinez-Martinez S, Lopez-Maderuelo MD, Nino-Moreno P, Orti L, Pantoja-Uceda D, Pineda-Lucena A, Cyert MS, Redondo JM (2009) A conserved docking surface on calcineurin mediates interaction with substrates and immunosuppressants. Mol Cell 33:616–626

Roy J, Li H, Hogan PG, Cyert MS (2007) A conserved docking site modulates substrate affinity for calcineurin, signaling output, and in vivo function. Mol Cell 25:889–901

Rudolph HK, Antebi A, Fink GR, Buckley CM, Dorman TE, LeVitre J, Davidow LS, Mao JI, Moir DT (1989) The yeast secretory pathway is perturbed by mutations in PMR1, a member of a Ca2+ ATPase family. Cell 58:133–145

Schlingmann KP, Gudermann T (2005) A critical role of TRPM channel-kinase for human magnesium transport. J Physiol 566:301–308

Schmitz C, Perraud AL, Johnson CO, Inabe K, Smith MK, Penner R, Kurosaki T, Fleig A, Scharenberg AM (2003) Regulation of vertebrate cellular Mg2+ homeostasis by TRPM7. Cell 114:191–200

Shaker JL, Deftos L (2000) Calcium and phosphate homeostasis (De Groot LJ, Beck-Peccoz P, Chrousos G, Dungan K, Grossman A, Hershman JM, Koch C, McLachlan R, New M, Rebar R, Singer F, Vinik A, Weickert MO, eds). MDText.com, Inc., South Dartmouth. PMID: 25905252

Sopko R, Huang D, Preston N, Chua G, Papp B, Kafadar K, Snyder M, Oliver SG, Cyert M, Hughes TR, Boone C, Andrews B (2006) Mapping pathways and phenotypes by systematic gene overexpression. Mol Cell 21:319–330

Sorin A, Rosas G, Rao R (1997) PMR1, a Ca2+-ATPase in yeast Golgi, has properties distinct from sarco/endoplasmic reticulum and plasma membrane calcium pumps. J Biol Chem 272:9895–9901

Starovasnik MA, Davis TN, Klevit RE (1993) Similarities and differences between yeast and vertebrate calmodulin: an examination of the calcium-binding and structural properties of calmodulin from the yeast *Saccharomyces cerevisiae*. Biochemistry 32:3261–3270

Stathopoulos AM, Cyert MS (1997) Calcineurin acts through the CRZ1/TCN1-encoded transcription factor to regulate gene expression in yeast. Genes Dev 11:3432–3444

Stathopoulos-Gerontides A, Guo JJ, Cyert MS (1999) Yeast calcineurin regulates nuclear localization of the Crz1p transcription factor through dephosphorylation. Genes Dev 13:798–803

Strayle J, Pozzan T, Rudolph HK (1999) Steady-state free Ca($^{2+}$) in the yeast endoplasmic reticulum reaches only 10 microM and is mainly controlled by the secretory pathway pump *pmr1*. EMBO J 18:4733–4743

Takita Y, Engstrom L, Ungermann C, Cunningham KW (2001) Inhibition of the Ca($^{2+}$)-ATPase Pmc1p by the v-SNARE protein Nyv1p. J Biol Chem 276:6200–6206

Tanida I, Takita Y, Hasegawa A, Ohya Y, Anraku Y (1996) Yeast Cls2p/Csg2p localized on the endoplasmic reticulum membrane regulates a non-exchangeable intracellular Ca2+ pool cooperatively with calcineurin. FEBS Lett 379:38–42

Tomar P, Sinha H (2014) Conservation of PHO pathway in ascomycetes and the role of Pho84. J Biosci 39:525–536

Viladevall L, Serrano R, Ruiz A, Domenech G, Giraldo J, Barcelo A, Arino J (2004) Characterization of the calcium-mediated response to alkaline stress in *Saccharomyces cerevisiae*. J Biol Chem 279:43614–43624

Wachek M, Aichinger MC, Stadler JA, Schweyen RJ, Graschopf A (2006) Oligomerization of the Mg2+-transport proteins Alr1p and Alr2p in yeast plasma membrane. FEBS J 273:4236–4249

Wei Y, Marchi V, Wang R, Rao R (1999) An N-terminal EF hand-like motif modulates ion transport by Pmr1, the yeast Golgi Ca(2+)/Mn(2+)-ATPase. Biochemistry 38:14534–14541

Wiesenberger G, Steinleitner K, Malli R, Graier WF, Vormann J, Schweyen RJ, Stadler JA (2007) $Mg^{2+}$ deprivation elicits rapid $Ca^{2+}$ uptake and activates $Ca^{2+}$/calcineurin signaling in *Saccharomyces cerevisiae*. Eukaryot Cell 6:592–599

Yoshimura H, Tada T, Iida H (2004) Subcellular localization and oligomeric structure of the yeast putative stretch-activated Ca2+ channel component Mid1. Exp Cell Res 293:185–195

Zhou Y, Meraner P, Kwon HT, Machnes D, Oh-hora M, Zimmer J, Huang Y, Stura A, Rao A, Hogan PG (2010a) STIM1 gates the store-operated calcium channel ORAI1 in vitro. Nat Struct Mol Biol 17:112–116

Zhou Y, Ramachandran S, Oh-hora M, Rao A, Hogan PG (2010b) Pore architecture of the ORAI1 store-operated calcium channel. Proc Natl Acad Sci U S A 107:4896–4901

# Chapter 8
# Potassium and Sodium Transport in Yeast

**Lynne Yenush**

**Abstract** As the proper maintenance of intracellular potassium and sodium concentrations is vital for cell growth, all living organisms have developed a cohort of strategies to maintain proper monovalent cation homeostasis. In the model yeast *Saccharomyces cerevisiae*, potassium is accumulated to relatively high concentrations and is required for many aspects of cellular function, whereas high intracellular sodium/potassium ratios are detrimental to cell growth and survival. The fact that *S. cerevisiae* cells can grow in the presence of a broad range of concentrations of external potassium (10 μM–2.5 M) and sodium (up to 1.5 M) indicates the existence of robust mechanisms that have evolved to maintain intracellular concentrations of these cations within appropriate limits. In this review, current knowledge regarding potassium and sodium transporters and their regulation will be summarized. The cellular responses to high sodium and potassium and potassium starvation will also be discussed, as well as applications of this knowledge to diverse fields, including antifungal treatments, bioethanol production and human disease.

**Keywords** Ion homeostasis • Potassium transport • Sodium transport • Trk1 • Ena1 • Nha1 • Pma1 • Yeast

## 8.1 Introduction

Ion homeostasis is a fundamental requirement for all organisms. Many different minerals are required for essential biochemical processes, but accumulation of these elements is toxic. As these elements are present as charged molecules in aqueous cellular environments, they are not able to freely diffuse across cell membranes. Thus, all living organisms have developed efficient systems to acquire and store these elements and robust mechanisms to maintain homeostatic concentrations to avoid toxicity.

L. Yenush (✉)
Instituto de Biología Molecular y Celular de Plantas (IBMCP), Universitat Politècnica de València-Consejo Superior de Investigaciones Científicas, Avd. de los Naranjos s/n, Valencia 46022, Spain
e-mail: lynne@ibmcp.upv.es

 
J. Ramos et al. (eds.), *Yeast Membrane Transport*, Advances in Experimental Medicine and Biology 892, DOI 10.1007/978-3-319-25304-6_8

*Saccharomyces cerevisiae* has been developed into a productive model to study many aspects of ion homeostasis based on its advantages as an experimental system and the high level of conservation throughout evolution of many proteins that transport ions (Saier 2000; Wolfe and Pearce 2006; Botstein and Fink 2011). Moreover, this model system is amenable to genome-level approaches, which have extensively characterized the yeast 'ionome' and defined genes and gene networks that contribute to its maintenance (Eide et al. 2005; Yu et al. 2012). Remarkably, in these studies, a relatively low number of genetic alterations were shown to have large effects on the mineral composition of yeast cells: approximately 5 % of the strains analyzed in rich media (212 of 4,358 knock-outs) and 9 % of the strains analyzed in minimal media (1065 of 11,890 haploid and diploid knock-outs and overexpression strains) showed significant differences in the relative concentrations of the 13–17 cations tested. These studies have revealed an important role for mitochondrial and vacuolar function and the ESCRT pathway (involved in vesicle trafficking) in the regulation of yeast ion homeostasis. Additionally, many of the strains identified displayed alterations in the accumulation of multiple elements. Only a scarce number of mutants were shown to be defective in only one element. These results indicate that the mechanisms that have evolved to maintain ion homeostasis are robust and in many cases act in a coordinated manner.

Potassium is a key monovalent cation necessary for multiple aspects of cell growth and survival, for example compensation of negative charges of macromolecules to maintain electroneutrality, cell turgor and volume, enzyme activity, protein synthesis, and maintenance of proper membrane potential and intracellular pH. In most cell types, potassium is accumulated against its concentration gradient to relatively high amounts, whereas sodium accumulation is actively avoided because of its toxicity. In many mammalian cell types, this low sodium/potassium ratio at the cellular level is actively maintained by P type $Na^+$, $K^+$ ATPases, which drive sodium out of the cell in exchange for potassium (Skou and Esmann 1992). The resulting sodium gradient is used for the coupled uptake of many ions and nutrients via secondary, sodium-coupled carriers. Essentially, yeast cells maintain low sodium/potassium ratios through efficient and selective potassium uptake (and not sodium), efficient efflux of excess sodium and efficient sequestration of sodium in the vacuole. In the majority of these transport processes, a proton motive force created by $H^+$-ATPases is required.

This review will focus on our current knowledge regarding potassium and sodium transport and how homeostasis of these ions is achieved and maintained in baker's yeast. As mentioned above, although not directly involved in transporting potassium or sodium themselves, $H^+$-ATPases are key regulators of these transport processes and so will be discussed first. General aspects of potassium and sodium uptake and efflux will be considered and our current knowledge regarding the structure and function of the implicated transporters will be presented. Our understanding of how potassium homeostasis is regulated and how yeast cells respond to both excess extracellular sodium and potassium and potassium starvation will be discussed. Finally, some applications of this knowledge to other fields will also be presented.

## 8.2 The Role of $H^+$-ATPases in Potassium and Sodium Transport

### *8.2.1 Pma1*

In *S. cerevisiae*, the plasma membrane $H^+$-ATPase encoded by the *PMA1* gene is largely responsible for creating the proton motive force across the plasma membrane. This proton gradient drives nutrient uptake by secondary, proton-coupled carriers (Barnett 2008). The *PMA1* gene is essential and it encodes a 100 kDa $P_2$-type ATPase that is highly stable and abundant in the yeast plasma membrane and has been estimated to consume at least 20 % of cellular ATP (Benito et al. 1991; Morsomme et al. 2000). The enzyme is activated by glucose and acidic internal pH and, not surprisingly, alterations in its activity have an important impact on intracellular pH and ion homeostasis (Serrano 1983; Perlin et al. 1988; Goossens et al. 2000). Mutants with partial loss of function of the *PMA1* gene are unable to grow at low external pH and display tolerance to cations due to alterations in the membrane potential that lead to a decrease in the uptake of positively charged molecules, such as Hygromycin B (McCusker et al. 1987; Perlin et al. 1988). The *S. cerevisiae* genome contains a second gene, *PMA2* which is approximately 90 % identical to *PMA1* (Schlesser et al. 1988). Although the Pma2 protein can pump protons and can substitute for Pma1 when expressed under the control of a strong promoter, in standard growth conditions, this gene is expressed at very low levels and therefore does not have an important impact on ion homeostasis (Supply et al. 1993).

Transcriptional regulation of *PMA1* (and in some cases *PMA2*) has been described in response to carbon source (mediated by the Rap1 and Gcr1 transcription factors), during the diauxic shift, entry into stationary phase and stress conditions (Rao et al. 1993; Portillo 2000; Fernandes and Sá-Correia 2003). As mentioned, on the protein level, decreased intracellular pH activates the enzyme, as does glucose addition. The mechanism of activation by acidic pH is not clear, but it may reflect the pH optimum of the enzyme that has been observed in reconstituted systems or post-translational modifications yet to be defined on the molecular level. Glucose activation of Pma1 rapidly results in an increase in the $V_{max}$ and a decrease in the affinity for ATP and is mediated, at least in part, by phosphorylation of the autoinhibitory C-terminal domain. Although the exact molecular mechanism has yet to be fully elucidated, several Pma1 phosphorylation sites have been implicated. Specifically, the phosphorylation of threonine 912 is required for glucose activation, but appears to be constitutive, while phosphorylation of serine 911 is induced by glucose addition and is also necessary for full Pma1 activation (Lecchi et al. 2007). The NPR family kinases Ptk2 and Hrk1 have been shown to positively regulate Pma1 activity (Goossens et al. 2000). Evidence has been presented suggesting that Ptk2 phosphorylates serine 899 of Pma1 (Eraso et al. 2006). Moreover, a role for the PP1-type phosphatase, Glc7 in the regulation of Pma1 activity has been proposed (Williams-Hart et al. 2002). In addition, the Yck1 and Yck2 casein kinases have been reported to negatively regulate Pma1 activity (Estrada et al. 1996). Other studies have suggested a role for calcium-dependent signaling in glucose-mediated Pma1 activation, although the mechanism is still unknown (Trópia et al. 2006; Pereira et al. 2008; Bouillet et al. 2012).

### *8.2.2 V-ATPase*

The vacuolar $H^+$-ATPase (V-ATPase) is also involved in determining the electrical potential across membranes of intracellular compartments and accordingly, it plays a crucial role in several physiological processes, including ion homeostasis (Kane 2007). The V-ATPase is a protein complex composed of a soluble, multi-subunit $V_1$ catalytic region and a membrane-embedded, multi-subunit $V_O$ region, whose structural organization is similar to the $F_1F_o$-ATPase (Nishi and Forgac 2002; Zhang et al. 2008). Two V-ATPase complexes have been identified. The first complex, which is present in vacuolar membranes contains the Vph1 subunit in the $V_o$ complex and is responsible for acidifying the vacuole. In the second complex, Stv1 substitutes Vph1 and this complex is responsible for the acidification of the Golgi apparatus/endosomes, where it is targeted (Tarsio et al. 2011). The V-ATPase is regulated on the level of complex formation/dissociation. This regulation seems to be conserved evolutionarily and is complex. For example, glucose starvation, decreasing intracellular pH, and poor nutrient conditions favor the dissociation and concomitant reduction in the activation of the V-ATPase, whereas glucose re-addition and increasing intracellular pH have the opposite effect (Kane 2012).

In *S. cerevisiae*, experimental evidence has been reported that shows that the Pma1 plasma membrane and the V-ATPases act coordinately to control cytosolic pH homeostasis (Martínez-Muñoz and Kane 2008). The electrogenic nature of their combined activities is a major determinate in the generation of not only plasma membrane, but also organellar membrane potential. As mentioned, this electro-chemical gradient is used for the uptake of nutrients from the cell environment by proton-coupled carriers (Barnett 2008). It also thought to play an important role in the ability of yeast cells to accumulate potassium against a steep concentration gradient and to enable the extrusion and organellar distribution of potassium and sodium via proton-coupled antiporters (Gaber 1992; Rodríguez-Navarro 2000; Arino et al. 2010).

## 8.3 Potassium Uptake and Efflux

Since as early as the 1940s, researchers proposed a relationship between potassium and proton transport in yeast and during the following years many aspects of these transport processes were characterized (Borst-Pauwels 1981). The steady state intracellular potassium concentration in yeast cells is maintained between 200 and 300 mM depending on the strain and growth conditions and is thought to depend on continuous uptake and efflux processes (Lapathitis and Kotyk 1998; Arino et al. 2010). As mentioned, the membrane potential generated by the plasma membrane $H^+$-ATPase is vital for potassium uptake in yeast. However, the coordination of potassium fluxes across the plasma membrane is also crucial to maintain proper membrane potential, as demonstrated by the hyperpolarization of mutants defective in high affinity uptake and the depolarization observed in mutants lacking potassium efflux systems (Madrid et al. 1998; Maresova and Sychrova 2005; Kinclova-Zimmermannova et al. 2006; Maresova et al. 2006). Thus, it is clear that the coordination of these processes is crucial for yeast cell growth and survival. In the next sections, the proteins responsible for mediating the uptake and efflux of potassium across the plasma membrane will be discussed.

### *8.3.1 Trk1 and Trk2*

In 1984, Rodriguez-Navarro and Ramos proposed a dual mode of potassium transport by showing that yeast displayed both high and low affinity potassium uptake depending on the growth history of the cells (Rodríguez-Navarro and Ramos 1984). In 1988, the first potassium transporter gene, *TRK1* was cloned on the basis of its ability to complement a yeast mutant defective in potassium uptake (Gaber et al. 1988). *TRK1* is a non-essential gene that encodes an integral membrane protein of 1235 amino acids (Fig. 8.1). Based on the structure of the KcsA $K^+$ channel from *Streptomyces lividans*, Trk1 has been proposed to be composed of four repetitions of an M1PM2 motif (Durell and Guy 1999). M1 and M2 are transmembrane segments that are connected by the P helix (Fig. 8.1). An extensive mutagenesis analysis has identified residues in the second transmembrane helix (M2) of the fourth M1PM2 repetition ($M2_D$) of Trk1 as being crucial for potassium transport (Haro and Rodríguez-Navarro 2003). It has been proposed that the four M1PM2 repetitions of the Trk1 monomer fold into a symmetric array and that four Trk1 monomers form a tetramer in the plasma membrane (Durell and Guy 1999). Although initial reports suggested that Trk1 is localized in plasma membrane lipid "rafts", further characterization of the protein distribution in the yeast plasma membrane shows that essentially all integral membrane proteins are found in two classes of microdomains that share biochemical properties with mammalian "rafts", but the overall organization and function of these microdomains appears to be quite different (Yenush et al. 2005; Malinsky et al. 2013).

Whereas wild type strains are able to grow in low micromolar potassium concentrations and exhibit high affinity and high velocity potassium uptake ($V_{max}$ 30 nmol/mg cells/min and $K_m$ of 0.024 mM), strains lacking *TRK1* are unable to grow in 0.1 mM KCl and show a marked reduction in potassium uptake kinetics, demonstrating that Trk1 is a major contributor to high affinity potassium uptake (Rodríguez-Navarro and Ramos 1984; Gaber et al. 1988). Each transporter has two cation binding sites and normally functions as a $K^+$ co-transporter, thought to be driven by the membrane potential created by the Pma1 $H^+$-ATPase. However, this

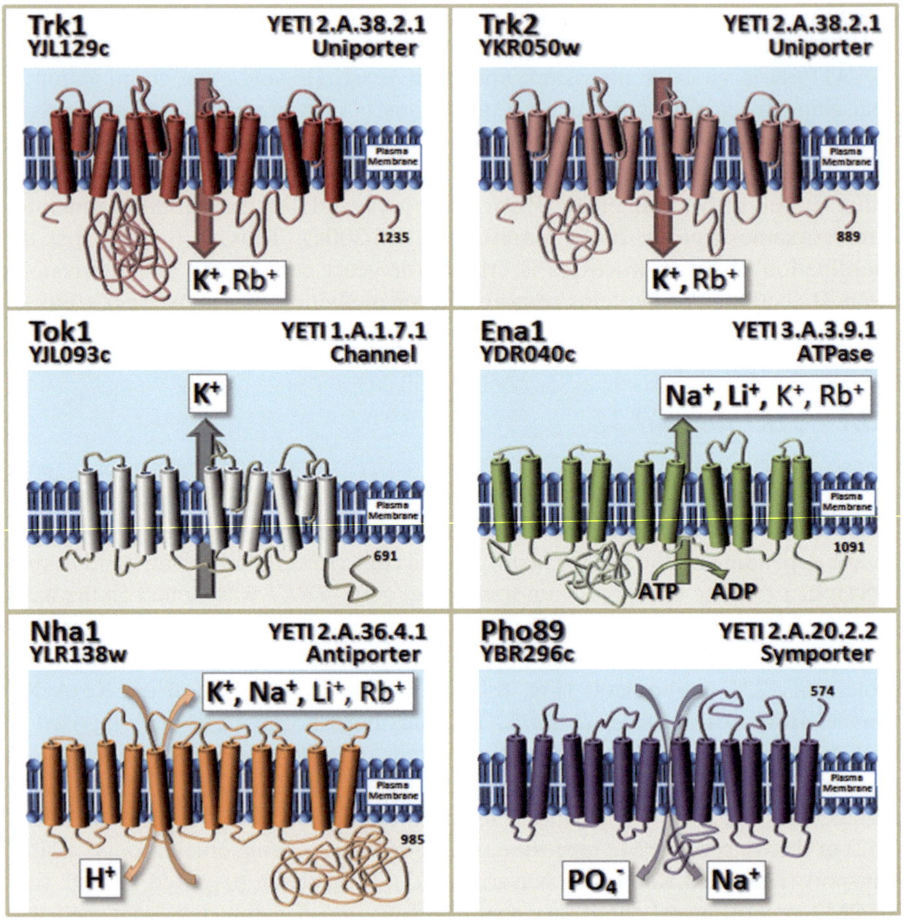


**Fig. 8.1** ***Saccharomyces cerevisiae*** **plasma membrane potassium and sodium transport proteins.** For each protein the standard name, systematic name, Yeast transporter information code based on the Transport classification system (YETI), transporter type, proposed topology and substrate specificity are shown (cations in bold are preferred substrates). Numbers at the end of each sequence represent the length of the protein. See text for more details and references

affirmation assumes a plasma membrane potential of −300 mV, which has not be confirmed experimentally in *S. cerevisiae*. Thus, other scenarios, such as Trk1 acting as a $K^+$-$Na^+$ symporter cannot be ruled out (reviewed in: (Arino et al. 2010)).

*TRK1* orthologues have been identified in other yeast, fungi and higher plants (Rodríguez-Navarro 2000). In fact, *S. cerevisiae* contains a second gene, *TRK2* which encodes a protein that is 55 % identical to Trk1 (Ko and Gaber 1991). The proposed topology is the same for Trk2, with the main structural difference residing in the length of the second cytosolic segment (Trk1 642 aa; Trk2 326 aa) (Fig. 8.1). Deletion of the *TRK2* gene has little effect on yeast growth, although the potassium requirements of the double *trk1 trk2* mutant increase 10-fold, as compared to the *trk1* simple mutant (Ko et al. 1990). Trk2 was initially proposed to mediate low affinity transport. However, later studies showed that Trk2, when expressed from a strong promoter, can mediate high/moderate affinity potassium uptake (Ramos et al. 1994; Michel et al. 2006). Thus, although Trk2 participates in potassium uptake, Trk1 is the dominant transporter, likely due to the higher expression of the *TRK1* gene. Interestingly, the Trk transporters have also been shown to mediate the efflux of anions such as $Cl^-$, $I^-$ and $Br^-$ and $SCN^-$ and $NO_3^-$, presumably through the pore created by the formation of the Trk1 or Trk2 tetramers (Kuroda et al. 2004; Rivetta et al. 2011). Although the physiological significance of this activity detected in electrophysiology experiments is not clear, it has been proposed to balance charges generated by Pma1 proton pumping activity (Rivetta et al. 2011).

As mentioned, Trk1 is the transporter responsible for potassium uptake and as such plays an important role in yeast physiology. Although there is no evidence for transcriptional regulation of either *TRK1* or *TRK2* in response to cation-related stresses, many proteins have been identified that affect the activity of this transporter, presumably at the post-translational level (Fig. 8.2). For example,

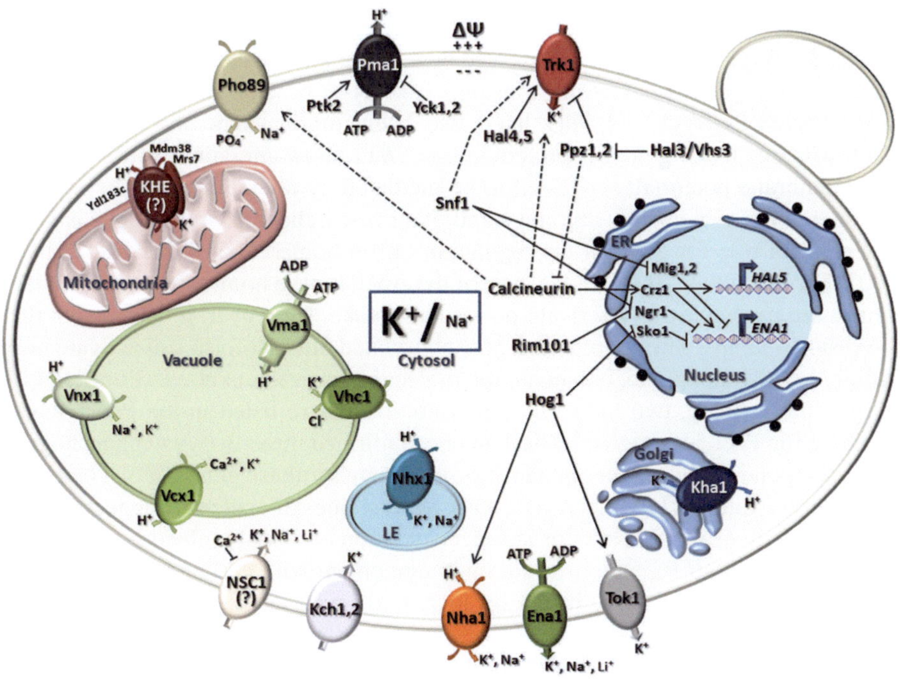


**Fig. 8.2 Schematic representation of transporters and regulators controlling potassium and sodium transport in *Saccharomyces cerevisiae*.** The major contributors to the known signal transduction pathways involved in potassium and sodium homeostasis are depicted. *Discontinuous lines* represent interactions that have not been fully documented. See text for details and references. (*LE* late endosome)

the functionally redundant protein kinases encoded by the *HAL4 (SAT4)* and *HAL5* genes were identified as positive regulators of Trk1 (Mulet et al. 1999). Overexpression of these genes confers tolerance to toxic concentrations of NaCl or LiCl and this phenotype requires the presence of the *TRK1* and *TRK2* genes. Moreover, the double *hal4 hal5* mutant presents defects in $Rb^+$ uptake and a slow growth phenotype in minimal media that can be ameliorated with increased external potassium. Evidence for direct phosphorylation of Trk1 by these kinases is lacking. However, it has been shown that the Hal4 and Hal5 kinases are required for Trk1 plasma membrane accumulation (Perez-Valle et al. 2007). The deletion of the last 35 amino acids of the Trk1 protein stabilizes the transporter in the plasma membrane, suggesting that this region is implicated in plasma membrane delivery and/or maintenance. Interestingly, several other nutrient transporters, in addition to Trk1, also fail to accumulate at the plasma membrane in *hal4 hal5* mutants leading to defects in both carbon and nitrogen metabolism, suggesting a more general role for the Hal4 and Hal5 kinases (Perez-Valle et al. 2010).

The Arl1 protein, which encodes a G protein of the Ras superfamily involved in protein trafficking, has been suggested to modulate Trk1 activity, as toxic cation sensitivity and a reduction in $Rb^+$ uptake has been documented in the mutant strain (Munson et al. 2004). Moreover, both *HAL4* and *HAL5* act as multi-copy suppressors of the *arl1* mutant strain. However, in this report no defect in Trk1 protein levels or trafficking was observed in *arl1* mutants, so the mechanism through which Arl1 regulates potassium transport has yet to be elucidated. Other protein kinases such as Sky1 and Snf1 have also been implicated in the regulation of Trk1. Mutants lacking the SR protein kinase *SKY1* show alterations in $Rb^+$ uptake and membrane potential, suggested to be mediated by alterations in Trk1 activity, although the mechanism is unknown and other researchers have described a Trk1-independent role for Sky1 in the regulation of ion homeostasis (Erez and Kahana 2002; Forment et al. 2002). In the case of the AMP kinase homologue Snf1, mutant strains are unable to fully activate potassium uptake. Moreover, it was shown that the residual kinase activity of a non-phosphorylated Snf1 isoform can activate high affinity potassium uptake, but again, the molecular basis is unknown (Portillo et al. 2005). Interestingly, two Snf1 phosphorylation sites are listed in the Phosphogrid database for Trk1, although they have not be confirmed directly (www.phosphogrid.org). The gene encoding the trehalose-6-phosphate synthase gene (*TPS1*) has been shown to activate Trk1 (Mulet et al. 2004). Several lines of evidence suggest that the mechanism involves the direct or indirect activation of Trk1 by glucose phosphates (Glc-1-P and Glc-6-P), which would be in agreement with earlier studies showing that potassium uptake is activated by increased levels of phosphorylated sugars (Alijo and Ramos 1993).

Protein phosphatases have also been reported to modulate Trk1 activity. First, early reports suggested that the $Ca^{2+}$/calmodulin-dependent calcineurin phosphatase is required for Trk1 to properly discriminate between potassium and sodium under conditions of salt stress (Mendoza et al. 1994). More recently, it was shown that the absence of calcineurin also affects high affinity potassium uptake in the absence of salt stress (Casado et al. 2010). The mechanism of this regulation is thought to involve the calcineurin-dependent regulation of the *HAL5* gene. Several lines of evidence suggest that a second protein phosphatase, Ppz1 is an important regulator of Trk1 activity. Strains lacking *PPZ1* and the related *PPZ2* gene are tolerant to toxic cations, as are strains that overexpress the Ppz1 regulatory subunit *HAL3* (Ferrando et al. 1995; Posas et al. 1995; de Nadal et al. 1998). In addition, strains lacking the *PPZ1* and *PPZ2* genes display increased turgor pressure and increased pH, due to excess potassium accumulation (Yenush et al. 2002). These phenotypes require the presence of the *TRK1* and *TRK2* genes. Furthermore, Ppz1 was shown to co-localize and physically interact with Trk1 and in *ppz1 ppz2* mutants an increase in Trk1 phosphorylation levels are observed (Yenush et al. 2005). Taken together, these data suggest that Ppz1 is a negative regulator of Trk1. Moreover, the interaction between Ppz1 and Hal3 is pH-dependent, leading to a model in which the Hal3-Ppz1 complex participates in the maintenance of internal potassium concentrations by responding to changes in internal pH. The kinase(s) responsible for Trk1 phosphorylation and the mechanism by which this class of post-translational modification alters the properties of the transporter still need to be defined.

An alternative approach that has been taken to identify regulators of Trk1 is high-throughput screening of the yeast mutant collection looking for genes whose disruption leads to increased or decreased tolerance to toxic cations, such as hygromycin B (Barreto et al. 2011; Fell et al. 2011). In these screens, 150–200 mutants encoding genes belonging to several functional groups, including protein kinases and phosphatases, transcription, cell cycle, and DNA processing were enriched. Some of the regulators identified in both screens have been mentioned above, such as Arl1, Sky1, Hal4, and Hal5. Interestingly, both screens also identified many mutants related to various aspects of vesicle trafficking, such as SNARE proteins and components of the CORVET and HOPS complexes. However, many of these mutants are not defective in Trk1 plasma membrane accumulation as might be expected, thus their participation in the regulation of potassium uptake remains to be defined.

### *8.3.2 Proteins Involved in Low Affinity Potassium Uptake*

As mentioned, *S. cerevisiae* cells display high and low affinity potassium uptake depending on the growth history of the cells and the media employed. Under normal growth conditions, where the potassium concentration is not limiting, Trk1 would mediate the majority of the so-called low affinity potassium uptake. When the extracellular potassium concentration decreases, Trk1 switches to a high affinity mode to mediate growth in the presence of as little as 10 $\mu$M $K^+$. Importantly, deletion of both *TRK1* and *TRK2* in *S. cerevisiae* is not lethal. These mutant strains display ectopic low affinity potassium uptake, indicating that additional mechanisms of potassium uptake must exist (Madrid et al. 1998). Electrophysiology studies revealed inward potassium currents in *trk1 trk2* mutants, whose activity is inhibited by calcium (Bihler et al. 1998, 2002). A putative channel was proposed to be responsible for these currents and named NSC1 (non-specific cation channel), but the protein responsible was not identified. It has been proposed that the "very low affinity" potassium uptake observed in *trk1 trk2* strains is mediated by multiple transport processes (reviewed in: (Arino et al. 2010)). Recently, two putative low affinity potassium transporter proteins were identified that may account for some of these currents. Kch1 and Kch2 (Prm6) were identified as necessary components of the pheromone-induced activation of the high affinity $Ca^{2+}$ influx system (HACS) (Stefan et al. 2013). These fungal-specific proteins are predicted to have several transmembrane segments and have been shown to localize to the yeast plasma membrane. The inward rectifying currents are notably reduced in strains lacking both *KCH1* and *KCH2* and overexpression of either gene improves the growth of *trk1 trk2* strains in low potassium medium, supporting a role for these proteins as potassium transporters or channels. However, under normal growth conditions, their activity appears to be eclipsed by much higher Trk1 activity. The fact that inward rectifying currents are still observed in strains lacking *trk1 trk2 kch1* and *kch2* indicate that additional mechanisms of potassium uptake are present. Candidates for these uptake systems include non-specific uptake by the Qdr2 drug/$H^+$ antiporter and sugar and amino acid permeases (Ko et al. 1993; Wright et al. 1997; Vargas et al. 2007). Finally, deletion of the gene encoding a small hydrophobic protein called Pmp3, which is highly conserved in yeast and plants, has been proposed to facilitate cation uptake in a Trk1,2-independent manner, via an unknown mechanism (Navarre and Goffeau 2000).

### *8.3.3 Tok1*

At least three different transporters contribute to potassium efflux in *S. cerevisiae*. Although both Ena1 and Nha1 can transport potassium, they were first identified based on their capacity for sodium efflux, and so will be discussed below. The third protein, Tok1, is an outwardly rectifying plasma membrane potassium channel and it is the only potassium-specific efflux system described in yeast (Gustin et al. 1986; Bertl et al. 1993; Ketchum et al. 1995; Zhou et al. 1995; Reid et al. 1996). The *TOK1* gene encodes a protein of 691 amino acids that contains eight transmembrane segments, the last four of which participate in the formation of two pore-forming P domains responsible for $K^+$ conductance (Fig. 8.1) (Ketchum et al. 1995; Martinac et al. 2008). The activity of the channel is regulated by both membrane potential and external potassium (Bertl et al. 1993; Vergani et al. 1997; Fairman et al. 1999). Accordingly, depolarization of the membrane leads to channel opening and potassium efflux, presumably to restore proper membrane potential. Gating of the channel is regulated by the carboxy terminal cytosolic segment, which prevents channel closure (Loukin and Saimi 2002). Although the electrophysiological data generated both in yeast and Xenopus oocytes clearly define the activity of the Tok1 channel, the physiological role of this potassium efflux activity remains unclear, as no growth-related phenotypes have been detected for the *tok1* mutant strain (Gustin et al. 1986; Ketchum et al. 1995; Zhou et al. 1995; Lesage et al. 1996; Reid et al. 1996; Loukin et al. 1997; Bertl et al. 1993, 1998, 2003). The function of Tok1 may involve plasma membrane potential maintenance as it has been shown that *tok1* mutant strains are depolarized, while strains overexpressing *TOK1* are hyperpolarized (Maresova et al. 2006). Tok1 has also been reported to be phosphorylated almost immediately upon sodium chloride treatment by the Hog1 MAP kinase (Proft and Struhl 2004). The HOG signaling pathway is a conserved Mitogen Activated Protein Kinase (MAPK) pathway, which in conditions of hyperosmotic stress leads to the activation of the Hog1 MAP kinase (reviewed in (de Nadal et al. 2002)). Although the functional consequences of this phosphorylation were not examined in detail in this study, mathematical modeling predicts that Hog1-mediated phosphorylation of Tok1 reduces $Na^+$ influx under NaCl stress (Ke et al. 2013).

## 8.4 Sodium Uptake and Efflux

As discussed earlier, due to its toxicity, sodium accumulation is actively avoided by yeast cells. Under normal laboratory growth conditions, the amount of intracellular sodium is very low. In the presence of high external concentrations, sodium is thought to enter the cell in various ways, principally by displacing potassium. For example, Trk1 and Trk2 can transport sodium, although the affinity is much lower than for potassium (Haro and Rodríguez-Navarro 2002). In fact, in the presence of high sodium, Trk1 is thought to undergo an undefined modification which improves its capability to discriminate between the two cations and thus favor potassium uptake (Mendoza et al. 1994). Sodium also enters through other non-specific, low-affinity potassium transporters, such as NSC1. These transporters do not appear to discriminate between these two cations, as *trk1 trk2* mutant strains, which depend on these low-affinity transport mechanisms, accumulate more sodium than the wild type strain (Gómez et al. 1996).

### *8.4.1 Pho89*

Interestingly, in *S. cerevisiae* one sodium-dependent nutrient transporter has been described. The *PHO89* gene encodes a sodium-phosphate co-transporter protein of 574 amino acids with twelve predicted membrane-spanning domains (Persson et al. 1999) (Fig. 8.1). Expression of the *PHO89* gene is induced by both phosphate limitation and alkaline pH (Martinez and Persson 1998; Serrano et al. 2002). The transporter, whose $K_m$ value for inorganic phosphate is 0.5 $\mu$M, is highly specific for sodium and maximum phosphate uptake is observed at 25 mM NaCl and pH 9.5 (Martinez and Persson 1998). Another related gene, *PHO84* encodes a proton-coupled phosphate transporter, which is responsible for phosphate uptake at acidic pH (Persson et al. 1999). Pho89 is the only known sodium-dependent secondary nutrient transporter in *S. cerevisiae*. Recent work shows the detectable accumulation of intracellular $Na^+$ as a result of Pho89 activity only in the absence of Ena1 (Serra-Cardona et al. 2014). Moreover, in this same study it was shown that the transcription of both *PHO89* and *ENA1* are coordinately regulated during alkaline stress. Thus, it appears that Ena1 activity is likely to suffice to avoid accumulation of toxic levels of intracellular sodium introduced via Pho89.

## *8.4.2 Nha1*

Two classes of transport proteins have been shown to be important for sodium efflux in *S. cerevisiae*, Nha1 and the Ena family of ATPases. A role for Nha1 in tolerance to toxic sodium concentrations was initially shown by its recovery in a screen to identify genes improving the growth of a salt sensitive strain (Prior et al. 1996). The *NHA1* gene encodes a protein of 985 amino acids, which is predicted to contain twelve transmembrane segments and a large cytosolic carboxy terminal domain (550 amino acids) (Fig. 8.1). The overall structure and transporter activity is conserved in all kingdoms of life, although diversity exists in the physiological function of this family of transporters. In the case of *S. cerevisiae*, Nha1 is localized to the plasma membrane and acts as a dimeric, electrogenic proton antiporter with similar affinity for both $K^+$ and $Na^+$ that is also capable of transporting $Rb^+$ and $Li^+$ (Bañuelos et al. 1998; Mitsui et al. 2005; Ohgaki et al. 2005). Thus, under acidic pH conditions, Nha1 is able to transport sodium out of the cell, although this is unlikely to be its most important physiological function. Accordingly, loss or increase of Nha1 function has been shown to influence cytosolic pH, membrane potential, Trk1-dependent potassium uptake and to be involved in the initial adaptation to both osmotic and alkaline pH stress (Prior et al. 1996; Sychrová et al. 1999; Bañuelos et al. 2002; Proft and Struhl 2004; Kinclova-Zimmermannova et al. 2006; Kinclova-Zimmermannova and Sychrova 2006).

The expression of the *NHA1* gene has not been found to be regulated under osmotic or pH stress conditions and thus is thought to represent a constitutively expressed housekeeping gene (Bañuelos et al. 1998). Extensive mutagenesis studies in several yeast species have identified many amino acids required for activity and substrate specificity (reviewed in (Arino et al. 2010)). Several functions have also been ascribed to the large carboxy terminal tail. For example, a short 16 amino acid sequence predicted start at the end of the last transmembrane segment and continue into the beginning of the large cytosolic domain is required for proper function and targeting to the plasma membrane, while amino acids 920–930 have been implicated in $Li^+$ transport (Kinclová et al. 2001; Mitsui et al. 2004a). In addition, regions of the Nha1 carboxy terminus have also been defined which are responsible for the ability of *NHA1* overexpression to rescue the synthetic lethality of a mutant strain lacking both the *SIT4* phosphatase gene and the *HAL3* gene encoding the regulatory subunit of the Ppz1 phosphatase (Simón et al. 2003). The *sit4 hal3* double mutant has been reported to have a defect in the $G_1/S$ transition of the cell cycle and the identification of *NHA1* as a multi-copy suppressor has led to the suggestion that Nha1 plays a role in cell cycle progression, although the mechanism of the cell cycle arrest of this mutant and the basis of the *NHA1*-mediated rescue are not known (Simón et al. 2001).

As mentioned above, Nha1 has been implicated in the initial adaptation to hyperosmotic stress. In addition to Tok1 (see above), upon salt stress, the Hog1 MAP kinase also very rapidly phosphorylates Nha1 on T765 and T876 (Proft and Struhl 2004). Experimental data presented by these authors show that, under certain conditions, this post-translational modification increases its ability to confer tolerance to NaCl and so was interpreted as activating Nha1 sodium extrusion activity. Subsequently, Kinclova-Zimmermannova and Sychrova showed that sorbitol treatment decreases Nha1 $K^+$ efflux activity in a Hog1-dependent manner (Kinclova-Zimmermannova and Sychrova 2006). Further experiments are required to definitively determine the function and molecular mechanism of this post-translational modification. In agreement with a role for multiple phosphorylation in the regulation of Nha1, the phospho-binding 14-3-3 protein, Bmh1, was found to interact with Nha1 and to influence toxic cation tolerance (Zahrádka et al. 2012). However, the Nha1-Bmh1 interaction does not require the presence of the Hog1 kinase and the mechanism by which this interaction may affect Nha1 activity is as yet undefined. An additional 12 phosphorylation sites are listed in the Phosphogrid database in the carboxy terminus of Nha1, some of which are suggested to respond to salt stress and may represent candidates for 14-3-3 protein interaction sites (www.phosphogrid.org). Another protein, named Cos3 has also been described to interact with Nha1 (Mitsui et al. 2004b). Gain or loss of function of this gene has been shown to alter salt resistance in a Nha1-dependent manner, although the mechanism by which it may regulate the antiporter is unclear, especially considering that it is localized mostly to the vacuolar membrane.

### *8.4.3 Ena1*

As mentioned, at acidic intracellular pH the Nha1 antiporter can extrude sodium, whereas at higher pH, the Ena1 transporter is principally responsible for sodium extrusion (Bañuelos et al. 1998). Chromosome IV of most yeast genomes contains 3–5 tandem copies encoding ENA P-type ATPases, which are classified in the fungal-specific IID subfamily (for reviews, see (Benito et al. 2002; Arino et al. 2010; Palmgren and Nissen 2011)). One exception is the CEN.PK strain and its derivatives that encode only one divergent *ENA* gene called *ENA6* (Daran-Lapujade et al. 2009). In the rest of the strains analyzed, the *ENA* genes encode identical or nearly identical proteins that are 1091 amino acids long and are predicted to contain ten transmembrane segments and a larger cytosolic nucleotide-binding domain between the fourth and fifth membrane helices (Fig. 8.1). ENA transporters are localized to the plasma membrane and form a typical phospho-enzyme intermediate, using the energy generated from ATP hydrolysis to transport $K^+$, $Na^+$ or $Li^+$ (with varying affinities) against their concentration gradient (Haro et al. 1991; Wieland et al. 1995; Benito et al. 1997). Lack of the *ENA* genes, either in the CEN.PK strains or by genetic manipulation deleting the complete cluster, leads to marked salt and alkaline pH sensitivity, confirming the role of these genes as important participants in sodium (and lithium) extrusion (Haro et al. 1991; Daran-Lapujade et al. 2009). At the post-translational level, very little is known regarding possible regulation of ENA proteins. Strains lacking the *SRO7* gene, which encodes a protein involved in exocytosis homologous to the Drosophila Lgl tumor suppressor gene, were shown to be salt sensitive and to display defects in the proper accumulation of Ena1 at the plasma membrane, although no further progress has been made (Larsson et al. 1998; Wadskog et al. 2006). By contrast, a considerable amount of information is available regarding the transcriptional regulation of the key component of this gene cluster, *ENA1*. Here, the major contributors will be discussed, but for more details, excellent reviews are available (Ruiz and Arino 2007; Arino et al. 2010).

Under standard growth conditions, the expression of the *ENA* genes is low, as observed for the rest of the transport proteins discussed above. However, in contrast to other transporters whose mRNA levels are generally unaltered by environmental conditions, expression of the *ENA1* gene is specifically and markedly increased in response to osmotic, saline and alkaline pH stress via the action of several signaling pathways (Mendoza et al. 1994; Márquez and Serrano 1996; Lamb et al. 2001) (Fig. 8.2). Under conditions of mild saline (0.3–0.4 M NaCl) and osmotic stress, the HOG pathway plays a dominant role in *ENA1* induction (Marquez and Serrano 1996). As mentioned above, among the first regulatory events to occur upon Hog1 activation is the phosphorylation of Nha1 and Tok1 (Proft and Struhl 2004). However, activated Hog1 quickly accumulates in the nucleus and mediates the induction of *ENA1* (and many other target genes) via several mechanisms (Ferrigno et al. 1998; Posas et al. 2000; Rep et al. 2000). First, Hog1 phosphorylates the bZip transcription factor Sko1 and converts the Sko1-Ssn6-Tup1 complex from a transcriptional repressor to an activator (Proft and Struhl 2002). In addition, the histone deacetylase complex Rpd3-Sin3 is recruited to the *ENA1* promoter in a Hog1-dependent manner, facilitating the association of RNA polymerase II and transcriptional activation (De Nadal et al. 2004). Finally, Hog1, like other MAP kinases, has also been shown to be involved in transcriptional elongation of many of its target genes under stress conditions, but whether this activity of Hog1 is involved in *ENA1* induction has not been reported (reviewed in (de Nadal and Posas 2011)).

Another important pathway regulating the induction of the *ENA1* gene under stress conditions is mediated by the protein phosphatase, calcineurin (Mendoza et al. 1994). Calcineurin is a calcium/calmodulin-dependent, PP2B-type heterodimeric phosphatase composed of one of two redundant catalytic subunits (Cna1 or Cna2) and the regulatory subunit encoded by the *CNB1* gene (Klee et al. 1988). Osmotic stress has been proposed to provoke a calcium burst responsible for the activation of the calcineurin pathway (Matsumoto et al. 2002). Induction of the expression of the gene encoding the $Na^+$-ATPase by calcineurin occurs mainly through the dephosphorylation of the transcription factor Crz1 which has been shown to bind to two calcineurin-dependent response elements (CDRE) in the *ENA1* promoter and activate transcription (Mendizabal et al. 2001). Mutations in genes encoding another protein phosphatase, Ppz1 and its regulatory subunit, Hal3, have also been shown to affect *ENA1* expression (Ferrando et al. 1995; Posas et al. 1995). As discussed above, Ppz1 is a negative regulator of Trk1. In the *ppz1* mutant, an increase in basal *ENA1* transcription is observed and it has been shown to be fully dependent on the calcineurin/Crz1 pathway, suggesting that Ppz1 is a negative regulator of calcineurin (Ruiz et al. 2003). Mutants lacking both the *PPZ1* and *PPZ2* genes display an increase in internal $K^+$ and a more alkaline cytosolic pH, which contribute to even higher basal levels of *ENA1* (Yenush et al. 2002). In this case, both the calcineurin/Crz1 pathway and a second alkaline responsive element in the *ENA1* promoter contribute to the higher mRNA levels (Ruiz et al. 2003).

Although it has been shown that the Hog1 and calcineurin pathways account for the vast majority of *ENA1* induction in response to saline and osmotic stress, other pathways have also been identified that contribute to the regulation of *ENA1* expression in response to different stresses (Marquez and Serrano 1996). For example, several studies have shown that the $C_2H_2$ family zinc finger transcription factor, Rim101 is important for *ENA1* induction in conditions of alkaline stress, in cooperation with the AMP kinase homologue, Snf1 (see below) and the calcineurin pathway (Lamb et al. 2001; Serrano et al. 2002; Platara et al. 2006). Rim101 acts as a negative regulator of the Ngr1 repressor. Thus, upon activation of Rim101, Ngr1-mediated repression is released, leading to transcriptional activation of *ENA1* (Lamb and Mitchell 2003). Mutants lacking *RIM101* are sensitive to toxic cations and this phenotype was initially attributed to defects in the induction of *ENA1* transcription. However, in response to moderate saline stress, *ENA1* induction is not affected in *rim101* mutants, likely due to the dominant role played by the Hog1 pathway. In this case, the Rim101 pathway is required for proper accumulation of the Ena1 protein (Marqués et al. 2015).

*ENA1* expression has also been shown to respond to nutrient availability. For example, *ENA1* expression is under glucose repression: expression is higher in media containing galactose, instead of glucose as the carbon source (Alepuz et al. 1997). This induction has been shown to require *ENA1* promoter sequences that are bound by the Mig1 and Mig2 transcriptional repressors and to be mediated by the Snf1 kinase (Alepuz et al. 1997; Proft and Serrano 1999). Mutants lacking the *snf1* gene are sensitive to toxic cation concentrations (Alepuz et al. 1997). In addition to its role as a regulator of Trk1, mentioned above, defects in *ENA1* induction have also been postulated to contribute to this *snf1* phenotype. However, in the case of salt stress, it appears that the Ngr1 repressor, and not Mig1, are involved in Snf1-mediated *ENA1* induction (Ye et al. 2008). Interestingly, in the case of alkaline stress, both MIG and Ngr1 promoter elements have been implicated in Snf1-dependent *ENA1* induction (Platara et al. 2006). Signal transduction routes responding to nitrogen source quality can also influence *ENA1* expression. More specifically, treatment with rapamycin, which inhibits the TORC1 signaling pathway, has been shown to lead to an increase in *ENA1* mRNA levels

(Crespo et al. 2001). The salt sensitivity of mutants in two TOR-regulated GATA transcription factors, Gln3 and Gat1, and the presence of GATA motifs in the *ENA1* promoter suggest that these proteins mediate rapamycin-dependent *ENA1* induction. However, additional studies showing the absence of *ENA1* regulation by the Sit4 phosphatase, a regulator of Gln3, and the cytoplasmic localization of Gln3 under salt stress conditions have called into question the validity of this straightforward model (Masuda et al. 2000; Tate and Cooper 2007). Finally, the Protein kinase A (PKA) pathway has been implicated in the inhibition of *ENA1* induction by controlling the subcellular localization and increasing the repressor activity of the Sko1 transcription factor and by antagonizing the calcineurin pathway, through the phosphorylation of Crz1 (Nakamura et al. 1993; Pascual-Ahuir et al. 2001; Proft et al. 2001; Kafadar and Cyert 2004).

## 8.5 Intracellular $K^+$/$Na^+$ Transport Proteins

One shortcoming of many of the approximations routinely used to study ion homeostasis in yeast is that the intracellular distribution of the different elements is not always considered. It has long been known that yeast cells accumulate many solutes, including cations, in the vacuole and this sequestration has been proposed to be important for both proper homeostasis and survival in response to ionic stress conditions (Okorokov et al. 1980; Perkins and Gadd 1993; Nass et al. 1997). The presence of ion transporters in the membranes of organelles indicates that subcellular compartmentalization and distribution of ions is actively maintained by the cell. The first attempts to measure the distribution of potassium in yeast was carried out in 1976 using energy-dispersive X-ray microanalysis (Roomans and Sevéus 1976). These authors concluded that the amount of potassium was similar in the cytoplasm and nucleus and that vacuoles contained half the amount of potassium found in the cytosol. For these experiments, the cells were incubated overnight in water, a treatment that is likely to distort the cation distribution as compared to cells that are actively growing. Several studies reported data estimating the cytosolic vs. vacuolar distribution in different yeast species by using treatments that specifically permeabilize the plasma membrane (Okorokov et al. 1980; Perkins and Gadd 1993; de Nadal et al. 1999; Montiel and Ramos 2007). Although informative, these approaches do not account for ion content in other compartments, as all of the non-cytoplasmic ion content is generally attributed to the vacuole. More recently, Herrera and co-workers used subcellular fractionation protocols and atomic emission spectrophotometry to better define the distribution of both potassium and sodium under different growth conditions (Herrera et al. 2013). While their results confirm the accumulation of potassium and sodium in the vacuole relative to the cytosol, they also show that the nucleus contains an important percentage of the total intracellular potassium (and sodium, if present) which is maintained constant under different growth conditions, consistent with the results reported by Roomans and Sevéus (Roomans and Sevéus 1976). The authors propose that potassium and sodium enter non-specifically through nuclear pores and act to neutralize the negative charges found in this organelle, analogous to that reported in mammalian cells (Strick et al. 2001). On the other hand, they show that the amount of potassium (and especially sodium when added to the medium) is relatively low in the cytosol and find that the amount of cytosolic potassium does not markedly change during potassium starvation, indicating mobilization from the vacuole under these conditions. The main characteristics of the transporters that contribute to this subcellular distribution of potassium and sodium will be presented below and are shown schematically in Fig. 8.2.

## *8.5.1 Vacuole*

### 8.5.1.1 Vnx1

The main proton-coupled antiporter mediating potassium or sodium transport across the vacuolar membrane is encoded by the *VNX1* gene (Cagnac et al. 2007). The protein encoded by this gene is 908 amino acids long and predicted to contain 13 transmembrane segments and a 242 amino acid amino terminal cytosolic domain. Vnx1 was identified in a functional screen of all antiporter mutants predicted to be localized to the vacuolar membrane or endosomes looking for alterations in $Na^+/H^+$ or $K^+/H^+$ exchange activity in purified vacuoles (Cagnac et al. 2007). Protein sequence alignments place Vnx1 in the CAX (calcium exchanger) family, but this protein shows no calcium transport activity. Instead, this transporter exchanges protons for potassium or sodium ions, having a higher affinity for the latter. Thus, Vnx1 uses the proton gradient generated by the Vma1 $H^+$-ATPase (see above) to mediate the transport of potassium (or sodium, if present) into the vacuole.

### 8.5.1.2 Vcx1 and Vch1

A second transporter, encoded by the *VCX1* gene was subsequently shown to be responsible for the residual potassium/$H^+$ exchange activity remaining in vacuoles purified from *vnx1* mutants (Cagnac et al. 2010). This transporter, which is 411 amino acids long with 11 transmembrane helices, was first characterized as a vacuolar $Ca^{2+}/H^+$ exchanger and this likely represents its main activity, although as stated, Vcx1 can also transport $K^+$ (Cunningham and Fink 1996; Pozos et al. 1996). The crystal structure of this protein was recently solved, which may aid in defining the molecular determinants of substrate specificity (Waight et al. 2013). A recent study has provided evidence that another transporter, encoded by the *VCH1* gene functions as a vacuolar $K^+/Cl^-$ co-transporter (Petrezselyova et al. 2013). Vch1 contains 1120 amino acids and 12 putative transmembrane segments. Although its transport activity has not been directly tested, based on sequence homology to other members of the cation-$Cl^-$ co-transporter (CCC) family, the subcellular localization and phenotypic data demonstrating a role in the proper maintenance of intracellular potassium and vacuolar morphology, Vch1 very likely mediates electroneutral symport of potassium and chloride ions into the vacuole (André and Scherens 1995; Petrezselyova et al. 2013).

## 8.5.2 *Endosomes/Golgi*

### 8.5.2.1 Nhx1

Among the organellar monovalent cation transport proteins, the endosomal $Na^+/H^+$ antiporter encoded by the *NHX1* gene was the first identified in yeast and may be the most extensively characterized (Nass et al. 1997). The Nhx1 antiporter has 12 predicted hydrophobic domains distributed over a total length of 633 amino acids. Not all of the reports in the literature are consistent regarding Nhx1 topology, but the observation that the carboxy terminal sequence of Nhx1 has been shown to interact with at least one regulatory protein (see below) suggests that it is likely that this region of the protein is cytosolic (Wells and Rao 2001; Ali et al. 2004). Several reports have established that this transporter localizes to the membrane of late endosomes (the pre-vacuolar compartment), as well as recycling endosomes and the trans-Golgi network, where it contributes to pH maintenance within vesicles by mediating potassium (or sodium, if present) sequestration in these compartments in exchange for protons (Nass and Rao 1998; Brett et al. 2005; Kojima et al. 2012). Disruption of the gene leads to several phenotypes, including sensitivity to low pH and high salt, a decrease in cytosolic pH and vesicle trafficking defects, a function shown to require the ion transporter capacity (Bowers et al. 2000; Brett et al. 2005; Mukherjee et al. 2006). Accordingly, a role for Nhx1 in osmotic shock adaptation and sequestration of toxic cations and surplus potassium has been documented (Nass and Rao 1999; Quintero et al. 2000). Nhx1 was also described to be necessary for the recruitment of the ESCRT-0 component Vps27 to endosomes necessary for multi-vesicular body (MVB) formation in a cell-free assay, although a second report, using a genetic approach, suggests a role for Nxh1 downstream of MVB formation (Kallay et al. 2011; Mitsui et al. 2011). Finally, a link between Nhx1 and vesicle fusion and a physical interaction between the carboxy terminus of Nhx1 and a Rab family GTPase- activating protein (Gyp6) has been reported (Ali et al. 2004; Qiu and Fratti 2010). Therefore, it appears that Nhx1 may be involved in several aspects of vesicle trafficking in yeast.

### 8.5.2.2 Kha1

Kha1 is the sodium or potassium-proton antiporter that shares the highest level of homology to bacterial antiporters (Ramírez et al. 1998). The *KHA1* gene encodes an 873 amino acid protein predicted to have 12 transmembrane segments, which, although initially thought to be a plasma membrane transporter, has been shown to localize to the membrane of the Golgi apparatus (Ramírez et al. 1998; Flis et al. 2005; Maresova and Sychrova 2005). The phenotypic characterization of the *kha1* mutant alone or in combination with other mutants suggests that it acts as a proton-coupled antiporter facilitating the accumulation of potassium in this organelle (Maresova and Sychrova 2005). Specifically, the alkaline pH sensitivity of this mutant can be ameliorated by high external potassium. Additional studies have provided evidence for a broad substrate specificity by showing that in strains lacking the Arl1 GTPase, Kha1 increases potassium, sodium and lithium tolerance (Marešová and Sychrová 2010). Although the transporter activity has not yet been tested directly, Kha1 is thought to participate in the regulation of potassium and pH homeostasis in the Golgi apparatus, likely in coordination with the Gef1 chloride channel (Flis et al. 2005).

### *8.5.3 Mitochondria*

Ion fluxes are especially important in the mitochondria. The respiration-dependent negative membrane potential of mitochondria facilitates the entry of cations such as potassium, which, if not counter-acted, would result in excessive accumulation and osmotic swelling (reviewed in (Bernardi 1999)). Potassium-proton exchange (KHE) in the inner mitochondrial membrane is an essential element of Peter Mitchell's chemiosmotic theory proposed in 1961 (Mitchell 1961). Although this activity has been well-documented in purified mitochondria from many different organisms (reviewed in (Bernardi 1999)), the identification of the protein(s) responsible for KHE has been elusive. In *S. cerevisiae*, three genes have been identified to play a role in KHE: *MDM38*, *YLH47* (*MRS7*) and *YDL183c* (Nowikovsky et al. 2004; Froschauer et al. 2005; Zotova et al. 2010). *MDM38*, which was first identified in a comprehensive screen for searching for genes that affect mitochondrial function and morphology, appears to play to most important role in KHE (Dimmer et al. 2002; Nowikovsky et al. 2004, 2007; Zotova et al. 2010). Mdm38 is the orthologue of the human protein Leucine zipper–EF-hand–containing transmembrane 1 (LETM1), which is thought to be responsible for the seizures observed in patients with Wolf-Hirschhorn syndrome (Endele et al. 1999; Rauch et al. 2001; Schlickum et al. 2004). Expression of this gene in yeast can rescue the mitochondrial function and morphology phenotypes of *mdm38* mutants (Nowikovsky et al. 2004). Although some authors suggest that LETM1 may be involved in mitochondrial $Ca^{2+}/H^{+}$ exchange, several lines of evidence suggest that the physiological function of Mdm38 and LETM1 is related to KHE (reviewed in (Nowikovsky and Bernardi 2014)). Ylh47 (Mrs7) is homologous to Mdm38, whereas the protein encoded by the *YDL183c* gene shares no sequence similarity. However overexpression of either *YLH47* or *YDL183c* can suppress *mdm38* mitochondrial dysfunction and the triple *mdm38 ylh47 ydl183c* mutant has more severe phenotypes than any of the single or double mutant combinations (Nowikovsky et al. 2004; Zotova et al. 2010). All three proteins are predicted to have a single membrane spanning domain and so are not likely to mediate KHE individually. However, all three proteins have been shown to be present in high molecular weight complexes and both Mdm38 and Ylh47 can oligomerize, leading to the hypothesis that these proteins are functionally redundant, necessary co-factors of an as yet unidentified KHE (Zotova et al. 2010) (Fig. 8.2).

## 8.6 Physiological Consequences and Cellular Responses to Alterations in Potassium and Sodium Concentrations

### *8.6.1 Saline Stress*

Perturbations in the extracellular and/or intracellular concentrations of sodium and potassium lead to diverse cellular responses. As discussed above, sodium is actively extruded from yeast cells, so that a physiological response to low sodium (assuming sufficient potassium is present) is not expected. However, in the case of exposure to high concentrations of sodium, yeast cells respond on several levels and the response varies according to the severity and duration of the treatment. High sodium concentrations present a dual toxicity; ionic stress and hyperosmotic stress. One factor contributing to ionic toxicity is the capacity of sodium to displace potassium or in some cases magnesium in the active sites of some enzymes. For example, the *HAL2* gene, which confers halotolerance upon overexpression, encodes for a nucleotidase that hydrolyses 3′-phosphoadenosine-5′-phosphate (PAP) to AMP which requires magnesium for catalysis (Murguía et al. 1996). Inhibition of this enzyme by low concentrations of lithium or sodium leads to the accumulation of toxic amounts of PAP and structural data suggests that lithium ions occupy a magnesium binding site necessary for proper catalysis (Albert et al. 2000).

To avoid sodium toxicity, yeast cells actively maintain a high $K^+/Na^+$ ratio. In response to saline stress, sodium extrusion, limitation of sodium entry and vacuolar sequestration are key processes, as discussed above. Ena1 and Nha1 are largely responsible for sodium extrusion under alkaline and acidic conditions, respectively and their activation represents one important physiological response to high sodium concentrations (Bañuelos et al. 1998). High salt concentrations also exert hyperosmotic shock and an essential component of the response to this class of stress is the metabolic adjustment toward production and accumulation of the compatible solute, glycerol to maintain water balance (for reviews see: (Blomberg 2000; Hohmann 2002)). Under these conditions, yeast cells also transiently arrest cell cycle progression and reduce both transcription and translation, presumably to provide time for adaptation, and Hog1 has been directly implicated in many of these processes (Teige et al. 2001; Proft and Struhl 2004; Clotet and Posas 2007; Melamed et al. 2008).

An important aspect of the salt stress response also involves remodeling of the gene expression profile. Several studies have examined the transcriptional response to high sodium concentrations and depending on the conditions employed, as many as 400 and 250 genes may be up-regulated or down-regulated, respectively (Posas et al. 2000; Rep et al. 2000; Causton et al. 2001; Yale and Bohnert 2001). Many of the genes whose mRNA levels are altered under saline stress are also regulated in a similar manner under a variety of stress conditions, and so represent a general stress response mediated in large part by Protein kinase A ((Hohmann et al. 2007) and references therein). The kinetics of the transcriptional regulation of individual genes during stress conditions varies widely, with many promoters responding quickly and transiently and others whose regulation is slower and in some cases prolonged, likely correlating with the function of the encoded protein in the acute response or long term adaptation, respectively. Hog1 is required for the regulation of a subset of genes in response to saline stress, including *ENA1*, as discussed above, and those necessary for glycerol production (Albertyn et al. 1994; Norbeck et al. 1996). Interestingly, the vast majority of the genes up-regulated in response to hyperosmotic stress are not required for cell survival under these conditions (Warringer et al. 2003). A recent report, using a novel signal rewiring approach, suggests that the Hog1-dependent induction of only the *GPD1 (glycerol-3-phosphate dehydrogenase-1)* and *GPP2 (glycerol-3-phosphatase-2)* genes, involved in glycerol biosynthesis, is necessary for osmoadaption (Babazadeh et al. 2014).

Analysis of gene expression has revealed many key features of stress responses. However, as mentioned, the alteration of the expression pattern of specific genes does not necessarily indicate an essential role for the encoded protein in stress adaptation. For example, even if an mRNA accumulates under certain stress conditions, the transcript must still be translated and the protein correctly processed, delivered and possibly activated in order to carry out its function. Several proteomics approaches have been undertaken to study changes in total protein accumulation under conditions of salt stress (reviewed in (Szopinska and Morsomme 2010)). Irrespective of the technique employed, all studies confirm the accumulation of key enzymes needed to shift metabolism towards glycerol production, underscoring the importance of this physiological response (Blomberg 1995; Norbeck and Blomberg 1996; Li et al. 2003; Soufi et al. 2009). A strong correlation between the subset of osmotic shock up-regulated proteins and their corresponding mRNA changes is observed in almost all cases, as would be expected. However, the overall relationship between the proteomic data and published mRNA changes are generally poor, indicating the complexity inherent in extrapolating from gene expression data, as mentioned above. One study analyzed specifically the plasma membrane proteins whose levels are affected during salt stress using a quantitative, gel-free iTRAQ labeling approach (Szopinska et al. 2011). Twelve plasma membrane proteins, including both eisosome components Lsp1 and Pil1, involved in endocytosis, were shown to accumulate, whereas 20 proteins, including Pma1 and ABC transporters, glucose and amino acid transporters, t-SNAREs, and proteins involved in cell wall biogenesis decreased during salt stress treatments. These data fit well with an increase in endocytosis of nutrient permeases in response to salt stress and are consistent with the decrease in amino acid uptake observed under these conditions (Norbeck and Blomberg 1998).

### *8.6.2 Increased Intracellular Potassium*

Like sodium, addition of high extracellular concentrations of potassium (>1 M) also leads to hyperosmotic stress and so in this aspect the cellular response will be similar to that discussed above for sodium. Due to the efficacy of the Ena1 and Nha1 extrusion systems and the reduction in Trk1 activity, wild type cells do not accumulate high internal concentrations of potassium, even in the presence of very high extracellular potassium. However, mutants lacking the ENA gene cluster, *NHA1* or both the *PPZ1* and *PPZ2* phosphatases are sensitive to high extracellular potassium, due to reduced extrusion or inability to inhibit uptake, respectively (Bañuelos et al. 1998; Yenush et al. 2002). The *ppz1 ppz2* mutant has been used as a tool to study some aspects of the physiological consequences of steady state increases in intracellular potassium. These strains were shown to have an increase in cell size and intracellular pH and to display plasma membrane depolarization and constitutive activation of the Slt2/Mpk1 cell wall integrity pathway, suggesting that the cell wall is reinforced to counteract the tugor pressure resulting from increased intracellular potassium (Yenush et al. 2002; Merchan et al. 2004). Interestingly, resistance to DNA damaging agents is also reduced in *ppz1 ppz2* mutants and these phenotypes are rescued by further disruption of the *TRK1* and *TRK2* genes or of the *SLT2/MPK1* gene and are phenocopied by overexpression of a constitutively active version of the Slt2/Mpk1 MAP kinase kinase, *MKK1* (Merchan et al. 2011). Thus, it appears that the constitutive activation of the MAP kinase pathway required for cell wall reinforcement in *ppz1 ppz2* mutants is detrimental for some aspects of DNA integrity.

### *8.6.3 Potassium Starvation*

Many studies have investigated various aspects of the physiological response to and consequences of lowering internal potassium concentrations, either by modifying the external media or by examining strains with genetic modifications that lead to reduced potassium uptake, namely the *trk1 trk2* and *hal4 hal5* mutants. Strains lacking the *TRK1* and *TRK2* genes are hyperpolarized and have a slightly decreased intracellular pH, even under non-limiting potassium conditions, despite the fact that the internal potassium concentration is not markedly different from the wild type control (Madrid et al. 1998; Navarrete et al. 2010). The hyperpolarization of the *trk1 trk2* mutants explains the general sensitivity to toxic cations, whereas the reduced intracellular pH has been attributed to decreased Pma1 activity, which fits well with the reduction in amino acid uptake also observed in these mutants (Yenush et al. 2002; Navarrete et al. 2010). Mutants lacking the genes encoding for the *HAL4* and *HAL5* kinases share many of the same phenotypes with the *trk1 trk2* mutants, such as acidic intracellular pH, decreased amino acid uptake and sensitivity to toxic cations, which is expected for strains lacking positive regulators of these potassium transporters (Perez-Valle et al. 2010). However, the *hal4 hal5* mutants appear to have additional, Trk1-independent defects which lead to a decrease in the accumulation of many nutrient transporters at the plasma membrane, although the molecular mechanism underlying this defect has yet to be defined (Perez-Valle et al. 2007, 2010). Despite the fact that different conditions were used, analysis of the gene expression profiles of both *hal4 hal5* and *trk1 trk2* mutants shows a strong correlation among the genes that are up- or down-regulated (correlation coefficient = 0.77) and indicate cellular processes that are altered in both mutants, such as methionine biosynthesis (Perez-Valle et al. 2010; Barreto et al. 2012). Part of this phenotype may be explained by the marked reduction observed in the accumulation of the high affinity methionine permease, Mup1, which correlates with reduced methionine uptake (Perez-Valle et al. 2010). Studies of these mutants highlight aspects of cell function that are affected in strains where high affinity potassium uptake is permanently disabled. Another physiological situation is the adaptation process that takes place in response to a sudden drop in external potassium concentrations.

Several approaches have been taken to analyze wild type and mutant strains either grown in or shifted to media with limiting potassium concentrations in order to characterize the changes produced by the starvation and the cellular responses that lead to the re-establishment of potassium homeostasis. After several hours of potassium starvation, wild type strains lose 70 % of their internal potassium, the cell volume decreases by about 20 % and cells become hyperpolarized, but the internal pH remains essentially the same (Navarrete et al. 2010). The transcriptional response to potassium starvation has been studied in two ways. In the first approach, cells were grown in chemostat cultures in the presence of limiting concentrations of potassium (Hess et al. 2006). The transcriptional response was moderate, with a total of approximately 110 different genes up- or down-regulated more than threefold in the two lowest potassium concentrations tested (0.65 and 1.3 mM), as compared with the non-limiting potassium control. The majority of the affected transcripts encode proteins involved in nitrogen metabolism. Subsequent experiments revealed ammonium toxicity under limiting potassium conditions and suggest that yeast cells respond to this toxicity by secreting amino acids (Hess et al. 2006). Ammonium was suggested and later proven to enter through the Trk potassium transporters as part of a second study investigating the transcriptional response to short-term potassium starvation (Barreto et al. 2012). In this study, cells were grown in the presence of non-limiting potassium and then shifted to essentially potassium-free media (15 μM) and the transcriptional profile was determined at a series of time points using microarrays. More than 800 genes were shown to be up-regulated at least at one time point, whereas more than 900 genes were shown to be down-regulated. The bulk of the transcriptional response was not observed until 60 mins.

Based on the transcriptional profile and further experiments, the shift to potassium-free media was shown to lead to a myriad of effects, including induction of oxidative stress, alterations in sulfur metabolism, phosphate starvation, pronounced reduction in genes necessary for ribosome biogenesis and translation, activation of the retrograde pathway, alteration of cell cycle-related gene and protein expression profiles and blockage of septin assembly. A similar study was also done using a different approach: Serial Analysis of Gene Expression (SAGE)-tag sequencing (Anemaet and van Heusden 2014). After 60 min of potassium starvation, mRNA levels of 105 and 172 genes were significantly up- or down-regulated, respectively. Although a lower number of genes were shown to be differentially expressed using this technique, there is a reasonable correlation between both studies, especially for genes related to the cell cycle and phosphate starvation. More recently, a detailed study confirmed and further characterized the phosphate deprivation response triggered by potassium starvation (Canadell et al. 2015). Proteomics approaches have also been employed to examine the changes at the level of protein accumulation in control and *trk1 trk2* mutants and in both non-limiting potassium and in response to potassium starvation (Curto et al. 2010; Gelis et al. 2012). Whereas, in the *trk1 trk2* mutants, no differentially expressed proteins were identified in non-limiting potassium medium, the studies using potassium-starved *trk1 trk2* cells showed a marked decrease in the total amount of protein recovered after prolonged potassium starvation. However, as stated by the authors, in both studies the protein recovery was sub-optimal and so key changes in individual protein accumulation of proteins outside the pI and molecular weight range and/or below the abundance threshold may have gone undetected in these experimental approaches.

A mathematical model has helped to determine key events required for effective adaptation to potassium starvation (Kahm et al. 2012). This approach has revealed a complex interplay between biophysical forces and molecular regulation facilitating potassium homeostasis by predicting that proton extrusion and an increased rate of the bicarbonate reaction are vital for cells to maintain a minimal concentration of intracellular potassium in response to sudden starvation. Upon shifting cells to potassium-free media, potassium loss proceeds in two phases; an initial rapid loss, followed by a longer and slower decrease in internal potassium. In *trk1 trk2* mutants, the second phase of potassium loss is much less pronounced than in the wild type cells, presumably due to the hyperpolarization of the membrane. This observation indicates that the lack of the high affinity transporters is not playing a pivotal role in net potassium loss during starvation. Using what is referred to as a reverse tracking algorithm, an initial burst of Pma1 activity and the bicarbonate reaction are predicted to be necessary to maintain the minimum amount of intracellular potassium required for viability. In both cases, this burst in activity will hyperpolarize the plasma membrane, but by two different mechanisms: Pma1 activation will lead to a decrease in the internal positive charge due to proton pumping outside the cell, whereas the bicarbonate reaction will lead to increased internal negative charge due to the accumulation of $HCO_3^-$ inside the cell. Importantly, the increase in Pma1 activity and transient activation of the bicarbonate reaction in response to potassium starvation predicted by the model were both confirmed experimentally. The mechanisms by which the cells sense and signal changes in the external potassium concentrations are still unknown, but this study highlights the usefulness of mathematical models to elucidate important aspects of cell physiology. These authors also present evidence showing that internal steady state potassium concentration is determined by the external concentration, thus indicating that potassium homeostasis is an example of non-perfect adaptation. A more recent study showed that the Trk1 and Trk2 transporters are required for the stabilization of intracellular potassium content by affecting the internal potassium concentrations attained at low extracellular potassium content (Herrera et al. 2014).

## 8.7 Extrapolations and Applications

As summarized above, a large number of laboratories have contributed to various aspects of the study of potassium and sodium transport in the model yeast *S. cerevisiae*. This information is important from a purely scientific point of view, but it also has many different applications, some of which will be mentioned here. For example, the experimental data generated has been used to construct mathematical models describing complex physiological processes, such as response to potassium starvation and to hyperosmotic shock (Klipp et al. 2005; Kahm et al. 2012; Ke et al. 2013). The predictive power of these models has confirmed the validity of these types of approaches and can serve as a framework for modeling processes in multicellular organisms.

On the other hand, the *S. cerevisiae* model system has been used as a point of reference to compare and contrast mechanisms of ion homeostasis in other yeast species, including those that cause disease in humans. Studies of the distribution and function of sodium and potassium transporters in non-conventional yeast species have been expertly reviewed (Ramos et al. 2011). Briefly, in most yeast species studied to date, surplus potassium and sodium are extruded via the joint participation of NHA antiporters, ENA ATPases and TOK potassium channels, whereas potassium uptake is mediated by various combinations of at least three types of systems unevenly spread among the yeast species: TRK and HAK (High Affinity $K^+$) transporters and the ACU (Alkali Cation Uptake) ATPases. Yeast HAK transporters are homologous to the Kup system of *Escherichia coli* and have been proposed to work as $K^+$–$H^+$ symporters with a high concentrative capacity (Rodríguez-Navarro 2000). Whereas HAK transporters are found in many species, including higher plants, functional ACU ATPases have been described only in non-conventional yeast, such as *Ustilago maydis*, *Pichia sorbitophila* and the extremely halotolerant and adaptable fungus, *Hortaea werneckii* (Rodríguez-Navarro 2000; Benito et al. 2004; Plemenitaš et al. 2014). Thus, it appears that many of the general aspects of sodium and potassium transport described above are well-conserved, but depending on the niche, alternative strategies for acquiring and maintaining potassium and sodium homeostasis have evolved.

A large body of evidence indicates that excessive potassium efflux and intracellular potassium depletion are key early steps in apoptosis in mammalian cells (Yu 2003). Several studies suggest that these changes are also implicated in cell death in yeast. For example, prolonged potassium starvation has been shown to lead to cell death through a process in which many of the biochemical markers associated with apoptosis in metazoan cells are detected, such as phosphatidylserine externalization, changes in chromatin condensation, DNA and vacuole fragmentation, as well as enhanced accumulation of reactive oxygen species (ROS) (Lauff and Santa-María 2010). Moreover, both potassium and proton fluxes were shown to influence glucose-induced cell death (Hoeberichts et al. 2010). Using a series of mutants defective for Pma1 activity or potassium uptake or efflux, it was shown that cells that had either reduced Pma1 activity or maintained higher internal potassium concentrations were less sensitive to cell death produced by glucose addition to starved cells, whereas those with lower internal potassium were more sensitive. These effects were also correlated with ROS production and the authors suggest that this is a key event in inducing cell death under these conditions.

Thus, it appears that in yeast, as in mammalian cells, internal potassium homeostasis is vital for cell survival and conditions which alter this balance can lead to cell death. This notion is further supported by studies demonstrating a connection between the fungicidal activities of killer toxin K1, Histatin 5 (Hst 5) and lactoferrin with potassium homeostasis (Ahmed et al. 1999; Sesti et al. 2001; Baev et al. 2003, 2004; Andrés et al. 2008). Although not all the data reported are consistent with this hypothesis, Tok1 has been proposed to be the target of the yeast viral killer toxin K1, which has been shown to bind to and activate the channel from both sides of the plasma membrane (Ahmed et al. 1999; Sesti et al. 2001; Breinig et al. 2002). Hst5, a histidine-rich cationic protein produced in human saliva, is a key component of the non-immune defense system of the oral cavity that possesses both fungistatic and fungicidal activities against several potentially pathogenic fungi, such as *Candida albicans, Candida glabrata*, *Candida krusei* and *Cryptococcus neoformans* (Tsai and Bobek 1997a, b). This toxin induces non-cytolytic efflux of cellular ATP, potassium, and magnesium, implicating these ion movements in the mechanism of Hst5 toxicity. Genetic approaches suggest that Tok1 modulates Hst5-mediated toxicity, whereas Trk1 was shown to be a critical effector of its fungicidal activity in *C. albicans* (Baev et al. 2003, 2004). Similarly, lactoferrin, a protein present in all mammalian mucosal secretions, exhibits antifungal and antibacterial activities through a mechanism that is still being defined (Farnaud and Evans 2003). Lactoferrin causes a rapid release of potassium from *C. albicans* cells and cell death can be inhibited by high extracellular potassium or by treatment with chloride or potassium channel blockers, suggesting a role for potassium channels in the mechanism of action of this fungal toxin (Viejo-Díaz et al. 2004a, b; Andrés et al. 2008).

As alluded to above, alterations in potassium homeostasis also affect plasma membrane potential, nutrient uptake and survival at alkaline pH, which in turn have been linked to flocculation, invasiveness and virulence. For example, hyper-activation of Ppz phosphatases results in alteration of potassium transport leading to Protein kinase A activation and increased expression of the flocculin-encoding *FLO11* gene, thus modulating flocculation and invasive growth in *S. cerevisiae* (González et al. 2013). Whether this mechanism is relevant in pathogenic fungi is still to be determined. A very relevant finding in this context is the identification of *ENA1* as a virulence gene in *Cryptococcus neoformans* (Idnurm et al. 2009). A subset of a library of signature-tagged insertion mutants of this human pathogenic fungus was screened in a murine inhalation model to identify genes required for virulence. Inactivation of the *ENA1* gene led to an avirulent phenotype, which was attributed to the reduced viability of this mutant under alkaline pH conditions. The ability of fungi to grow in slightly alkaline conditions is essential for pathogenesis because, in general, many human host environments have a relatively high pH (reviewed in: (Davis 2009)). This also explains the important role of the alkaline-responsive PacC/Rim101 pathway in pathogenic yeast virulence in found in several species (reviewed in (Cornet and Gaillardin 2014)).

Sensitivity to antifungal drugs is also affected in mutants with alterations in potassium homeostasis. Studies in *S. cerevisiae* have revealed that treatment with ketoconazole, miconazole or amiodarone leads to potassium efflux, similar to that observed for Hst5 and lactoferrin (Peña et al. 2009; Calahorra et al. 2011). Ketoconazole and miconazole are members of the azole class of antifungal drugs, whose main mechanism of toxicity is the inhibition of the biosynthesis of the fungal specific sterol, ergosterol by directly binding and inactivating cytochrome P-450, thus leading to alterations in the properties of the yeast plasma membrane (reviewed in: (Saag and Dismukes 1988)). Amiodarone is a cationic amphipathic drug that interacts preferentially with lipid membranes and has been used clinically as an anti-arrhythmic agent for many years (Mason 1987). It was shown to have broad fungicidal activity (Courchesne 2002). Studies aimed at determining the mechanism of action in yeast showed that mutation of *PMA1*, *TOK1* or *ENA1-4* protected against aminodarone toxicity, suggesting that initial drug-induced hyperpolarization is important in the mechanism of antifungal activity and this was confirmed by decreasing the membrane potential by glucose removal or addition of salts (Maresova et al. 2009). This transient hyperpolarization is followed by depolarization, $Ca^{2+}$ and $H^{+}$ influx and loss in cell viability (Courchesne and Ozturk 2003; Maresova et al. 2009; Peña et al. 2009). Other observations consistent with a role for the determinants of potassium homeostasis in fungal drug sensitivity of pathogenic yeast include studies showing that disruption of *ENA1* and *NHA1* in *C. neoformans* alter membrane potential and the sensitivity to several antifungal drugs (Jung et al. 2012). Finally, in the case of *C. glabrata*, treatment with another azole drug, fluconazole was shown to lead to membrane hyperpolarization and increased sensitivity to cationic drugs (Elicharova and Sychrova 2014). Moreover, in strains lacking *ENA1* and/ or the *CNH1* cation ATPase this combined treatment was even more effective.

Taken together, these observations clearly indicate that the proteins involved in determining and maintaining plasma membrane potential through the modulation of potassium homeostasis represent promising targets for complimentary antifungal treatments. The fact that, for example, the *TRK1* gene, which has no homologues in mammalian cells, is present as a single copy in the *C. albicans* genome and the sequence of the MPM segments (see Sect. 8.3.1 on Trk1) is highly conserved among fungal species suggests that inhibitors of this protein have the potential to be broad-spectrum antifungal treatments with potentially low toxicity (Miranda et al. 2009).

*S. cerevisiae* is also used in a wide range of industrial processes. The impact of potassium homeostasis on nutrient uptake and cell survival will obviously affect many aspects of yeast performance in industrial fermentations. One poignant example was recently reported showing the direct effect of both external and internal potassium and pH on ethanol tolerance in conditions relevant for the industrial production of bioethanol (Lam et al. 2014). These authors showed that increasing the potassium concentration and the pH of the media leads to important improvements in both ethanol tolerance and production under high-glucose and high–cell-density conditions, essentially by boosting cell viability. These results were confirmed using yeast mutants that were engineered to have increased internal potassium and pH by increasing Trk1 activity (*ppz1 ppz2* mutants) and ectopically overexpressing *PMA1*. The ethanol production of these modified laboratory strains was superior to industrial stains currently in use. The authors suggest that this protective effect of augmenting potassium and proton fluxes is due to the counteraction of the dissipation of the potassium and proton gradients caused by partial permeabilization of the plasma membrane in the presence of high ethanol concentrations. Their work shows that these complex, but genetically determined biophysical parameters may be key points of intervention for the development of yeast strains capable of higher bioethanol production in industrial settings.

The knowledge and reagents generated in *S. cerevisiae* have also served as a platform for discovery and characterization of ion transporters from both plants and mammals. For example, in the case of plant ion transporters, yeast mutants defective in high affinity potassium transport (*trk1 trk2*) were used to identify and determine structure/function relationships for the plant potassium channels KAT1 and AKT1 (Anderson et al. 1992, 1994; Schachtman et al. 1992; Sentenac et al. 1992; Uozumi et al. 1995; Nakamura et al. 1997). The *Arabidopsis thaliana* SOS (Salt Overly Sensitive) pathway, consisting of the SOS1 sodium ATPase, the SOS2 protein kinase and the SOS3 $Ca^{2+}$ sensor was functionally reconstituted in yeast strains devoid of *ENA1-4*, *NHA1* and *NHX1* (Quintero et al. 2002). The *Arabidopsis thaliana* CHX17 gene was shown to complement the *S. cerevisiae kha1* mutant phenotypes, suggesting that this transporter can function as a $K^+/H^+$ exchanger in the Golgi of yeast (Maresova and Sychrova 2006). Finally, several studies have used *S. cerevisiae nhx1* or *ena1-4 nha1 nhx1* mutants to study plant intracellular $Na^+/H^+$ exchangers from both rice and *Arabidopsis thaliana* (Gaxiola et al. 1999; Quintero et al. 2000; Fukuda et al. 2004; Kinclová-Zimmermannová et al. 2004). These studies demonstrate the level of conservation that exists between yeast and plants and confirms the utility of yeast model systems for the study of higher organisms.

Not surprisingly, this same complementation approach has been used for the characterization of mammalian ion transporters as well. For example, several functional studies of inward rectifying potassium channels have been done using *trk1 trk2* mutants (for examples see: (Tang et al. 1995; Minor et al. 1999; Hasenbrink et al. 2005; Haass et al. 2007; Schwarzer et al. 2008; D'Avanzo et al. 2010)). These heterologous expression systems have also been used for high-throughput screenings searching for small molecule modulators of potassium channels (Zaks-Makhina et al. 2004, 2009; Bagriantsev et al. 2013). The heterotrimeric ENaC sodium channel has also been functionally expressed in yeast and shown to increase salt sensitivity (Gupta and Canessa 2000). Mutations in this sodium channel have been linked to an inherited form of hypertension called Liddle's Syndrome (Shimkets et al. 1994). Other transporters, such as the $Na^+$,$K^+$-ATPase, the CFTR (cystic fibrosis transmembrane conductance regulator) chloride channel and a $Na^+$-phosphate co-transporter have also been studied by heterologous expression in yeast (for a review see: (Kolb et al. 2011)). Using *ena1-4 nha1 nhx1* mutants, several mammalian $Na^+/H^+$ exchangers have also been characterized (Montero-Lomelí and Okorokova Façanha 1999; Flegelova et al. 2006; Xiang et al. 2007). One very interesting study used the yeast model system to characterize mutations in the human NHE9 $Na^+/H^+$ antiporter that have been associated with autism (Kondapalli et al. 2013). In this study, equivalent mutations found in autistic patients were introduced into the *NHX1* gene and functional studies of these modified transporters showed that two of them led to a loss of Nhx1 function. Finally, as discussed earlier, another example of the utility of yeast to contribute to the definition of the function of the proteins encoded by disease related genes is the discovery that the LETM1 gene, responsible for seizures associated with some forms of Wolf-Hirschhorn Syndrome, can functionally complement the *mdm38* yeast mutant mitochondrial KHE phenotypes (see above) (Nowikovsky et al. 2004).

In conclusion, our knowledge regarding potassium and sodium transport in yeast is quite extensive, but far from complete. Although most of the major transporters have been identified and extensively characterized, some fluxes, such as the low affinity potassium uptake NSC1 activity and the mitochondrial $K^+/H^+$ exchange across the inner membrane, await molecular characterization. Moreover, several ORFs with weak sequence homology to mammalian ion transporters still have unknown functions and may help to complete the picture, especially in the case of the intracellular distribution of potassium and sodium. Our understanding of the regulation of many of these transporters, especially on the post-translational level is also very limited. Integration of the signals leading to the establishment and maintenance of ion homeostasis in response to changing environments is an area where progress still needs to be made. Given the importance of this field to basic science and its applications ranging from industrial processes to plant salt and drought tolerance and mammalian physiology, disease states and drug discovery, new advances in the study of yeast potassium and sodium transport are likely to bring new insight with both expected and novel impacts in the future.

**Acknowledgements** L.Y. is funded by grant BFU2011-30197-C03-03 from the Spanish Ministry of Science and Innovation (Madrid, Spain) and EUI2009-04147 [Systems Biology of Microorganisms (SysMo2) European Research Area-Network (ERA-NET)].

## References


Ahmed A, Sesti F, Ilan N, Shih TM, Sturley SL et al (1999) A molecular target for viral killer toxin: TOK1 potassium channels. Cell 99:283–291

Albert A, Yenush L, Gil-Mascarell MR, Rodriguez PL, Patel S et al (2000) X-ray structure of yeast Hal2p, a major target of lithium and sodium toxicity, and identification of framework interactions determining cation sensitivity. J Mol Biol 295:927–938

Albertyn J, Hohmann S, Thevelein JM, Prior BA (1994) GPD1, which encodes glycerol-3-phosphate dehydrogenase, is essential for growth under osmotic stress in Saccharomyces cerevisiae, and its expression is regulated by the high-osmolarity glycerol response pathway. Mol Cell Biol 14:4135–4144

Alepuz PM, Cunningham KW, Estruch F (1997) Glucose repression affects ion homeostasis in yeast through the regulation of the stress-activated ENA1 gene. Mol Microbiol 26:91–98

Ali R, Brett CL, Mukherjee S, Rao R (2004) Inhibition of sodium/proton exchange by a Rab-GTPase-activating protein regulates endosomal traffic in yeast. J Biol Chem 279:4498–4506

Alijo R, Ramos J (1993) Several routes of activation of the potassium uptake system of yeast. Biochim Biophys Acta 1179:224–228

Anderson JA, Huprikar SS, Kochian LV, Lucas WJ, Gaber RF (1992) Functional expression of a probable Arabidopsis thaliana potassium channel in Saccharomyces cerevisiae. Proc Natl Acad Sci U S A 89:3736–3740

Anderson JA, Nakamura RL, Gaber RF (1994) Heterologous expression of K+ channels in Saccharomyces cerevisiae: strategies for molecular analysis of structure and function. Symp Soc Exp Biol 48:85–97

André B, Scherens B (1995) The yeast YBR235w gene encodes a homolog of the mammalian electroneutral Na(+)-(K+)-C1- cotransporter family. Biochem Biophys Res Commun 217:150–153

Andrés MT, Viejo-Díaz M, Fierro JF (2008) Human lactoferrin induces apoptosis-like cell death in Candida albicans: critical role of K+-channel-mediated K+ efflux. Antimicrob Agents Chemother 52:4081–4088

Anemaet IG, van Heusden GP (2014) Transcriptional response of Saccharomyces cerevisiae to potassium starvation. BMC Genomics 15:1040

Arino J, Ramos J, Sychrova H (2010) Alkali metal cation transport and homeostasis in yeasts. Microbiol Mol Biol Rev 74:95–120

Babazadeh R, Furukawa T, Hohmann S, Furukawa K (2014) Rewiring yeast osmostress signalling through the MAPK network reveals essential and non-essential roles of Hog1 in osmoadaptation. Sci Rep 4:4697

Baev D, Rivetta A, Li XS, Vylkova S, Bashi E et al (2003) Killing of Candida albicans by human salivary histatin 5 is modulated, but not determined, by the potassium channel TOK1. Infect Immun 71:3251–3260

Baev D, Rivetta A, Vylkova S, Sun JN, Zeng GF et al (2004) The TRK1 potassium transporter is the critical effector for killing of Candida albicans by the cationic protein, Histatin 5. J Biol Chem 279:55060–55072

Bagriantsev SN, Ang KH, Gallardo-Godoy A, Clark KA, Arkin MR et al (2013) A high-throughput functional screen identifies small molecule regulators of temperature- and mechano-sensitive K2P channels. ACS Chem Biol 8:1841–1851

Bañuelos MA, Sychrová H, Bleykasten-Grosshans C, Souciet JL, Potier S (1998) The Nha1 antiporter of Saccharomyces cerevisiae mediates sodium and potassium efflux. Microbiology 144(Pt 10):2749–2758

Bañuelos MA, Ruiz MC, Jiménez A, Souciet JL, Potier S et al (2002) Role of the Nha1 antiporter in regulating K(+) influx in Saccharomyces cerevisiae. Yeast 19:9–15

Barnett JA (2008) A history of research on yeasts 13. Active transport and the uptake of various metabolites. Yeast 25:689–731

Barreto L, Canadell D, Petrezselyova S, Navarrete C, Maresova L et al (2011) A genomewide screen for tolerance to cationic drugs reveals genes important for potassium homeostasis in Saccharomyces cerevisiae. Eukaryot Cell 10:1241–1250

Barreto L, Canadell D, Valverde-Saubí D, Casamayor A, Ariño J (2012) The short-term response of yeast to potassium starvation. Environ Microbiol 14:3026–3042

Benito B, Moreno E, Lagunas R (1991) Half-life of the plasma membrane ATPase and its activating system in resting yeast cells. Biochim Biophys Acta 1063:265–268

Benito B, Quintero FJ, Rodríguez-Navarro A (1997) Overexpression of the sodium ATPase of Saccharomyces cerevisiae: conditions for phosphorylation from ATP and Pi. Biochim Biophys Acta 1328:214–226

Benito B, Garciadeblás B, Rodríguez-Navarro A (2002) Potassium- or sodium-efflux ATPase, a key enzyme in the evolution of fungi. Microbiology 148:933–941

Benito B, Garciadeblás B, Schreier P, Rodríguez-Navarro A (2004) Novel p-type ATPases mediate high-affinity potassium or sodium uptake in fungi. Eukaryot Cell 3:359–368

Bernardi P (1999) Mitochondrial transport of cations: channels, exchangers, and permeability transition. Physiol Rev 79:1127–1155

Bertl A, Slayman CL, Gradmann D (1993) Gating and conductance in an outward-rectifying K+ channel from the plasma membrane of Saccharomyces cerevisiae. J Membr Biol 132:183–199

Bertl A, Bihler H, Reid JD, Kettner C, Slayman CL (1998) Physiological characterization of the yeast plasma membrane outward rectifying K+ channel, DUK1 (TOK1), in situ. J Membr Biol 162:67–80

Bertl A, Ramos J, Ludwig J, Lichtenberg-Fraté H, Reid J et al (2003) Characterization of potassium transport in wild-type and isogenic yeast strains carrying all combinations of trk1, trk2 and tok1 null mutations. Mol Microbiol 47:767–780

Bihler H, Slayman CL, Bertl A (1998) NSC1: a novel high-current inward rectifier for cations in the plasma membrane of Saccharomyces cerevisiae. FEBS Lett 432:59–64

Bihler H, Slayman CL, Bertl A (2002) Low-affinity potassium uptake by Saccharomyces cerevisiae is mediated by NSC1, a calcium-blocked non-specific cation channel. Biochim Biophys Acta 1558:109–118

Blomberg A (1995) Global changes in protein synthesis during adaptation of the yeast Saccharomyces cerevisiae to 0.7 M NaCl. J Bacteriol 177:3563–3572

Blomberg A (2000) Metabolic surprises in Saccharomyces cerevisiae during adaptation to saline conditions: questions, some answers and a model. FEMS Microbiol Lett 182:1–8

Borst-Pauwels GW (1981) Ion transport in yeast. Biochim Biophys Acta 650:88–127

Botstein D, Fink GR (2011) Yeast: an experimental organism for 21st Century biology. Genetics 189:695–704

Bouillet LE, Cardoso AS, Perovano E, Pereira RR, Ribeiro EM et al (2012) The involvement of calcium carriers and of the vacuole in the glucose-induced calcium signaling and activation of the plasma membrane H(+)-ATPase in Saccharomyces cerevisiae cells. Cell Calcium 51:72–81

Bowers K, Levi BP, Patel FI, Stevens TH (2000) The sodium/proton exchanger Nhx1p is required for endosomal protein trafficking in the yeast Saccharomyces cerevisiae. Mol Biol Cell 11:4277–4294

Breinig F, Tipper DJ, Schmitt MJ (2002) Kre1p, the plasma membrane receptor for the yeast K1 viral toxin. Cell 108:395–405

Brett CL, Tukaye DN, Mukherjee S, Rao R (2005) The yeast endosomal Na+K+/H+ exchanger Nhx1 regulates cellular pH to control vesicle trafficking. Mol Biol Cell 16:1396–1405

Cagnac O, Leterrier M, Yeager M, Blumwald E (2007) Identification and characterization of Vnx1p, a novel type of vacuolar monovalent cation/H+ antiporter of Saccharomyces cerevisiae. J Biol Chem 282:24284–24293

Cagnac O, Aranda-Sicilia MN, Leterrier M, Rodriguez-Rosales MP, Venema K (2010) Vacuolar cation/H+ antiporters of Saccharomyces cerevisiae. J Biol Chem 285:33914–33922

Calahorra M, Lozano C, Sánchez NS, Peña A (2011) Ketoconazole and miconazole alter potassium homeostasis in Saccharomyces cerevisiae. Biochim Biophys Acta 1808:433–445

Canadell D, González A, Casado C, Ariño J (2015) Functional interactions between potassium and phosphate homeostasis in Saccharomyces cerevisiae. Mol Microbiol 95:555–572

Casado C, Yenush L, Melero C, del Carmen Ruiz M, Serrano R et al (2010) Regulation of Trk-dependent potassium transport by the calcineurin pathway involves the Hal5 kinase. FEBS Lett 584:2415–2420

Causton HC, Ren B, Koh SS, Harbison CT, Kanin E et al (2001) Remodeling of yeast genome expression in response to environmental changes. Mol Biol Cell 12:323–337

Clotet J, Posas F (2007) Control of cell cycle in response to osmostress: lessons from yeast. Methods Enzymol 428:63–76

Cornet M, Gaillardin C (2014) pH signaling in human fungal pathogens: a new target for antifungal strategies. Eukaryot Cell 13:342–352

Courchesne WE (2002) Characterization of a novel, broad-based fungicidal activity for the antiarrhythmic drug amiodarone. J Pharmacol Exp Ther 300:195–199

Courchesne WE, Ozturk S (2003) Amiodarone induces a caffeine-inhibited, MID1-dependent rise in free cytoplasmic calcium in Saccharomyces cerevisiae. Mol Microbiol 47:223–234

Crespo JL, Daicho K, Ushimaru T, Hall MN (2001) The GATA transcription factors GLN3 and GAT1 link TOR to salt stress in Saccharomyces cerevisiae. J Biol Chem 276:34441–34444

Cunningham KW, Fink GR (1996) Calcineurin inhibits VCX1-dependent H+/Ca2+ exchange and induces Ca2+ ATPases in Saccharomyces cerevisiae. Mol Cell Biol 16:2226–2237

Curto M, Valledor L, Navarrete C, Gutiérrez D, Sychrova H et al (2010) 2-DE based proteomic analysis of Saccharomyces cerevisiae wild and K+ transport-affected mutant (trk1,2) strains at the growth exponential and stationary phases. J Proteomics 73:2316–2335

D'Avanzo N, Cheng WW, Xia X, Dong L, Savitsky P et al (2010) Expression and purification of recombinant human inward rectifier K+ (KCNJ) channels in Saccharomyces cerevisiae. Protein Expr Purif 71:115–121

Daran-Lapujade P, Daran JM, Luttik MA, Almering MJ, Pronk JT et al (2009) An atypical PMR2 locus is responsible for hypersensitivity to sodium and lithium cations in the laboratory strain Saccharomyces cerevisiae CEN.PK113-7D. FEMS Yeast Res 9:789–792

Davis DA (2009) How human pathogenic fungi sense and adapt to pH: the link to virulence. Curr Opin Microbiol 12:365–370

de Nadal E, Posas F (2011) Elongating under stress. Genet Res Int 2011:326286

de Nadal E, Clotet J, Posas F, Serrano R, Gomez N et al (1998) The yeast halotolerance determinant Hal3p is an inhibitory subunit of the Ppz1p Ser/Thr protein phosphatase. Proc Natl Acad Sci U S A 95:7357–7362

de Nadal E, Calero F, Ramos J, Ariño J (1999) Biochemical and genetic analyses of the role of yeast casein kinase 2 in salt tolerance. J Bacteriol 181:6456–6462

de Nadal E, Alepuz PM, Posas F (2002) Dealing with osmostress through MAP kinase activation. EMBO Rep 3:735–740

De Nadal E, Zapater M, Alepuz PM, Sumoy L, Mas G et al (2004) The MAPK Hog1 recruits Rpd3 histone deacetylase to activate osmoresponsive genes. Nature 427:370–374

Dimmer KS, Fritz S, Fuchs F, Messerschmitt M, Weinbach N et al (2002) Genetic basis of mitochondrial function and morphology in Saccharomyces cerevisiae. Mol Biol Cell 13:847–853

Durell SR, Guy HR (1999) Structural models of the KtrB, TrkH, and Trk1,2 symporters based on the structure of the KcsA K(+) channel. Biophys J 77:789–807

Eide DJ, Clark S, Nair TM, Gehl M, Gribskov M et al (2005) Characterization of the yeast ionome: a genome-wide analysis of nutrient mineral and trace element homeostasis in Saccharomyces cerevisiae. Genome Biol 6:R77

Elicharova H, Sychrova H (2014) Fluconazole affects the alkali-metal-cation homeostasis and susceptibility to cationic toxic compounds of Candida glabrata. Microbiology 160:1705–1713

Endele S, Fuhry M, Pak SJ, Zabel BU, Winterpacht A (1999) LETM1, a novel gene encoding a putative EF-hand Ca(2+)-binding protein, flanks the Wolf-Hirschhorn syndrome (WHS) critical region and is deleted in most WHS patients. Genomics 60:218–225

Eraso P, Mazón MJ, Portillo F (2006) Yeast protein kinase Ptk2 localizes at the plasma membrane and phosphorylates in vitro the C-terminal peptide of the H+-ATPase. Biochim Biophys Acta 1758:164–170
Erez O, Kahana C (2002) Deletions of SKY1 or PTK2 in the Saccharomyces cerevisiae trk1Deltatrk2Delta mutant cells exert dual effect on ion homeostasis. Biochem Biophys Res Commun 295:1142–1149
Estrada E, Agostinis P, Vandenheede JR, Goris J, Merlevede W et al (1996) Phosphorylation of yeast plasma membrane H+-ATPase by casein kinase I. J Biol Chem 271:32064–32072
Fairman C, Zhou X, Kung C (1999) Potassium uptake through the TOK1 K+ channel in the budding yeast. J Membr Biol 168:149–157
Farnaud S, Evans RW (2003) Lactoferrin – a multifunctional protein with antimicrobial properties. Mol Immunol 40:395–405
Fell GL, Munson AM, Croston MA, Rosenwald AG (2011) Identification of yeast genes involved in k homeostasis: loss of membrane traffic genes affects k uptake. G3 (Bethesda) 1:43–56
Fernandes AR, Sá-Correia I (2003) Transcription patterns of PMA1 and PMA2 genes and activity of plasma membrane H+-ATPase in Saccharomyces cerevisiae during diauxic growth and stationary phase. Yeast 20:207–219
Ferrando A, Kron SJ, Rios G, Fink GR, Serrano R (1995) Regulation of cation transport in Saccharomyces cerevisiae by the salt tolerance gene HAL3. Mol Cell Biol 15:5470–5481
Ferrigno P, Posas F, Koepp D, Saito H, Silver PA (1998) Regulated nucleo/cytoplasmic exchange of HOG1 MAPK requires the importin beta homologs NMD5 and XPO1. EMBO J 17:5606–5614
Flegelova H, Haguenauer-Tsapis R, Sychrova H (2006) Heterologous expression of mammalian Na/H antiporters in Saccharomyces cerevisiae. Biochim Biophys Acta 1760:504–516
Flis K, Hinzpeter A, Edelman A, Kurlandzka A (2005) The functioning of mammalian ClC-2 chloride channel in Saccharomyces cerevisiae cells requires an increased level of Kha1p. Biochem J 390:655–664
Forment J, Mulet JM, Vicente O, Serrano R (2002) The yeast SR protein kinase Sky1p modulates salt tolerance, membrane potential and the Trk1,2 potassium transporter. Biochim Biophys Acta 1565:36–40
Froschauer E, Nowikovsky K, Schweyen RJ (2005) Electroneutral K+/H+ exchange in mitochondrial membrane vesicles involves Yol027/Letm1 proteins. Biochim Biophys Acta 1711:41–48
Fukuda A, Nakamura A, Tagiri A, Tanaka H, Miyao A et al (2004) Function, intracellular localization and the importance in salt tolerance of a vacuolar Na(+)/H(+) antiporter from rice. Plant Cell Physiol 45:146–159
Gaber RF (1992) Molecular genetics of yeast ion transport. Int Rev Cytol 137:299–353
Gaber RF, Styles CA, Fink GR (1988) TRK1 encodes a plasma membrane protein required for high-affinity potassium transport in Saccharomyces cerevisiae. Mol Cell Biol 8:2848–2859
Gaxiola RA, Rao R, Sherman A, Grisafi P, Alper SL et al (1999) The Arabidopsis thaliana proton transporters, AtNhx1 and Avp1, can function in cation detoxification in yeast. Proc Natl Acad Sci U S A 96:1480–1485
Gelis S, Curto M, Valledor L, González A, Ariño J et al (2012) Adaptation to potassium starvation of wild-type and K(+)-transport mutant (trk1,2) of Saccharomyces cerevisiae: 2-dimensional gel electrophoresis-based proteomic approach. Microbiologyopen 1:182–193
Gómez MJ, Luyten K, Ramos J (1996) The capacity to transport potassium influences sodium tolerance in Saccharomyces cerevisiae. FEMS Microbiol Lett 135:157–160
González A, Casado C, Petrezsélyová S, Ruiz A, Ariño J (2013) Molecular analysis of a conditional hal3 vhs3 yeast mutant links potassium homeostasis with flocculation and invasiveness. Fungal Genet Biol 53:1–9
Goossens A, de La Fuente N, Forment J, Serrano R, Portillo F (2000) Regulation of yeast H(+)-ATPase by protein kinases belonging to a family dedicated to activation of plasma membrane transporters. Mol Cell Biol 20:7654–7661
Gupta SS, Canessa CM (2000) Heterologous expression of a mammalian epithelial sodium channel in yeast. FEBS Lett 481:77–80

Gustin MC, Martinac B, Saimi Y, Culbertson MR, Kung C (1986) Ion channels in yeast. Science 233:1195–1197
Haass FA, Jonikas M, Walter P, Weissman JS, Jan YN et al (2007) Identification of yeast proteins necessary for cell-surface function of a potassium channel. Proc Natl Acad Sci U S A 104:18079–18084
Haro R, Rodríguez-Navarro A (2002) Molecular analysis of the mechanism of potassium uptake through the TRK1 transporter of Saccharomyces cerevisiae. Biochim Biophys Acta 1564:114–122
Haro R, Rodríguez-Navarro A (2003) Functional analysis of the M2(D) helix of the TRK1 potassium transporter of Saccharomyces cerevisiae. Biochim Biophys Acta 1613:1–6
Haro R, Garciadeblas B, Rodríguez-Navarro A (1991) A novel P-type ATPase from yeast involved in sodium transport. FEBS Lett 291:189–191
Hasenbrink G, Schwarzer S, Kolacna L, Ludwig J, Sychrova H et al (2005) Analysis of the mKir2.1 channel activity in potassium influx defective Saccharomyces cerevisiae strains determined as changes in growth characteristics. FEBS Lett 579:1723–1731
Herrera R, Álvarez MC, Gelis S, Ramos J (2013) Subcellular potassium and sodium distribution in Saccharomyces cerevisiae wild-type and vacuolar mutants. Biochem J 454:525–532
Herrera R, Alvarez MC, Gelis S, Kodedová M, Sychrová H et al (2014) Role of Saccharomyces cerevisiae Trk1 in stabilization of intracellular potassium content upon changes in external potassium levels. Biochim Biophys Acta 1838:127–133
Hess DC, Lu W, Rabinowitz JD, Botstein D (2006) Ammonium toxicity and potassium limitation in yeast. PLoS Biol 4:e351
Hoeberichts FA, Perez-Valle J, Montesinos C, Mulet JM, Planes MD et al (2010) The role of K+ and H+ transport systems during glucose- and H2O2-induced cell death in Saccharomyces cerevisiae. Yeast 27:713–725
Hohmann S (2002) Osmotic stress signaling and osmoadaptation in yeasts. Microbiol Mol Biol Rev 66:300–372
Hohmann S, Krantz M, Nordlander B (2007) Yeast osmoregulation. Methods Enzymol 428:29–45
Idnurm A, Walton FJ, Floyd A, Reedy JL, Heitman J (2009) Identification of ENA1 as a virulence gene of the human pathogenic fungus Cryptococcus neoformans through signature-tagged insertional mutagenesis. Eukaryot Cell 8:315–326
Jung KW, Strain AK, Nielsen K, Jung KH, Bahn YS (2012) Two cation transporters Ena1 and Nha1 cooperatively modulate ion homeostasis, antifungal drug resistance, and virulence of Cryptococcus neoformans via the HOG pathway. Fungal Genet Biol 49:332–345
Kafadar KA, Cyert MS (2004) Integration of stress responses: modulation of calcineurin signaling in Saccharomyces cerevisiae by protein kinase A. Eukaryot Cell 3:1147–1153
Kahm M, Navarrete C, Llopis-Torregrosa V, Herrera R, Barreto L et al (2012) Potassium starvation in yeast: mechanisms of homeostasis revealed by mathematical modeling. PLoS Comput Biol 8:e1002548
Kallay LM, Brett CL, Tukaye DN, Wemmer MA, Chyou A et al (2011) Endosomal Na+(K+)/H+ exchanger Nhx1/Vps44 functions independently and downstream of multivesicular body formation. J Biol Chem 286:44067–44077
Kane PM (2007) The long physiological reach of the yeast vacuolar H+-ATPase. J Bioenerg Biomembr 39:415–421
Kane PM (2012) Targeting reversible disassembly as a mechanism of controlling V-ATPase activity. Curr Protein Pept Sci 13:117–123
Ke R, Ingram PJ, Haynes K (2013) An integrative model of ion regulation in yeast. PLoS Comput Biol 9:e1002879
Ketchum KA, Joiner WJ, Sellers AJ, Kaczmarek LK, Goldstein SA (1995) A new family of outwardly rectifying potassium channel proteins with two pore domains in tandem. Nature 376:690–695
Kinclová O, Ramos J, Potier S, Sychrová H (2001) Functional study of the Saccharomyces cerevisiae Nha1p C-terminus. Mol Microbiol 40:656–668

Kinclova-Zimmermannova O, Sychrova H (2006) Functional study of the Nha1p C-terminus: involvement in cell response to changes in external osmolarity. Curr Genet 49:229–236
Kinclová-Zimmermannová O, Flegelová H, Sychrová H (2004) Rice Na+/H+-antiporter Nhx1 partially complements the alkali-metal-cation sensitivity of yeast strains lacking three sodium transporters. Folia Microbiol (Praha) 49:519–525
Kinclova-Zimmermannova O, Gaskova D, Sychrova H (2006) The Na+, K+/H+ -antiporter Nha1 influences the plasma membrane potential of Saccharomyces cerevisiae. FEMS Yeast Res 6:792–800
Klee CB, Draetta GF, Hubbard MJ (1988) Calcineurin. Adv Enzymol Relat Areas Mol Biol 61:149–200
Klipp E, Nordlander B, Krüger R, Gennemark P, Hohmann S (2005) Integrative model of the response of yeast to osmotic shock. Nat Biotechnol 23:975–982
Ko CH, Gaber RF (1991) TRK1 and TRK2 encode structurally related K+ transporters in Saccharomyces cerevisiae. Mol Cell Biol 11:4266–4273
Ko CH, Buckley AM, Gaber RF (1990) TRK2 is required for low affinity K+ transport in Saccharomyces cerevisiae. Genetics 125:305–312
Ko CH, Liang H, Gaber RF (1993) Roles of multiple glucose transporters in Saccharomyces cerevisiae. Mol Cell Biol 13:638–648
Kojima A, Toshima JY, Kanno C, Kawata C, Toshima J (2012) Localization and functional requirement of yeast Na+/H+ exchanger, Nhx1p, in the endocytic and protein recycling pathway. Biochim Biophys Acta 1823:534–543
Kolb AR, Buck TM, Brodsky JL (2011) Saccharomyces cerivisiae as a model system for kidney disease: what can yeast tell us about renal function? Am J Physiol Renal Physiol 301:F1–F11
Kondapalli KC, Hack A, Schushan M, Landau M, Ben-Tal N et al (2013) Functional evaluation of autism-associated mutations in NHE9. Nat Commun 4:2510
Kuroda T, Bihler H, Bashi E, Slayman CL, Rivetta A (2004) Chloride channel function in the yeast TRK-potassium transporters. J Membr Biol 198:177–192
Lam FH, Ghaderi A, Fink GR, Stephanopoulos G (2014) Biofuels. Engineering alcohol tolerance in yeast. Science 346:71–75
Lamb TM, Mitchell AP (2003) The transcription factor Rim101p governs ion tolerance and cell differentiation by direct repression of the regulatory genes NRG1 and SMP1 in Saccharomyces cerevisiae. Mol Cell Biol 23:677–686
Lamb TM, Xu WJ, Diamond A, Mitchell AP (2001) Alkaline response genes of Saccharomyces cerevisiae and their relationship to the RIM101 pathway. J Biol Chem 276:1850–1856
Lapathitis G, Kotyk A (1998) Univalent cation fluxes in yeast. Biochem Mol Biol Int 44:371–380
Larsson K, Böhl F, Sjöström I, Akhtar N, Strand D et al (1998) The Saccharomyces cerevisiae SOP1 and SOP2 genes, which act in cation homeostasis, can be functionally substituted by the Drosophila lethal(2)giant larvae tumor suppressor gene. J Biol Chem 273:33610–33618
Lauff DB, Santa-María GE (2010) Potassium deprivation is sufficient to induce a cell death program in Saccharomyces cerevisiae. FEMS Yeast Res 10:497–507
Lecchi S, Nelson CJ, Allen KE, Swaney DL, Thompson KL et al (2007) Tandem phosphorylation of Ser-911 and Thr-912 at the C terminus of yeast plasma membrane H+-ATPase leads to glucose-dependent activation. J Biol Chem 282:35471–35481
Lesage F, Guillemare E, Fink M, Duprat F, Lazdunski M et al (1996) A pH-sensitive yeast outward rectifier K+ channel with two pore domains and novel gating properties. J Biol Chem 271:4183–4187
Li J, Steen H, Gygi SP (2003) Protein profiling with cleavable isotope-coded affinity tag (cICAT) reagents: the yeast salinity stress response. Mol Cell Proteomics 2:1198–1204
Loukin SH, Saimi Y (2002) Carboxyl tail prevents yeast K(+) channel closure: proposal of an integrated model of TOK1 gating. Biophys J 82:781–792
Loukin SH, Vaillant B, Zhou XL, Spalding EP, Kung C et al (1997) Random mutagenesis reveals a region important for gating of the yeast K+ channel Ykc1. EMBO J 16:4817–4825

Madrid R, Gómez MJ, Ramos J, Rodríguez-Navarro A (1998) Ectopic potassium uptake in trk1 trk2 mutants of Saccharomyces cerevisiae correlates with a highly hyperpolarized membrane potential. J Biol Chem 273:14838–14844

Malinsky J, Opekarová M, Grossmann G, Tanner W (2013) Membrane microdomains, rafts, and detergent-resistant membranes in plants and fungi. Annu Rev Plant Biol 64:501–529

Maresova L, Sychrova H (2005) Physiological characterization of Saccharomyces cerevisiae kha1 deletion mutants. Mol Microbiol 55:588–600

Maresova L, Sychrova H (2006) Arabidopsis thaliana CHX17 gene complements the kha1 deletion phenotypes in Saccharomyces cerevisiae. Yeast 23:1167–1171

Marešová L, Sychrová H (2010) Genetic interactions among the Arl1 GTPase and intracellular Na(+) /H(+) antiporters in pH homeostasis and cation detoxification. FEMS Yeast Res 10:802–811

Maresova L, Urbankova E, Gaskova D, Sychrova H (2006) Measurements of plasma membrane potential changes in Saccharomyces cerevisiae cells reveal the importance of the Tok1 channel in membrane potential maintenance. FEMS Yeast Res 6:1039–1046

Maresova L, Muend S, Zhang YQ, Sychrova H, Rao R (2009) Membrane hyperpolarization drives cation influx and fungicidal activity of amiodarone. J Biol Chem 284:2795–2802

Márquez JA, Serrano R (1996) Multiple transduction pathways regulate the sodium-extrusion gene PMR2/ENA1 during salt stress in yeast. FEBS Lett 382:89–92

Marqués MC, Zamarbide-Forés S, Pedelini L, Llopis-Torregrosa V, Yenush L (2015) A functional Rim101 complex is required for proper accumulation of the Ena1 Na+-ATPase protein in response to salt stress in Saccharomyces cerevisiae. FEMS Yeast Res 15(4):fov017

Martinac B, Saimi Y, Kung C (2008) Ion channels in microbes. Physiol Rev 88:1449–1490

Martinez P, Persson BL (1998) Identification, cloning and characterization of a derepressible Na+-coupled phosphate transporter in Saccharomyces cerevisiae. Mol Gen Genet 258:628–638

Martínez-Muñoz GA, Kane P (2008) Vacuolar and plasma membrane proton pumps collaborate to achieve cytosolic pH homeostasis in yeast. J Biol Chem 283:20309–20319

Mason JW (1987) Amiodarone. N Engl J Med 316:455–466

Masuda CA, Ramírez J, Peña A, Montero-Lomelí M (2000) Regulation of monovalent ion homeostasis and pH by the Ser-Thr protein phosphatase SIT4 in Saccharomyces cerevisiae. J Biol Chem 275:30957–30961

Matsumoto TK, Ellsmore AJ, Cessna SG, Low PS, Pardo JM et al (2002) An osmotically induced cytosolic Ca2+ transient activates calcineurin signaling to mediate ion homeostasis and salt tolerance of Saccharomyces cerevisiae. J Biol Chem 277:33075–33080

McCusker JH, Perlin DS, Haber JE (1987) Pleiotropic plasma membrane ATPase mutations of Saccharomyces cerevisiae. Mol Cell Biol 7:4082–4088

Melamed D, Pnueli L, Arava Y (2008) Yeast translational response to high salinity: global analysis reveals regulation at multiple levels. RNA 14:1337–1351

Mendizabal I, Pascual-Ahuir A, Serrano R, de Larrinoa IF (2001) Promoter sequences regulated by the calcineurin-activated transcription factor Crz1 in the yeast ENA1 gene. Mol Genet Genomics 265:801–811

Mendoza I, Rubio F, Rodriguez-Navarro A, Pardo JM (1994) The protein phosphatase calcineurin is essential for NaCl tolerance of Saccharomyces cerevisiae. J Biol Chem 269:8792–8796

Merchan S, Bernal D, Serrano R, Yenush L (2004) Response of the Saccharomyces cerevisiae Mpk1 mitogen-activated protein kinase pathway to increases in internal turgor pressure caused by loss of Ppz protein phosphatases. Eukaryot Cell 3:100–107

Merchan S, Pedelini L, Hueso G, Calzada A, Serrano R et al (2011) Genetic alterations leading to increases in internal potassium concentrations are detrimental for DNA integrity in Saccharomyces cerevisiae. Genes Cells 16:152–165

Michel B, Lozano C, Rodríguez M, Coria R, Ramírez J et al (2006) The yeast potassium transporter TRK2 is able to substitute for TRK1 in its biological function under low K and low pH conditions. Yeast 23:581–589

Minor DL, Masseling SJ, Jan YN, Jan LY (1999) Transmembrane structure of an inwardly rectifying potassium channel. Cell 96:879–891

Miranda M, Bashi E, Vylkova S, Edgerton M, Slayman C et al (2009) Conservation and dispersion of sequence and function in fungal TRK potassium transporters: focus on Candida albicans. FEMS Yeast Res 9:278–292

Mitchell P (1961) Coupling of phosphorylation to electron and hydrogen transfer by a chemi-osmotic type of mechanism. Nature 191:144–148

Mitsui K, Kamauchi S, Nakamura N, Inoue H, Kanazawa H (2004a) A conserved domain in the tail region of the Saccharomyces cerevisiae Na+/H+ antiporter (Nha1p) plays important roles in localization and salinity-resistant cell-growth. J Biochem 135:139–148

Mitsui K, Ochi F, Nakamura N, Doi Y, Inoue H et al (2004b) A novel membrane protein capable of binding the Na+/H+ antiporter (Nha1p) enhances the salinity-resistant cell growth of Saccharomyces cerevisiae. J Biol Chem 279:12438–12447

Mitsui K, Yasui H, Nakamura N, Kanazawa H (2005) Oligomerization of the Saccharomyces cerevisiae Na+/H+ antiporter Nha1p: implications for its antiporter activity. Biochim Biophys Acta 1720:125–136

Mitsui K, Koshimura Y, Yoshikawa Y, Matsushita M, Kanazawa H (2011) The endosomal Na(+)/H(+) exchanger contributes to multivesicular body formation by regulating the recruitment of ESCRT-0 Vps27p to the endosomal membrane. J Biol Chem 286:37625–37638

Montero-Lomelí M, Okorokova Façanha AL (1999) Expression of a mammalian Na+/H+ antiporter in Saccharomyces cerevisiae. Biochem Cell Biol 77:25–31

Montiel V, Ramos J (2007) Intracellular Na and K distribution in Debaryomyces hansenii. Cloning and expression in Saccharomyces cerevisiae of DhNHX1. FEMS Yeast Res 7:102–109

Morsomme P, Slayman CW, Goffeau A (2000) Mutagenic study of the structure, function and biogenesis of the yeast plasma membrane H(+)-ATPase. Biochim Biophys Acta 1469:133–157

Mukherjee S, Kallay L, Brett CL, Rao R (2006) Mutational analysis of the intramembranous H10 loop of yeast Nhx1 reveals a critical role in ion homoeostasis and vesicle trafficking. Biochem J 398:97–105

Mulet JM, Leube MP, Kron SJ, Rios G, Fink GR et al (1999) A novel mechanism of ion homeostasis and salt tolerance in yeast: the Hal4 and Hal5 protein kinases modulate the Trk1-Trk2 potassium transporter. Mol Cell Biol 19:3328–3337

Mulet JM, Alejandro S, Romero C, Serrano R (2004) The trehalose pathway and intracellular glucose phosphates as modulators of potassium transport and general cation homeostasis in yeast. Yeast 21:569–582

Munson AM, Haydon DH, Love SL, Fell GL, Palanivel VR et al (2004) Yeast ARL1 encodes a regulator of K+ influx. J Cell Sci 117:2309–2320

Murguía JR, Bellés JM, Serrano R (1996) The yeast HAL2 nucleotidase is an in vivo target of salt toxicity. J Biol Chem 271:29029–29033

Nakamura T, Liu Y, Hirata D, Namba H, Harada S et al (1993) Protein phosphatase type 2B (calcineurin)-mediated, FK506-sensitive regulation of intracellular ions in yeast is an important determinant for adaptation to high salt stress conditions. EMBO J 12:4063–4071

Nakamura RL, Anderson JA, Gaber RF (1997) Determination of key structural requirements of a K+ channel pore. J Biol Chem 272:1011–1018

Nass R, Rao R (1998) Novel localization of a Na+/H+ exchanger in a late endosomal compartment of yeast. Implications for vacuole biogenesis. J Biol Chem 273:21054–21060

Nass R, Rao R (1999) The yeast endosomal Na+/H+ exchanger, Nhx1, confers osmotolerance following acute hypertonic shock. Microbiology 145(Pt 11):3221–3228

Nass R, Cunningham KW, Rao R (1997) Intracellular sequestration of sodium by a novel Na+/H+ exchanger in yeast is enhanced by mutations in the plasma membrane H+-ATPase. Insights into mechanisms of sodium tolerance. J Biol Chem 272:26145–26152

Navarre C, Goffeau A (2000) Membrane hyperpolarization and salt sensitivity induced by deletion of PMP3, a highly conserved small protein of yeast plasma membrane. EMBO J 19:2515–2524

Navarrete C, Petrezselyova S, Barreto L, Martinez JL, Zahradka J et al (2010) Lack of main K plus uptake systems in Saccharomyces cerevisiae cells affects yeast performance in both potassium-sufficient and potassium-limiting conditions. FEMS Yeast Res 10:508–517

Nishi T, Forgac M (2002) The vacuolar (H+)-ATPases – nature's most versatile proton pumps. Nat Rev Mol Cell Biol 3:94–103
Norbeck J, Blomberg A (1996) Protein expression during exponential growth in 0.7 M NaCl medium of Saccharomyces cerevisiae. FEMS Microbiol Lett 137:1–8
Norbeck J, Blomberg A (1998) Amino acid uptake is strongly affected during exponential growth of Saccharomyces cerevisiae in 0.7 M NaCl medium. FEMS Microbiol Lett 158:121–126
Norbeck J, Pâhlman AK, Akhtar N, Blomberg A, Adler L (1996) Purification and characterization of two isoenzymes of DL-glycerol-3-phosphatase from Saccharomyces cerevisiae. Identification of the corresponding GPP1 and GPP2 genes and evidence for osmotic regulation of Gpp2p expression by the osmosensing mitogen-activated protein kinase signal transduction pathway. J Biol Chem 271:13875–13881
Nowikovsky K, Bernardi P (2014) LETM1 in mitochondrial cation transport. Front Physiol 5:83
Nowikovsky K, Froschauer EM, Zsurka G, Samaj J, Reipert S et al (2004) The LETM1/YOL027 gene family encodes a factor of the mitochondrial K+ homeostasis with a potential role in the Wolf-Hirschhorn syndrome. J Biol Chem 279:30307–30315
Nowikovsky K, Reipert S, Devenish RJ, Schweyen RJ (2007) Mdm38 protein depletion causes loss of mitochondrial K+/H+ exchange activity, osmotic swelling and mitophagy. Cell Death Differ 14:1647–1656
Ohgaki R, Nakamura N, Mitsui K, Kanazawa H (2005) Characterization of the ion transport activity of the budding yeast Na+/H+ antiporter, Nha1p, using isolated secretory vesicles. Biochim Biophys Acta 1712:185–196
Okorokov LA, Lichko LP, Kulaev IS (1980) Vacuoles: main compartments of potassium, magnesium, and phosphate ions in Saccharomyces carlsbergenis cells. J Bacteriol 144:661–665
Palmgren MG, Nissen P (2011) P-type ATPases. Annu Rev Biophys 40:243–266
Pascual-Ahuir A, Posas F, Serrano R, Proft M (2001) Multiple levels of control regulate the yeast cAMP-response element-binding protein repressor Sko1p in response to stress. J Biol Chem 276:37373–37378
Peña A, Calahorra M, Michel B, Ramírez J, Sánchez NS (2009) Effects of amiodarone on K+, internal pH and Ca2+ homeostasis in Saccharomyces cerevisiae. FEMS Yeast Res 9:832–848
Pereira MB, Tisi R, Fietto LG, Cardoso AS, França MM et al (2008) Carbonyl cyanide m-chlorophenylhydrazone induced calcium signaling and activation of plasma membrane H(+)-ATPase in the yeast Saccharomyces cerevisiae. FEMS Yeast Res 8:622–630
Perez-Valle J, Jenkins H, Merchan S, Montiel V, Ramos J et al (2007) Key role for intracellular K+ and protein kinases Sat4/Hal4 and Hal5 in the plasma membrane stabilization of yeast nutrient transporters. Mol Cell Biol 27:5725–5736
Perez-Valle J, Rothe J, Primo C, Martinez Pastor M, Arino J et al (2010) Hal4 and Hal5 protein kinases are required for general control of carbon and nitrogen uptake and metabolism. Eukaryot Cell 9:1881–1890
Perkins J, Gadd GM (1993) Accumulation and intracellular compartmentation of lithium ions in Saccharomyces cerevisiae. FEMS Microbiol Lett 107:255–260
Perlin DS, Brown CL, Haber JE (1988) Membrane potential defect in hygromycin B-resistant pma1 mutants of Saccharomyces cerevisiae. J Biol Chem 263:18118–18122
Persson BL, Petersson J, Fristedt U, Weinander R, Berhe A et al (1999) Phosphate permeases of Saccharomyces cerevisiac: structure, function and regulation. Biochim Biophys Acta 1422:255–272
Petrezselyova S, Kinclova-Zimmermannova O, Sychrova H (2013) Vhc1, a novel transporter belonging to the family of electroneutral cation-Cl(−) cotransporters, participates in the regulation of cation content and morphology of Saccharomyces cerevisiae vacuoles. Biochim Biophys Acta 1828:623–631
Platara M, Ruiz A, Serrano R, Palomino A, Moreno F et al (2006) The transcriptional response of the yeast Na(+)-ATPase ENA1 gene to alkaline stress involves three main signaling pathways. J Biol Chem 281:36632–36642
Plemenitaš A, Lenassi M, Konte T, Kejžar A, Zajc J et al (2014) Adaptation to high salt concentrations in halotolerant/halophilic fungi: a molecular perspective. Front Microbiol 5:199

Portillo F (2000) Regulation of plasma membrane H(+)-ATPase in fungi and plants. Biochim Biophys Acta 1469:31–42

Portillo F, Mulet JM, Serrano R (2005) A role for the non-phosphorylated form of yeast Snf1: tolerance to toxic cations and activation of potassium transport. FEBS Lett 579:512–516

Posas F, Camps M, Ariño J (1995) The PPZ protein phosphatases are important determinants of salt tolerance in yeast cells. J Biol Chem 270:13036–13041

Posas F, Chambers JR, Heyman JA, Hoeffler JP, de Nadal E et al (2000) The transcriptional response of yeast to saline stress. J Biol Chem 275:17249–17255

Pozos TC, Sekler I, Cyert MS (1996) The product of HUM1, a novel yeast gene, is required for vacuolar Ca2+/H+ exchange and is related to mammalian Na+/Ca2+ exchangers. Mol Cell Biol 16:3730–3741

Prior C, Potier S, Souciet JL, Sychrova H (1996) Characterization of the NHA1 gene encoding a Na+/H+-antiporter of the yeast Saccharomyces cerevisiae. FEBS Lett 387:89–93

Proft M, Serrano R (1999) Repressors and upstream repressing sequences of the stress-regulated ENA1 gene in Saccharomyces cerevisiae: bZIP protein Sko1p confers HOG-dependent osmotic regulation. Mol Cell Biol 19:537–546

Proft M, Struhl K (2002) Hog1 kinase converts the Sko1-Cyc8-Tup1 repressor complex into an activator that recruits SAGA and SWI/SNF in response to osmotic stress. Mol Cell 9:1307–1317

Proft M, Struhl K (2004) MAP kinase-mediated stress relief that precedes and regulates the timing of transcriptional induction. Cell 118:351–361

Proft M, Pascual-Ahuir A, de Nadal E, Ariño J, Serrano R et al (2001) Regulation of the Sko1 transcriptional repressor by the Hog1 MAP kinase in response to osmotic stress. EMBO J 20:1123–1133

Qiu QS, Fratti RA (2010) The Na+/H+ exchanger Nhx1p regulates the initiation of Saccharomyces cerevisiae vacuole fusion. J Cell Sci 123:3266–3275

Quintero FJ, Blatt MR, Pardo JM (2000) Functional conservation between yeast and plant endosomal Na(+)/H(+) antiporters. FEBS Lett 471:224–228

Quintero FJ, Ohta M, Shi H, Zhu JK, Pardo JM (2002) Reconstitution in yeast of the Arabidopsis SOS signaling pathway for Na+ homeostasis. Proc Natl Acad Sci U S A 99:9061–9066

Ramírez J, Ramírez O, Saldaña C, Coria R, Peña A (1998) A Saccharomyces cerevisiae mutant lacking a K+/H+ exchanger. J Bacteriol 180:5860–5865

Ramos J, Alijo R, Haro R, Rodriguez-Navarro A (1994) TRK2 is not a low-affinity potassium transporter in Saccharomyces cerevisiae. J Bacteriol 176:249–252

Ramos J, Ariño J, Sychrová H (2011) Alkali-metal-cation influx and efflux systems in nonconventional yeast species. FEMS Microbiol Lett 317:1–8

Rao R, Drummond-Barbosa D, Slayman CW (1993) Transcriptional regulation by glucose of the yeast PMA1 gene encoding the plasma membrane H(+)-ATPase. Yeast 9:1075–1084

Rauch A, Schellmoser S, Kraus C, Dörr HG, Trautmann U et al (2001) First known microdeletion within the Wolf-Hirschhorn syndrome critical region refines genotype-phenotype correlation. Am J Med Genet 99:338–342

Reid JD, Lukas W, Shafaatian R, Bertl A, Scheurmann-Kettner C et al (1996) The S. cerevisiae outwardly-rectifying potassium channel (DUK1) identifies a new family of channels with duplicated pore domains. Recept Channels 4:51–62

Rep M, Krantz M, Thevelein JM, Hohmann S (2000) The transcriptional response of Saccharomyces cerevisiae to osmotic shock. Hot1p and Msn2p/Msn4p are required for the induction of subsets of high osmolarity glycerol pathway-dependent genes. J Biol Chem 275:8290–8300

Rivetta A, Kuroda T, Slayman C (2011) Anion currents in yeast K+ transporters (TRK) characterize a structural homologue of ligand-gated ion channels. Pflugers Arch 462:315–330

Rodríguez-Navarro A (2000) Potassium transport in fungi and plants. Biochim Biophys Acta 1469:1–30

Rodríguez-Navarro A, Ramos J (1984) Dual system for potassium transport in Saccharomyces cerevisiae. J Bacteriol 159:940–945

Roomans GM, Sevéus LA (1976) Subcellular localization of diffusible ions in the yeast Saccharomyces cerevisiae: quantitative microprobe analysis of thin freeze-dried sections. J Cell Sci 21:119–127

Ruiz A, Arino J (2007) Function and regulation of the Saccharomyces cerevisiae ENA sodium ATPase system. Eukaryot Cell 6:2175–2183

Ruiz A, Yenush L, Ariño J (2003) Regulation of ENA1 Na(+)-ATPase gene expression by the Ppz1 protein phosphatase is mediated by the calcineurin pathway. Eukaryot Cell 2:937–948

Saag MS, Dismukes WE (1988) Azole antifungal agents: emphasis on new triazoles. Antimicrob Agents Chemother 32:1–8

Saier MH (2000) A functional-phylogenetic classification system for transmembrane solute transporters. Microbiol Mol Biol Rev 64:354–411

Schachtman DP, Schroeder JI, Lucas WJ, Anderson JA, Gaber RF (1992) Expression of an inward-rectifying potassium channel by the Arabidopsis KAT1 cDNA. Science 258:1654–1658

Schlesser A, Ulaszewski S, Ghislain M, Goffeau A (1988) A second transport ATPase gene in Saccharomyces cerevisiae. J Biol Chem 263:19480–19487

Schlickum S, Moghekar A, Simpson JC, Steglich C, O'Brien RJ et al (2004) LETM1, a gene deleted in Wolf-Hirschhorn syndrome, encodes an evolutionarily conserved mitochondrial protein. Genomics 83:254–261

Schwarzer S, Kolacna L, Lichtenberg-Fraté H, Sychrova H, Ludwig J (2008) Functional expression of the voltage-gated neuronal mammalian potassium channel rat ether à go-go1 in yeast. FEMS Yeast Res 8:405–413

Sentenac H, Bonneaud N, Minet M, Lacroute F, Salmon JM et al (1992) Cloning and expression in yeast of a plant potassium ion transport system. Science 256:663–665

Serra-Cardona A, Petrezsélyová S, Canadell D, Ramos J, Ariño J (2014) Coregulated expression of the Na+/phosphate Pho89 transporter and Ena1 Na+-ATPase allows their functional coupling under high-pH stress. Mol Cell Biol 34:4420–4435

Serrano R (1983) In vivo glucose activation of the yeast plasma membrane ATPase. FEBS Lett 156:11–14

Serrano R, Ruiz A, Bernal D, Chambers JR, Arino J (2002) The transcriptional response to alkaline pH in Saccharomyces cerevisiae: evidence for calcium-mediated signalling. Mol Microbiol 46:1319–1333

Sesti F, Shih TM, Nikolaeva N, Goldstein SA (2001) Immunity to K1 killer toxin: internal TOK1 blockade. Cell 105:637–644

Shimkets RA, Warnock DG, Bositis CM, Nelson-Williams C, Hansson JH et al (1994) Liddle's syndrome: heritable human hypertension caused by mutations in the beta subunit of the epithelial sodium channel. Cell 79:407–414

Simón E, Clotet J, Calero F, Ramos J, Ariño J (2001) A screening for high copy suppressors of the sit4 hal3 synthetically lethal phenotype reveals a role for the yeast Nha1 antiporter in cell cycle regulation. J Biol Chem 276:29740–29747

Simón E, Barceló A, Ariño J (2003) Mutagenesis analysis of the yeast Nha1 Na+/H+ antiporter carboxy-terminal tail reveals residues required for function in cell cycle. FEBS Lett 545:239–245

Skou JC, Esmann M (1992) The Na, K-ATPase. J Bioenerg Biomembr 24:249–261

Soufi B, Kelstrup CD, Stoehr G, Fröhlich F, Walther TC et al (2009) Global analysis of the yeast osmotic stress response by quantitative proteomics. Mol Biosyst 5:1337–1346

Stefan CP, Zhang N, Sokabe T, Rivetta A, Slayman CL et al (2013) Activation of an essential calcium signaling pathway in Saccharomyces cerevisiae by Kch1 and Kch2, putative low-affinity potassium transporters. Eukaryot Cell 12:204–214

Strick R, Strissel PL, Gavrilov K, Levi-Setti R (2001) Cation-chromatin binding as shown by ion microscopy is essential for the structural integrity of chromosomes. J Cell Biol 155:899–910

Supply P, Wach A, Goffeau A (1993) Enzymatic properties of the PMA2 plasma membrane-bound H(+)-ATPase of Saccharomyces cerevisiae. J Biol Chem 268:19753–19759

Sychrová H, Ramírez J, Peña A (1999) Involvement of Nha1 antiporter in regulation of intracellular pH in Saccharomyces cerevisiae. FEMS Microbiol Lett 171:167–172

Szopinska A, Morsomme P (2010) Quantitative proteomic approaches and their application in the study of yeast stress responses. OMICS 14:639–649

Szopinska A, Degand H, Hochstenbach JF, Nader J, Morsomme P (2011) Rapid response of the yeast plasma membrane proteome to salt stress. Mol Cell Proteomics 10:M111.009589

Tang W, Ruknudin A, Yang WP, Shaw SY, Knickerbocker A et al (1995) Functional expression of a vertebrate inwardly rectifying K+ channel in yeast. Mol Biol Cell 6:1231–1240

Tarsio M, Zheng H, Smardon AM, Martínez-Muñoz GA, Kane PM (2011) Consequences of loss of Vph1 protein-containing vacuolar ATPases (V-ATPases) for overall cellular pH homeostasis. J Biol Chem 286:28089–28096

Tate JJ, Cooper TG (2007) Stress-responsive Gln3 localization in Saccharomyces cerevisiae is separable from and can overwhelm nitrogen source regulation. J Biol Chem 282:18467–18480

Teige M, Scheikl E, Reiser V, Ruis H, Ammerer G (2001) Rck2, a member of the calmodulin-protein kinase family, links protein synthesis to high osmolarity MAP kinase signaling in budding yeast. Proc Natl Acad Sci U S A 98:5625–5630

Trópia MJ, Cardoso AS, Tisi R, Fietto LG, Fietto JL et al (2006) Calcium signaling and sugar-induced activation of plasma membrane H(+)-ATPase in Saccharomyces cerevisiae cells. Biochem Biophys Res Commun 343:1234–1243

Tsai H, Bobek LA (1997a) Human salivary histatin-5 exerts potent fungicidal activity against Cryptococcus neoformans. Biochim Biophys Acta 1336:367–369

Tsai H, Bobek LA (1997b) Studies of the mechanism of human salivary histatin-5 candidacidal activity with histatin-5 variants and azole-sensitive and -resistant Candida species. Antimicrob Agents Chemother 41:2224–2228

Uozumi N, Gassmann W, Cao Y, Schroeder JI (1995) Identification of strong modifications in cation selectivity in an Arabidopsis inward rectifying potassium channel by mutant selection in yeast. J Biol Chem 270:24276–24281

Vargas RC, García-Salcedo R, Tenreiro S, Teixeira MC, Fernandes AR et al (2007) Saccharomyces cerevisiae multidrug resistance transporter Qdr2 is implicated in potassium uptake, providing a physiological advantage to quinidine-stressed cells. Eukaryot Cell 6:134–142

Vergani P, Miosga T, Jarvis SM, Blatt MR (1997) Extracellular K+ and Ba2+ mediate voltage-dependent inactivation of the outward-rectifying K+ channel encoded by the yeast gene TOK1. FEBS Lett 405:337–344

Viejo-Díaz M, Andrés MT, Fierro JF (2004a) Effects of human lactoferrin on the cytoplasmic membrane of Candida albicans cells related with its candidacidal activity. FEMS Immunol Med Microbiol 42:181–185

Viejo-Díaz M, Andrés MT, Fierro JF (2004b) Modulation of in vitro fungicidal activity of human lactoferrin against Candida albicans by extracellular cation concentration and target cell metabolic activity. Antimicrob Agents Chemother 48:1242–1248

Wadskog I, Forsmark A, Rossi G, Konopka C, Oyen M et al (2006) The yeast tumor suppressor homologue Sro7p is required for targeting of the sodium pumping ATPase to the cell surface. Mol Biol Cell 17:4988–5003

Waight AB, Pedersen BP, Schlessinger A, Bonomi M, Chau BH et al (2013) Structural basis for alternating access of a eukaryotic calcium/proton exchanger. Nature 499:107–110

Warringer J, Ericson E, Fernandez L, Nerman O, Blomberg A (2003) High-resolution yeast phenomics resolves different physiological features in the saline response. Proc Natl Acad Sci U S A 100:15724–15729

Wells KM, Rao R (2001) The yeast Na+/H+ exchanger Nhx1 is an N-linked glycoprotein. Topological implications. J Biol Chem 276:3401–3407

Wieland J, Nitsche AM, Strayle J, Steiner H, Rudolph HK (1995) The PMR2 gene cluster encodes functionally distinct isoforms of a putative Na+ pump in the yeast plasma membrane. EMBO J 14:3870–3882

Williams-Hart T, Wu X, Tatchell K (2002) Protein phosphatase type 1 regulates ion homeostasis in Saccharomyces cerevisiae. Genetics 160:1423–1437

Wolfe DM, Pearce DA (2006) Channeling studies in yeast: yeast as a model for channelopathies? Neuromolecular Med 8:279–306

Wright MB, Ramos J, Gomez MJ, Moulder K, Scherrer M et al (1997) Potassium transport by amino acid permeases in Saccharomyces cerevisiae. J Biol Chem 272:13647–13652

Xiang M, Feng M, Muend S, Rao R (2007) A human Na+/H+ antiporter sharing evolutionary origins with bacterial NhaA may be a candidate gene for essential hypertension. Proc Natl Acad Sci U S A 104:18677–18681

Yale J, Bohnert HJ (2001) Transcript expression in Saccharomyces cerevisiae at high salinity. J Biol Chem 276:15996–16007

Ye T, Elbing K, Hohmann S (2008) The pathway by which the yeast protein kinase Snf1p controls acquisition of sodium tolerance is different from that mediating glucose regulation. Microbiology 154:2814–2826

Yenush L, Mulet JM, Arino J, Serrano R (2002) The Ppz protein phosphatases are key regulators of K+ and pH homeostasis: implications for salt tolerance, cell wall integrity and cell cycle progression. EMBO J 21:920–929

Yenush L, Merchan S, Holmes J, Serrano R (2005) pH-responsive, posttranslational regulation of the Trk1 potassium transporter by the type 1-related Ppz1 phosphatase. Mol Cell Biol 25:8683–8692

Yu SP (2003) Regulation and critical role of potassium homeostasis in apoptosis. Prog Neurobiol 70:363–386

Yu D, Danku JM, Baxter I, Kim S, Vatamaniuk OK et al (2012) High-resolution genome-wide scan of genes, gene-networks and cellular systems impacting the yeast ionome. BMC Genomics 13:623

Zahrádka J, van Heusden GP, Sychrová H (2012) Yeast 14-3-3 proteins participate in the regulation of cell cation homeostasis via interaction with Nha1 alkali-metal-cation/proton antiporter. Biochim Biophys Acta 1820:849–858

Zaks-Makhina E, Kim Y, Aizenman E, Levitan ES (2004) Novel neuroprotective K+ channel inhibitor identified by high-throughput screening in yeast. Mol Pharmacol 65:214–219

Zaks-Makhina E, Li H, Grishin A, Salvador-Recatala V, Levitan ES (2009) Specific and slow inhibition of the kir2.1 K+ channel by gambogic acid. J Biol Chem 284:15432–15438

Zhang Z, Zheng Y, Mazon H, Milgrom E, Kitagawa N et al (2008) Structure of the yeast vacuolar ATPase. J Biol Chem 283:35983–35995

Zhou XL, Vaillant B, Loukin SH, Kung C, Saimi Y (1995) YKC1 encodes the depolarization-activated K+ channel in the plasma membrane of yeast. FEBS Lett 373:170–176

Zotova L, Aleschko M, Sponder G, Baumgartner R, Reipert S et al (2010) Novel components of an active mitochondrial K(+)/H(+) exchange. J Biol Chem 285:14399–14414

# Chapter 9
# Carboxylic Acids Plasma Membrane Transporters in *Saccharomyces cerevisiae*

**Margarida Casal, Odília Queirós, Gabriel Talaia, David Ribas, and Sandra Paiva**

**Abstract** This chapter covers the functionally characterized plasma membrane carboxylic acids transporters Jen1, Ady2, Fps1 and Pdr12 in the yeast *Saccharomyces cerevisiae,* addressing also their homologues in other microorganisms, as filamentous fungi and bacteria. Carboxylic acids can either be transported into the cells, to be used as nutrients, or extruded in response to acid stress conditions. The secondary active transporters Jen1 and Ady2 can mediate the uptake of the anionic form of these substrates by a $H^+$-symport mechanism. The undissociated form of carboxylic acids is lipid-soluble, crossing the plasma membrane by simple diffusion. Furthermore, acetic acid can also be transported by facilitated diffusion *via* Fps1 channel. At the cytoplasmic physiological pH, the anionic form of the acid prevails and it can be exported by the Pdr12 pump. This review will highlight the mechanisms involving carboxylic acids transporters, and the way they operate according to the yeast cell response to environmental changes, as carbon source availability, extracellular pH and acid stress conditions.

**Keywords** Carboxylic acids • Plasma membrane transporters • Yeast • Jen1 • Ady2 • Pdr12 • Fps1

M. Casal (✉) • G. Talaia • D. Ribas • S. Paiva
CBMA-Centre of Molecular and Environmental Biology, Department of Biology, University of Minho, Campus de Gualtar, 4710-057 Braga, Portugal
e-mail: mcasal@bio.uminho.pt

O. Queirós
CBMA-Centre of Molecular and Environmental Biology, Department of Biology, University of Minho, Campus de Gualtar, 4710-057 Braga, Portugal

CESPU, Instituto de Investigação e Formação Avançada em Ciências e Tecnologias da Saúde, Rua Central de Gandra, 1317, 4585-116 Gandra, PRD, Portugal

 
J. Ramos et al. (eds.), *Yeast Membrane Transport*, Advances in Experimental Medicine and Biology 892, DOI 10.1007/978-3-319-25304-6_9

## 9.1 Introduction

Carboxylic acids are ubiquitous organic compounds that can be used by the cell machinery as natural substrates, or found as final products or by-products of fermentative processes. As weak acids, they can partially dissociate according to their p$K_a$ and to the pH of the aqueous solution, following the Henderson–Hasselbalch equation [$pH = pK_a + \log (A^-/HA)$]. One of the most important factors influencing carboxylic acids transport across biological membranes is the environmental pH. When the pH is below the p$K_a$ of the acid, the protonated undissociated form predominates and it is able to cross biological membranes by passive diffusion. However, when the pH is above p$K_a$ of the acid, the charged anionic form predominates, requiring a transporter to cross the biological membrane (reviewed by Casal et al. 2008).

The first evidence for a carboxylic acid transporter in yeasts was described in *Schizosaccharomyces pombe*, associated with the uptake of malic acid (Osothsilp and Subden 1986). The gene encoding this activity (*MAE1*) was also the first carboxylate permease gene identified in yeasts (Grobler et al. 1995). According to the Transport Classification Database (TCDB, www.tcdb.org), Mae1 displays ten predicted transmembrane segments (TMS), belonging to the Tellurite-resistance/Dicarboxylate Transporter family (TC 2.A.16). Evidences for the existence of a transporter mediating the uptake of lactic acid were found for the first time in the yeasts *Candida utilis* (Leão and van Uden 1986) and *Saccharomyces cerevisiae* (Cassio et al. 1987) and the associated gene was named *JEN1* (Casal et al. 1999). This transporter is a member of the Major Facilitator Superfamily (MFS) belonging to the lactate/pyruvate:$H^+$ sub-family (TC 2.A.1.12.2). Some years later, the Ady2 acetate-propionate-formate transporter was described in *S. cerevisiae.* This transporter is expressed in the presence of non-fermentable carbon sources and subjected to glucose repression (Paiva et al. 2004). Ady2 displays six predicted TMS and belongs to the Acetate Uptake Transporter (AceTr) family (TC2.A.96.1.4). Additional roles have been attributed to this protein, namely in ammonia export (Palkova et al. 2002) and *ascus* formation (Rabitsch et al. 2001).

Jen1 and Ady2, involved in the uptake of the anionic form of monocarboxylic acids, are so far the only secondary active transporters assigned in *S. cerevisiae.* At lower pH values, when the undissociated form of the acid prevails, carboxylic acids can also cross the plasma membrane by simple diffusion. However, it has been found that the aquaglyceroporin Fps1, a channel primarily involved in osmoadaptation, can mediate the entry of undissociated acetic acid by facilitated diffusion (Mollapour and Piper 2007). Both simple and facilitated diffusion correspond to energy-independent transport mechanisms. Once in the cytosol, at neutral pH values, the acids dissociate, releasing toxic counter-anions and protons, inducing an intracellular acidification that affects cell homeostasis. However, *S. cerevisiae* can still grow in the presence of high concentration of weak carboxylic acids at low pH values, though presenting increased lag phase duration and lower biomass yields (Piper et al. 1997). Pdr12, an ATP-binding cassette (ABC) transporter, member of the Pleiotropic Drug Resistance (PDR) family, was demonstrated to be essential to the acquisition of tolerance to weak acid stress, being involved in the extrusion of the carboxylate anions and participating in cellular detoxification (Piper et al. 1998).

The importance of carboxylic acids transporters in yeast relies thus in two fundamental processes, the uptake of the acids to be used as nutrients and their extrusion in response to acid stress conditions. The present review will focus on the role of the currently known plasma membrane carboxylic acids transporters in *S. cerevisiae* Jen1, Ady2, Fps1 and Pdr12, addressing also their homologues found in other microorganisms, as filamentous fungi and bacteria.

## 9.2 The Carboxylic Acids Transporter Jen1, a Member of the Major Facilitator Superfamily

Jen1 was the first monocarboxylic acids transporter described in fungi (Casal et al. 1999). Besides its role in the uptake of lactate, pyruvate, acetate and propionate (Casal et al. 1999), it also transports the micronutrient selenite (McDermott et al. 2010) and the antitumor compound 3-bromopyruvate (Lis et al. 2012). The transport of the substrate is bidirectional, being Jen1 also involved in the acids efflux (van Maris et al. 2004; Pacheco et al. 2012). The kinetic parameters estimated for this transporter, concerning lactate uptake in *S. cerevisiae* W303-1A lactic acid-grown cells are: $V_{max}$ of 0.40 nmol of lactic acid $s^{-1}$ mg of dry weight$^{-1}$ and $K_m$ of 0.69 mM lactic acid (Casal et al. 1999). When overexpressed in the host *Pichia pastoris* a fivefold increase in $V_{max}$ was achieved ($V_{max} = 2.15 \pm 0.14$ nmol of lactic acid $\cdot s^{-1}$ mg of dry weight$^{-1}$), whereas the $K_m$ was of the same order of magnitude ($K_m = 0.54 \pm 0.08$ mM) (Soares-Silva et al. 2003). The reconstitution of Jen1 in *P. pastoris* vesicles demonstrated that the proton motive force is necessary for transport, confirming the proton-symport mechanism. Kinetic properties of the reconstituted transporter were found to be similar to the ones reported for *S. cerevisiae* intact cells (Soares-Silva et al. 2003). These evidences clearly demonstrated Jen1 as a fully functional lactate permease.

In *Kluyveromyces lactis* two Jen1 homologues were identified, one encoding a monocarboxylate transporter (KlJen1), for lactate and pyruvate, and the other a dicarboxylate transporter (KlJen2) for malate and succinate (Lodi et al. 2004; Queiros et al. 2007). A similar situation was found in *Candida albicans* where CaJen1 transports monocarboxylic acids, such as lactate, whereas CaJen2 transports the dicarboxylic acids succinate and malate. The phylogenetic analysis of these Jen1 homologues suggested the existence of two functional clusters, Jen1 and Jen2, comprising the functionally characterized monocarboxylate and dicarboxylate transporters, respectively (Casal et al. 2008). The *Yarrowia lipolytica* genome encodes six Jen1 homologues. The YlJen1 encoding genes have their expression increased in the presence of different carboxylic acids. Recently they were associated with mono-, di- and tricarboxylic acids transport, such as lactate, pyruvate, fumarate, malate, succinate and citrate, suggesting the ability to transport these acids (Dulermo et al. 2015; Guo et al. 2015).

Using the information that resulted from the sequencing of various yeast and fungi genomes it has been possible to trace the evolution of the Jen1 family members (Lodi et al. 2007; Casal et al. 2008; Dulermo et al. 2015; Sá-Pessoa et al. 2015; Soares-Silva et al. accepted for publication). Since YlJen homologues represent a separated cluster, they were designated to their own subfamily called Jen3 (Dulermo et al. 2015). YlJen1 and YlJen5 were considered ancestors of Jen1, therefore also known as preJen1 proteins. These proteins are considered an evolutionary intermediate between Jen2 (the ancestral homologue) and Jen1 members (Lodi et al. 2007; Dulermo et al. 2015). Jen1 transporters differentiated from preJen1 in *K. lactis* and they are present as single copy in *Saccharomyces* species. Post-WGD (Whole Genome Duplication) species (such as *Saccharomyces* genus) in general lack Jen2, indicating that this precursor was evolutionary lost prior to the WGD (Lodi et al. 2007).

### *9.2.1 Expression and Regulation*

In *S. cerevisiae* the regulation of Jen1 expression has also been studied over the past years at the transcriptional, post-transcriptional and post-translational levels. In lactic acid, pyruvic acid, acetic acid or glycerol-grown cells *JEN1* is highly expressed, whereas in glucose, formic and propionic acid-grown cells it is undetectable (Casal et al. 1999). *JEN1* glucose repression involves the transcription factors Mig1 and Mig2 (Bojunga and Entian 1999). When glucose is withdrawn, Hap2/3/4/5 complex, Adr1 and Cat8 transcription factors act together (Fig. 9.1), upregulating numerous genes involved in the utilization of non-fermentable carbon sources, including *JEN1* and *ADY2* (Bojunga and Entian 1999; Haurie et al. 2001; Lodi et al. 2002; Ratnakumar et al. 2009). The kinase Snf1, a AMPK homologue, is also involved in the release of glucose repression of both genes. It phosphorylates Mig1 resulting in its translocation from the nucleus, activating Cat8 and Adr1 (Lodi et al. 2002; Tachibana et al. 2005; Abate et al. 2012). It was also demonstrated that *SNF1* deletion inhibits H3 acetylation at the *ADY2* promoter, impairing chromatin remodelling and *ADY2* transcription, in a process independent of Adr1 and Cat8 (Abate et al. 2012).

The influence of the carbon source in *JEN1* expression has also been highlighted at the post-transcriptional level, particularly in mRNA turnover (Andrade and Casal 2001). The transcriptional map of yeast genome revealed two *JEN1* transcripts with 2200 and 1900 nucleotides (Richard et al. 1997). It has been found that, in *S. cerevisiae* W303-1A lactic-grown cells, the long-mRNA decay is triggered by a pulse of glucose. This was in contrast to what was observed for CEN.PK2-1C

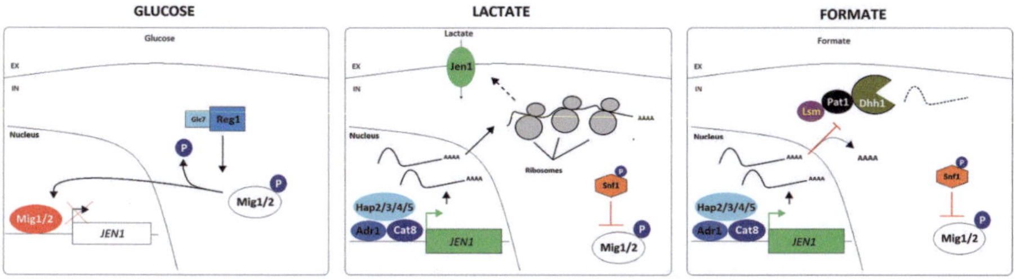


**Fig. 9.1 Post-transcriptional regulation of *JEN1*.** In **glucose**, the transcription of *JEN1* is repressed by Mig1/2 transcription factors. In the cytoplasm, Mig1/2 are activated by dephosphorylation through a Reg1/Glc7 phosphatase and, then, imported into the nucleus (Carlson 1999). In **lactate**, as sole carbon source, the transcriptional factors Cat8, Adr1 and Hap2/3/4/5 are able to activate *JEN1* expression. The inactivation of Mig1/2 by phosphorylation, through Snf1 protein kinase, leads to their export to the cytoplasm. Therefore, in these conditions, the formation and maturation of mRNA is unaffected which ultimately leads to the expression of the lactate transporter at the plasma membrane. Cat8, Adr1 and Hap2/3/4/5 are transcriptional activators associated to non-fermentable carbon sources growth (Hedges et al. 1995; Olesen and Guarente 1990; Haurie et al. 2001), as lactate. Interestingly, *CAT8* is repressed at the DNA level by Mig1/2 proteins (Randez-Gil et al. 1997). In **formate**, as sole carbon source, the transcription of *JEN1* is also active, however the early synthetized mRNA is targeted for degradation, consequently, no Jen1 protein is produced. Dhh1 RNA helicase, Pat1 and Lsm are involved in *JEN1* mRNA degradation, in this condition

cells, where the long-mRNA retained a half-life time of 17–20 min, independently of the addition of glucose to the culture medium. However, when CEN.PK2-1C cells were grown on ethanol, the *JEN1* long-mRNA half-life time was reduced from 26 to 7 min. upon the pulse of glucose. Mapping of the *JEN1* 5′ and 3′ UTR transcripts revealed multiple transcription start sites located at positions -51 (for the long transcript, translated), and +391 or +972 (for the short transcripts, not translated). It was demonstrated that when *JEN1*(+391) small transcript is present, it works as a glucose sensor, promoting *JEN1*(-51) protein coding mRNA rapid decay (Andrade et al. 2005).

The Dhh1 RNA helicase and the Pat1-Lsm decapping enhancers were found to play a crucial role in the mechanisms underlying the regulation of *JEN1* 5′-3′mRNA decay pathway (Mota et al. 2014). Dhh1 together with Pat1-Lsm complex controls the translation initiation, by targeting mRNAs to the P-bodies, contributing to the recruitment of the mRNA decapping machinery (reviewed by Olszewska et al. 2102). When cells were incubated in formic acid *JEN1* mRNA accumulates in *dhh1*, *pat1* or *lsm* mutants, contrarily to what was found in the wild-type strain (Mota et al. 2014). In this condition it was found that *JEN1* mRNA, although transcriptionally active, was targeted for degradation *via* Dhh1 and Pat1-Lsm complex (Fig. 9.1). Microarray analysis revealed that this mechanism of regulation is also shared by other genes involved in non-fermentable carbon source metabolism, as *ADY2*, *CAT8* and *HAP4*. Besides its general effect in the global cytoplasmic mRNA decay, Dhh1 has additional roles in the post-transcriptional or transcriptional regulation of several genes in response to environmental stimuli (Mota et al. 2014).

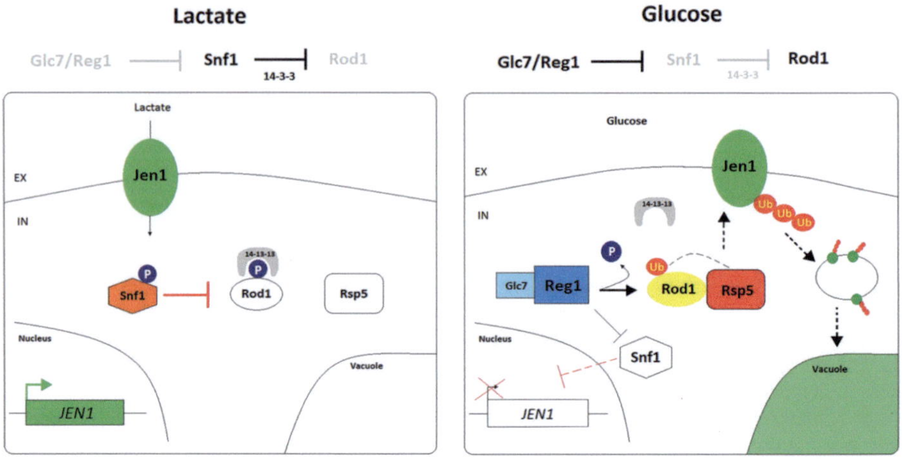


**Fig. 9.2** **Jen1 inactivation by glucose.** In **lactate**, as only carbon source, Jen1 is expressed at the plasma membrane. In this condition, Rsp5 ubiquitin ligase is unable to target Jen1 for degradation *via* the arrestin-like protein Rod1. Rod1 phosphorylation is dependent on Snf1 protein kinase. Phosphorylated Rod1 interacts with 14-3-3 proteins, becoming inaccessible to Rsp5, thus impairing Jen1 degradation. Upon a pulse of **glucose**, Jen1 is rapidly removed from the plasma membrane. Reg1/Glc7 phosphatase is involved in Rod1 activation and Snf1 inactivation, through dephosphorylation. As a consequence, Rod1 is released from a phospho-dependent interaction with 14-3-3 proteins and ubiquitylated by Rsp5. Finally, ubiquitylated-Rod1-Rsp5 complex ubiquitylates Jen1 resulting in its endocytosis and vacuolar degradation

Another level of Jen1 regulation involves protein trafficking and turnover. The addition of a pulse of glucose to lactic acid-grown cells very rapidly triggered the loss of Jen1 activity and endocytosis, followed by vacuolar degradation (Paiva et al. 2002). The HECT E3 ubiquitin ligase Rsp5, the unique member of the Nedd4 family in yeast, modifies Jen1 at the cell surface by polyubiquitylation (Paiva et al. 2009). Jen1 has been reported as one of the first examples where endocytic internalization and sorting at multivesicular bodies (MVBs) require ubiquitin-K63 linked chain(s). It has also been demonstrated that the yeast Rod1, a protein of the ART family (arrestin-related trafficking adaptors), is essential for glucose-induced Jen1 ubiquitylation and endocytosis. In lactic acid-grown cells Snf1 protein kinase inactivates Rod1 by phosphorylation. Upon a pulse of glucose, the PP1 phosphatase Glc7/Reg1 activates Rod1, which also dephosphorylates Snf1 inactivating it (Becuwe et al. 2012) (Fig. 9.2). Therefore, Rod1 serves as a relay between glucose signalling and endocytosis.

## 9.2.2 *Jen1 Structure*

Jen1 shares the common topology of the MFS members (Fig. 9.3), known as MFS fold, which comprises 12 TMS organized in 6 + 6 folded domains in the

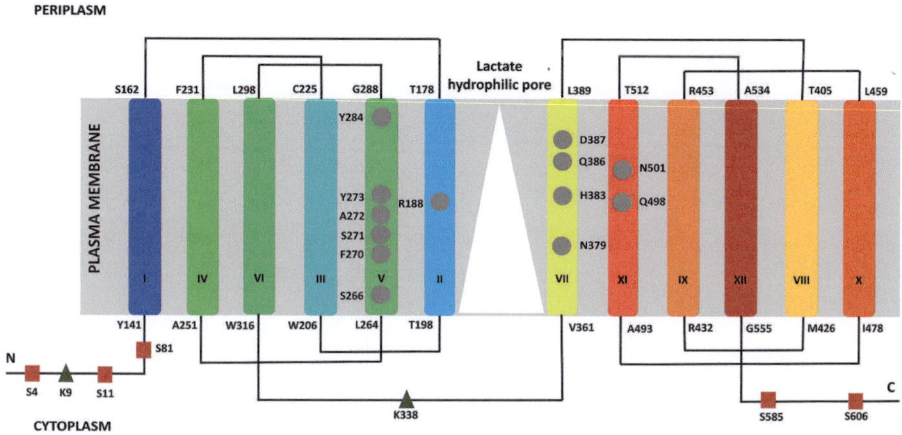


**Fig. 9.3 ScJen1 overall topology.** The predicted transmembrane segments are colored in rainbow spectrum, from the I TMS to XII TMS. Residues involved in substrate binding, specificity and translocation are shown as circles: II TMS (R188), V TMS (S266, F270, S271, A272, Y273, Y284) VII TMS (N379, H383, Q386, D387) and XI TMS (Q498, N501). Residues marked with squares (S4, S11, S81, S585, S606) are predicted targets for phosphorylation. Ubiquitylated residues (K9, K338) are shown as triangles. The hydrophilic lactate pore is designated by a white triangle between the II TMS and VII TMS, pointing the transport of lactate molecule from periplasm to cytoplasm. The N-terminus and C-terminus of the protein are annotated as N and C, respectively

N- and C-termini, separated by a central cytoplasmic loop (Marger and Saier 1993). Currently, several members of this superfamily, covering six subfamilies, have their structure solved, namely, the lactose/$H^+$ symporter LacY (Abramson et al. 2003), the glycerol-3-phosphate/Pi antiporter GlpT (Huang et al. 2003), the multidrug/$H^+$ antiporter EmrD (Yin et al. 2006), the L-fucose/$H^+$ symporter FucP (Dang et al. 2010), the oligopeptide/$H^+$ symporters PepTSo and PepTSt (Newstead et al. 2011), and the D-xylose/$H^+$ symporter XylE (Solcan et al. 2012).

Jen1 structural-functional relationships have been elucidated by a rational mutational analysis of conserved amino acid residues. The conserved sequence $^{379}$**NXX**[S/T]**HX**[S/T]**QD**$^{387}$, located towards the periplasmic side of putative transmembrane segment seven (7-TMS), is part of the substrate translocation pathway (Soares-Silva et al. 2007). The residue N379 was found to be irreplaceable, and crucial for protein activity. Residues H383 and D387 affect both the transport capacity and the specificity. On the other hand, Q386N substitution reduces the binding affinity for all Jen1 substrates, while Q386A increases the affinity for pyruvate (Soares-Silva et al. 2007). In addition, the conserved residues F270 (5-TMS) and Q498 (11-TMS) are essential for the substrate specificity, as they are involved in the distinction between mono- and dicarboxylates. The residues N501 (11-TMS) and R188 (2-TMS) are important players in the protein function as they are irreplaceable for Jen1 activity (Soares-Silva et al. 2011) (Fig. 9.3).

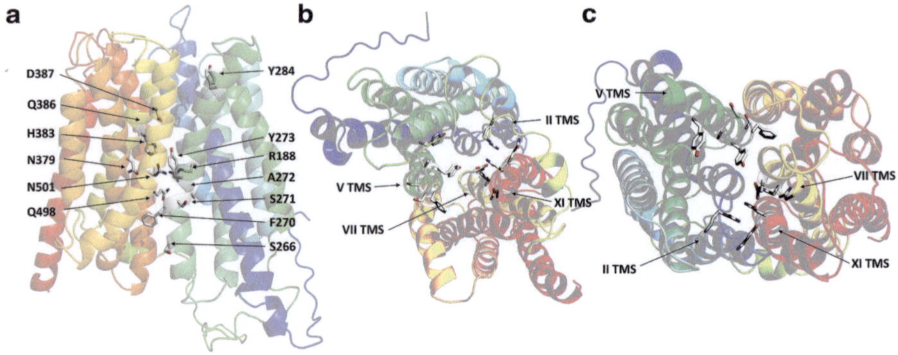


**Fig. 9.4 Jen1 predicted 3D structural model**: (**a**) transversal, (**b**) cytoplasmic and (**c**) periplasmic view. Residues critical for substrate binding, specificity and trajectory are shown as stick molecular structures and identified by black arrows: II TMS (R188), V TMS (S266, F270, S271, A272, Y273, Y284) VII TMS (N379, H383, Q386, D387) and XI TMS (Q498, N501). The 12 TMS are colored in rainbow spectrum from the N-terminus in blue to the C-terminus in red. The model in ribbon representations was obtained with Modeller software (Version 9.14, Ben Web, CA, USA) based on GlpT transporter X-ray structure. The images were obtained with the PyMOL Molecular Graphics System (Version 1.5.0.4 Schrödinger, LLC)

The predicted structural similarity of Jen1 with the GlpT permease revealed that all the polar residues above mentioned, namely R188 (2-TMS), S266, F270, S271, A272, Y273 (5-TMS), N379, H383, D387 (7-TMS) and Q498, N501 (11-TMS), are perfectly aligned in an imaginary axis that lies parallel to the protein pore (Fig. 9.4). Furthermore, docking calculations revealed a 'trajectory-like' substrate displacement along the Jen1 pore, where R188 plays a major dynamic role mediating the orderly relocation of the substrate by subsequent $H^+$-bond interactions involving also residues H383, N501 and Q498 (Soares-Silva et al. 2011). The association between structural-functional studies and prediction models provided extremely valid information on the structure properties and features of Jen1. The conserved amino acids residues, in particular the polar and charged present in TMS, were found essential for binding and translocation of both the substrate (the anionic form of the acid) and the co-substrate (the proton) (Soares-Silva et al. 2011).

## 9.3 The Carboxylic Acids Transporter Ady2, a Member of the AceTr Family

Members of the Acetate Uptake Transporter (AceTr) Family (TC 2.A.96.1.4) are found in archaea, eubacteria as well as in simple or complex eukaryotes. They share several conserved motifs namely the amino acid residues sequences

NP(A/V/G)P(L/F/V)GL and (Y/F)G(X)FW (Augstein et al. 2003). The NPA sequence found in the first motif is typical of the Major Intrinsic Protein (MIP) family, also known as aquaporins. The AceTr members have a similar size and contain five to six predicted TMS, characteristic of the MIP family. However, while the characteristic NPA motif is present twice in MIP proteins, between the second and the third, and between the fifth and sixth TMS (Agre et al. 1995), in AceTr proteins it is present only once in the N-terminal part at the first predicted TMS (Augstein et al. 2003).

Gpr1 (TC 2.A.96.1.2) from *Yarrowia lipolytica* was the first AceTr family member identified in yeasts (Kujau et al. 1992), shown to be involved in acetic acid sensitivity, cell and colony morphology, yeast-to-hyphae transition and cell life span (Tzschoppe et al. 1999). *GPR1* transcription is induced or derepressed in acetic acid or ethanol supplemented medium, compared to glucose-grown cells. Its localization at the plasma membrane is supported by subcellular fractionation studies and by fluorescence microscopy analysis (Augstein et al. 2003).

The *S. cerevisiae ADY2* (TC 2.A.96.1.4), along with *ATO3* (TC 2.A.96.1.5) and *FUN34* (TC 2.A.96.1.5) genes are homologous of *Y. lipolytica GPR1* gene (Tzschoppe et al. 1999). *ADY2* was first characterized as a gene required for proper *ascus* formation on a sporulation medium in which acetate is the main carbon source (Rabitsch et al. 2001). Other reports associate *ADY2* and the two homologue genes *ATO3* and *FUN34* with ammonium export (Palkova et al. 2002). Microarray analysis suggested the involvement of *ADY2* in acetate utilization and its subsequent disruption abolished active acetate transport (Paiva et al. 2004). Despite these evidences, the assignment of Ady2 as a carboxylate transporter held some controversy due to its multifunctional activity. However, the identification of Ady2 homologues as acetate transporters, namely AcpA (TC 2.A.96.1.3) from *Aspergillus nidulans* (Robellet et al. 2008) and SatP (TC 2.A.96.1.1) from *E. coli* (Sa-Pessoa et al. 2013), strongly supported this assignment for *S. cerevisiae* Ady2.

AcpA is a transporter essential for the uptake and use of acetate as sole carbon source in *A. nidulans* (Robellet et al. 2008), with high-affinity and high-capacity for acetate, and low capacity for other monocarboxylates. The AcpA transporter has an unusual expression profile showing activity in resting *A. nidulans* conidiospores, consistent with a role in spore maintenance or homoeostasis (Sá-Pessoa et al. 2015). The early activity of AcpA in resting conidiospores might be justified since transporters are efficient scavengers that can supply metabolites avoiding the cost of biosynthesis. In addition, the non-repressibility of AcpA by a primary carbon source such as glucose, shows that it serves cellular needs for acetate accumulation other than acting as a carbon source supplier (Sá-Pessoa et al. 2015). The AcpB *A. nidulans* was found to be responsible for residual acetate transport in mycelia and no function was found for AcpC. These results are in agreement with the lack of detectable transcription and minimal evolutionary conservation in fungi (Sá-Pessoa et al. 2015). The AlcS (TC 2.A.96.2.1), another *A. nidulans* plasma membrane protein of the AceTr family, is induced by ethanol and repressed by glucose (Fillinger and Felenbok 1996), however it is not involved in the transport of carboxylic acids (Flipphi et al. 2006).

The SatP protein from *E. coli*, previously known as YaaH, is a succinate-acetate/$H^+$ symporter, differing from the other family members. The ability to transport both a monocarboxylic acid and a dicarboxylic acid was reported for the first time in this family. Before this, other AceTr family members were associated exclusively to the transport of monocarboxylates (Sá-Pessoa et al. 2013). From a physiological point of view, the SatP transporter was found to play an important role for *E. coli* cells prior to the acetate switch momentum. When cells are grown aerobically in glucose, along with sugar consumption, the excretion of acetate occurs. However when glucose is depleted, cells switch the metabolism to use acetate as carbon source. SatP is highly expressed and active during the exponential phase of growth, prior to the acetate switch, most likely being involved in acetate efflux (Sá-Pessoa et al. 2013). Besides SatP, *E. coli* cells express another acetate transporter, the ActP, which is transcribed along with acetyl-CoA synthetase. This transporter belongs to the Sodium: Solute Symporter family and is highly specific for short-chain aliphatic monocarboxylates, namely acetate, glycolate and propionate (Gimenez et al. 2003). When *E. coli* cells are grown aerobically in glucose, ActP is mostly expressed and active after the acetate switch phase, which suggests its involvement in acetate uptake. SatP contributes to the acetate–succinate intracellular balance, although it is not crucial for cells to grow on glucose or acetic acid (Sá-Pessoa et al. 2013).

In the archaea *Methanosarcina acetivorans*, *MA4008* gene was found by quantitative transcription analysis of acetic acid *versus* methanol-grown cells, with 125-fold induction (Rohlin and Gunsalus 2010). This high level of expression is only achieved in the presence of acetate, as the only carbon source. Methanol + acetic acid grown cells display lower levels of expression than cells grown in sole methanol, which is the preferred carbon source of this species. In addition, the expression of *MA4008* was similar to acetate kinase (*ack*) and acetyl-CoA phosphotransferase (*pta*) encoding genes which are required for acetate utilization in bacteria. This indicates that the *MA4008* gene is expressed only in the absence of a rich carbon source, suggesting its putative role in acetate uptake (Rohlin and Gunsalus 2010).

A phylogenetic tree of the AceTr family members functionally analysed is presented in Fig. 9.5. The *S. cerevisiae* and *Y. lipolytica* members form a distinct clade with similar evolutionary distance from the *A. nidulans* AcpA and AcpB clades. The AcpA clade includes members from 21 species of *Aspergillus* genus with known genomes and most ascomycetes. A second clade, which includes AcpB, close to the AcpA proteins, is also partially conserved in *Aspergillus* sp. or other dikarya. The AcpC *A. nidulans* clade includes few sequences and is significantly more distant from AcpA, AcpB or the yeast clades (Sá-Pessoa et al. 2015). The AceTr bacteria homologues, are the most distant members from the fungi species, possibly due to evolutionary distance factors, such as substrate specificity. The SatP transporter accepts both acetate and succinate (Sá-Pessoa et al. 2013) in contrast to known fungi homologues, which do not.

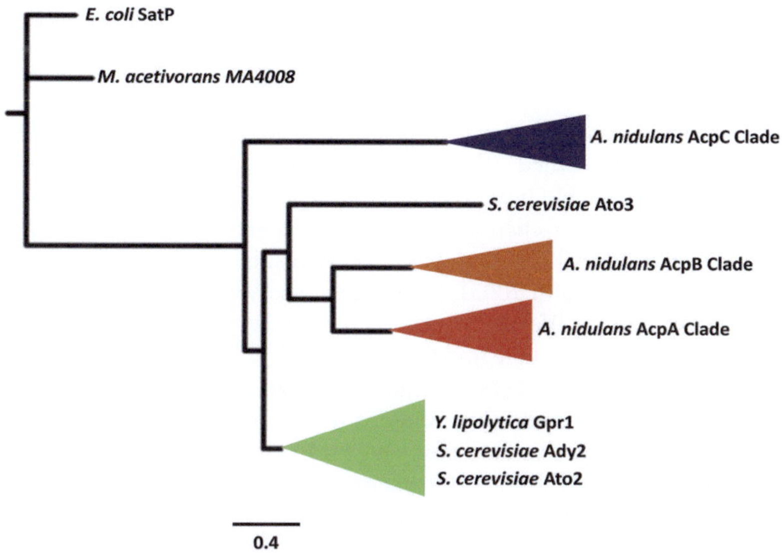


**Fig. 9.5 Phylogenetic tree of AceTr family.** A phylogenetic tree was constructed from alignments of bacteria and fungi AceTr homologous protein sequences, using the Nearest-Neighbour-Joining algorithm. NCBI Protein accession numbers are shown next to the species name. Microbial species shown in the tree and protein clades are as follows: *Aspergillus sp.* (AcpA, AcpB and AcpC clades); *Saccharomyces cerevisiae; Yarrowia lipolytica; Escherichia coli; Metanosarcina acetirovans*. The SatP transporter from *E. coli* was used as outgroup. The scale bar indicates de number of amino acid substitution per site. Adapted from Sá-Pessoa et al. (2015)

The Ady2 transporter has potential biotechnological applications for the efflux of carboxylic acids, such as in lactate secreting *S. cerevisiae* cell factories (Pacheco et al. 2012). In *S. cerevisiae* strains, engineered to produce lactate from glucose, the overexpression of Ady2 promotes the increase of lactic acid accumulation in the extracellular medium. Upon glucose exhaustion, a switch in the overall cell metabolism occurs, and the extracellular acid is again transported inside the cell *via* Ady2 (Pacheco et al. 2012). In order to uncover novel lactate transporters in *S. cerevisiae* a laboratory evolution approach was carried out in a *jen1* deleted strain. As a result, two independent evolved mutants associated with a gain of function for lactate uptake were selected. The whole-genome resequencing identified two single-nucleotide changes both located in *ADY2*, the L219V and A252G (de Kok et al. 2012). The two previous amino acid residues were equivalently mutated in the bacteria homologue SatP, specifically in L131 and A164. Both mutations enhanced the lactate uptake (Sá-Pessoa et al. 2013). Overall, these results demonstrate the relevance of carboxylate transporters as modulators of the production and secretion of carboxylic acids by microbial cells. The molecular engineering of these transporters can improve microbial cell factories for organic acids bio-production.

## 9.4 Role of Plasma Membrane Transporters in Carboxylic Acid Stress Resistance

Short-chain carboxylic acids are important biotechnological compounds with large application in the so called Bio-based economy for obtaining building-block chemicals from sustainable biomass (Sauer et al. 2008). Additionally, in food and beverages industries, short-chain monocarboxylic acids like acetic, sorbic, benzoic or propionic acids are widely used as preservatives, in order to prevent microbial spoilage (Russell and Gould 2003). It has been observed that some yeasts are acid resistant and able to proliferate in acidic environments (Fleet 2007; Piper et al. 1998). This property is desirable for industrial carboxylic acids production, avoiding neutralizing agents' addition to the medium and expensive downstream processes to regenerate undissociated carboxylic acids, as it happens when less tolerant bacteria are used (Sauer et al. 2010). However, concerning the use of carboxylic acids as food and beverage preservatives, the capacity to resist to an acidic environment can lead to the development of contaminant yeasts and food spoilage (Piper et al. 1998), requiring higher concentrations of the acids with deleterious effects in human health. In this way, a deep understanding of the mechanisms underlying acid resistance and stress response is fundamental to control these biotechnological processes.

The maintenance of the intracellular pH (pHi) and the development of a stress response is crucial for yeast cell survival in an acidic environment. Intracellular acidification can affect the redox homeostasis, enzymatic activities (consequently, altering metabolic pathways and energy yield) or nutrient transport across the cell (Goffeau and Slayman 1981; Casal et al. 2008). On the other hand, the accumulation of the carboxylic acid counter-ion ($RCOO^-$) induces toxic effects such as high internal turgor pressure, oxidative stress and lipid peroxidation (Mollapour and Piper 2008; Mira et al. 2010). Furthermore, organic acids can cause changes in membrane and cell wall structure and composition (Piper et al. 2001; Ullah et al. 2013). Carboxylic acids stress response in yeast includes (i) a general and pleiotropic stress response that affects the expression of several genes under the control of Msn2 and Msn4 transcription factors (Martinez-Pastor et al. 1996; Schmitt and McEntee 1996) and (ii) a specific response that involves the efflux of the organic anions by the pump Pdr12 and the efflux of protons by the pump Pma1 (Piper et al. 1998; Holyoak et al. 1999; Bauer et al. 2003). The rate of entrance of the undissociated form of carboxylic acids into the cell (by simple diffusion or *via* Fps1) and the efflux of the organic anions (*via* Pdr12) are key events in yeast adaptation to the acidic environment (Fig. 9.6).

### *9.4.1 The Pleiotropic Drug Resistance Pdr12 Pump*

Pdr12 is a member of a large ATP-binding cassette (ABC) transporters PDR subfamily. Contrarily to other members of this subfamily that have a broad range of substrates, Pdr12 is specifically involved in the efflux of carboxylic acids (Seret et al. 2009).

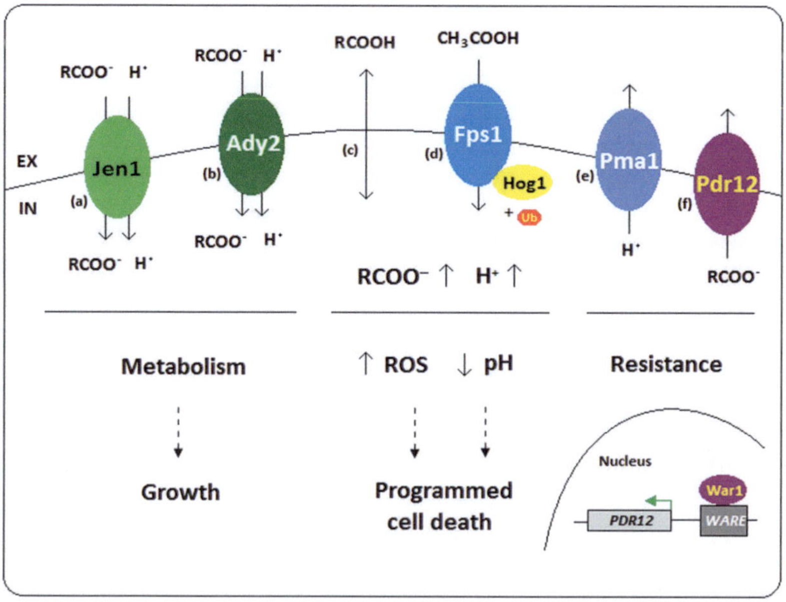


**Fig. 9.6 Mechanisms of carboxylic acids transport in*S. cerevisiae*.** Jen1 and Ady2 are able to mediate the transport of carboxylates ($RCOO^-$, (**a**) = lactate, pyruvate, acetate or propionate; (**b**) = acetate, propionate, formate or lactate, respectively) into the cell, where they are used as carbon and energy sources. When the undissociated form of the acid (RCOOH) prevails, it enters by simple diffusion (**c**) and in the case of acetic acid ($CH_3COOH$) also through the Fps1 channel (**d**). The accumulation of protons ($H^+$) and anions ($RCOO^-$) may cause severe acidic conditions (low pH and high levels of ROS), which eventually lead to programmed cell death. Acid stress triggers the efflux of $H^+$ by Pma1 (**e**) and of carboxylates by Pdr12 (**f**) = sorbate, propionate, benzoate or levulinate). To increase the expression of Pdr12, cell activates the War1 transcription factor which induces the transcription of *PDR12*, through *WARE* cis-acting response element. Furthermore, the ubiquitylation and degradation of Fps1, promoted by Hog1 MAPK pathway, serves as a mechanism to avoid acetic acid facilitated diffusion

The PDR subfamily is characterized by the presence of nuclear binding domains (NBD), that contain three conserved motifs: Walker A, Walker B and ABC signature or C loop, alternating with TMS containing six transmembrane segments, in the following pattern NBD-TMS-NBD-TMS, a reverse topology of the classic ABC transporters (Decottignies and Goffeau 1997; Seret et al. 2009; Lamping et al. 2010). The two NBD work in tandem forming the ATP binding site, while TMS bind the substrate and translocate it across the plasma membrane (Martinez and Falson 2014).

In *S. cerevisiae* the PDR subfamily comprises the following members: Pdr5, Pdr10, Pdr15, Snq2, Pdr12, Aus1, Pdr11, Pdr18, Adp1 and YOL075c (Rea 1999; Prasad and Goffeau 2012). An extensive phylogenetic analysis of 349 PDR proteins from 55 fungal species identified nine clusters (A, B, C, D, E, F, G, H1, H2), being Pdr12-like pumps associated with cluster D, which also includes Snq2. This cluster, as well as the clusters A (Pdr5-like) and E (Aus1/Pdr11-like), only contains Saccharomycotina members (Lamping et al. 2010).

Another phylogenetic study, based on Génolevures database comprising nine hemiascomycetous species, identified a family (GL3C0025) containing all PDR proteins of *S. cerevisiae*, with 62 members clustered in 5 groups (A to E). Cluster A grouped Pdr12 together with eight other members from *Candida glabrata*, *Zygosaccharomyces rouxii*, *Saccharomyces kluyvery*, *K. lactis* and *Y. lipolytica*. In this analysis, the close homologue Snq2 was represented in cluster B (Seret et al. 2009). Although the function of Pdr12 and Snq2 diverge (short chain carboxylic acids efflux and antifungal azoles or other hydrophobic compounds extrusion, respectively), they share high sequence identity (46 %), what may suggest a common ancestral to both clusters (Seret et al. 2009).

Pdr12 is located at the plasma membrane, being essential to yeast adaptation to weak acid stress (Piper et al. 1998). Its correct expression, trafficking and/or functionality is controlled by Pdr10, another member of ABC subfamily that shares 36 % identity with Pdr12. Specifically, Pdr10 controls the local environment of Pdr12 in the membrane and its partition in lipid rafts (Rockwell et al. 2009).

Pdr12 is specifically involved in the efflux of propionic, sorbic and benzoic acids, increasing cells resistance to these acids (Piper et al. 1998; Holyoak et al. 1999). Carboxylic acids with aliphatic chain lengths higher than C7 (thus, with high degree of hydrophobicity), or highly hydrophilic (lactic, acetic and formic) acids, are not exported by Pdr12 (Holyoak et al. 1999). In accordance to this, it has been observed that the presence of propionic, sorbic or benzoic acids, as well as acidic pH, strongly induces *PDR12* expression, differently to what is observed with acetic and formic acids (Piper et al. 2001; Hatzixanthis et al. 2003).

In contrast with a wild type strain, *pdr12* mutant cells present higher sensitivity to sorbic, benzoic, propionic and levulinic acids, as well as to other carboxylic acids ranging from C1 to C7 (Piper et al. 1998; Holyoak et al. 1999; Nygård et al. 2014). The opposite is observed when overexpressing *PDR12* (Nygård et al. 2014). The influence of Pdr12 in carboxylic acid stress response depends on the degree of the acid hydrophobicity, as *PDR12* expression decreases the tolerance to the hydrophilic formic, acetic, lactic and glycolic acids, in contrast to what happens for the moderately hydrophobic sorbic, propionic and levulinic acids (Nygard 2014).

The release of $H^+$ and of the charged anion ($RCOO^-$) due to the acid dissociation alters the plasma membrane electrochemical potential ($Z\Delta pH$). Pma1 is able to mediate the efflux of the protons, neutralizing the cytoplasm acidification. But there is still a charge imbalance. Thus, the extrusion of the anionic form of the acid by Pdr12 avoids its toxic accumulation and restores the charge balance at expenses of intracellular ATP (Piper et al. 1998; Holyoak et al. 1999).

The carboxylic acid stress response is regulated by different pathways, namely through the transcription factors Msn2 and Msn4, Haa1, Rim1 and War1 (reviewed by Mira et al. 2010). *PDR12* promoter contains a weak-acid response element (*WARE*), recognized by the zinc-finger protein War1, exclusively involved in *PDR12* expression regulation (Kren et al. 2003). Deletion of *WAR1* leads to a failure of Pdr12 accumulation at the plasma membrane and to a decreased resistance to the moderately hydrophobic carboxylic acids. This phenotype is restored by the constitutive expression of *PDR12* in the *war1* null mutant (Kren et al. 2003; Schuller et al. 2004). War1 is activated by carboxylic acids, putatively through the modification of the phosphorylation status of two serine residues at the positions 923 and 930 (Mollapour and Piper 2012). It has been observed that the carboxylic acid stress induces a conformational change of War1, enhancing its association to *PDR12* promoter (Gregori et al. 2008). In fact, gel electrophoresis assays demonstrated a mobility shift of War1 in carboxylic acid stressed cells, due to changes in its phosphorylation status. However, a War1 kinase has not been identified so far (Kren et al. 2003). War1 is conserved among several yeast species and an orthologue found in *C. albicans* was shown to mediate sorbic acid tolerance, like in *S. cerevisiae* (Lebel et al. 2006).

In contrast to what happens with War1, Pdr12 is negatively regulated by Cmk1, a calmodulin protein kinase (Holyoak et al. 2000; Nygård et al. 2014). Like the strain overexpressing *PDR12*, the deletion of *CMK1* increases the resistance to levulinic and propionic acids and the sensitivity to formic, acetic, glycolic and lactic acids. However, the role of Cmk1 is not limited to Pdr12 negative regulation, as improved tolerance to the acetic and formic acids observed in *pdr12* mutant strain, was further enhanced by *CMK1* deletion, indicating an additional role of Cmk1 in the carboxylic acids resistance capacity (Nygård et al. 2014).

Phylogenetic studies indicate the existence of Pdr12 homologues in other yeast species. In *C. glabrata* and *C. albicans*, like in *S. cerevisiae*, an increased resistance to sorbic and benzoic acids is accompanied by an increased expression of Pdr12, mediated by the activation of War1 (Mundy and Cormack 2009; Piper 2011). Furthermore, in *C. glabrata* Hog1 is required for full activation of *PDR12* (Jandric et al. 2013). In *K. lactis*, Pdr12 is associated with the efflux of the 4-methylthio-oxobutyric acid, a fusel acid derived from methionine (Hebert et al. 2011), a feature well described in *S. cerevisiae* (Hazelwood et al. 2006), suggesting an analogue function for Pdr12 proteins in these two species. Although *PDR12* orthologues have been annotated in several yeast genomes, a functional role has only been studied in detail in *S. cerevisiae.*

### 9.4.2 The Aquaglyceroporin Fps1

The yeast aquaglyceroporin Fps1 is able to mediate the uptake of the undissociated form of acetic acid into the cell (Mollapour and Piper 2007). *S. cerevisiae* genome encodes another member of aquaglyceroporins, Yfl045c, whose function is still unclarified (Ahmadpour et al. 2014). Both Fps1 and Yfl045c belong to the MIP family, a group of integral membrane proteins present in all organisms. These transporters form channels or pores through biological membranes, mediating the transport of small uncharged molecules, such as water or glycerol (Hohmann et al. 2000; Tamas et al. 1999). Fps1, in contrast to what is described for other aquaglyceroporins, does not act as a monomer, but most probably as a homo-tetramer, as described for aquaporins (Beese-Sims et al. 2011).

MIP channels have two symmetrical halves and are characterized by the presence of 6 TMS and two important loops, B and E, containing the highly conserved NPA motif. However, in Fps1-like proteins, these loops are less conserved presenting the NPX and NXA motifs (Pettersson et al. 2005). The loops B and E are located between second and third TMS and fifth and sixth TMS respectively, consisting in half-helices entering the membrane in opposite sites, forming a channel pore (Gustavsson et al. 2005).

More than 400 MIPs have been found in fungal available genomes, phylogenetically classified according to protein similarity and substrate selectivity (Verma et al. 2014). According to this classification, besides the traditional and dominant clusters of aquaporins (water channels) and aquaglyceroporins (water and small uncharged molecules channels), two new clusters have been identified, XIP (X-intrinsic proteins) and SIP (small and basic intrinsic proteins) groups, which are also present in plants. Aquaglyceroporins cluster contains at least seven subfamilies in fungi, including Fps1-like subfamily which is found invariably in yeast Ascomycota (Verma et al. 2014).

Fps1 was firstly described as being involved in glycerol efflux in response to osmotic stress (Luyten et al. 1995). However, Fps1 was shown to be also involved in the influx of arsenite, antimonite and the undissociated form of acetic acid (Mollapour and Piper 2007; Wysocki et al. 2001). Regulation of Fps1 expression is dependent on the Hog1 MAPK pathway (Tamas et al. 1999). Hog1 is activated by hyperosmotic stress, binding in these conditions to the N-terminal of Fps1, leading to the channel closure and preventing glycerol outflow (Lee et al. 2013). In hypoosmotic conditions, water enters into the cell causing turgor pressure, inducing the opening of the channel and glycerol efflux through Fps1 (Tamas et al. 1999). Acetic acid stress activates Hog1 MAPK pathway leading to Fps1 decay. Hog1 promotes Fps1 phosphorylation, ubiquitylation, endocytosis and degradation in the vacuole, impairing acetic acid uptake and therefore, rendering the cells more resistant to the acidic stress (Mollapour and Piper 2007; Piper et al. 2011) (Fig. 9.6).

*FPS1* orthologues have been found in *Zygosaccharomyces rouxii*, *K. lactis*, *Kluyveromyces marxianus* and *Pichia angusta* genomes but not in *C. albicans*, *Candida tropicalis*, *D. hansenii* and *Pichia sorbitophila*. The heterologous expression of *FPS1* from the first three above mentioned yeast species in a *S. cerevisiae fps1* mutant complements its phenotype regarding the capacity to recover from a hypoosmotic shock, indicating their conserved role in glycerol efflux (Neves et al. 2004). In *C. glabrata* two orthologues were found: CgFps1 and CgFps2, both regulated by the Hog1 pathway. Like in *S. cerevisiae*, they are associated with glycerol efflux and tolerance to osmotic stress where CgFps1 has a more important role than CgFps2 (Beese-Sims et al. 2012). Only *CaFPS1* was able to complement the *fps1* mutant phenotype of *S. cerevisiae*, reflecting the higher degree of homology between ScFps1 and CaFps1 (Beese-Sims et al. 2012). Despite the similarity of functions of *S. cerevisiae FPS1* yeast orthologues concerning glycerol transport and osmotic stress response, no role concerning carboxylic acid stress, and more precisely, concerning the response to acetic acid stress, has yet been attributed to them.

## 9.5 Conclusions

Carboxylic acids are intermediates of central metabolic pathways like glycolysis and tricarboxylic acid cycle (TCA), thus playing a key role in the metabolism of all cells. The study of the metabolism of carboxylic acids in yeasts is also important from a biotechnological point of view, particularly regarding the production and preservation of foods and beverages, but also in the pharmaceutical and biomedical industries.

*S. cerevisiae* can use a wide range of nutrients as carbon and energy sources, with preference for carbohydrates, in particular glucose or other related rapidly fermentable sugars, like fructose and mannose (review by Conrad et al. 2014). The transition from a fermentative to an oxidative metabolism requires sugar depletion, and is associated with a metabolic reprogramming, namely of TCA, oxidative phosphorylation, glyoxylate cycle and gluconeogenesis, and membrane transporters (Gombert et al. 2001; Soontorngun et al. 2007; Casal et al. 2008). In this chapter we reviewed the role of carboxylic acid transporters in the capacity of the yeast cell to adapt and respond to environmental changes (Fig. 9.6). On one hand, Jen1 and Ady2, expressed only in the absence of glucose, are involved in the metabolic use of monocarboxylic acids. On the other hand, Fps1 and Pdr12 are related to acid stress responses. The entrance of the undissociated acid in the cell, either by simple diffusion or *via* Fps1, in the case of acetic acid, imposes a severe stress: the anion accumulation is toxic and the released protons acidify the intracellular pH. Pumps specific for protons (Pma1) and for monocarboxylate anions (Pdr12) are crucial for the cells to recover and essential to maintain an equilibrated intracellular environment and cellular homeostasis.

**Acknowledgements** This work was supported by FCT I.P. through the strategic funding UID/BIA/04050/2013 and the project PTDC/BIA-MIC/5184/2014. D.R and G.T. acknowledge FCT for the SFRH/BD/96166/2013 and SFRH/BD/86221/2012, respectively.

## References

Abate G, Bastonini E, Braun KA, Verdone L, Young ET, Caserta M (2012) Snf1/AMPK regulates Gcn5 occupancy, H3 acetylation and chromatin remodelling at *S. cerevisiae ADY2* promoter. Biochim Biophys Acta 1819:419–427

Abramson J, Smirnova I, Kasho V, Verner G, Kaback HR, Iwata S (2003) Structure and mechanism of the lactose permease of *Escherichia coli*. Science 301:610–615

Agre P, Brown D, Nielsen S (1995) Aquaporin water channels: unanswered questions and unresolved controversies. Curr Opin Cell Biol 7:472–483

Ahmadpour D, Geijer C, Tamas MJ, Lindkvist-Petersson K, Hohmann S (2014) Yeast reveals unexpected roles and regulatory features of aquaporins and aquaglyceroporins. Biochim Biophys Acta 1840:1482–1491

Andrade RP, Casal M (2001) Expression of the lactate permease gene *JEN1* from the yeast *Saccharomyces cerevisiae*. Fungal Genet Biol 32:105–111

Andrade RP, Kotter P, Entian KD, Casal M (2005) Multiple transcripts regulate glucose-triggered mRNA decay of the lactate transporter *JEN1* from *Saccharomyces cerevisiae*. Biochem Biophys Res Commun 332:254–262

Augstein A, Barth K, Gentsch M, Kohlwein SD, Barth G (2003) Characterization, localization and functional analysis of Gpr1p, a protein affecting sensitivity to acetic acid in the yeast *Yarrowia lipolytica*. Microbiology 149:589–600

Bauer BE, Rossington D, Mollapour M, Mamnun Y, Kuchler K, Piper PW (2003) Weak organic acid stress inhibits aromatic amino acid uptake by yeast, causing a strong influence of amino acid auxotrophies on the phenotypes of membrane transporter mutants. Eur J Biochem 270:3189–3195

Becuwe M, Vieira N, Lara D, Gomes-Rezende J, Soares-Cunha C, Casal M, Haguenauer-Tsapis R, Vincent O, Paiva S, Leon S (2012) A molecular switch on an arrestin-like protein relays glucose signaling to transporter endocytosis. J Cell Biol 196:247–259

Beese-Sims SE, Lee J, Levin DE (2011) Yeast Fps1 glycerol facilitator functions as a homotetramer. Yeast 28:815–819

Beese-Sims SE, Pan SJ, Lee J, Hwang-Wong E, Cormack BP, Levin DE (2012) Mutants in the *Candida glabrata* glycerol channels are sensitized to cell wall stress. Eukaryot Cell 11:1512–1519

Bojunga N, Entian KD (1999) Cat8p, the activator of gluconeogenic genes in *Saccharomyces cerevisiae*, regulates carbon source-dependent expression of NADP-dependent cytosolic isocitrate dehydrogenase (Idp2p) and lactate permease (Jen1p). Mol Gen Genet 262:869–875

Carlson M (1999) Glucose repression in yeast. Curr Opin Microbiol 2(2):202–207

Casal M, Paiva S, Andrade RP, Gancedo C, Leao C (1999) The lactate-proton symport of *Saccharomyces cerevisiae* is encoded by *JEN1*. J Bacteriol 181:2620–2623

Casal M, Paiva S, Queiros O, Soares-Silva I (2008) Transport of carboxylic acids in yeasts. FEMS Microbiol Rev 32:974–994

Cassio F, Leao C, van Uden N (1987) Transport of lactate and other short-chain monocarboxylates in the yeast *Saccharomyces cerevisiae*. Appl Environ Microbiol 53:509–513

Conrad M, Schothorst J, Kankipati HN, Van Zeebroeck G, Rubio-Texeira M, Thevelein JM (2014) Nutrient sensing and signaling in the yeast *Saccharomyces cerevisiae*. FEMS Microbiol Rev 38:254–299

Dang S, Sun L, Huang Y, Lu F, Liu Y, Gong H, Wang J, Yan N (2010) Structure of a fucose transporter in an outward-open conformation. Nature 467:734–738

de Kok S, Nijkamp JF, Oud B, Roque FC, de Ridder D, Daran JM, Pronk JT, van Maris AJ (2012) Laboratory evolution of new lactate transporter genes in a jen1Delta mutant of *Saccharomyces cerevisiae* and their identification as *ADY2* alleles by whole-genome resequencing and transcriptome analysis. FEMS Yeast Res. doi:10.1111/j.1567-1364.2012.00787.x

Decottignies A, Goffeau A (1997) Complete inventory of the yeast ABC proteins. Nat Genet 15:137–145

Dulermo R, Gamboa-Melendez H, Michely S, Thevenieau F, Neuveglise C, Nicaud JM (2015) The evolution of Jen3 proteins and their role in dicarboxylic acid transport in Yarrowia. MicrobiologyOpen 4:100–120

Fillinger S, Felenbok B (1996) A newly identified gene cluster in *Aspergillus nidulans* comprises five novel genes localized in the alc region that are controlled both by the specific transactivator AlcR and the general carbon-catabolite repressor CreA. Mol Microbiol 20:475–488

Fleet GH (2007) Yeasts in foods and beverages: impact on product quality and safety. Curr Opin Biotechnol 18:170–175

Flipphi M, Robellet X, Dequier E, Leschelle X, Felenbok B, Velot C (2006) Functional analysis of alcS, a gene of the alc cluster in *Aspergillus nidulans*. Fungal Genet Biol 43:247–260

Gimenez R, Nunez MF, Badia J, Aguilar J, Baldoma L (2003) The gene yjcG, cotranscribed with the gene acs, encodes an acetate permease in *Escherichia coli*. J Bacteriol 185:6448–6455

Goffeau A, Slayman CW (1981) The proton-translocating ATPase of the fungal plasma membrane. Biochim Biophys Acta 639:197–223

Gombert AK, Moreira dos Santos M, Christensen B, Nielsen J (2001) Network identification and flux quantification in the central metabolism of *Saccharomyces cerevisiae* under different conditions of glucose repression. J Bacteriol 183:1441–1451

Gregori C, Schuller C, Frohner IE, Ammerer G, Kuchler K (2008) Weak organic acids trigger conformational changes of the yeast transcription factor War1 *in vivo* to elicit stress adaptation. J Biol Chem 283:25752–25764

Grobler J, Bauer F, Subden RE, Van Vuuren HJ (1995) The mae1 gene of *Schizosaccharomyces pombe* encodes a permease for malate and other C4 dicarboxylic acids. Yeast 11:1485–1491

Guo H, Liu P, Madzak C, Du G, Zhou J, Chen J (2015) Identification and application of keto acids transporters in *Yarrowia lipolytica*. Sci Rep 5:8138

Gustavsson S, Lebrun A-S, Nordén K, Chaumont F, Johanson U (2005) A novel plant major intrinsic protein in physcomitrella patens most similar to bacterial glycerol channels. Plant Physiol 139:287–295

Hatzixanthis K, Mollapour M, Seymour I, Bauer BE, Krapf G, Schüller C, Kuchler K, Piper PW (2003) Moderately lipophilic carboxylate compounds are the selective inducers of the *Saccharomyces cerevisiae* Pdr12p ATP-binding cassette transporter. Yeast 20:575–585

Haurie V, Perrot M, Mini T, Jeno P, Sagliocco F, Boucherie H (2001) The transcriptional activator Cat8p provides a major contribution to the reprogramming of carbon metabolism during the diauxic shift in *Saccharomyces cerevisiae*. J Biol Chem 276:76–85

Hazelwood LA, Tai SL, Boer VM, de Winde JH, Pronk JT, Daran JM (2006) A new physiological role for Pdr12p in *Saccharomyces cerevisiae*: export of aromatic and branched-chain organic acids produced in amino acid catabolism. FEMS Yeast Res 6:937–945

Hebert A, Forquin-Gomez MP, Roux A, Aubert J, Junot C, Loux V, Heilier JF, Bonnarme P, Beckerich JM, Landaud S (2011) Exploration of sulfur metabolism in the yeast Kluyveromyces lactis. Appl Microbiol Biotechnol 91:1409–1423

Hedges D, Proft M, Entian KD (1995) CAT8, a new zinc cluster-encoding gene necessary for derepression of gluconeogenic enzymes in the yeast Saccharomyces cerevisiae. Mol Cell Biol 15:1915–1922

Hohmann I, Bill RM, Kayingo I, Prior BA (2000) Microbial MIP channels. Trends Microbiol 8:33–38

Holyoak CD, Bracey D, Piper PW, Kuchler K, Coote PJ (1999) The *Saccharomyces cerevisiae* weak-acid-inducible ABC transporter Pdr12 transports fluorescein and preservative anions from the cytosol by an energy-dependent mechanism. J Bacteriol 181:4644–4652

Holyoak CD, Thompson S, Ortiz Calderon C, Hatzixanthis K, Bauer B, Kuchler K, Piper PW, Coote PJ (2000) Loss of Cmk1 Ca(2+)-calmodulin-dependent protein kinase in yeast results in constitutive weak organic acid resistance, associated with a post-transcriptional activation of the Pdr12 ATP-binding cassette transporter. Mol Microbiol 37:595–605

Huang YF, Lemieux MJ, Song JM, Auer M, Wang DN (2003) Structure and mechanism of the glycerol-3-phosphate transporter from Escherichia coli. Science 301:616–620

Jandric Z, Gregori C, Klopf E, Radolf M, Schüller C (2013) Sorbic acid stress activates the Candida glabrata high osmolarity glycerol MAP kinase pathway. Front Microbiol 4:350

Kren A, Mamnun YM, Bauer BE, Schuller C, Wolfger H, Hatzixanthis K, Mollapour M, Gregori C, Piper P, Kuchler K (2003) War1p, a novel transcription factor controlling weak acid stress response in yeast. Molecular and cellular biology 23:1775–1785

Kujau M, Weber H, Barth G (1992) Characterization of mutants of the yeast *Yarrowia lipolytica* defective in acetyl-coenzyme A synthetase. Yeast 8:193–203

Lamping E, Baret PV, Holmes AR, Monk BC, Goffeau A, Cannon RD (2010) Fungal PDR transporters: phylogeny, topology, motifs and function. Fungal Genet Biol 47:127–142

Leão C, van Uden N (1986) Transport of lactate and other short-chain monocarboxylates in the yeast Candida utilis. Appl Microbiol Biotechnol 23:389

Lebel K, MacPherson S, Turcotte B (2006) New tools for phenotypic analysis in Candida albicans: the WAR1 gene confers resistance to sorbate. Yeast 23:249–259

Lee J, Reiter W, Dohnal I, Gregori C, Beese-Sims S, Kuchler K, Ammerer G, Levin DE (2013) MAPK Hog1 closes the S. cerevisiae glycerol channel Fps1 by phosphorylating and displacing its positive regulators. Genes Dev 27:2590–2601

Lis P, Zarzycki M, Ko YH, Casal M, Pedersen PL, Goffeau A, Ulaszewski S (2012) Transport and cytotoxicity of the anticancer drug 3-bromopyruvate in the yeast Saccharomyces cerevisiae. J Bioenerg Biomembr 44:155–161

Lodi T, Diffels J, Goffeau A, Baret PV (2007) Evolution of the carboxylate Jen transporters in fungi. FEMS Yeast Res 7:646–656

Lodi T, Fontanesi F, Ferrero I, Donnini C (2004) Carboxylic acids permeases in yeast: two genes in Kluyveromyces lactis. Gene 339:111–119

Lodi T, Fontanesi F, Guiard B (2002) Co-ordinate regulation of lactate metabolism genes in yeast: the role of the lactate permease gene JEN1. Mol Genet Genom 266:838–847

Luyten K, Albertyn J, Skibbe WF, Prior BA, Ramos J, Thevelein JM, Hohmann S (1995) Fps1, a yeast member of the MIP family of channel proteins, is a facilitator for glycerol uptake and efflux and is inactive under osmotic stress. EMBO J 14:1360–1371

Marger MD, Saier MH Jr (1993) A major superfamily of transmembrane facilitators that catalyse uniport, symport and antiport. Trends Biochem Sci 18:13–20

Martinez L, Falson P (2014) Multidrug resistance ATP-binding cassette membrane transporters as targets for improving oropharyngeal candidiasis treatment. Adv Cell Mol Otalaryngol 2:23955

Martinez-Pastor MT, Marchler G, Schuller C, Marchler-Bauer A, Ruis H, Estruch F (1996) The Saccharomyces cerevisiae zinc finger proteins Msn2p and Msn4p are required for transcriptional induction through the stress response element (STRE). EMBO J 15:2227–2235

McDermott JR, Rosen BP, Liu Z (2010) Jen1p: a high affinity selenite transporter in yeast. Mol Biol cell 21:3934–3941

Mira NP, Becker JD, Sa-Correia I (2010) Genomic expression program involving the Haa1p-regulon in Saccharomyces cerevisiae response to acetic acid. Omics 14:587–601

Mollapour M, Piper PW (2007) Hog1 mitogen-activated protein kinase phosphorylation targets the yeast Fps1 aquaglyceroporin for endocytosis, thereby rendering cells resistant to acetic acid. Mol Cell Biol 27:6446–6456

Mollapour, M., & Piper, P. W. (2008). Chapter 10. In Simon MS, Avery V, Pieter Van W (eds) Weak organic acid resistance of spoilage yeasts. British Mycological Society symposia series. Academic Press, New York, pp 143–155

Mollapour M, Piper PW (2012) Activity of the yeast zinc-finger transcription factor War1 is lost with alanine mutation of two putative phosphorylation sites in the activation domain. Yeast 29:39–44

Mota S, Vieira N, Barbosa S, Delaveau T, Torchet C, Le Saux A, Garcia M, Pereira A, Lemoine S, Coulpier F et al (2014) Role of the DHH1 gene in the regulation of monocarboxylic acids transporters expression in Saccharomyces cerevisiae. PloS One 9, e111589

Mundy RD, Cormack B (2009) Expression of Candida glabrata Adhesins after exposure to chemical preservatives. J Infect Dis 199:1891–1898

Neves L, Oliveira R, Lucas C (2004) Yeast orthologues associated with glycerol transport and metabolism. FEMS Yeast Res 5:51–62

Newstead S, Drew D, Cameron AD, Postis VL, Xia X, Fowler PW, Ingram JC, Carpenter EP, Sansom MS, McPherson MJ et al (2011) Crystal structure of a prokaryotic homologue of the mammalian oligopeptide-proton symporters, PepT1 and PepT2. EMBO J 30:417–426

Nygard Y (2014). Production of D-xylonate and organic acid tolerance in yeast. PhD thesis, Aalto University School of Chemical Technology, Espoo, Finland

Nygård Y, Mojzita D, Toivari M, Penttilä M, Wiebe MG, Ruohonen L (2014) The diverse role of Pdr12 in resistance to weak organic acids. Yeast 31:219–232

Olesen JT, Guarente L (1990) The HAP2 subunit of yeast CCAAT transcriptional activator contains adjacent domains for subunit association and DNA recognition: model for the HAP2/3/4 complex. Genes Dev 4(10):1714–1729

Olszewska M, Bujarski JJ, Kurpisz M (2012) P-bodies and their functions during mRNA cell cycle: mini-review. Cell Biochem Funct 30:177–182

Osothsilp C, Subden RE (1986) Malate transport in Schizosaccharomyces pombe. J Bacteriol 168:1439–1443

Pacheco A, Talaia G, Sa-Pessoa J, Bessa D, Goncalves MJ, Moreira R, Paiva S, Casal M, Queiros O (2012) Lactic acid production in Saccharomyces cerevisiae is modulated by expression of the monocarboxylate transporters Jen1 and Ady2. FEMS Yeast Res 12:375–381

Paiva S, Devaux F, Barbosa S, Jacq C, Casal M (2004) Ady2p is essential for the acetate permease activity in the yeast Saccharomyces cerevisiae. Yeast 21:201–210

Paiva S, Kruckeberg AL, Casal M (2002) Utilization of green fluorescent protein as a marker for studying the expression and turnover of the monocarboxylate permease Jen1p of Saccharomyces cerevisiae. Biochem J 363:737–744

Paiva S, Vieira N, Nondier I, Haguenauer-Tsapis R, Casal M, Urban-Grimal D (2009) Glucose-induced ubiquitylation and endocytosis of the yeast Jen1 transporter: role of lysine 63-linked ubiquitin chains. Journal Biol Chem 284:19228–19236

Palkova Z, Devaux F, Icicova M, Minarikova L, Le Crom S, Jacq C (2002) Ammonia pulses and metabolic oscillations guide yeast colony development. Mol Biol Cell 13:3901–3914

Pettersson N, Filipsson C, Becit E, Brive L, Hohmann S (2005) Aquaporins in yeasts and filamentous fungi. Biol Cell 97:487–500

Piper P, Calderon CO, Hatzixanthis K, Mollapour M (2001) Weak acid adaptation: the stress response that confers yeasts with resistance to organic acid food preservatives. Microbiology 147:2635–2642

Piper P, Mahe Y, Thompson S, Pandjaitan R, Holyoak C, Egner R, Muhlbauer M, Coote P, Kuchler K (1998) The pdr12 ABC transporter is required for the development of weak organic acid resistance in yeast. EMBO J 17:4257–4265

Piper PW (2011) Chapter 4 - Resistance of yeasts to weak organic acid food preservatives. In Allen SS, Laskin I, Geoffrey MG (eds) Advances in applied microbiology, vol 77. Academic Press, pp 97–113

Piper PW, Ortiz-Calderon C, Holyoak C, Coote P, Cole M (1997) Hsp30, the integral plasma membrane heat shock protein of Saccharomyces cerevisiae, is a stress-inducible regulator of plasma membrane H(+)-ATPase. Cell Stress Chaperones 2:12–24

Prasad R, Goffeau A (2012) Yeast ATP-binding cassette transporters conferring multidrug resistance. Annu Rev Microbiol 66:39–63

Queiros O, Pereira L, Paiva S, Moradas-Ferreira P, Casal M (2007) Functional analysis of Kluyveromyces lactis carboxylic acids permeases: heterologous expression of KlJEN1 and KlJEN2 genes. Curr Genet 51:161–169

Rabitsch KP, Toth A, Galova M, Schleiffer A, Schaffner G, Aigner E, Rupp C, Penkner AM, Moreno-Borchart AC, Primig M et al (2001) A screen for genes required for meiosis and spore formation based on whole-genome expression. Curr Biol 11:1001–1009

Randez-Gil F, Bojunga N, Proft M, Entian KD (1997) Glucose derepression of gluconeogenic enzymes in Saccharomyces cerevisiae correlates with phosphorylation of the gene activator Cat8p. Mol Cell Biol 17:2502–2510

Ratnakumar S, Kacherovsky N, Arms E, Young ET (2009) Snf1 controls the activity of adr1 through dephosphorylation of Ser230. Genetics 182:735–745

Rea PA (1999) MRP subfamily ABC transporters from plants and yeast. J Exp Bot 50:895–913

Richard GF, Fairhead C, Dujon B (1997) Complete transcriptional map of yeast chromosome XI in different life conditions. J Mol Biol 268:303–321

Robellet X, Flipphi M, Pegot S, Maccabe AP, Velot C (2008) AcpA, a member of the GPR1/FUN34/YaaH membrane protein family, is essential for acetate permease activity in the hyphal fungus Aspergillus nidulans. Biochem J 412:485–493

Rockwell N, Wolfger H, Kuchler K, Thorner J (2009) ABC transporter Pdr10 regulates the membrane microenvironment of Pdr12 in *Saccharomyces cerevisiae*. J Membr Biol 229:27–52

Rohlin L, Gunsalus RP (2010) Carbon-dependent control of electron transfer and central carbon pathway genes for methane biosynthesis in the Archaean, Methanosarcina acetivorans strain C2A. BMC Microbiol 10:62

Russell NJ, Gould G (2003) Acidulants and low pH. In: Russell NJ, Gould G (eds) Food preservatives. Springer, New York, pp 25–47

Sá-Pessoa J, Amillis S, Casal M, Diallinas G (2015) Expression and specificity profile of the major acetate transporter AcpA in Aspergillus nidulans. Fungal Genet Biol 76:93–103

Sa-Pessoa J, Paiva S, Ribas D, Silva IJ, Viegas SC, Arraiano CM, Casal M (2013) SATP (YaaH), a succinate-acetate transporter protein in *Escherichia coli*. Biochem J 454:585–595

Sauer M, Porro D, Mattanovich D, Branduardi P (2008) Microbial production of organic acids: expanding the markets. Trends Biotechnol 26:100–108

Sauer M, Porro D, Mattanovich D, Branduardi P (2010) 16 years research on lactic acid production with yeast – ready for the market? Biotechnol Genet Eng Rev 27:229–256

Schmitt AP, McEntee K (1996) Msn2p, a zinc finger DNA-binding protein, is the transcriptional activator of the multistress response in *Saccharomyces cerevisiae*. Proc Natl Acad Sci U S A 93:5777–5782

Schuller D, Valero E, Dequin S, Casal M (2004) Survey of molecular methods for the typing of wine yeast strains. FEMS Microbiol Lett 231:19–26

Seret M-L, Diffels J, Goffeau A, Baret P (2009) Combined phylogeny and neighborhood analysis of the evolution of the ABC transporters conferring multiple drug resistance in hemiascomycete yeasts. BMC Genomics 10:459

Soares-Silva I, Paiva S, Diallinas G, Casal M (2007) The conserved sequence NXX[S/T]HX[S/T]QDXXXT of the lactate/pyruvate:H(+) symporter subfamily defines the function of the substrate translocation pathway. Mol Membr Biol 24:464–474

Soares-Silva I, Sa-Pessoa J, Myrianthopoulos V, Mikros E, Casal M, Diallinas G (2011) A substrate translocation trajectory in a cytoplasm-facing topological model of the monocarboxylate/H(+) symporter Jen1p. Mol Microbiol 81:805–817

Soares-Silva I, Schuller D, Andrade RP, Baltazar F, Cassio F, Casal M (2003) Functional expression of the lactate permease Jen1p of *Saccharomyces cerevisiae* in *Pichia pastoris*. Biochem J 376:781–787

Soares-Silva I, Ribas D, Foskolou Z, Barata B, Bessa D, Queiros O, Paiva S, Casal (2015) The Debaryomyces hansenii carboxylate transporters Jen1 homologues are functional in Saccharomyces cerevisiae. FEMS Yeast Res. http://dx.doi.org/10.1093/femsyr/fov094

Solcan N, Kwok J, Fowler PW, Cameron AD, Drew D, Iwata S, Newstead S (2012) Alternating access mechanism in the POT family of oligopeptide transporters. EMBO J 31:3411–3421

Soontorngun N, Larochelle M, Drouin S, Robert F, Turcotte B (2007) Regulation of gluconeogenesis in *Saccharomyces cerevisiae* is mediated by activator and repressor functions of Rds2. Mol Cell Biol 27:7895–7905

Tachibana C, Yoo JY, Tagne JB, Kacherovsky N, Lee TI, Young ET (2005) Combined global localization analysis and transcriptome data identify genes that are directly coregulated by Adr1 and Cat8. Mol Cell Biol 25:2138–2146

Tamas MJ, Luyten K, Sutherland FC, Hernandez A, Albertyn J, Valadi H, Li H, Prior BA, Kilian SG, Ramos J et al (1999) Fps1p controls the accumulation and release of the compatible solute glycerol in yeast osmoregulation. Mol Microbiol 31:1087–1104

Tzschoppe K, Augstein A, Bauer R, Kohlwein SD, Barth G (1999) Trans-dominant mutations in the *GPR1* gene cause high sensitivity to acetic acid and ethanol in the yeast Yarrowia lipolytica. Yeast 15:1645–1656

Ullah A, Chandrasekaran G, Brul S, Smits GJ (2013) Yeast adaptation to weak acids prevents futile energy expenditure. Front Microbiol 4:142

van Maris AJ, Konings WN, van Dijken JP, Pronk JT (2004) Microbial export of lactic and 3-hydroxypropanoic acid: implications for industrial fermentation processes. Metab Eng 6:245–255

Verma RK, Prabh ND, Sankararamakrishnan R (2014) New subfamilies of major intrinsic proteins in fungi suggest novel transport properties in fungal channels: implications for the host-fungal interactions. BMC Evol Biol 14:173

Wysocki R, Chéry CC, Wawrzycka D, Van Hulle M, Cornelis R, Thevelein JM, Tamás MJ (2001) The glycerol channel Fps1p mediates the uptake of arsenite and antimonite in *Saccharomyces cerevisiae*. Mol Microbiol 40:1391–1401

Yin Y, He X, Szewczyk P, Nguyen T, Chang G (2006) Structure of the multidrug transporter EmrD from *Escherichia coli*. Science 312:741–744

# Chapter 10
# Inorganic Phosphate and Sulfate Transport in *S. cerevisiae*

**D.R. Samyn and B.L. Persson**

**Abstract** Inorganic ions such as phosphate and sulfate are essential macronutrients required for a broad spectrum of cellular functions and their regulation. In a constantly fluctuating environment microorganisms have for their survival developed specific nutrient sensing and transport systems ensuring that the cellular nutrient needs are met. This chapter focuses on the *S. cerevisiae* plasma membrane localized transporters, of which some are strongly induced under conditions of nutrient scarcity and facilitate the active uptake of inorganic phosphate and sulfate. Recent advances in studying the properties of the high-affinity phosphate and sulfate transporters by means of site-directed mutagenesis have provided further insight into the molecular mechanisms contributing to substrate selectivity and transporter functionality of this important class of membrane transporters.

**Keywords** Phosphate • Sulfate • Transport • Regulation • *S. cerevisiae*

## 10.1 Inorganic Phosphate Transport in *S. cerevisiae*

### *10.1.1 Introduction*

In the periodic table of elements, phosphorus is found in Group 15 (pnictogens) amongst other well-known elements like nitrogen and arsenic. It was discovered by the German alchemist Hennig Brand while searching for the mythical substance that has the capabilities of converting base metals into silver or gold. In his search, which involved human urine as a source (phosphorus is now primarily obtained from phosphate rock, $Ca_3(PO_4)_2$), he came across a waxy white solid that had properties of glowing in the dark and bursting into flames when removed from water. In the seventeenth century the element was given its current name, coming from the Greek phos, "light", and phoros, "bringing".

D.R. Samyn (✉) • B.L. Persson
Department of Chemistry and Biomedical Sciences, Centre for Biomaterials Chemistry, Linnaeus University, 391 82 Kalmar, Sweden
e-mail: dieter.samyn@lnu.se



J. Ramos et al. (eds.), *Yeast Membrane Transport*, Advances in Experimental Medicine and Biology 892, DOI 10.1007/978-3-319-25304-6_10

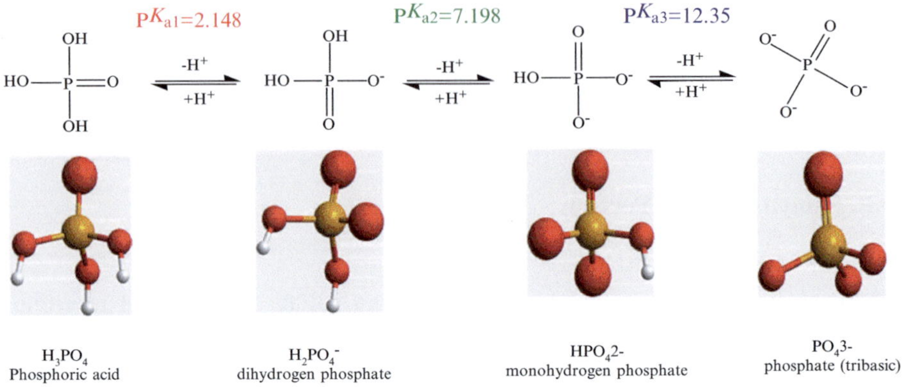


**Fig. 10.1** The protonation states of inorganic phosphates, with the corresponding pKa's

**Table 10.1** Similarities between phosphate and carbon utilization in *S. cerevisiae*

| |
|---|
| Phosphate and carbon can be stored intracellular in polymeric forms (i.e. polyphosphate and glycogen, respectively) |
| Multiple genes encode enzymes for phosphate and carbon utilization |
| Glucose represses genes for carbon utilization while phosphate similarly represses genes for phosphate utilization |
| Cells require large amounts; carbon for energy and biosynthesis, phosphate for nucleic acids and phospholipids (primarily) |

*Ortho*phosphate forms the basic anionic unit of all phosphates and is a tetrahedral structure of one phosphorus atom surrounded by four oxygen atoms ($PO_4^{3-}$). The anion is tribasic and the three valences can be satisfied by hydrogen, inorganic (metal) ions, organic ions or a combination of the three (Ellinger 1972). Inorganic phosphate (Pi) occurs when the three valences are satisfied with hydrogen (*ortho*phosphoric acid), inorganic ions or combinations of both. Because of this, inorganic phosphates have four distinct protonation states (see Fig. 10.1). Due to the different pKa values for each phosphate ion entity, the intra- and extracellular pH will dictate the significance presence of distinct ions. Under physiological growth conditions (i.e. acidic-neutral pH range), Pi ions will occur as $H_2PO_4^-$ or to a lesser extent $HPO_4^{2-}$. In biological systems, Pi is identified as the mono- or divalent anion of phosphoric acid [$HPO_4^{-2}$, $H_2PO_4^-$]. This makes Pi highly improbable to cross membranes, such as the plasma membrane, by passive transport. In order to circumvent this physical limitation, *S. cerevisiae* utilizes secondary active transport to drive the transport of Pi across membranes. Pi is required for cellular functions such as DNA and membrane lipid synthesis, intracellular signalling and the generation of high-energy phosphate esters like ATP. In order to emphasize the importance and the ubiquitous character of phosphate in yeast metabolism a correlation can be made between phosphate and carbon utilization as shown in Table 10.1.

Because of the central role of Pi, it is important that the cell is able to counter any environmental fluctuations which might lead to a disturbance in cellular phosphate homeostasis. This requires the cell to ensure a rapid cellular response to any fluctuation in external and internal Pi levels.

## 10.1.2 PHO Regulon

*S. cerevisiae* makes use of a dual-transporter system to ensure that the cell is more capable to handle phosphate starvation and to adapt to changes in the external environment. This dual-transporter system is under the control of a phosphate-responsive signaling system (known as the *PHO* pathway) (Toh-e et al. 1973, 1988; Auesukaree et al. 2004; Persson et al. 1999; Mouillon and Persson 2006). This permits S. *cerevisiae* to sense and respond to variations in the concentration of inorganic phosphate in the surrounding environment. This pathway regulates the expression of so-called "*PHO*" genes, which will enable the cell to scavenge and take up the Pi from the surroundings (Lenburg and O'Shea 1996). The *PHO* regulon consists of several proteins, which possess either a *PHO*-specific regulatory or responsive function. The regulatory system is composed of at least five proteins: the transcriptional activators Pho2 and Pho4, the Pho80-Pho85 cyclin-cyclin dependent protein kinase (CDK) complex (Huang et al. 2007), and the Pho81 CDK inhibitor (Johnston and Carlson 1992). The relationship between Pho4 and the Pho80-Pho85 complex is such that the latter is able to phosphorylate the former. Due to the fact that the *PHO4* gene is constitutively expressed independent of the concentration of extra- and intracellular Pi, the regulation of its transcriptional factor capacity is not only regulated by means of phosphorylation events. More recently, a novel YTH (YT521-B homology)-containing protein Pho92 has been identified to play a role in the regulation of the *PHO4* gene expression (Kang et. al. 2014). The Pho92 exerts its function in stabilizing the *PHO4* mRNA under high Pi conditions. This results in a regulation of the high-affinity phosphate metabolism by inhibiting *PHO4* expression at its post-translational level.

### 10.1.2.1 High Phosphate Conditions

Under high phosphate conditions (see Fig. 10.2) the Pho80-Pho85 kinase complex will be able to hyper-phosphorylate Pho4. This will lead to Pho4 firstly exhibiting a lower affinity for Pho2 and the nuclear importer protein Pse1, and secondly becoming a preferred substrate for the nuclear exporter protein Msn5. These actions result in Pho4 being localized extra-nuclear and thus not being able to activate gene expression.

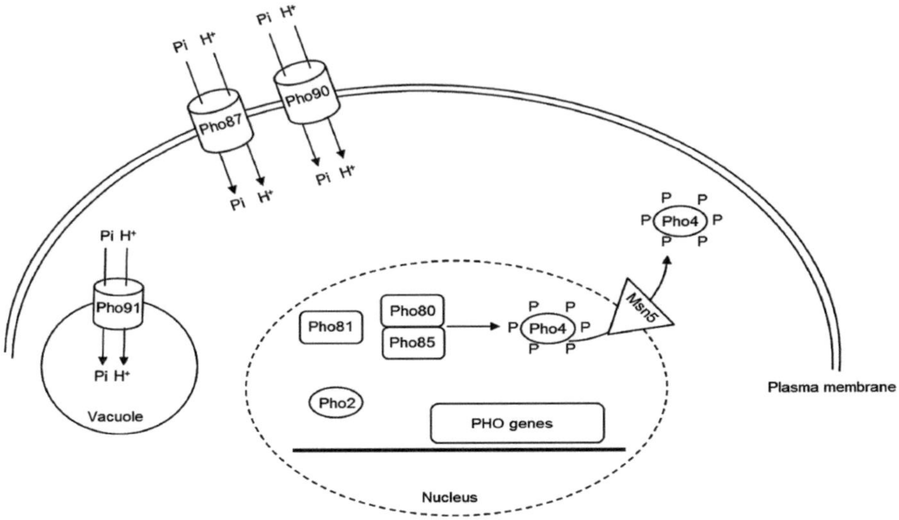


**Fig. 10.2** *PHO* regulon operational under high Pi conditions. Inactivation of the *PHO*-responsive genes is dependent on external phosphate concentrations and the localization of the transcription factor Pho4. Under high Pi conditions the Pho4 will be hyperphosphorylated by the Pho80-Pho85 CDK complex and reside in the cytoplasm and not be able to induce the expression of *PHO*-responsive genes

#### 10.1.2.2 Low Phosphate Conditions

Under low phosphate conditions, the intracellular levels of the metabolite inositol heptakisphosphate (IP7) are elevated, resulting in the allosteric modulation of Pho81 (Lee et al. 2007). Pho81 will now inhibit the Pho80-Pho85 kinase activity, leading to the hypo-phosphorylated status of Pho4 (Ogawa et al. 1995; Lee et al. 2009; 2011). This will lead to in an elevated affinity of Pho4 for the Pse1 nuclear import receptor, resulting in a translocation of Pho4 out of the cytoplasm into the nucleus (Kaffman et al. 1998). Once Pho4 resides in the nucleus, it will interact with Pho2, leading to an activation of phosphate-responsive genes and their corresponding proteins (Komeili and O'Shea 1999) (see Fig. 10.3) such as:

(i) the high-affinity Pi transporters Pho84 (Bun-ya et al. 1991) and Pho89 (Martinez and Persson 1998);
(ii) the Pho5 repressible acidic phosphatase, together with its homologues Pho11 and Pho12. These phosphatases will be secreted into the periplasm, where they will liberate *ortho*phosphate from phospho-ester substrates (Persson et al. 2003);
(iii) the Pho8, a vacuolar alkaline phosphatase (Kaneko et al. 1987);
(iv) the Pho86, which is required for endoplasmatic reticulum exit of the Pho84 (Bun-ya et al. 1996; Lau et al. 2000);
(v) the Pho81, which is the negative regulator of the Pho80-Pho85 kinase complex;

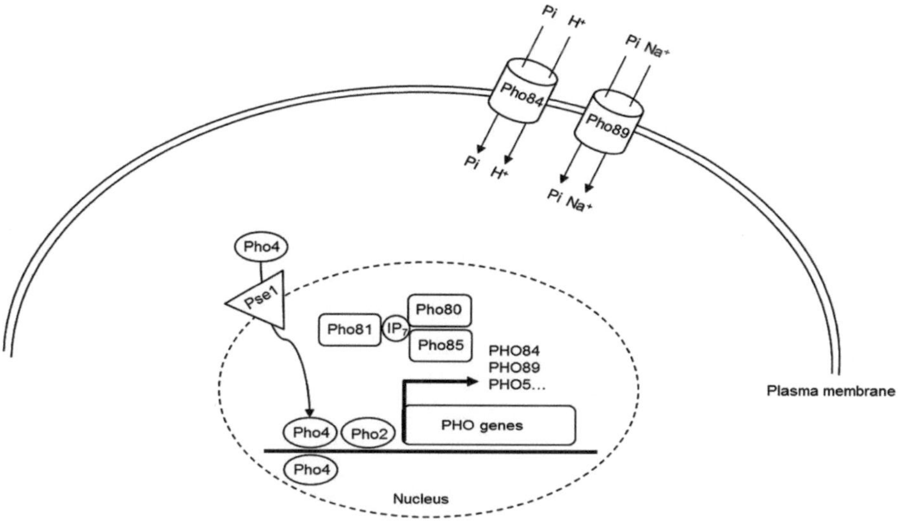


**Fig. 10.3** *PHO* regulon operational under low phosphate conditions. Activation of the *PHO*-responsive genes is dependent on external phosphate concentrations and the localization of the transcription factor Pho4. Under low Pi conditions the Pho81 inhibits the Pho80-Pho85 CDK complex and Pho4 will be hyporphosphorylated resulting in the Pho4 being located in the nucleus where it will induced the expression of *PHO*-responsive genes

(vi) the Spl2 (Wykoff et al. 2007), which acts as a negative regulator of Pho87 (Ghillebert et al. 2011) and possibly also of Pho90 (Hürlimann et al. 2009).

The phosphorylation events will determine the activity and localization of the transcription factor Pho4. In order for the cell to adapt to phosphate fluctuations, it must be capable of not only of switching the expression of the genes corresponding to the phosphate transporters on and off, but also possess some more fine-tuned intermediate responses (Springer et al. 2003). The magnitude and duration of gene activation is dependent on the degree of phosphorylation. Pho4 harbors five phosphorylation sites (O'Neill et al. 1996). Depending on which site is phosphorylated, Pho4 is either (i) exported out of the nucleus (Kaffman et al. 1998), (ii) inhibited in its interaction with Pse1 and thus blocking nuclear import, or (iii) inhibited in the interaction with Pho2 (Komeili and O'Shea 1999).

Furthermore, it has recently been shown that *S. cerevisiae* displays temporal gene induction upon phosphate limitation/starvation conditions (Vardi et al. 2014). This dynamic response enables the cell to correct potential limitations in both a rapid, yet gradual way so that there is more control concerning the cellular fitness. As a result of external Pi uptake by yeast a decrease of the external phosphate levels will occur. Once the external Pi level decreases below the Kd ($\sim$220 $\mu$M) of the low-affinity transport system the *PHO* regulon is induced. This Pi condition will trigger the first wave of a subset of genes that are induced instantaneously and encompass the high-affinity transporter *PHO84*, *PHO86* and the phosphatases *PHO12* and *PHO11* resulting in an increase in the internal Pi. Furthermore, the *PHM1-4* (which will direct intracellular Pi to the vacuole to be stored as polyphosphates), *SPL2* (which is responsible for the down-regulation of the low-affinity transporters) and *PHO81* genes will be induced resulting in a decrease of intracellular Pi. This first wave of gene induction will establish a positive feedback loop maintaining Pho4-dependent gene expression. A second, delayed, wave of gene inductions seems to occur when the in-flux of Pi by means of the high-affinity transporters decreases and the external Pi level drops below the low-affinity transporter Kd (~220 μM). Genes that are induced during this wave are the secreted phosphatase *PHO5*, *PHM8*, *PHO8* and *DDP1* which will act as phosphatases on intracellular substrates, *PHO89* high-affinity Pi transporter and *KCS1* which will reduce IP7 concentrations (Vardi et al. 2014).

## *10.1.3 Inorganic Phosphate Transporters*

### 10.1.3.1 Low-affinity Inorganic Phosphate Transporters

General Properties

The transporters Pho87 (2.A.47.2.1), Pho90 (2.A.47.2.3) and Pho91 (2.A.47.2.2) all belong to the 2.A.47 divalent Anion:$Na^+$ Symporter (DASS) family (also referred to as the SLC13 family). They all have $K_m$ values in the range of 200 μM–1 mM (Wykoff and O'Shea 2001), with an acidic pH optimum and thus using the proton gradient as a driving force for translocation of Pi ions in a symport manner. The low-affinity transporters are expressed during conditions where there is an ample amount of external Pi available and are considered to be "household" transporter proteins.

Pho87 and Pho90

The primary sequences of Pho87 (Bun-ya et al. 1996) and Pho90 are more than 60 % identical. This is partly due to the 375 amino acid residues in the amino terminal SPX domain, harbored by both transporters, while being absent in the high-affinity Pi transporters. The SPX domain is thought to have regulatory functions and is present in several other eukaryote proteins (Barabote et al. 2006). In Pho87 and Pho90 this domain is involved in the regulation of Pi uptake but it is not important for the uptake *per se* since a truncation does not affect protein level or membrane localization at high Pi conditions. Moreover, the SPX-truncated Pho90 displays an increased Pi transport compared to the corresponding wild-type protein (Hürlimann et al. 2009). Starvation for glucose and treatment with rapamycin, a condition known to mimic amino acid and nitrogen starvation, triggers endocytosis and vacuolar degradation of both transporters, a mechanism that was shown to require the SPX domain in both Pho87 and Pho90 (Ghillebert et al. 2011). Under low Pi conditions both transporters are down-regulated. For the Pho87 down-regulation it has been shown that Spl2 is required, but whether or not Spl2 is also involved in the down-regulation of Pho90 is still under debate. It has previously been shown that Pho87, in the absence of Pho84, is able to act as a nutrient sensor, signalling the presence of Pi for activation of the Protein Kinase A (PKA) pathway, a function normally provided by Pho84 (Giots et al. 2003).

Pho91

Pho91 also harbors an N-terminal SPX domain, but the importance of the domain in this transporter has not been addressed. Pho91 is the least investigated phosphate transporter, and its precise localization is still uncertain. Previously it was shown that Pho91 allows for vigorous growth when over-expressed under the ADH1 promoter in a strain deleted of all the other transporters (Wykoff and O'Shea 2001) and expression under its native promoter allowed for minimal growth, implying a plasma membrane localization. In contrast with the above-mentioned localization, a more recent study (Hurlimann et al. 2007) showed that the N-terminal GFP-tagged Pho91 was detected in the vacuolar membrane irrespective of the weak CYC1 promoter or the strong ADH1 promoter was used. Most recently it was shown that when over-expressed from the ADH1 promoter Pho91 not only transports Pi, but also supports strong glycerol-3-phosphate uptake, which again implies plasma membrane localization (Popova et al. 2010).

#### 10.1.3.2 High-affinity Inorganic Phosphate Transporters

During conditions where the external environment is depleted of Pi, cells are in need of transporters that are able to scavenge the little amount of Pi available. This is done by means of having transporters expressed at the plasma membrane level that exhibit a high-affinity towards the substrate. *S. cerevisiae* has two Pi transporters that fulfill this function.

Pho84, a Pi:$H^+$-Coupled Symporter

*General Properties*

The Pho84 protein is considered the main transporter of Pi into the cell during Pi limitations. The Pho84 transporter is at its maximum operational capacity at pH 4.5 (Berhe et al. 2001; Bun-ya et al. 1991) and displays a strong affinity for Pi with a $K_m$ in the range of 8–58 μM (Tamai et al. 1985; Wykoff and O'Shea 2001). The transport of one Pi ion is driven in a symport manner by 2–3 protons, using the energy from the proton motive force (Borst-Pauwels 1993). In addition to transporting Pi ions, Pho84 has been shown to function as a low-affinity metal translocator in the transport of manganese (Jensen et al. 2003) and selenite (Lazard et al. 2010). Several other phosphate esters have been shown to either bind to or get transported by Pho84 (Pratt et al. 2004; Popova et al. 2010). The Pho84 high-affinity Pi:$H^+$ (2.A.1.9.1) transporter is classified amongst the phosphate:$H^+$ symporter family, belonging to the major facilitator superfamily. As a member of this superfamily, Pho84 displays typical characteristics (Pao et al. 1998).

*Mechanisms in Pho84 Dynamics*

In order to control the number of transporters, the cell has adopted mechanisms that act at three different levels: (i) gene transcription (explained in Sect. 1.2.2), (ii) alteration of the intrinsic transport activity and (iii) intracellular membrane trafficking. Almost a hundred plasma membrane nutrient transporters have been identified in *S. cerevisiae* (Lauwers et al. 2010), and all these proteins share a common strategy when it comes to their regulation of intracellular membrane trafficking. The presence or absence of nutrient transporters in the plasma membrane appear closely related to nutrient availability. A key player in intracellular trafficking is the highly conserved protein, ubiquitin (Ub), which determines the removal route of plasma membrane proteins (Haguenauer-Tsapis and André 2004). Prior to being removed from the plasma membrane, Ub will be covalently attached to a target protein, and it is the only known signal to trigger the internalization of plasma membrane nutrient transporters. One of the first acknowledgments on the involvement of Ub in membrane trafficking was made for Ste6 (Kölling and Hollenberg 1994) where it was shown that mutations reducing the Ste6-ubiquitination impaired delivery of the protein from the plasma membrane to the vacuole. For the covalent "labeling" of these plasma membrane transporters the cell needs three different types of enzymes (together with several co-factors), which differ from each other in function, moment of action and amount.

It has been shown that for the Pho84 transporter an environmental change from a Pi shortage to excess will trigger removal of the transporter from the plasma membrane and result in its degradation in the vacuole (Lundh et al. 2009). Moreover, it has been shown that in some cases it is required that the membrane protein is phosphorylated prior to Rsp5-mediated ubiquitination (Marchal et al. 1998; Kelm et al. 2004). This seems also to be the case for the Pho84 transporter (Lundh et al. 2009) (see Fig. 10.4).

*Molecular Mechanism of Pi Transport*

An *in silico* model of the Pho84 protein (Lagerstedt et al. 2004), modeled on the glycerol-3-phosphate transporter of *E. coli*, has recently been used as a basis for selecting residues that are involved in several steps in the Pi transport (Samyn et al. 2012).

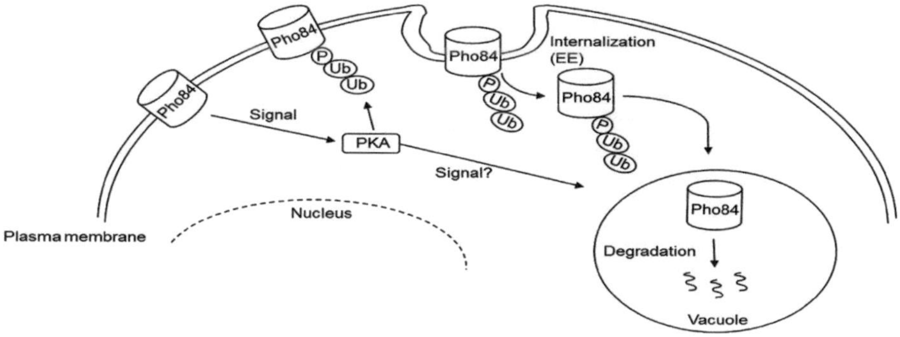


**Fig. 10.4** Schematic overview of the down-regulation of Pho84 (EE: Early endosome). When external Pi conditions are elevated the high-affinity Pi transporter Pho84 is removed from the plasma membrane. This removal is a consequence of a phosphorylation and ubiquitination event which leads to the degradation of the Pho84 in the vacuole

A simplified structural overview of the inwards-facing confirmation ($C_i$) of the transporter with the key sections that are involved in phosphate transport is indicated in Fig. 10.5. Three distinct sections can be found in the protein:

Section I is located towards the extracellular side of the transporter. This section harbors the residues Phe81 and Phe361 that seem to contribute to what can be interpreted as a "lid", sealing of the transportation pathway (Fig. 10.5, detailed structure section I). In a recently solved crystal structure of a fungal (*Piriformospora indica*) high-affinity Pi transporter, PiPT, the Phe50 and Phe327 residues are predicted to have the same function (Pedersen et al. 2013). In contrary to a previous performed SCAM (Substituted Cysteine Accessibility Method) analysis (Popova et al. 2010), kinetics studies on site-directed mutations have indicated that Arg168 is not immediately involved in the binding and transport of Pi (Samyn et al. 2012). Although the effects observed when replacing the Arg168 by an Ala or Glu are not drastic, there might be an influence on the local positive electrostatic charge.

Section II harbors residues that are proposed to play a role in the "shuttling" of protons. The binding and release of protons is crucial for the symporter mechanism. Three mechanistic features can be observed in a proton-driven uptake system: (i) coupling between substrate binding/dissociation and conformational changes in the transporter; (ii) coupling between the substrate binding/dissociation and protonation/deprotonation of residues; and (iii) coupling between protonation/deprotonation and conformational changes in the transporter (Abramson et al. 2004; Kaback et al. 2001). Most frequently these are Glu/Asp/His residues, and to a lesser extent Lys/Arg/Tyr. In section II Asp79 and Asp76 (Fig. 10.5, detailed structure section II) might form a relay system that shuttles protons through the transportation channel and eventually away from the binding site. This mechanism has been proposed in the PiPT and has led to the postulation of a "cytoplasmatic proton-tunnel" which is responsible for the needed deprotonation

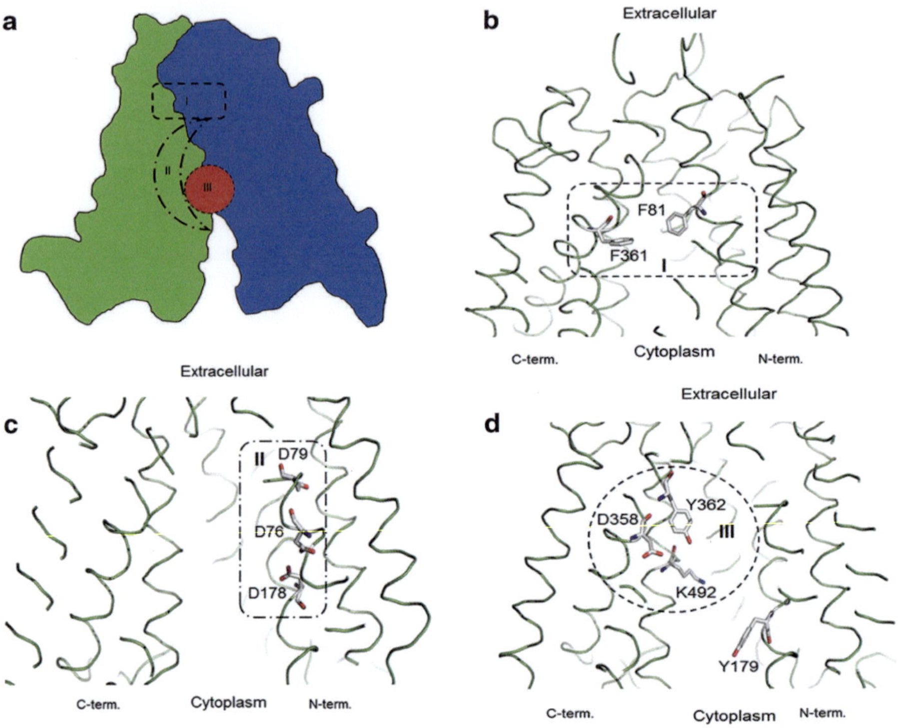


**Fig. 10.5** Overview of sections and residues playing a role in the Pho84 transport mechanism. Panel **A**: Cartoon overview of the Pho84. Green represents the N-terminal portion and blue represents the C-terminal half of the protein. Section that harbor residues that play a role in different steps of the transport mechanism are indicated with Roman numerals (I-III). Panel **B-D** represents a more detailed ribbon view (made with PyMol) of the transporter sections and the residues within each section are indicated as stick representations

step crucial to the release of Pi from the binding site into the cytosol (Pedersen et al. 2013). Although Asp178 has been shown to play a part in the binding site the kinetics rather points to a role in the release step of Pi and more so in the deprotonation event (Samyn et al. 2012).

Section III of the protein represents the proposed binding site where two positively charged residues play a role in binding Pi: Lys492 and Asp358 (Fig. 10.5, detailed structure section III). Substrate-binding to the Pho84 occurs optimally under acidic conditions (pH 4.5) during which Pi mainly exist as the monobasic ion, $H_2PO_4^-$. Thus, it is to be expected that the substrate-binding site have a positive surface electrostatic potential. The Lys492 together with Asp358 are localized in the same plane that has been indicated as the substrate-binding site in the glycerol-3-phosphate transporter (GlpT) (Huang et al. 2003) and the lactose transporter (LacY) (Abramson et al. 2003). Because of this favorable location it is considered to be the binding site of the Pho84 transporter. Kinetic studies on site-directed mutants of Lys492 and Asp358 have shown the importance of these residues (Samyn et al. 2012). Lys492 is important for the affinity ($K_m$) towards the Pi, whereas Asp358 is crucial for the activity ($V_{max}$) of the transporter.

### *Pho84 as a Transceptor*

When cells are deprived of Pi the cell growth will be arrested, lead to an increased level of stored carbohydrates (e.g. trehalose) and develop high stress tolerance. These features are due to a downregulation of the PKA pathway (Thevelein and de Winde 1999). Supplementing the cells with Pi will prompt a rapid switch to a high PKA phenotype resulting in an activation of the PKA phosphorylation target trehalase (Giots et al. 2003; Mouillon and Persson 2005; Popova et al. 2010). Because of the double function as transporter and receptor Pho84 is recognized as a transceptor (Holsbeeks et al. 2004; Thevelein and Voordeckers 2009). Furthermore, mutagenesis of Asp358 in Pho84 abolished transport but left signaling largely unaffected (Samyn et al. 2012). This shows that mutagenesis of Asp358 does not significantly compromise the localization and the sensing/signaling function of Pho84 confirming that substrate transport is not required for signaling by the Pho84 transceptor.

## Pho89, a Pi:$Na^+$-Coupled Symporter

Despite belonging to the high-affinity phosphate transporter group the Pho89 exhibits properties different than those of the Pho84 transporter. The Pho89 transporter (2.A.20.2.2) is classified amongst the Pi transporter family (PiT; 2.A.20) (Saier et al. 2000) and shows a high degree of homology with the type III $Na^+$:Pi (SLC20 family) co-transporter in mammals (Collins et al. 2004). The contribution to Pi uptake in yeast by Pho89 displays a 100-fold lower transport activity than Pho84 (Pattison-Granberg and Persson 2000). The transport of Pi occurs in a symport manner with cations as co-solutes, preferably $Na^+$ but also with $Li^+$ and $K^+$, and the optimum pH for transport is 9.5 (Zvygailskaya et al. 2008; Martinez and Persson 1998). The high pH optimum coincides with that Pho89 also have been shown to have a Pho2/Pho4-dependent up-regulation in response to alkaline conditions. The $Na^+$-ATPase ENA1 is also upregulated in response to alkaline conditions and extruding $Na^+$ out of the cell contributing to a $Na^+$ motive force across the plasma membrane favouring the $Na^+$ dependent Pi transport by Pho89 (Serrano et al. 2002). A signature sequence for the PiT family, with the consensus amino acid sequence GANDVANA, has been proposed (Saier et al. 2000). This signature sequence has been identified in more than a hundred proteins with homology to the PiT family. The GANDVANA sequence is present in two copies in human PiT2. Mutational analysis of the conserved Asp residue to an Asn in any of the two sequences confers knockout of Pi transport (Bottger and Pedersen 2005). Pho89 also harbors two copies of the signature sequence located in the N- and C-terminal domain at similar positions. Whether these sequences have a similar importance as shown in the human PiT2 is currently unknown (Salaün et al. 2001). Multiple sequence alignments of PiT family members representing all kingdoms of life show that two Glu residues are highly conserved, corresponding to Glu55 and Glu491 in Pho89 (Bottger and Pedersen 2002). In PiT1 and Pit2 these residues were shown to be crucial for Pi transport activity. A replacement of any of these Glu residues in Pho89 with Lys abolishes transport activity. So does a replacement of Glu55 by a Gln, whereas a replacement of Glu491 by a Gln retained some of its activity (Andersson et al. 2012). This similarity suggests that these Glu residues have a functional role in the transport mechanism. In contrast to Pho84 and Pho87 (Giots et al. 2003), the Pho89 (Zvyagilskaya et al. 2008) is not able to sense and signal the availability of Pi for rapid activation of the PKA pathway in Pi starved cells.

Recently, the recombinant full-length Pho89 transporter has been heterologously expressed in *P. pastoris* (Sengottaiyan et al. 2013a) and functionally reconstituted into proteoliposomes. Uptake experiments performed on the proteoliposomal system clearly show a $Na^+$ gradient dependency. Furthermore, circular dichroism experiments confirm that the Pho89, in agreement with the predicted structure is composed of 12 transmembrane segments (Sengottaiyan et al. 2013b).

## 10.2 Sulfate Transport in *S. cerevisiae*

### *10.2.1 Introduction*

Sulfur is considered to be a major macronutrient (Maw 1963) and is taken up by means of specific membrane transporters. Once inorganic sulfur ($SO_4^{2-}$) is taken up into the cell it will undergo a series of enzymatic reactions (Sulfate Assimilation Pathways; SAP) to form sulfide and homocysteine, and be incorporated into S-containing compounds (glutathione, S-adensoylmethionine, cysteine, etc.). Sulfate transporters can be found in all organisms ranging from prokaryotes to higher eukaryotes (Mount and Romero 2004).

### *10.2.2 The High-Affinity Sulfate Transporters Sul1 and Sul2 in S. cerevisiae*

#### 10.2.2.1 General Properties

Cellular uptake of $SO_4^{2-}$ in *S. cerevisiae* is facilitated by two SLC26-type gene products, Sul1 and Sul2 (McCready and Din 1974; Roomans et al. 1979). Both transporters mediate $SO_4^{2-}$ transport in a symport manner by utilizing the $H^+$-gradient (Breton and Surdin-Kerjan 1977; Cherest et al. 1997). Sul1 and Sul2 are considered as high-affinity sulfate transporter, with Km values towards sulfate of 4.5 μM and 10 μM respectively (Cherest et al. 1997). Furthermore, a when both high-affinity transporters are absent cells are still able to grow under high sulfate conditions. This implies that in addition to the high-affinity sulfate transport system *S. cerevisiae* also has unidentified low-affinity sulfate transport system.

#### 10.2.2.2 Mechanisms in Sul2 Dynamics

During conditions where extracellular sulfate is limited, Sul2 accounts for the major influx. The expression of both sulfate transporters is strongly induced during sulfate limiting conditions (Boer et al. 2003), whereas replenishing the cell with ample amounts of sulfate results in endocytosis (Jennings and Cui 2012) of the transporters and a swift reduction in sulfate uptake is observed. The inactivation and degradation of Sul2 has been shown to be dependent on extracellular rather than cytosolic $SO_4^{2-}$ concentrations (Jennings et al. 2012). Moreover, rather than the substrate dictating the degradation of Sul2 it is the transport process itself that triggers this event, and thus can be seen as an autoregulatory inactivation. This rapid downregulation would then only require a conformational state during the catalytic cycle where the transporter undergoes inactivation and degradation, which is similar to the *S. cerevisiae* general amino acid transport 1 (GAP1) and high-affinity Pi transporter Pho84 (Lundh et al. 2009; Cain and Kaiser 2011).

#### 10.2.2.3 Molecular Mechanism of Sulfate Transport

To date there is no crystal structure or in silico 3D model available of neither Sul1 nor Sul2. This poses a limitation to the understanding of the molecular mechanisms underlying the transport mechanism of these transporters. A recent study by Kankipati et al. 2015 investigated several amino acid residues in their importance to the molecular transport mechanism. This was done by means of a rational approach based on sequence alignments and restricted amino acid side chain selection criteria, similar to the study of the high-affinity Pi transporter Pho84 (Samyn et al. 2012). In this study Asp124, D483, Glu427 and Glu538 in Sul1, and Asp140, Asp305 and Glu443 in Sul2 were selected as targets for site directed mutagenesis. These residues were selected based on their side chain properties which are believed to be putative $H^+$ -binding sites. In Sul1 residues Asp124 and Glu427 are crucial for functional transport of sulfate. In Sul2 the residues Asp140 and Glu443 are important for functional transport of sulfate. The involvement of the Asp124 in Sul1 and Asp 140 in Sul2 residues in the transport is that these residues undergo a protonation resulting in a decreased repulsion of the negatively charged sulfate molecule and thus facilitating the binding to the transporter. Furthermore, the binding of the sulfate ion results in a conformational changes followed by transferring the $H^+$ to the next binding site harboring the Glu427 in Sul1 and the Glu443 in Sul2. The latter would explain the importance of these Glu residues which also could undergo a protonation/deprotonation cycle in order for the sulfate ion to be transported (Kankipati et al. 2015).

### 10.2.2.4 Sul1 and Sul2 as a Transceptor

Depriving cells of $SO_4^{2-}$ will result in the downregulation of the PKA pathway (Thevelein and de Winde 1999). Upon supplementing the cells with $SO_4^{2-}$ a rapid switch to a high PKA phenotype resulting in an activation of the PKA phosphorylation target trehalase will occur (Kankipati et al. 2015). Because of the double function as transporter and receptor Sul1 and Sul2 are recently added to the ever growing list of transceptor. Furthermore, mutagenesis of Glu427 in Sul1 and Glu443 in Sul2 abolished transport of $SO_4^{2-}$ but left signaling unaffected (Kankipati et al. 2015) confirming that substrate transport is not required for signaling by the Sul1 and Sul2 transceptors.

## References


Abramson J, Smirnova I, Kasho V, Verner G, Kaback HR, Iwata S (2003) Structure and mechanism of the lactose permease of *Escherichia coli*. Science 301:610–615

Andersson MR, Samyn DR, Persson BL (2012) Mutational analysis of conserved glutamic acids of Pho89, a *Saccharomyces cerevisiae* high-affinity inorganic phosphate:$Na^+$ symporter. Biologia 67(6):1056–1061

Auesukaree C, Homma T, Tochio H, Shirakawa M, Kaneko Y, Harashima S (2004) Intracellular phosphate serves as a signal for the regulation of the PHO pathway in *Saccharomyces cerevisiae*. J Biol Chem 279:17289–17294

Barabote RD, Tamang DG, Abeywardena SN, Fallah NS, Fu JY, Lio JK, Mirhosseini P, Pezeshk R, Podell S, Salampessy ML, Thever MD, Saier MH Jr (2006) Extra domains in secondary transport carriers and channel proteins. Biochim Biophys Acta 1758:1557–1579

Berhe A, Zvyagilskaya R, Lagersted JO, Pratt JR, Persson BL (2001) Properties of the cysteine-less Pho84 phosphate transporter of *Saccharomyces cerevisiae*. Biochem Biophys Res Comm 287:837–842

Boer VM, de Winde JH, Pronk JT, Piper MD (2003) The genome-wide transcriptional responses of *Saccharomyces cerevisiae* grown on glucose in aerobic chemostat cultures limited for carbon, nitrogen, phosphorus, or sulfur. J Biol Chem 278:3265–3274

Borst-Pauwels GWFH (1993) Mutual interaction of ion uptake and membrane-potential. Biochim Biophys Acta 1145:15–24

Bottger P, Pedersen L (2002) Two highly conserved glutamate residues critical for type III sodium-dependent phosphate transport revealed by uncoupling transport function from retroviral receptor function. J Biol Chem 277:42741–42747

Bottger P, Pedersen L (2005) Evolutionary and experimental analyses of inorganic phosphate transporter PiT family reveals two related sigNature sequences harboring highly conserved aspartic acids critical for sodium-dependent phosphate transport function of human PiT2. Febs J 272:3060–3074

Breton A, Surdin-Kerjan Y (1977) Sulfate uptake in *Saccharomyces cerevisiae*: biochemical and genetic study. J Bacteriol 132:224–232

Bun-ya M, Nishimura M, Harashima S, Oshima Y (1991) The Pho84 gene of *Saccharomyces cerevisiae* encodes an inorganic-phosphate transporter. Mol Cell Biol 11:3229–3238

Bun-ya M, Shikata K, Nakade S, Yompakdee C, Harashima S, Oshima Y (1996) Two new genes, PHO86 and PHO87, involved in inorganic phosphate uptake in *Saccharomyces cerevisiae*. Curr Genetics 29:344–351

Cain NE, Kaiser CA (2011) Transport activity-dependent intracellular sorting of the yeast general amino acid permease. Mol Biol Cell 22:1919–1929

Cherest H, Davidian JC, Thomas D, Benes V, Ansorge W, Surdin-Kerjan Y (1997) Molecular characterization of two high affinity sulfate transporters in *Saccharomyces cerevisiae*. Genetic 145:627–635

Collins JF, Bai LQ, Ghishan FK (2004) The SLC20 family of proteins: dual functions as sodium-phosphate cotransporters and viral receptors. Pflugers Archiv-European J Physiol 447:647–652

Ellinger RH (1972) Phosphates in food processing. In: Furia TE (ed) CRC Handbook of food additives. CRC Press, Cleveland

Ghillebert RSE, De Snijder P, Smets B, Winderickx J (2011) Differential roles for the low-affinity phosphate transporters Pho87 and Pho90 in *Saccharomyces cerevisiae*. Biochem J 434:243–251

Giots F, Donaton MCV, Thevelein JM (2003) Inorganic phosphate is sensed by specific phosphate carriers and acts in concert with glucose as a nutrient signal for activation of the protein kinase A pathway in the yeast *Saccharomyces cerevisiae*. Mol Microbiol 47:1163–1181

Haguenauer-Tsapis R, Andre B (2004) Membrane trafficking of yeast transporters: mechanisms and physiological control of downregulation. Topics in Current Genetics (2004) ed Berlin-Heidelberg: Springer

Holsbeeks I, Lagatie O, Van Nuland A, Van de Velde S, Thevelein JM (2004) The eukaryotic plasma membrane as a nutrient-sensing device. Trends Biochem Sci 29:556–564

Huang YF, Lemieux MJ, Song JM, Aue M, Wang DN (2003) Structure and mechanism of the glycerol-3-phosphate transporter from *Escherichia coli*. Science 301:616–620

Huang K, Ferrin-O'Connell I, Zhang W, Leonard GA, O'Shea EK, Quiocho FA (2007) Structure of the Pho85-Pho80 CDK-cyclin complex of the phosphate-responsive signal transduction pathway. Mol Cell 28:614–623

Hürlimann BC, Stadler-Waibel M, Werner TP, Freimoser FM (2007) Pho91 is a vacuolar phosphate phosphate and polyphosphate *Saccharomyces cerevisiae*. Mol Biol Cell 18:4438–4445

Hürlimann HC, Pinson B, Stadler-Waibel M, Zeeman SC, Freimoser FM (2009) The SPX domain of the yeast low-affinity phosphate transporter Pho90 regulates transport activity. Embo Rep 10:1003–1008

Jennings ML, Cui J (2012) Inactivation of *Saccharomyces cerevisiae* sulfate transporter Sul2p: use it and lose it. Biophys J 102:768–776

Jensen LT, Ajua-Alemanji M, Culotta VC (2003) The *Saccharomyces cerevisiae* high affinity phosphate transporter encoded by *PHO84* also functions in manganese homeostasis. J Biol Chem 278:42036–42040

Johnston M, Carlson M (1992) Regulation of carbon and phosphate utilization. In: Jones EW, Pringle JR, Broach JR (eds) The molecular and cellular biology of the yeast *Saccharomyces*: gene expression. Cold Spring Harbor Laboratory Press, Cold Spring Harbor, pp 193–281

Kaback HR, Sahin-Tóth M, Weinglass AB (2001) The kamikaze approach to membrane transport. Nat Rev Mol Cell Biol 2:610–620

Kaffman A, Rank NM, O'Neill EM, Huang LS, O'Shea EK (1998) The receptor Msn5 exports the phosphorylated transcription factor Pho4 out of the nucleus. Nature 396:482–486

Kaneko Y, Hayashi N, Toh-e A, Banno I, Oshima Y (1987) Structural characteristics of the Pho8 gene encoding repressible alkaline-phosphatase in *Saccharomyces-cerevisiae*. Gene 58:137–148

Kang HJ, Jeong SJ, Kim KN, Baek IJ, Chang M, Kang CM, Park YS, Yun CW (2014) A novel protein, Pho92, has a conserved YTH domain and regulates phosphate metabolism by decreasing the mRNA stability of PHO4 in *Saccharomyces cerevisiae*. Biochem J 457:391–400

Kankipati HN, Rubio-Texeira M, Castermans D, Diallinas G, Thevelein JM (2015) Sul1and Sul2 sulfate transceptors signal to protein kinase A upon exit of sulfur starvation. J Biol Chem. pii: jbc.M114.629022. [Epub ahead of print]

Kelm KB, Huyer G, Huang JC, Michaelis S (2004) The internalization of yeast Ste6p follows an ordered series of events involving phosphorylation, ubiquitination, recognition and endocytosis. Traffic 5:165–180

Kölling R, Hollenberg CP (1994) The ABC-transporter Ste6 accumulates in the plasma membrane in a ubiquitinated form in endocytosis mutants. EMBO J 13:3261–3271

Komeili A, O'Shea EK (1999) Roles of phosphorylation sites in regulating activity of the transcription factor Pho4. Science 284:977–980

Lagerstedt JO, Voss JC, Wieslander A, Persson BL (2004) Structural modeling of dual-affinity purified Pho84 phosphate transporter. Febs Lett 578:262–268

Lau WTW, Howson RW, Malkus P, Schekman R, O'Shea EK (2000) Pho86p, an endoplasmic reticulum (ER) resident protein in *Saccharomyces cerevisiae*, is required for ER exit of the high-affinity phosphate transporter Pho84p. PNAS 97:1107–1112

Lauwers E, Erpapazoglou Z, Haguenauer-Tsapis R, Andre B (2010) The ubiquitin code of yeast permease trafficking. Trends in Cell Biol 20:196–204

Lazard M, Blanquet S, Fisicaro P, Labarraque G, Plateau P (2010) Uptake of selenite by *Saccharomyces cerevisiae* involves the high and low affinity orthophosphate transporters. J Biol Chem 285:32029–32037

Lee YS, Mulugu S, York JD, O'Shea EK (2007) Regulation of a cyclin-CDK-CDK inhibitor complex by inositol pyrophosphates. Science 316:109–112

Lee PS, Greenwell PW, Dominska M, Gawel M, Hamilton M, Petes TD (2009) A fine-structure map of spontaneous mitotic crossovers in the yeast *Saccharomyces cerevisiae*. PLoS Gen 5, e1000410

Lee MV, Topper SE, Hubler SL, Hose J, Wenger CD, Coon JJ, Gasch AP (2011) A dynamic model of proteome changes reveals new roles for transcript alteration in yeast. Mol Syst Biol 7:514

Lenburg ME, OShea EK (1996) Signaling phosphate starvation. Trends Biochem Sci 21:383–387

Lundh F, Mouillon JM, Samyn D, Stadler K, Popova Y, Lagerstedt JO, Thevelein JM, Persson BL (2009) Molecular mechanisms controlling phosphate-induced downregulation of the yeast Pho84 phosphate transporter. Biochem 48:4497–4505

Marchal C, Haguenauer-Tsapis R, Urban-Grimal D (1998) A PEST-like sequence mediates phosphorylation and efficient ubiquitination of yeast uracil permease. Mol Cell Biol 18:314–321

Martinez P, Persson BL (1998) Identification, cloning and characterization of a derepressible $Na^{+}$-coupled phosphate transporter in *Saccharomyces cerevisiae*. Mol General Gen 258:628–638

Maw GA (1963) The uptake of inorganic sulfate by a brewer's yeast. Folia Microbiol (Praha) 40:325–332

McCready RG, Din GA (1974) Active sulfate transport in *Saccharomyces cerevisiae*. FEBS Lett 38:361–363

Mouillon JM, Persson BL (2005) Inhibition of the protein kinase A alters the degradation of the high-affinity phosphate transporter Pho84 in *Saccharomyces cerevisiae*. Curr Gen 48:226–234

Mouillon JM, Persson BL (2006) New aspects on phosphate sensing and signalling in *Saccharomyces cerevisiae*. Fems Yeast Res 6:171–176

Mount DB, Romero MF (2004) The SLC26 gene family of multifunctional anion exchangers. Pflugers Arch 447:710–721

Ogawa N, Noguchi KI, Sawai H, Yamashita Y, Yompakdee C, Oshima Y (1995) Functional domains of Pho81p, an Inhibitor of Pho85p Protein-Kinase, in the transduction pathway of P-I signals in *Saccharomyces cerevisiae*. Mol Cell Biol 15:997–1004

ONeill EM, Kaffman A, Jolly ER, OShea EK (1996) Regulation of *PHO4* nuclear localization by the *PHO80-PHO85* cyclin-CDK complex. Science 271:209–212

Pao SS, Paulsen IT, Saier MH Jr (1998) Major facilitator superfamily. Microbiol Mol Biol Rev 62:1–34

Pattison-Granberg J, Persson BL (2000) Regulation of cation-coupled high-affinity phosphate uptake in the yeast *Saccharomyces cerevisiae*. J Bact 182:5017–5019

Pedersen BP, Kumar H, Waight AB, Risenmay AJ, Roe-Zurz Z, Chau BH, Schlessinger A, Bonomi M, Harries W, Sali A, Johri AK, Stroud RM (2013) Crystal structure of a eukaryotic phosphate transporter. Nature 496:533–536

Persson BL, Petersson J, Fristedt U, Weinander R, Berhe A, Pattison J (1999) Phosphate permeases of *Saccharomyces cerevisiae*: structure, function and regulation. Biochim Biophys Acta Rev Biomembr 1422:255–272

Persson BL, Lagerstedt JO, Pratt JR, Pattison-Granberg J, Lundh K, Shokrollahzadeh S, Lundh F (2003) Regulation of phosphate acquisition in *Saccharomyces cerevisiae*. Curr Gen 43:225–244

Popova Y, Thayumanavan P, Lonati E, Agrochao M, Thevelein JM (2010) Transport and signaling through the phosphate-binding site of the yeast Pho84 phosphate transceptor. PNAS 107:2890–2895

Pratt JR, Mouillon JM, Lagerstedt JO, Pattison-Granberg J, Lundh KI, Persson BL (2004) Effects of methylphosphonate, a phosphate analogue, on the expression and degradation of the high-affinity phosphate transporter Pho84, in *Saccharomyces cerevisiae*. Biochemistry 43:14444–14453

Roomans GM, Kuypers GA, Theuvenet AP, Borst-Pauwels GW (1979) Kinetics of sulfate uptake by yeast. Biochim Biophys Acta 551:197–206

Saier MH, Beatty JT, Goffeau A, Harley KT, Heijne WHM, Huang SH, Jack DL, Jahn PS, Lew K, Liu J, Pao SS, Paulsen IT, Tseng TT, Virk PS (2000) The major facilitator superfamily. J Mol Microbiol Biotech 2:255–279

Salaün C, Rodrigues P, Heard JM (2001) Transmembrane topology of PiT-2, a phosphate transporter-retrovirus receptor. J Virol 75:5584–5592

Samyn DR, Ruiz-Pávon L, Andersson MR, Popova Y, Thevelein JM, Persson BL (2012) Mutational analysis of putative phosphate- and proton-binding sites in the *Saccharomyces cerevisiae* Pho84 phosphate:H(+) transceptor and its effect on signalling to the PKA and PHO pathways. Biochem J 445:413–422

Sengottaiyan P, Ruiz-Pavón L, Persson BL (2013a) Functional expression, purification and reconstitution of the recombinant phosphate transporter Pho89 of *Saccharomyces cerevisiae*. FEBS J 280:965–975

Sengottaiyan P, Petrlova J, Lagerstedt JO, Ruiz-Pavon L, Budamagunta MS, Voss JC, Persson BL (2013b) Characterization of the biochemical and biophysical properties of the *Saccharomyces cerevisiae* phosphate transporter Pho89. Biochem Biophys Res Commun 436:551–556

Serrano R, Ruiz A, Bernal D, Chambers JR, Arino J (2002) The transcriptional response to alkaline pH in Saccharomyces cerevisiae: evidence for calcium-mediated signaling. Mol Microbiol 46:1319–1333

Springer M, Wykoff DD, Miller N, O'Shea EK (2003) Partially phosphorylated Pho4 activates transcription of a subset of phosphate-responsive genes. Plos Biol 1:261–270

Tamai Y, Toh-e A, Oshima Y (1985) Regulation of inorganic-phosphate transport-systems in *Saccharomyces cerevisiae*. J Bacteriol 164:964–968

Thevelein JM, Voordeckers K (2009) Functioning and evolutionary significance of nutrient transceptors. Mol Biol Evol 26:2407–2414

Thevelein JM, de Winde JH (1999) Novel sensing mechanisms and targets for the cAMP-protein kinase A pathway in the yeast *Saccharomyces cerevisiae*. Mol Microbiol 33:904–918

Toh-e A, Ueda Y, Kakimoto SI, Oshima Y (1973) Isolation and characterization of acid-phosphatase mutants in *Saccharomyces cerevisiae*. J Bacteriol 113:727–738

Toh-e A, Tanaka K, Uesono Y, Wickner RB (1988) Pho85, a negative regulator of the Pho system, is a homolog of the protein-kinase gene, Cdc28, of *Saccharomyces cerevisiae*. Mol Geneal Gen 214:162–164

Vardi N, Levy S, Gurvich Y, Polacheck T, Carmi M, Jaitin D, Amit I, Barkai N (2014) Sequential feedback induction stabilizes the phosphate starvation response in budding yeast. Cell Rep 9:1122–1134

Wykoff DD, O'Shea EK (2001) Phosphate transport and sensing in *Saccharomyces cerevisiae*. Genetics 159:1491–1499

Wykoff DD, Rizvi AH, Raser JM, Margolin B, O'Shea EK (2007) Positive feedback regulates switching of phosphate transporters in *S cerevisiae*. Mol Cell 27:1005–1013

Zvyagilskaya RA, Lundh F, Samyn D, Pattison-Granberg J, Mouillon JM, Popova Y, Thevelein JM, Persson BL (2008) Characterization of the Pho89 phosphate transporter by functional hyperexpression in *Saccharomyces cerevisiae*. FEMS Yeast Res 8:685–696

# Chapter 11
# Interactions Between Monovalent Cations and Nutrient Homeostasis

**David Canadell and Joaquín Ariño**

**Abstract** Maintenance of appropriate fluxes of monovalent cation is a requirement for growth and survival. In the budding yeast *Saccharomyces cerevisiae* an electrochemical gradient of $H^+$ is fundamental for the uptake of diverse cations, such as $K^+$, and of many other nutrients. In spite of early work suggesting that alterations in monovalent cation fluxes impact on the uptake and utilization of nutrients, such as phosphate anions, only recently this important aspect of the yeast physiology has been addressed and characterized in some detail. This chapter provides a historical background and summarizes the latest findings.

**Keywords** Potassium homeostasis • Phosphate uptake • Ammonium assimilation • *Saccharomyces cerevisiae*

## 11.1 How Yeast Cells Obtain Nutrients?

The yeast plasma membrane represents the primary selective barrier for nutrient trafficking in the cell, since the cell wall is considered a semi selective porous structure and does not represent an absolute obstacle for solutes (De Nobel and Barnett 1991). Nutrients and ions move across the membranes by mean of free or facilitated diffusion, and by active transport. Free diffusion is limited to lipid-solute/uncharged molecules and implies a favorable gradient. Similarly, facilitated diffusion relies on permeases or channels to transport molecules via the differential solute concentration inside and outside the cell. For example, sugars like glucose can be uptaked by permeases (Ozcan and Johnston 1999), whereas the passive efflux of potassium is mediated by the voltage-gated $K^+$-specific channel Tok1, which is activated by membrane depolarization (Ketchum et al. 1995).

In yeast cells, active transport is responsible for the uptake of most nutrients as well as cations like potassium. The proton motive force enables uptake of

D. Canadell • J. Ariño (✉)
Departament de Bioquímica i Biologia Molecular & Institut de Biotecnologia i Biomedicina, Universitat Autónoma de Barcelona, Bellaterra, 08193 Barcelona, Spain
e-mail: Joaquin.Arino@uab.es

 
J. Ramos et al. (eds.), *Yeast Membrane Transport*, Advances in Experimental Medicine and Biology 892, DOI 10.1007/978-3-319-25304-6_11

nutrients/cations in symport with protons (and their efflux as an antiport mechanism). The majority of the transporters of nutrients necessary for the growth of yeast cells are dependent on the proton gradient. These include some sugar transporters, all amino acids transporters, organic acid transporters, purines, pyrimidines and vitamin transporters, anion transporters, such as phosphate and sulfate, and some monovalent (i.e. potassium) or divalent cations (i.e. magnesium, zinc) and others transport systems.

### *11.1.1 Pma1 as Energy Source for Nutrient Transport*

The plasma membrane $H^+$-ATPase, encoded by the essential *PMA1* gene, belongs to the widely distributed family of P2-type ATPases and is the main primary yeast transporter (Ambesi et al. 2000; Serrano et al. 1986). Pma1 pumps protons through the plasma membrane out of the cell and, therefore, creates an electrochemical gradient of protons across the plasma membrane that is indispensable for all secondary active symporters and antiporters as well as for cytosolic pH regulation. The electrochemical gradient of protons created by Pma1 activity also controls the accumulation of essential $K^+$ ions (Rodriguez-Navarro 2000).

The activity of Pma1 is tightly regulated according to the metabolic activity and physiological conditions of cells. Addition of glucose to starved cells results in a powerful stimulus leading to a rapid and strong Pma1 activation, and Pma1 activity is also positively regulated in response to decreased intracellular pH or increased potassium uptake (Seto-Young and Perlin 1991). Pma1 expression and activity is also tightly regulated by carbon or nitrogen sources like glucose or amino acid levels (see Sects. 11.3.2 and 11.4 for details).

### *11.1.2 Relationship Between Pma1 and Cation Transport*

Uptake of potassium is the main consumer of the electrochemical gradient generated by Pma1 (Madrid et al. 1998). High-affinity potassium uptake is mediated by the plasma membrane transporters Trk1 and Trk2 (Gaber et al. 1988; Ko et al. 1990; Ko and Gaber 1991), where Trk1 is the most physiologically relevant. Yeast cells can concentrate potassium up to 200–300 mM into the cell, even if the external concentration is very low (<1 mM). Consequently, a *trk1 trk2* double mutant strain shows an increased requirement for external potassium and a hyperpolarized plasma membrane (Madrid et al. 1998). A low-affinity potassium transport was described in the *trk1 trk2* mutant strain, supported by a nonspecific cation "channel" named NSC1 (although the gene encoding this activity is not known) (Bihler et al. 1998, 2002) and, more recently, the calcium-related transporters Kch1 and Kch2 have been found to sustain growth of *trk1 trk2* mutant cells in low $K^+$ environments, suggesting they promote $K^+$ uptake (Stefan et al. 2013). In addition, the ability to transport potassium have been attributed to the drug antiporter Qdr2 (Vargas et al. 2007), as well as to some amino acid transporters (Wright et al. 1997).

Potassium transport into yeast cells results in a depolarization of the plasma membrane, stimulating the Pma1 activity and a simultaneous cytosolic alkalization (Rodriguez-Navarro 2000). The regulation of the activities of both Pma1 and Trk1 is fundamental for the modulation of the electrochemical gradient used by the other nutrient symporters. The balance between $K^+$ and $H^+$ influx and efflux sets the level of the plasma-membrane potential (Ke et al. 2013; Mulet et al. 1999; Navarrete et al. 2010). The efflux of potassium is mediated by the above mentioned voltage-gated $K^+$-specific channel Tok1 (Ketchum et al. 1995), by the Nha1 $Na^+(K^+)/H^+$ antiporter (Banuelos et al. 1998; Prior et al. 1996) and by the cluster of *ENA* genes (*ENA1-ENAn,* where *n* may differ from strain to strain) coding a P-type ATPase important for the efflux of $Na^+/K^+/Li^+$ (Benito et al. 2002; Garciadeblas et al. 1993; Haro et al. 1991). *ENA1* is the most relevant member of the cluster. It is barely expressed under standard growth conditions, but is rapidly induced by osmotic stress, salt stress or by shifting cells to alkaline pH (Ruiz and Arino 2007). It has been proposed that Ena1 could play another role controlling the membrane potential allowing the maintenance of normal uptake of nutrients through secondary transporters under stress conditions in which the proton gradient is reduced (Ke et al. 2013).

### *11.1.3 Intracellular (Organelle) Nutrient and Cation Transport*

Similarly to the plasma membrane, nutrient or cation transport in the vacuole and other acidic intracellular membrane organelles, such as endosomes or the Golgi apparatus, is driven by facilitated diffusion and active transport. In this case, the proton gradient is generated by V-ATPases pumping $H^+$ into the vacuole or organelle, a necessary process to maintain intracellular pH and the acidification of the vacuole (Martinez-Munoz and Kane 2008). The V-ATPase is a large, conserved protein complex made up by at least 14 gene products and organized into two subcomplexes: an integral membrane complex $V_0$ and the ATPase complex $V_1$ (Kane 2006). Two isoforms of the V-ATPase coexist, depending on the localization, one including the *VPH1* gene product, which targets the V-ATPase to the vacuole, and the other with Stv1, targeting the V-ATPase to the Golgi apparatus/endosome (Kawasaki-Nishi et al. 2001; Manolson et al. 1994).

The electrochemical gradient across the vacuolar membrane drives transport into and out of the intracellular organelles. $Ca^{2+}$ accumulates in the vacuole partially via the $Ca^{2+}/H^+$ antiporter, Vcx1 (Miseta et al. 1999). Similarly, transport of heavy metals, amino acids, and polyphosphate accumulation depend on the proton gradient across the vacuolar membrane (reviewed in Li and Kane 2009).

Three intracellular $K^+(Na^+)/H^+$ antiporters with different localizations have been described and characterized in *S. cerevisiae* cells. Nhx1 is located in the late endosomal/prevacuolar membranes (Brett et al. 2005), Kha1 in the membranes of the Golgi apparatus (Maresova and Sychrova 2005), and Vnx1 in vacuolar and endoplasmic-reticulum membranes (Cagnac et al. 2007). Recently, a Vhc1 vacuolar membrane cation-chloride cotransporter (CCC) that mediates $K^+$ and $Cl^-$ cotransport into the vacuole has been reported (Petrezselyova et al. 2013). Evidence gathered in the last few years indicates that V-ATPase function can be modulated by regulated assembly/disassembly processes. Similarly to the activation of Pma1, V-ATPase assembly is regulated by glucose and the nutrient state of the cell (Li and Kane 2009; Parra et al. 2014). Glucose is the main stimulus for V-ATPase assembly or disassembly. Addition of glucose to glucose-deprived cells causes V-ATPase re-assembly through the activation of PKA and alkalization of cytosol. At the same time, intracellular alkalinization triggers the activation of Pma1 at the plasma membrane and, together with the assembly of V-ATPase in vacuoles, controls intracellular pH. Otherwise, cells starved for glucose or subjected to less preferred carbon sources tend to V-ATPase disassembling (Parra et al. 2014).

The electrochemical gradient generated by the V-ATPase drives the entry of cations into the vacuole through different cation transporter systems. This allows storage of potassium to ensure optimal cytosolic potassium concentrations, diminishing the amounts of toxic sodium or lithium cations in the cytosol, or compensating the negative charges of vacuolar polyphosphates (Klionsky et al. 1990). Ion transport is also necessary to maintain V-ATPase activity, since it is an electrogenic pump and needs anion transport to provide charge compensation (Arai et al. 1989). Transport of ions consumes the gradient generated by the V-ATPase and, as it opposes to the constant $H^+$ pumping, it is crucial to control intracellular pH. The regulation of the gradient facilitates intracellular nutrient transport and endosomal trafficking (Ali et al. 2004; Arai et al. 1989). In addition, transport of cations into the vacuole served to control cellular turgor and for detoxification the cytosol of diverse toxic compounds.

## 11.2 The Relationship Between Phosphate and Cation Homeostasis

Inorganic phosphate ($P_i$) is the most abundant anion in *S. cerevisiae* cells. It is necessary for the biosynthesis of cellular components such as nucleic acids, nucleoproteins, and phospholipids, and is involved in many metabolic or signaling pathways. Transport of $P_i$ through the cell membrane is mediated by low-affinity $H^+/P_i$ symport ($K_M \sim 1$ mM), involving Pho87 and Pho90, which is sufficient for growth in standard external $P_i$ levels (Bun-ya et al. 1996; Ghillebert et al. 2011; Wykoff and O'Shea 2001), and by a high affinity ($K_M \sim 10$ μM) transport system, consisting of Pho84 and Pho89 transporters that are induced under phosphate-limiting conditions (Bun-ya et al. 1991; Martinez and Persson 1998; Persson et al. 1999).

Expression of *PHO84* and *PHO89* transporter-encoding genes, as well as that of other genes required for acquisition and utilization of $P_i$, such as the acid phosphatases *PHO5* and *PHO12*, or the *VTC1-4* genes, encoding proteins involved in polyphosphate metabolism, is increased in response to $P_i$ starvation. This response is regulated by the *PHO* signaling pathway, which controls the localization of the transcription factor Pho4 depending on phosphate availability (Auesukaree et al. 2003; Ogawa et al. 1995; Persson et al. 2003). Under normal phosphate conditions Pho4 is phosphorylated by the Pho85–Pho80 cyclin-dependent kinase complex and excluded from the nucleus genes (Kaffman et al. 1994; Toh-e A et al. 1988). When phosphate levels decrease, the levels of inositol heptakisphosphate ($IP_7$) increase and together with Pho81 inhibit the Pho85-Pho80 complex, thus preventing the phosphorylation of Pho4 (Lee et al. 2007, 2008). Unphosphorylated Pho4 then accumulates in the nucleus and binds to diverse phosphate-responsive gene promoters triggering their transcription.

Pho84 cotransports phosphate with $H^+$, and it is responsible for the vast majority of the high-affinity Pi uptake under normal acidic growth conditions when phosphate is scarce (Pattison-Granberg and Persson 2000). Pho84 has been defined as a phosphate transceptor, a concept related to transporters that also act as nutrient sensing proteins. Pho84 transports phosphate and signals to activate the protein kinase A (PKA) pathway when phosphate is added to starved cells (Popova et al. 2010).

## 11.2.1 The Role of Potassium in Phosphate Homeostasis

A possible relationship between potassium availability and normal phosphate uptake in yeast was described many years ago (Goodman and Rothstein 1957; Schmidt et al. 1949). These authors observed that $K^+$ enhanced phosphate uptake in yeast when the medium was adjusted to acidic pH, but decreased it at high pH, and proposed that the effect of $K^+$ upon phosphate uptake is due to an increase in intracellular pH, which accompanies $K^+$ uptake. However, these experiments were performed with yeast cultures genetically undefined and under non-physiological conditions (transport measurements were made in water or different buffers). It has not been until 2010 that the relationship between phosphate accumulation and metal homeostasis has been investigated in some detail (Rosenfeld et al. 2010). These authors showed that *pho80* mutants, who are unable to down regulate the *PHO* pathway and, consequently, accumulate high levels of intracellular phosphate, also display high levels of diverse metals, including potassium and sodium.

On the other hand, alterations in potassium homeostasis affect phosphate metabolism. Elimination of potassium from the medium or perturbation of the normal influx of the cation triggers a wide transcriptional response that includes the activation of a set of genes typically induced under phosphate starvation conditions (Barreto et al. 2012). This response also depends on the integrity of the *PHO* signaling pathway (Canadell et al. 2015). The *trk1 trk2* mutant, which lacks the high-affinity potassium transporter, growing under limiting potassium amounts, shows abnormally high Pho84 expression and the same phenomenon is observed for mutations in the regulatory potassium uptake system that leads to decreased high affinity potassium uptake. Similarly, mutants such as *ptk2* or *brp1* that show decrease Pma1 activity and, consequently, present defective potassium uptake (Barreto et al. 2011), also display an Trk1-dependent accumulation of Pho84 under potassium limitation.

Polyphosphates (PolyP) are linear polymers of phosphoanhydride-linked phosphate residues ubiquitously distributed in all organisms and living cells (Kornberg et al. 1999). PolyP have a variety of different functions, and in yeast cells it is of utmost importance as phosphate store and for buffering fluctuations of $P_i$ levels (Thomas and O'Shea 2005). Although polyP has been identified in diverse subcellular localizations, the yeast vacuole appears to be the major reservoir of the polymer (Kornberg 1995; Saito et al. 2005; Urech et al. 1978). Recently, Canadell and coworkers (Canadell et al. 2015) showed that decreased potassium uptake results in rapid degradation of PolyP with only limited changes in free $P_i$ levels. In correlation with polyP degradation, a decrease of 4 to 5-fold in the amount of vacuolar potassium was also observed after potassium starvation (Herrera et al. 2013) suggesting a charge balance effect, in which potassium (the most abundant cation in yeast cells) would counteract negatively charges of PolyP chains. Potassium starvation also caused a decline in the levels of ATP and ADP and in the overall level of adenine nucleotides. The decay in ATP levels could be attributed, at least in part, to the increased activity of the Pma1 ATPase, which is immediately activated after removal of potassium from the medium (Kahm et al. 2012). Remarkably, it has been proposed that the level of adenine nucleotides could function as intracellular signal for phosphate starvation, thus regulating the *PHO* pathway (Gauthier et al. 2008). This signal could be also important for adapting the *PHO* pathway under conditions of potassium limitation.

### *11.2.2 Sodium/Phosphate Cotransport: PHO89 and ENA1 Understand the Same Messages*

As mentioned above, uptake of $P_i$ under normal (acidic) growth conditions is mediated by the Pho84 transporter. However, *S. cerevisiae* also encodes a second high-affinity Pi transporter, Pho89, which shows marked differences with Pho84. Whereas Pho84 is a $H^+/P_i$ symporter, Pho89 is a phosphate/cation symporter that works most efficiently under alkaline conditions (optimum pH of 9.5). Phosphate transport through Pho89 requires the existence of an alkali-metal cation gradient, being $Na^+$ markedly preferred over $K^+$ or $Li^+$ (Martinez and Persson 1998; Zvyagilskaya et al. 2008) . Remarkably, whereas in mammals several $Na^+/P_i$,

$Na^+$/glucose, $Na^+$/amino acids or other cotransporters are described, Pho89 seems to be the only membrane monovalent cation/nutrient transporter in *S. cerevisiae*.

*PHO89* is induced under $P_i$ limitation, although this response is somewhat limited and is delayed in comparison with that of *PHO84* (Martinez and Persson 1998; Serra-Cardona et al. 2014; Zvyagilskaya et al. 2008). This response is essentially under the control of Pho4. In addition, as other members of the *PHO* regulon, *PHO89* and *PHO84* are induced by alkalinization of the medium (Serrano et al. 2002). Interestingly, in this case the response of *PHO89* is faster than that of *PHO84* (Serra-Cardona et al. 2014; Serrano et al. 2002; Viladevall et al. 2004). Serrano and coworkers demonstrated that a significant part of the positive response of the *PHO89* promoter in response to alkali stress was dependent on calcineurin activation and was mediated by the Crz1 transcription factor (Serrano et al. 2002). Very recently, Serra-Cardona and coworkers (Serra-Cardona et al. 2014) demonstrated that *PHO89* expression is also under the control of both the Snf1 protein kinase, through the Mig2 and Nrg1/2 repressors, and the Rim101 transcription factor, which represses the expression of Nrg1 (Lamb and Mitchell 2003). Remarkably, this regulatory network is the same that the previously elucidated for the $Na^+$-ATPase *ENA1* in response to alkalinization of the medium (Platara et al. 2006).

Uptake of $P_i$ through the Pho89 transporter, which becomes relevant at alkaline pH (Zvyagilskaya et al. 2008), has a forced trade-off: the intake of toxic $Na^+$ cations. Whereas maximum activity of Pho89 is reached with external concentrations of 15–20 mM $Na^+$ (Martinez and Persson 1998), more recent measurements under conditions of optimum Pho89 activity yielded a $K_M$ value for $Na^+$ coupling relatively low (~0.7 mM) and a Vmax of ~2 mmol $min^{-1}$ g $cells^{-1}$ (Zvyagilskaya et al. 2008). The existence of a common regulatory network in response to alkalinization of the medium for *ENA1* and *PHO89* suggested that both gene products could be functionally coupled. Indeed, kinetic studies show that in response to alkalinization of the medium *ENA1* mRNA peaks shortly after *PHO89* mRNA does, and both proteins reach a maximum of accumulation roughly at the same time (Serra-Cardona et al. 2014). These authors also showed that, when Pho89 is the only high-affinity $P_i$ transporter, the presence of Ena1 becomes necessary for growth at acidic pH, when $P_i$ is limiting, or at alkaline pH irrespective of the availability of Pi, and that *pho84 ena1* cells dramatically accumulate $Na^+$ cations when shifted to alkaline pH, whereas *pho84* cells do not.

It has been proposed that Ena1 plays important roles in establishing the membrane potential, thus maintaining normal uptake of nutrients through secondary transporters under certain stress conditions, such alkaline pH stress (Ke et al. 2013). Whereas this could hold true, in this case the role of Ena1 would be more specific. The requirement for Ena1-mediated efflux of $Na^+$ would be twofold: (1) necessary to avoid the accumulation of $Na^+$ above the toxic threshold and (2) to circumvent the abolition or even reversal of the extracellular/intracellular $Na^+$ gradient that is necessary to support effective $P_i$ uptake.

## 11.3 The Nitrogen Sources and Cation Homeostasis

Nitrogen is needed in yeast cells for biosynthesis of amino acids, nucleotides and a variety of nitrogen-containing compounds. *S. cerevisiae* can use a diversity of amino acids or other organic nitrogen compounds as nitrogen sources, but the standard laboratory yeast strains prefer glutamine and ammonium over other nitrogen sources (Zaman et al. 2008). Growth in a preferred nitrogen source leads to repression of genes required for catabolism of other nitrogen sources. The main nitrogen source used in a standard laboratory synthetic media is ammonium because it supports growth of yeast cells at an optimal rate. Ammonium is transported into the yeast cells by three ammonium transporters, Mep1, Mep2 and Mep3, which are under the nitrogen catabolite repression regulation (Marini et al. 1997). A strain lacking all three Mep proteins cannot grow on media containing less than 5 mM $NH_4^+$, but grows normally on high $NH_4^+$, demonstrating the existence of additional, low-affinity ammonium transport systems. In yeasts grown in glucose, ammonia can be assimilated by incorporation into the amino group of glutamate by two anabolic pathways: the reductive amination of 2-ketoglutarate, catalyzed by glutamate dehydrogenase (*GDH1* and *GDH3*), for which NADPH serves as the source of electrons, or by the ATP-dependent synthesis of glutamine from glutamate and ammonia, catalyzed by glutamine synthase (*GLN1*). Nitrogen-containing compounds such as amino acids or nucleotides are synthetized from glutamate or glutamine (Magasanik 2003).

### *11.3.1 Ammonium and Potassium Look Alike*

The bulk of intracellular negative charges due to the different anionic compounds are preferentially neutralized by potassium ions in yeast cells. Under situations of potassium starvation, sodium can replace potassium for some functions (Rodriguez-Navarro 2000). Yeast cells have a very different uptake capability for both cations, with a $K^+/Na^+$ $K_M$ ratio of 1:700 when high-affinity potassium uptake is working. At the same extracellular concentrations potassium is taken up much more efficiently than sodium and, therefore, intracellular concentrations of the latter are lower (Gomez et al. 1996; Ramos et al. 1985). Measurement of potassium transport is usually done by using the $Rb^+$ analogue, due to the very similar transport activity of potassium transporters for both cations (Rodriguez-Navarro 2000). Another cation that mimics potassium is the ammonium ion ($NH_4^+$), which actually has an ionic radius (1.43 Å) very similar to $K^+$ (1.33 Å), and in fact even closer that $Rb^+$ (1.47 Å) or $Na^+$ (0.95 Å) ones.

Replacement of potassium by ammonium was described in bacteria in potassium-limited chemostat cultures, mainly at alkaline pH (Buurman et al. 1989). Hess and collaborators (Hess et al. 2006) suggested that if ammonium is present in high concentrations in potassium-limited cultures, ammonium ions can enter the cell via the Trk potassium channels. Then, the unregulated influx would create an excess of internal ammonium that would become toxic, leading to production and excretion of amino acids to reduce internal ammonium levels. Barreto and coworkers (Barreto et al. 2012) demonstrated that, in fact, the lack of potassium causes a fast and persistent increase in intracellular ammonium concentration, that the ammonium influx is mediated by Trk1, and that this effect is abolished when the cell express a version of Trk1 unable to transport potassium. In addition, a *S. cerevisiae* natural strain resistant to high concentration of ammonium (K12) was isolated in sake breweries. This strain has a *TRK1* allele with greater affinity for potassium than the standard allele of *TRK1* found in *Saccharomyces* strains (Reisser et al. 2013), which led the authors to hypothesize that, since greater-affinity transporters are often associated with lower flux rates, this would imply reduced flux of ammonium into the yeast cells under conditions of ammonium toxicity. On the basis of data reported in the literature it is tempting to speculate that the Trk system could be the responsible for the unidentified low affinity ammonium uptake detected in *mep* mutants.

In cells lacking functional mitochondria the tricarboxylic acid cycle (TCA) is not able to work as a full cycle (due to the lack of succinate dehydrogenase activity), thus limiting the production of α-ketoglutarate necessary for ammonium fixation (Jazwinski 2013). As a response, the mitochondrial retrograde pathway activates retrograde genes encoding the enzymes needed to synthesize α-ketoglutarate and to maintain glutamate levels needed for anabolic reactions (Butow and Avadhani 2004; Liu and Butow 2006). Thus, the retrograde genes (*CIT2*, *ACO1* and *IDH1/2*), which encode early enzymes in the TCA cycle and are under the control of the transcription factors Rtg1/3, remain active, allowing the generation of isocitrate, necessary for both the glyoxylate cycle and the synthesis of α-ketoglutarate (Fig. 11.1). An increase in intracellular ammonium could elicit the retrograde response and induce *CIT2* and *DLD3* expression (Tate and Cooper 2003). In potassium-starved yeast cells the induction of retrograde response elements was observed. This induction was dependent on the presence of Rtg1/3 and occurred in the absence of any evident mitochondrial problem (Barreto et al. 2012). Therefore, it is likely that the intracellular accumulation of ammonium could be the triggering factor for the retrograde response observed in limited potassium conditions.

The metabolic data on amino acid secretion under potassium limitation or excess of ammonia obtained by Hess and coworkers (Hess et al. 2006) was analyzed using a metabolic network model (Mo et al. 2009). The potassium limitation and high ammonium conditions shared a common pattern, with high levels of amino acids and precursors of amino acid biosynthesis such as pyruvate, α-ketoglutarate, glutamine, glutamate, etc. The mRNA profiles of *GDH1*, *GDH2* and *GLN1* in potassium-starved cells and under high levels of ammonium are very similar, showing a decrease of *GDH1* and increase of both *GDH2* and *GLN1* (Fig. 11.1) (Barreto et al. 2012; ter Schure et al. 1995). The opposite behavior of *GDH1* and *GDH2*,

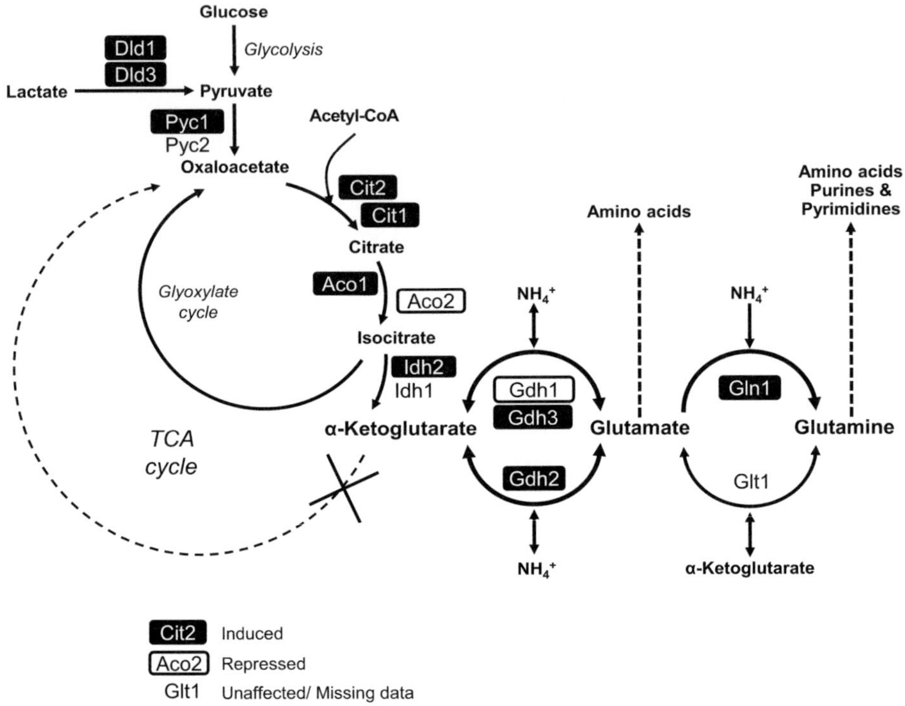


**Fig. 11.1 A possible link between retrograde response and ammonium toxicity in potassium deprived cells**. Lack of $K^+$ triggers induction of retrograde genes such as *CIT2* and *DLD3*. The activation of the retrograde pathway could serve to ensure sufficient levels of α-ketoglutarate to neutralize toxic amounts of ammonium by forming Glu or Gln. The format given to the gene names refers to their response to potassium starvation according to Barreto et al. (2012). Induction or repression indicates changes ≥2-fold

although remarkable, it is not completely unexpected since it has been proposed that both genes are subjected to different transcriptional regulation (ter Schure et al. 2000). Similarly, a strong repression of the general amino acid permease *GAP1* was detected in both conditions, highlighting once again the impact of potassium in controlling intracellular ammonium levels (Barreto et al. 2012; Hess et al. 2006; ter Schure et al. 1995). Mo and co-workers also studied the metabolic pattern of a strain in which the *GDH1* gene was deleted and *GDH2* was overexpressed. This strain, under standard growth conditions, also shows an altered metabolic profile related to amino acid biosynthesis and, interestingly, displays increased methylglyoxal and lactate accumulation, as it is observed under potassium starvation conditions (Barreto et al. 2012; Mo et al. 2009) (see also Sect. 11.4). All these results reinforce the notion that ammonium poisoning is behind many of the metabolic and transcriptional changes observed in cells exposed to lack of potassium.

### *11.3.2 The Role of Potassium in Amino Acid and Sulfate Transport*

As it has been mentioned previously, normal potassium uptake is needed to maintain the electrochemical proton gradient across the membrane generated by Pma1. Aside from other regulatory mechanisms, *PMA1* expression is controlled by *GLN3* (Cox et al. 1999). Gln3 acts as a transcriptional activator of genes required for the uptake and utilization of poor nitrogen sources, such as proline. Because amino acid uptake is mediated by proton symport, the Gln3-dependent activation of *PMA1* expression in yeast may reflect a need for efficient amino acid transport in nitrogen poor media. Interestingly, it has been shown that a single amino acid change can transform an amino acid permease such as Bap2 or Hip1 in a likely potassium transporter, able to partially suppress the increased $K^+$ requirement of *trk1 trk2* cells (Wright et al. 1997).

A relationship between sulfate transport and potassium was observed many years ago, proposing the release of a $K^+$ cation for every sulfate imported using the proton motive force as energy source (Borst-Pauwels 1981). However, as it happens with early phosphate uptake experiments, measurements were performed under non-physiological conditions and were subjected to experimental limitations, thus complicating their interpretation. More recently, working in standard growth conditions, Barreto and coworkers showed that lack of potassium promoted the induction of many genes related to the metabolism of methionine and its derivatives (Barreto et al. 2012), including the entire pathway for sulfur assimilation and Met uptake and biosynthesis, similarly to the profile observed in the response in sulfur-limited chemostat cultures (Boer et al. 2003). In fact, the response to potassium starvation likely reflects a shortage of the availability of sulfur-containing amino acids, since the levels of Met and Cys showed an almost immediate depletion in cells transferred to $K^+$-free medium, and increasing the amount of Met in the medium specifically attenuated the induction of genes relevant for the biosynthesis of sulfur-containing amino acids derivatives (Barreto et al. 2012). It must be noted that these results were observed in the BY4741 strain, which carries the *met17* mutation and, therefore, since it is unable to incorporate sulfate into homocysteine, it is auxotrophic for Met. However, the analysis of *lacZ* reporters for the genes *MET10*, *MUP3* and *STR3* in the *MET17* strains W303-1A and DBY746 confirmed that these promoters are still activated in response to lack of potassium, albeit to a lesser degree. Therefore, the *met17* mutation probably potentiated the impact of $K^+$ starvation on sulfur-related genes but it is not at the origin of the observed effect.

Remarkably, the increased expression of several genes encoding proteins required for sulfur utilization was also observed in *trk1 trk2* or in *hal4 hal5* cells growing in the presence of plenty of potassium (Barreto et al. 2012; Perez-Valle et al. 2010). Hal4 and Hal5 protein kinases regulate the localization of Trks transporters in the plasma membrane, as well as of other nutrient transporters. The *hal4 hal5* double mutant strain requires potassium supplementation for maximal growth and displays a marked defect in rubidium uptake (Mulet et al. 1999;

Perez-Valle et al. 2007). Therefore, the sulfur-amino acid related transcriptional profile of the *hal4 hal5* mutant could be caused, at least in part, by a poor potassium uptake of this strain. Proteomic studies confirmed that certain sulfur-amino acid biosynthesis-related proteins (Met22, Sam4, Sam1) are present in higher amounts in the *trk1 trk2* mutant strain during potassium starvation (Gelis et al. 2012).

## 11.4 Sugars Regulating Potassium and Proton Transport

As it is mentioned above, glucose is an important signal for Pma1 regulation that invokes multiple signaling pathways collaborating to allow proper activation of the ATPase by this sugar. Initially, Pma1 regulation by glucose was described at two different levels: a rapid 5- to 10-fold increase in ATPase catalytic activity (Serrano 1983), affecting both $K_M$ and Vmax, and a second phase involving a slow and moderate increase in the expression of the gene in a process that involves the Rap1 and Gcr1 transcription factors (Portillo 2000). The rapid response to glucose addition could be due to glucose-dependent phosphorylations of Pma1 identified some years ago. Phosphorylation of Ser-899 (Goossens et al. 2000) and tandem phosphorylation of Ser-911 and Thr-912 (Lecchi et al. 2007) was found to be mediated by the protein kinase Ptk2 and resulted in increased Pma1 activity. In the other hand, the addition of glucose prevents the inhibitory phosphorylation of Pma1 by yeast casein kinase I (encoded by the *YCK1* and *YCK2* genes), thus generating an increase in the Pma1 activity (Estrada et al. 1996). As a result, glucose activation of Pma1 is the consequence of downregulation of the inhibitory casein kinase I and upregulation of the activating protein kinase Ptk2. Other activating effects of glucose on Pma1 are mediated by the increase in calcium concentrations (Tropia et al. 2006) and by the disruption of the complex of Pma1 with acetylated tubulin that follows the addition of glucose (Campetelli et al. 2005). This implies that glucose triggers both up- and downregulation of signaling pathways with the main objective of activating the yeast plasma membrane ATPase. Similarly, expression of *ENA1*, encoding the $Na^+(K^+)$ ATPase, is affected by the availability of carbon sources. In comparison with galactose or raffinose, *ENA1* expression is repressed on glucose (Alepuz et al. 1997). Since activation of Pma1 by glucose favors the influx of potassium, the negative regulation of *ENA1* by glucose seems necessary to avoid a futile cycle in which potassium could be excluded from the cell with additional energy expense.

Similarly to amino acids transporters, the hexose transporter genes (*HXT1*, *HXT3* and *GAL2*) could give rise to mutations capable of suppressing the *trk1 trk2* phenotype at limiting potassium (Ko et al. 1993; Liang and Gaber 1996) in a way that is not obligatorily coupled to the binding or transport of glucose.

A remarkable link between carbohydrate metabolism and potassium transport is highlighted by the observation that *TPS1*, a gene encoding trehalose-6-phosphate synthase and known to modulate glucose metabolism, activates Trks and decreases the sensitivity of yeast cells to many toxic cations (Mulet et al. 2004). These effects are independent of Trk1 regulators such as Hal4, Hal5, or the phosphatase calcineurin. Mutants defective in *PGM2* or *HXK2* exhibited *trk1* related phenotypes similar to those of *tps1* mutants. The authors observed a positive correlation between the levels of glucose phosphates (Glc-1-P and Glc-6-P) caused by these mutations and Trk activity, leading to the proposal that these metabolites, directly or indirectly, activate potassium uptake via the Trk system (Mulet et al. 2004). These findings would support the early observation that potassium uptake in yeast is activated by glucose and other fermentable sugars and that phosphorylation of the sugar was sufficient to trigger the activating pathway (Alijo and Ramos 1993). The induction of genes related to trehalose biosynthesis and degradation was also observed in potassium-starved cells (Barreto et al. 2012). The authors suggested that besides of the function of phosphorylated glucose on Trk1 activation, the induction of trehalose related genes could be useful for generation of a futile cycle based in trehalose synthesis and degradation, aiming to down-regulate carbon flux through the upper part of the glycolysis and to decrease the generation of methylglyoxal (Fig. 11.2), as it has been suggested for other stresses (Barreto et al. 2012; Parrou et al. 1997). Methylglyoxal is a by-product of glycolysis, and an excess of methylglyoxal has been related with the production of ROS and activation of oxidative stress response genes (Desai et al. 2010; Maeta et al. 2004). Under potassium starvation accumulation of methylglyoxal, as well as induction of several genes needed for its degradation, accumulation of ROS, and activation of oxidative stress response genes was observed (Barreto et al. 2012). Detoxification of methylglyoxal leads to accumulation of lactate, which can be transformed to pyruvate by the product of *DLD3* (Fig. 11.2), a retrograde up-regulated gene, linking the detoxification of ammonium in potassium starved cells with methylglyoxal degradation.

An additional link between potassium and glucose was observed at the level of Snf1, a protein kinase central to adaptation to glucose depletion. It has been proposed that Snf1 acts as an additional regulator of potassium influx via Trk1 and/or Trk2 (Portillo et al. 2005). Mutant *snf1* cells were found to be unable to fully activate $K^+$ import, and genetic analysis suggested that the weak kinase activity of the non-phosphorylated form of Snf1 activates high-affinity $K^+$ uptake in the presence of glucose. It was proposed that the effect of Snf1 on the activation of Trk1 could probably be mediated by the Snf1-regulated transcription factor Sip4 controlling the expression of an unknown gene or genes (Portillo et al. 2005).

The collective evidence indicates that, besides other essential cellular roles, the availability of potassium impacts in the acquisition of several nutrients (such as phosphate, ammonium or sulfate-containing amino acids). It is worth noting that invasive growth has been related in yeasts to nutrient starvation (particularly of sugar and nitrogen sources). Notably, cells lacking Trk1,2 display an invasive phenotype that is abolished by increasing the potassium concentration in the medium (Gonzalez et al. 2013). A similar effect is observed after hyperactivation of the Ppz phosphatases, which results in blockage of potassium influx. The effect on invasiveness (and flocculation) provoked by decreased potassium availability

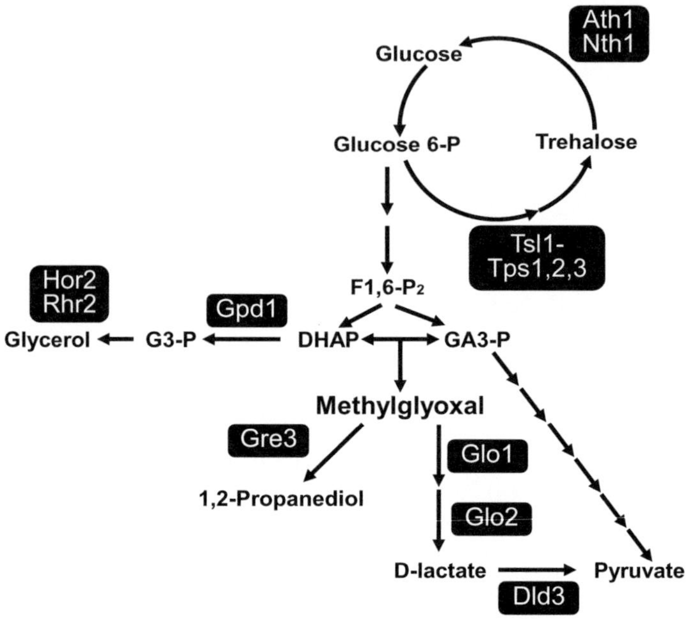


**Fig. 11.2 The generation of methylglyoxal and its link to trehalose metabolism.** Cells subjected to $K^+$ depletion show increased expression of genes required for detoxification of methylglyoxal to lactate (*GLO1* and *GLO2*) or to propanediol (*GRE3*). Similarly, a time-dependent increase in the expression of genes required for trehalose synthesis (*TSL1*, *TPS1*, *TPS2* and *TPS3*) and degradation (*NTH1*, *ATH1*) is observed (Barreto et al. 2012). This would generate a futile cycle based in trehalose synthesis and degradation that might down-regulate carbon flux through the upper part of glycolysis thus decreasing the generation of methylglyoxal. F1-6-$P_2$, fructose-1,6-bisphosphate; GA3-P, glyceraldehyde-3P; DHAP, dihydroxyacetone-P; G3-P, glycerol-3P

appears to be mediated by activation of the Tpk2 isoform of protein kinase A and the transcription factor *FLO11* (Gonzalez et al. 2013). These results further emphasize the link between potassium and nutrient homeostasis in yeast cells.

**Acknowledgments** Work in the laboratory of the authors was supported by grants BFU2011-30197-C3-01, BFU2014-54591-C2-1-P, and EUI2009-04147 (SysMo2) to J.A. (Ministry of Science and Innovation, Spain, and Fondo Europeo de Desarrollo Regional [FEDER]). J.A. is recipient of an Ajut 2014SGR-4 and was awarded an Institució Catalana de Recerca i Estudis Avançats (ICREA) Academia 2009 Award (Generalitat de Catalunya). D.C. was recipient of a predoctoral fellowship from the Spanish Ministry of Education.

## References

Alepuz PM, Cunningham KW, Estruch F (1997) Glucose repression affects ion homeostasis in yeast through the regulation of the stress-activated ENA1 gene. Mol Microbiol 26:91–98

Ali R, Brett CL, Mukherjee S et al (2004) Inhibition of sodium/proton exchange by a Rab-GTPase-activating protein regulates endosomal traffic in yeast. J Biol Chem 279:4498–4506

Alijo R, Ramos J (1993) Several routes of activation of the potassium uptake system of yeast. Biochim Biophys Acta 1179:224–228

Ambesi A, Miranda M, Petrov VV et al (2000) Biogenesis and function of the yeast plasma-membrane H(+)-ATPase. J Exp Biol 203:155–160

Arai H, Pink S, Forgac M (1989) Interaction of anions and ATP with the coated vesicle proton pump. Biochemistry 28:3075–3082

Auesukaree C, Homma T, Kaneko Y et al (2003) Transcriptional regulation of phosphate-responsive genes in low-affinity phosphate-transporter-defective mutants in Saccharomyces cerevisiae. Biochem Biophys Res Commun 306:843–850

Banuelos MA, Sychrova H, Bleykasten-Grosshans C et al (1998) The Nha1 antiporter of Saccharomyces cerevisiae mediates sodium and potassium efflux. Microbiology 144(Pt 10):2749–2758

Barreto L, Canadell D, Petrezselyova S et al (2011) A genomewide screen for tolerance to cationic drugs reveals genes important for potassium homeostasis in Saccharomyces cerevisiae. Eukaryot Cell 10:1241–1250

Barreto L, Canadell D, Valverde-Saubi D et al (2012) The short-term response of yeast to potassium starvation. Environ Microbiol 14:3026–3042

Benito B, Garciadeblas B, Rodriguez-Navarro A (2002) Potassium- or sodium-efflux ATPase, a key enzyme in the evolution of fungi. Microbiology 148:933–941

Bihler H, Slayman CL, Bertl A (1998) NSC1: a novel high-current inward rectifier for cations in the plasma membrane of Saccharomyces cerevisiae. FEBS Lett 432:59–64

Bihler H, Slayman CL, Bertl A (2002) Low-affinity potassium uptake by Saccharomyces cerevisiae is mediated by NSC1, a calcium-blocked non-specificity cation channel. Biochim Biophys Acta 1558:109–118

Boer VM, de Winde JH, Pronk JT et al (2003) The genome-wide transcriptional responses of Saccharomyces cerevisiae grown on glucose in aerobic chemostat cultures limited for carbon, nitrogen, phosphorus, or sulfur. J Biol Chem 278:3265–3274

Borst-Pauwels GW (1981) Ion transport in yeast. Biochim Biophys Acta 650:88–127

Brett CL, Tukaye DN, Mukherjee S et al (2005) The yeast endosomal Na+K+/H+ exchanger Nhx1 regulates cellular pH to control vesicle trafficking. Mol Biol Cell 16:1396–1405

Bun-ya M, Nishimura M, Harashima S et al (1991) The PHO84 gene of Saccharomyces cerevisiae encodes an inorganic phosphate transporter. Mol Cell Biol 11:3229–3238

Bun-ya M, Shikata K, Nakade S et al (1996) Two new genes, PHO86 and PHO87, involved in inorganic phosphate uptake in Saccharomyces cerevisiae. Curr Genet 29:344–351

Butow RA, Avadhani NG (2004) Mitochondrial signaling: the retrograde response. Mol Cell 14:1–15

Buurman ET, Pennock J, Tempest DW et al (1989) Replacement of potassium ions by ammonium ions in different micro-organisms grown in potassium-limited chemostat culture. Arch Microbiol 152:58–63

Cagnac O, Leterrier M, Yeager M et al (2007) Identification and characterization of Vnx1p, a novel type of vacuolar monovalent cation/H+ antiporter of Saccharomyces cerevisiae. J Biol Chem 282:24284–24293

Campetelli AN, Previtali G, Arce CA et al (2005) Activation of the plasma membrane H-ATPase of Saccharomyces cerevisiae by glucose is mediated by dissociation of the H(+)-ATPase-acetylated tubulin complex. FEBS J 272:5742–5752

Canadell D, Gonzalez A, Casado C et al (2015) Functional interactions between potassium and phosphate homeostasis in *Saccharomyces cerevisiae*. Mol Microbiol 95(3):555–572. doi:10.1111/mmi.12886

Cox KH, Pinchak AB, Cooper TG (1999) Genome-wide transcriptional analysis in S. cerevisiae by mini-array membrane hybridization. Yeast 15:703–713

De Nobel JG, Barnett JA (1991) Passage of molecules through yeast cell walls: a brief essay-review. Yeast 7:313–323

Desai KM, Chang T, Wang H et al (2010) Oxidative stress and aging: is methylglyoxal the hidden enemy? Can J Physiol Pharmacol 88:273–284

Estrada E, Agostinis P, Vandenheede JR et al (1996) Phosphorylation of yeast plasma membrane H+-ATPase by casein kinase I. J Biol Chem 271:32064–32072
Gaber RF, Styles CA, Fink GR (1988) TRK1 encodes a plasma membrane protein required for high-affinity potassium transport in Saccharomyces cerevisiae. Mol Cell Biol 8:2848–2859
Garciadeblas B, Rubio F, Quintero FJ et al (1993) Differential expression of two genes encoding isoforms of the ATPase involved in sodium efflux in Saccharomyces cerevisiae. Mol Gen Genet 236:363–368
Gauthier S, Coulpier F, Jourdren L et al (2008) Co-regulation of yeast purine and phosphate pathways in response to adenylic nucleotide variations. Mol Microbiol 68:1583–1594
Gelis S, Curto M, Valledor L et al (2012) Adaptation to potassium starvation of wild-type and K(+)-transport mutant (trk1,2) of Saccharomyces cerevisiae: 2-dimensional gel electrophoresis-based proteomic approach. MicrobiologyOpen 1:182–193
Ghillebert R, Swinnen E, De SP et al (2011) Differential roles for the low-affinity phosphate transporters Pho87 and Pho90 in Saccharomyces cerevisiae. Biochem J 434:243–251
Gomez MJ, Luyten K, Ramos J (1996) The capacity to transport potassium influences sodium tolerance in *Saccharomyces cerevisiae*. FEMS Microbiol Lett 135:157–160
Gonzalez A, Casado C, Petrezselyova S et al (2013) Molecular analysis of a conditional hal3 vhs3 yeast mutant links potassium homeostasis with flocculation and invasiveness. Fungal Genet Biol 53:1–9
Goodman J, Rothstein A (1957) The active transport of phosphate into the yeast cell. J Gen Physiol 40:915–923
Goossens A, de la Fuente N, Forment J et al (2000) Regulation of yeast H(+)-ATPase by protein kinases belonging to a family dedicated to activation of plasma membrane transporters. Mol Cell Biol 20:7654–7661
Haro R, Garciadeblas B, Rodriguez-Navarro A (1991) A novel P-type ATPase from yeast involved in sodium transport. FEBS Lett 291:189–191
Herrera R, Alvarez MC, Gelis S et al (2013) Subcellular potassium and sodium distribution in Saccharomyces cerevisiae wild type and vacuolar mutants. Biochem J 454:525–532
Hess DC, Lu W, Rabinowitz JD et al (2006) Ammonium toxicity and potassium limitation in yeast. PLoS Biol 4, e351
Jazwinski SM (2013) The retrograde response: when mitochondrial quality control is not enough. Biochim Biophys Acta 1833:400–409
Kaffman A, Herskowitz I, Tjian R et al (1994) Phosphorylation of the transcription factor PHO4 by a cyclin-CDK complex, pHO80–PHO85. Science 263:1153–1156
Kahm M, Navarrete C, Llopis-Torregrosa V et al (2012) Potassium starvation in yeast: mechanisms of homeostasis revealed by mathematical modeling. PLoS Comput Biol 8, e1002548
Kane PM (2006) The where, when, and how of organelle acidification by the yeast vacuolar H+-ATPase. Microbiol Mol Biol Rev 70:177–191
Kawasaki-Nishi S, Bowers K, Nishi T et al (2001) The amino-terminal domain of the vacuolar proton-translocating ATPase a subunit controls targeting and in vivo dissociation, and the carboxyl-terminal domain affects coupling of proton transport and ATP hydrolysis. J Biol Chem 276:47411–47420
Ke R, Ingram PJ, Haynes K (2013) An integrative model of ion regulation in yeast. PLoS Comput Biol 9, e1002879
Ketchum KA, Joiner WJ, Sellers AJ et al (1995) A new family of outwardly rectifying potassium channel proteins with two pore domains in tandem. Nature 376:690–695
Klionsky DJ, Herman PK, Emr SD (1990) The fungal vacuole: composition, function, and biogenesis. Microbiol Rev 54:266–292
Ko CH, Gaber RF (1991) TRK1 and TRK2 encode structurally related K+ transporters in Saccharomyces cerevisiae. Mol Cell Biol 11:4266–4273
Ko CH, Buckley AM, Gaber RF (1990) TRK2 is required for low affinity K+ transport in Saccharomyces cerevisiae. Genetics 125:305–312
Ko CH, Liang H, Gaber RF (1993) Roles of multiple glucose transporters in Saccharomyces cerevisiae. Mol Cell Biol 13:638–648

Kornberg A (1995) Inorganic polyphosphate: toward making a forgotten polymer unforgettable. J Bacteriol 177:491–496

Kornberg A, Rao NN, Ault-Riché D (1999) Inorganic polyphosphate: a molecule of many functions. Annu Rev Biochem 68:89–125

Lamb TM, Mitchell AP (2003) The transcription factor Rim101p governs ion tolerance and cell differentiation by direct repression of the regulatory genes NRG1 and SMP1 in Saccharomyces cerevisiae. Mol Cell Biol 23:677–686

Lecchi S, Nelson CJ, Allen KE et al (2007) Tandem phosphorylation of Ser-911 and Thr-912 at the C terminus of yeast plasma membrane H+-ATPase leads to glucose-dependent activation. J Biol Chem 282:35471–35481

Lee YS, Mulugu S, York JD et al (2007) Regulation of a cyclin-CDK-CDK inhibitor complex by inositol pyrophosphates. Science 316:109–112

Lee YS, Huang K, Quiocho FA et al (2008) Molecular basis of cyclin-CDK-CKI regulation by reversible binding of an inositol pyrophosphate. Nat Chem Biol 4:25–32

Li SC, Kane PM (2009) The yeast lysosome-like vacuole: endpoint and crossroads. Biochim Biophys Acta 1793:650–663

Liang H, Gaber RF (1996) A novel signal transduction pathway in Saccharomyces cerevisiae defined by Snf3-regulated expression of HXT6. Mol Biol Cell 7:1953–1966

Liu Z, Butow RA (2006) Mitochondrial retrograde signaling. Annu Rev Genet 40:159–185

Madrid R, Gomez MJ, Ramos J et al (1998) Ectopic potassium uptake in trk1 trk2 mutants of Saccharomyces cerevisiae correlates with a highly hyperpolarized membrane potential. J Biol Chem 273:14838–14844

Maeta K, Izawa S, Okazaki S et al (2004) Activity of the Yap1 transcription factor in Saccharomyces cerevisiae is modulated by methylglyoxal, a metabolite derived from glycolysis. Mol Cell Biol 24:8753–8764

Magasanik B (2003) Ammonia assimilation by Saccharomyces cerevisiae. Eukaryot Cell 2:827–829

Manolson MF, Wu B, Proteau D et al (1994) STV1 gene encodes functional homologue of 95-kDa yeast vacuolar H(+)-ATPase subunit Vph1p. J Biol Chem 269:14064–14074

Maresova L, Sychrova H (2005) Physiological characterization of Saccharomyces cerevisiae kha1 deletion mutants. Mol Microbiol 55:588–600

Marini AM, Soussi-Boudekou S, Vissers S et al (1997) A family of ammonium transporters in Saccharomyces cerevisiae. Mol Cell Biol 17:4282–4293

Martinez P, Persson BL (1998) Identification, cloning and characterization of a derepressible Na+-coupled phosphate transporter in Saccharomyces cerevisiae. Mol Gen Genet 258:628–638

Martinez-Munoz GA, Kane P (2008) Vacuolar and plasma membrane proton pumps collaborate to achieve cytosolic pH homeostasis in yeast. J Biol Chem 283:20309–20319

Miseta A, Kellermayer R, Aiello DP et al (1999) The vacuolar Ca2+/H+ exchanger Vcx1p/Hum1p tightly controls cytosolic Ca2+ levels in S. cerevisiae. FEBS Lett 451:132–136

Mo ML, Palsson BO, Herrgard MJ (2009) Connecting extracellular metabolomic measurements to intracellular flux states in yeast. BMC Syst Biol 3:37

Mulet JM, Leube MP, Kron SJ et al (1999) A novel mechanism of ion homeostasis and salt tolerance in yeast: the Hal4 and Hal5 protein kinases modulate the Trk1-Trk2 potassium transporter. Mol Cell Biol 19:3328–3337

Mulet JM, Alejandro S, Romero C et al (2004) The trehalose pathway and intracellular glucose phosphates as modulators of potassium transport and general cation homeostasis in yeast. Yeast 21:569–582

Navarrete C, Petrezselyova S, Barreto L et al (2010) Lack of main K+ uptake systems in Saccharomyces cerevisiae cells affects yeast performance in both potassium-sufficient and potassium-limiting conditions. FEMS Yeast Res 10:508–517

Ogawa N, Saitoh H, Miura K et al (1995) Structure and distribution of specific cis-elements for transcriptional regulation of PHO84 in Saccharomyces cerevisiae. Mol Gen Genet 249:406–416

Ozcan S, Johnston M (1999) Function and regulation of yeast hexose transporters. Microbiol Mol Biol Rev 63:554–569

Parra KJ, Chan CY, Chen J (2014) Saccharomyces cerevisiae vacuolar H+-ATPase regulation by disassembly and reassembly: one structure and multiple signals. Eukaryot Cell 13:706–714

Parrou JL, Teste MA, Francois J (1997) Effects of various types of stress on the metabolism of reserve carbohydrates in Saccharomyces cerevisiae: genetic evidence for a stress-induced recycling of glycogen and trehalose. Microbiology 143(Pt 6):1891–1900

Pattison-Granberg J, Persson BL (2000) Regulation of cation-coupled high-affinity phosphate uptake in the yeast Saccharomyces cerevisiae. J Bacteriol 182:5017–5019

Perez-Valle J, Jenkins H, Merchan S et al (2007) Key role for intracellular K+ and protein kinases Sat4/Hal4 and Hal5 in the plasma membrane stabilization of yeast nutrient transporters. Mol Cell Biol 27:5725–5736

Perez-Valle J, Rothe J, Primo C et al (2010) Hal4 and Hal5 protein kinases are required for general control of carbon and nitrogen uptake and metabolism. Eukaryot Cell 9:1881–1890

Persson BL, Petersson J, Fristedt U et al (1999) Phosphate permeases of Saccharomyces cerevisiae: structure, function and regulation. Biochim Biophys Acta 1422:255–272

Persson BL, Lagerstedt JO, Pratt JR et al (2003) Regulation of phosphate acquisition in Saccharomyces cerevisiae. Curr Genet 43:225–244

Petrezselyova S, Kinclova-Zimmermannova O, Sychrova H (2013) Vhc1, a novel transporter belonging to the family of electroneutral cation-Cl(-) cotransporters, participates in the regulation of cation content and morphology of Saccharomyces cerevisiae vacuoles. Biochim Biophys Acta 1828:623–631

Platara M, Ruiz A, Serrano R et al (2006) The transcriptional response of the yeast Na+-ATPase ENA1 gene to alkaline stress involves three main signaling pathways. J Biol Chem 281:36632–36642

Popova Y, Thayumanavan P, Lonati E et al (2010) Transport and signaling through the phosphate-binding site of the yeast Pho84 phosphate transceptor. Proc Natl Acad Sci U S A 107:2890–2895

Portillo F (2000) Regulation of plasma membrane H(+)-ATPase in fungi and plants. Biochim Biophys Acta 1469:31–42

Portillo F, Mulet JM, Serrano R (2005) A role for the non-phosphorylated form of yeast Snf1: tolerance to toxic cations and activation of potassium transport. FEBS Lett 579:512–516

Prior C, Potier S, Souciet JL et al (1996) Characterization of the NHA1 gene encoding a Na+/H+-antiporter of the yeast Saccharomyces cerevisiae. FEBS Lett 387:89–93

Ramos J, Contreras P, Rodriguez-Navarro A (1985) A potassium transport mutant of Saccharomyces cerevisiae. Arch Microbiol 143:88–93

Reisser C, Dick C, Kruglyak L et al (2013) Genetic basis of ammonium toxicity resistance in a sake strain of yeast: a Mendelian case. G3 (Bethesda) 3(4):733–740

Rodriguez-Navarro A (2000) Potassium transport in fungi and plants. Biochim Biophys Acta 1469:1–30

Rosenfeld L, Reddi AR, Leung E et al (2010) The effect of phosphate accumulation on metal ion homeostasis in Saccharomyces cerevisiae. J Biol Inorg Chem 15:1051–1062

Ruiz A, Arino J (2007) Function and regulation of the Saccharomyces cerevisiae ENA sodium ATPase system. Eukaryot Cell 6:2175–2183

Saito K, Ohtomo R, Kuga-Uetake Y et al (2005) Direct labeling of polyphosphate at the ultrastructural level in Saccharomyces cerevisiae by using the affinity of the polyphosphate binding domain of Escherichia coli exopolyphosphatase. Appl Environ Microbiol 71:5692–5701

Schmidt G, Hecht TL, Thannhauser SJ (1949) The effect of potassium ions on the absorption of orthophosphate and the formation of metaphosphate by bakers' yeast. J Biol Chem 178(2):733–742

Serra-Cardona A, Petrezselyova S, Canadell D et al (2014) Coregulated expression of the Na+/Phosphate Pho89 transporter and Ena1 Na+-ATPase allows their functional coupling under high-pH stress. Mol Cell Biol 34:4420–4435

Serrano R (1983) In vivo glucose activation of the yeast plasma membrane ATPase. FEBS Lett 156:11–14

Serrano R, Kielland-Brandt MC, Fink GR (1986) Yeast plasma membrane ATPase is essential for growth and has homology with (Na+ + K+), K+- and Ca2+-ATPases. Nature 319:689–693

Serrano R, Ruiz A, Bernal D et al (2002) The transcriptional response to alkaline pH in *Saccharomyces cerevisiae*: evidence for calcium-mediated signalling. Mol Microbiol 46:1319–1333

Seto-Young D, Perlin DS (1991) Effect of membrane voltage on the plasma membrane H(+)-ATPase of Saccharomyces cerevisiae. J Biol Chem 266:1383–1389

Stefan CP, Zhang N, Sokabe T et al (2013) Activation of an essential calcium signaling pathway in Saccharomyces cerevisiae by Kch1 and Kch2, putative low-affinity potassium transporters. Eukaryot Cell 12:204–214

Tate JJ, Cooper TG (2003) Tor1/2 regulation of retrograde gene expression in Saccharomyces cerevisiae derives indirectly as a consequence of alterations in ammonia metabolism. J Biol Chem 278:36924–36933

ter Schure EG, Sillje HH, Verkleij AJ et al (1995) The concentration of ammonia regulates nitrogen metabolism in Saccharomyces cerevisiae. J Bacteriol 177:6672–6675

ter Schure EG, van Riel NA, Verrips CT (2000) The role of ammonia metabolism in nitrogen catabolite repression in Saccharomyces cerevisiae. FEMS Microbiol Rev 24:67–83

Thomas MR, O'Shea EK (2005) An intracellular phosphate buffer filters transient fluctuations in extracellular phosphate levels. Proc Natl Acad Sci U S A 102:9565–9570

Toh-e A, Tanaka K, Uesono Y et al (1988) PHO85, a negative regulator of the PHO system, is a homolog of the protein kinase gene, CDC28, of Saccharomyces cerevisiae. Mol Gen Genet 214:162–164

Tropia MJ, Cardoso AS, Tisi R et al (2006) Calcium signaling and sugar-induced activation of plasma membrane H(+)-ATPase in Saccharomyces cerevisiae cells. Biochem Biophys Res Commun 343:1234–1243

Urech K, Durr M, Boller T et al (1978) Localization of polyphosphate in vacuoles of Saccharomyces cerevisiae. Arch Microbiol 116:275–278

Vargas RC, Garcia-Salcedo R, Tenreiro S et al (2007) Saccharomyces cerevisiae multidrug resistance transporter Qdr2 is implicated in potassium uptake, providing a physiological advantage to quinidine-stressed cells. Eukaryot Cell 6:134–142

Viladevall L, Serrano R, Ruiz A et al (2004) Characterization of the calcium-mediated response to alkaline stress in *Saccharomyces cerevisiae*. J Biol Chem 279:43614–43624

Wright MB, Ramos J, Gomez MJ et al (1997) Potassium transport by amino acid permeases in Saccharomyces cerevisiae. J Biol Chem 272:13647–13652

Wykoff DD, O'Shea EK (2001) Phosphate transport and sensing in Saccharomyces cerevisiae. Genetics 159:1491–1499

Zaman S, Lippman SI, Zhao X et al (2008) How Saccharomyces responds to nutrients. Annu Rev Genet 42:27–81

Zvyagilskaya RA, Lundh F, Samyn D et al (2008) Characterization of the Pho89 phosphate transporter by functional hyperexpression in Saccharomyces cerevisiae. FEMS Yeast Res 8:685–696

# Chapter 12
# Mathematical Modelling of Cation Transport and Regulation in Yeast

**Matthiasé Kahm and Maik Kschischo**

**Abstract** Mathematical modelling of ion transport is a strategy to understand the complex interplay between various ionic species and their transporters. Such models should provide new insights and suggest new interesting experiments. Two essential variables in models for ion transport and control are the membrane potential and the intracellular pH, which generates an additional layer of complexity absent from many other models of biochemical reaction pathways. The aim of this text is to introduce the reader to the basic principles and assumptions of modelling in this field. A simplified model of potassium transport will be used as an example and will be derived in a step by step manner. This forms the basis for understanding the advantages and limitations of more complex models. These are briefly reviewed at the end of this chapter.

**Keywords** Systems biology • Mathematical modelling • Membrane potential • Ion homeostasis • Biological thermodynamics

## 12.1 Why Mathematical Modelling of Cation Transport?

Cation transport in yeast is a well established field (Arino et al. 2010; Cyert and Philpott 2013). As reviewed in this monograph, the most important transport proteins have been characterised and their regulation has been studied with great success. Genome wide transcriptomics and proteomics studies revealed that transport of different ions is highly interdependent and also coupled to various stress responses and metabolic processes (Barreto et al. 2011; Merchan et al. 2011). Nevertheless, the field continuous to throw up surprises (Canadell et al.

M. Kahm
University of Applied Sciences Koblenz, RheinAhrCampus, Joseph-Rovan-Allee 2, 53424 Remagen, Germany
e-mail: matthias.kahm@googlemail.com

M. Kschischo (✉)
Department of Mathematics and Technology, University of Applied Sciences Koblenz, RheinAhrCampus, Remagen, Germany
e-mail: kschischo@rheinahrcampus.de

 
J. Ramos et al. (eds.), *Yeast Membrane Transport*, Advances in Experimental Medicine and Biology 892, DOI 10.1007/978-3-319-25304-6_12

2015). One intriguing example is the recent observation, that potassium availability has a great impact on ethanol production (Lam et al. 2014). These, and other findings, indicate that a detailed mechanistic understanding of phenotypes and their underlying biological processes is only possible if the membrane transport proteins are considered as components of a larger system (Hodgkin and Huxley 1952; Tosteson and Hoffman 1960). This complexity is difficult to accommodate by verbal models alone. The aim of this article is to review recent attempts towards a more holistic understanding of cation transport by means of mathematical modelling.

There are two basic approaches to building mathematical models in molecular systems biology (Snoep and Westerhoff 2005). Bottom-up models start from individual chemical reactions and integrate them into a larger system in order to study the emergent properties generated by the interactions between the different components. Top-down systems biology starts from systems wide data like cell growth and omics measurements and integrates them into phenomenological models to characterise the cells function. An intermediate approach is provided by constraint based models (Bordbar et al. 2014; Mo et al. 2009), which provide a means to integrate metabolomics, transcriptomics and proteomomics data with reconstructed metabolic networks. There is a already a large literature reviewing these modelling approaches (Keener and Sneyd 2004; Kremling 2013; Snoep and Westerhoff 2005). However, two fundamental biophysical variables, namely the membrane potential and the intracellular pH are often ignored. These are crucial to many cellular processes, but essential for the transport of ions.

Incorporating pH and membrane potential adds an additional layer of complexity to biochemical reaction models. This is illustrated in this text by deriving a minimal model for potassium regulation in *Saccharomyces cerevisiae*. This minimal model is not realistic, but it illustrates the structure of mechanistic bottom up models for ion regulation. Understanding this basic model provides a basis for understanding more advanced models, which will briefly be discussed thereafter. To make this paper more self contained, I will start with a very brief summary of the relevant biological knowledge, see also the articles of Lynne Yenusch (Chap. 8) and David Canadell and Joaquín Ariño (Chap. 11) in this monograph.

## 12.2 Potassium Transport Systems in *Saccharomyces Cerevisiae*

Intracellular potassium is essential to many biological processes including protein translation, pyruvate synthesis and regulation of the cell cycle (Arino et al. 2010; Cyert and Philpott 2013; Rodriguez-Navarro 2000). In many environments the potassium concentration is in the micromolar range (Rodriguez-Navarro 2000), whilst intracellular concentrations have to be sufficiently high for survival or even cell growth. How can the cell maintain this minimum intracellular concentration despite a very strong concentration gradient? As it turns out, there is currently no exhaustive answer to this ostensibly simple question.

The transporters involved in potassium transport are well characterised (Arino et al. 2010). The main route for $K^+$ uptake under normal physiological conditions is via the high affinity plasma membrane transporters Trk1 and Trk2, with Trk1 being the more important and higher expressed protein under physiological conditions (Gaber et al. 1988; Ko et al. 1990). In addition, there is a low affinity mode for potassium uptake which is even active in *trk1 trk2* double mutants (Bihler et al. 1998, 2002). This nonspecific cation channel with affinity in the millimolar range is called Nsc1, but the corresponding protein is currently unknown.

Uptake of potassium is only possible, if the net thermodynamic force is inwardly directed. The membrane potential counteracts the concentration gradient and thus facilitates uptake. Most of the energy stored in the membrane potential is generated by the extrusion of protons via the plasma membrane ATPase Pma1 (Serrano 1978). Pma1 is an essential protein, since the membrane potential is needed for many other transport mechanisms. Pma1 is also involved in cytosolic pH regulation.

Potassium can be exported from the cell via the plasma membrane proton antiporter Nha1, which is active under acidic pH and which is also responsible for the extrusion of toxic sodium ions (Martinez-Muñoz and Peña 2005). An alternative route to extrude $K^+$ is provided by the Ena1 transporter, which is only activated under sodium or osmotic stress and especially by alkaline pH (Haro et al. 1991). Ena1 can also export $Na^+$ and facilitates sodium detoxification under these conditions. In addition, potassium can be exported via Tok1, an outward rectifying potassium channel, which is active at positive and low negative membrane voltages (Bertl et al. 1993). Potassium can also be stored in various intracellular membrane organelles by consuming an electrical membrane potential difference between the cytoplasma and the interior of the compartment, which is generated by proton pumping via V-ATPases (Martinez-Muñoz and Peña 2005).

Potassium transport is coupled to proton transport, which in turn is coupled to energy metabolism and intracellular pH regulation. In addition, there is also a surprising link between potassium and phosphate metabolism, which despite of being known for many decades (Schmidt et al. 1949), has only recently been characterised under controlled experimental conditions (Canadell et al. 2015). This is also reviewed in Chap. 11. Thus, potassium regulation is an excellent example where individual membrane transport proteins are well characterised and still the complexity of the various couplings and interactions is not sufficiently understood. The proposition of systems biology is that this complexity can be handled by mathematical models. But, how can we write down equations incorporating the most important aspects of potassium transport?

## 12.3 A Minimal Model for Potassium

The main steps to develop a mechanistic model for ion regulation are now illustrated by means of a minimal model for potassium transport in *Saccharomyces cerevisae*. This minimal model is too simple to describe the whole complexity of potassium transport, but it provides the basic structure of more realistic models reviewed thereafter.

### 12.3.1 The Balance Equations

When modelling potassium regulation, the state variable of primary interest is the intracellular concentration $\left[K^{+}\right]_{i}$ of potassium ions. The systems border in this model is the plasma membrane and extracellular concentrations are treated as parameters or external inputs. The cell is considered as a single compartment and no distinction is made between ions in the cytoplasm and ions in intracellular compartments.

Changes of intracellular potassium are determined by potassium fluxes. This is quantitatively described by the differential equation

$$\frac{d}{dt}\left[K^{+}\right]_{i} = -J_{K},$$

where $J_K$ is the total flux of potassium across the plasma membrane. Here, we adopt the convention that a negative flux ($J_K < 0$) indicates uptake and a positive flux ($J_K > 0$) export. The total potassium flux $J_K$ is given by the sum of the potassium fluxes mediated by the different potassium membrane transport proteins. The model is designed for acidic extracellular conditions, for which we know from biological experience, that the Trk1,2 transporter system, the Nha1 transporter and the Tok1 gated channel are the most important potassium transport routes (Arino et al. 2010). Accordingly, $J_K$ is replaced by

$$\frac{d}{dt}\left[K^{+}\right]_{i} = -\left(J_{K}^{Trk1,2} + J_{K}^{Nha1} + J_{K}^{Tok1} + J_{K}^{Leak}\right). \tag{12.1}$$

Note, that Trk1 and Trk2 are treated as a single system and no distinction is made between these two proteins. The Ena1 transporter is assumed to be inactive under the conditions above (Martinez-Muñoz and Peña 2005). However, there might be alternative routes of potassium transport which can not be modelled in sufficient detail. The term $J_{K}^{Leak}$ is added to mimic these additional potassium fluxes which can not be attributed to any of the three systems Trk1,2, Nha1 or Tok1. All the fluxes depend on the activity of the different transporters as well as on ion concentrations and the membrane potential. This is discussed in Sect. 12.3.2.

Potassium transport is associated with proton transport. Instead of modelling intracellular proton concentrations, we model the dynamic changes of intracellular pH as Boron (2004) and Grabe and Oster (2001)

$$\frac{d}{dt}pH_{i} = \frac{1}{\beta}\left(J_{H}^{Pma1} + J_{H}^{Nha1} + J_{H}^{Leak}\right). \tag{12.2}$$

This equation incorporates the proton fluxes mediated by the proton pump Pma1 and the antiporter Nha1. Protons are involved in the transport of many other substances and nutrients and it would be hopeless to model them all explicitly. Instead, a proton leak current $J_{H}^{Leak}$ as a proxy for these alternative transport routes is added.

The quantity $\beta$ is the proton buffering capacity of the intracellular medium. It is a measure for the change in free proton concentration per change in pH.[1] Although the buffering capacity $\beta$ is in principle itself a function of pH, it can approximately be treated as constant for a large range of physiological intracellular pH values. The reason is that there are many different cellular buffers with different buffer capacities and $\beta$ represents the effective buffering capacity generated by the superposition of buffering effects (Boron 2004).

The plasma membrane potential $V_m$ contributes to the thermodynamic driving force for the transport of ions across the plasma membrane. The membrane potential is generated by charge differences between the outer and inner side of the membrane. If the membrane is considered as a capacitance storing electrical energy we have at every instance

$$c_m V_m = F\mathscr{R}\left(-\left[K^+\right]_i + \beta pH_i - X\right). \tag{12.3}$$

Here, $c_m = 1\frac{\mu F}{cm^2}$ is the specific membrane capacitance, $F$ is the Faraday constant and $\mathscr{R}$ is the ratio of cell volume and cell surface area. The equation simply states the total charge stored on the membrane capacitor is equal to the net charge stored inside the cell. In addition to the charges carried by potassium ions and protons, there are many other charge carriers inside the cell. Their net concentration is denoted by $X$.

It should be noted that Eq. 12.3 is only an approximation and does not imply that the intra- or extracellular medium is charged (Weiss 1996). In fact, electroneutrality is only violated in a very thin layer (on the order 1 nm in width) around the membrane. This means, that the above equation is strictly correct only when the concentrations in the bracket are defined as the total amount of intracellular ions divided by the cell volume. If we consider the concentrations away from the surface charge layer, a more exact description requires the incorporation of surface ionic densities (Mori et al. 2008). However, for our purpose the approximation of Eq. 12.3 is sufficiently accurate.

Concentration differences across membranes generate an osmotic pressure which drives water in the direction to reduce the concentration gradient. Typically, the concentration of solvent is much higher inside the cell than outside. To prevent the cell from bursting, the turgor pressure counteracts the osmotic pressure. The internal osmotic pressure $\Pi_i$, the external osmotic pressure $\Pi_o$ and turgor pressure $\Pi_t$ determine the rate of change of the cell volume $v$ via

$$\frac{dv}{dt} = L_p A\left(\Pi_i - \Pi_o - \Pi_t\right). \tag{12.4}$$

The parameter $L_P$ is called the hydraulic conductivity and describes the resistance of the cell membrane to changes of pressure (Weiss 1996). The turgor pressure can be approximated as a linear function of the cell volume (Gennemark et al. 2006). The

[1]Mathematically, the buffering capacity is defined as $\beta = \frac{d[H^+]}{d(pH)}$.

internal osmotic pressure $\Pi_i$ is a function of the intracellular solute concentration, whereas $\Pi_o$ depends on the extracellular concentrations (Box 12.1).

---

**Box 12.1 Osmotic pressure and turgor pressure**

The relationship between the concentrations $c_l$ of all solutes in the medium and the osmotic pressure $\Pi$ is given by the van't Hoff law (Katchalsky and Curran 1967)

$$\Pi = RT \sum_l n_l c_l.$$

The sum runs over all solutes contributing to the osmotic pressure ($T$ is the absolute temperature, $R$ is the gas constant). The factor $n_l$ is the number of moieties the substance dissociates in. For example, the solute potassium chloride dissociates into the two moieties $K^+$ and $Cl^-$. The equation is a variant of the ideal gas law. For our purposes we can use the expressions

$$\Pi_o = RT\left(\left[K^+\right]_o + \left[H^+\right]_o + [Cl^-]_o + [Z]_o\right)$$

$$\Pi_i = RT\left(\left[K^+\right]_i + \left[H^+\right]_i + [Z]_i\right) \cdot \frac{1}{1 - f_v \frac{v(0)}{v}}.$$

Here, the parameters $Z_o$ and $Z_i$ take the concentration of other intracellular and extracellular osmotic active solutes into account. The factor $f_V \approx \frac{1}{3}$ corrects for the fact that approximately one third of the cell is occupied by cellular structures which are not osmotically active and $v(0)$ is the reference cell volume at the initial time $t = 0$.
The relationship between turgor pressure and cell volume can be modelled by

$$\Pi_t = \frac{\frac{v}{v(0)} - k}{1 - k}.$$

The dimensionless parameter $k \approx 0.99$ is a measure of how sensitive the turgor pressure is against changes in the volume (Gennemark et al. 2006).

---

In summary, the minimal model (Fig. 12.1) incorporates the four dynamic state variables $[K^+]_i$, $pH_i$, $V_m$ and $v$. Equations (12.1), (12.2), (12.3) and (12.4) describe the dynamics of these four state variables as function of time. Mathematically, we have a differential-algebraic system consisting of the three differential equations (12.1), (12.2) and (12.4) complemented by one algebraic equation (12.3). In addition, the model requires initial conditions for the four state variables. To completely specify the model, we need to express the right hand sides of these equations as function of the state variables. This is the subject of the next section.

### *12.3.2 Modelling Individual Transport Systems*

Transport proteins have a three dimensional structure which determines their specificity for the transported substances or ions. Structural information could in principle be used to develop detailed kinetic models of the transport proteins.

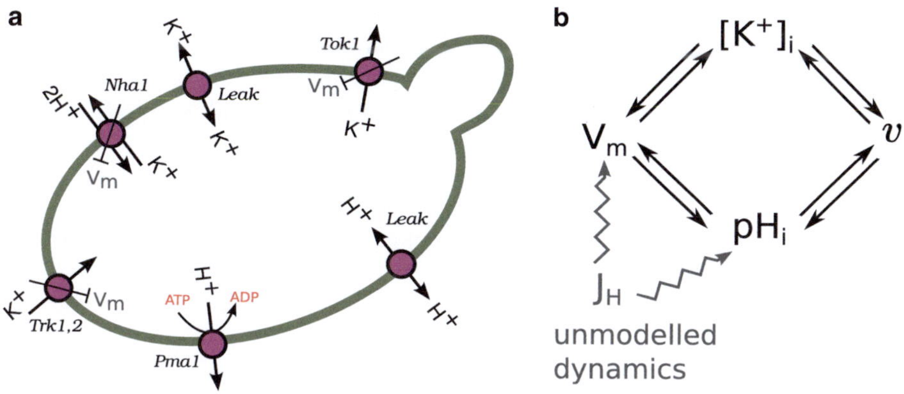


**Fig. 12.1** The minimal model for potassium transport. (**a**) The model incorporates the transport proteins Pma1, Trk1,2, Nha1 and Tok1. The potassium transporters Trk1 and Trk2 are considered as a single system. Proton or potassium fluxes mediated by other systems are modelled as leaks. (**b**) The dynamic variables of the minimal model are the intracellular potassium concentration $[K^+]_i$, the intracellular ph $pH_i$, the membrane voltage $V_m$ and the cell volume $v$. The graph illustrates the interactions between these variables. Processes not covered by the minimal model are shown in opaque. These processes are proton fluxes activated under potassium starvation conditions, see the discussion in Sect. 12.4

However, this approach is often not feasible, because either the detailed molecular mechanisms of transport are not known or the parameters of such a model are not easily accessible. Thus, we have to rely on much simpler phenomenological models. Fortunately, the basic laws of thermodynamics provide a useful constraint to derive approximate descriptions of membrane transport systems.

The thermodynamic driving force for the $j$-th ion is given by the gradient of the electrochemical potential (Katchalsky and Curran 1967)

$$\mu_j = RT \ln\left(\frac{[j]_i}{[j]_0}\right) + z_j F V_m \tag{12.5}$$

relative to the cells exterior. The external and internal concentration of the ion $j$ is denoted by $[j]_o$ and $[j]_i$ respectively, $z_j$ is the valency and $RT$ is the ideal gas constant times absolute temperature. The electrochemical potential is related to the Gibbs free energy

$$\Delta G_j = q_e \frac{\mu_j}{z_j F} = q_e(V_m - E_j), \tag{12.6}$$

which can be interpreted as the work performed by transporting the ion $j$ across the membrane (Endresen et al. 2000; Katchalsky and Curran 1967). Here, $q_e$ is the elementary charge and

$$E_j = \frac{RT}{z_j F} \ln\left(\frac{[j]_i}{[j]_0}\right) \tag{12.7}$$

denotes the equilibrium potential. The electrochemical potential (12.5) and the Gibbs free energy (12.6) for this ion $j$ are both zero for $V_m = E_j$.

In our basic model above, only protons and potassium ions are considered. Thus, we assume that all the fluxes for potassium ions ($j = K^+$) and protons ($j = H^+$) are functions of the electrochemical potential for potassium $\mu_K$ and protons $\mu_H$. Equivalently, we can replace the electrochemical potential by the Gibbs Free energy expressions of Eq. 12.5. If the transport protein is an uniporter for the ion $j$, then the flux depends only on $\Delta G_j$. We make this strong modelling assumption (Arino et al. 2010) for the case of the Trk1,2 potassium transporter. Accordingly we have

$$J_K^{Trk1,2} = J_K^{Trk1,2}(\Delta G_K),$$

with

$$J_K^{Trk1,2}(0) = 0.$$

The last equation represents the constraint that the transport of potassium requires a thermodynamic driving force and that the flux is zero if this force is zero. The simplest possible function having this property is a proportional relationship of the form $J_K^{Trk1,2}(\Delta G_K) \propto \Delta G_K$. Equivalently, we can express this as

$$J_K^{Trk1,2} \propto (V_m - E_K). \tag{12.8}$$

It is customary to specify this relation for the electrical current $I_K^{Trk1,2}$ instead of the mass flux $J_K^{Trk1,2}$. Both are related by a simple factor

$$I_K^{Trk1,2} = z_K \cdot F \cdot \frac{v}{A} \cdot J_K^{Trk1,2} = g^{Trk1,2} \cdot (V_m - E_K), \tag{12.9}$$

which depends on the valency $z_K = +1$, on the cell volume $v$ and on the surface area $A$ of the cell. The quantity $g^{Trk1,2}$ is the electrical resistance of the Trk1,2 system. If it were constant, then the electrical or mass current would change sign depending on the sign of $(V_m - E_K)$. However, electrophysiological measurements indicate that the current for $Trk1,2$ is almost unidirectional (Kuroda et al. 2004). This inward rectification can be incorporated by considering $g_{Trk1,2}$ as a function of the membrane voltage $V_m$. It is possible to fit the electrophysiological data (Kahm et al. 2012; Kuroda et al. 2004) in by an equation of the form [2]

$$I_K^{Trk1,2} = \underbrace{\frac{\gamma^{Trk1,2}}{1 + \exp[d \cdot (V_m - U_{1/2}^{Trk1,2})]}}_{g^{Trk1,2}(V_m)} (V_m - E_K) \tag{12.10}$$

[2]This shape can be motivated by a simple two state model of voltage dependent switching, see e.g. for a derivation (Keener and Sneyd 2004; Weiss 1996).

$$E_K = \frac{RT}{zF} \ln\left(\frac{[K^+]_i}{[K^+]_o}\right). \tag{12.11}$$

The function $g^{Trk1,2}(V_m)$ has a sigmoidal shape and generates a current voltage relationship of the form depicted in the left upper corner of Fig. 12.2. The details depend on the three positive parameters $d$, $U_{1/2}^{Trk1,2}$ and $\gamma^{Trk1,2}$. In any case, the current $I_K^{Trk1,2}(V_m)$ is almost zero for membrane voltages $V_m \gg U_{1/2}^{Trk1,2}$ and asymptotically linear for negative values $V_m \to -\infty$. Thus, the transport system *Trk*1, 2 is modelled as an inwardly directed voltage gated rectifier. For the potassium current $I_K^{Tok1}$ of the outward rectifying Tok1 channel (Bertl et al. 1993) it is reasonable to assume the same functional form, but with different parameters and in particular a negative $U_{1/2}^{Tok1}$, see Fig. 12.2.

The same idea can also be applied to the other passive or secondary active transport systems by using an equation of the general form

$$I = z \cdot F \cdot \frac{v}{A} \cdot J = g \cdot (V_m - E). \tag{12.12}$$

Here, $I$ and $J$ are the electrical and the corresponding mass current respectively; and $z$ is the valency of the ion. The equilibrium potential $E$ depends on the different ions transported and on the stoichiometry of the transport reaction. For symporters and antiporters, $E$ is a function of both $\Delta G_H$ and $\Delta G_K$. For example, *Nha*1 is believed to extrude one potassium ion in exchange for two protons. The total work is the sum of the work performed by exporting one potassium ion (the stoichiometric coefficient is $+1$) and importing two protons (the stoichiometric coefficient is $-2$). Thus,

$$\Delta G^{Nha1} = 1 \cdot \Delta G_k + (-2) \cdot \Delta G_H \tag{12.13}$$

$$= q_e(V_m - E_H) - 2q_e(V_m - E_K). \tag{12.14}$$

By inserting the expressions for $E_H$ and $E_K$ using Eq. (12.6) we obtain

$$\Delta G^{Nha1} = q_e(V_m - E_{Nha1}) \tag{12.15}$$

$$E_{Nha1} = \frac{RT}{F} \ln \frac{[H^+]_o^2 \cdot [K^+]_i}{[H^+]_i^2 \cdot [K^+]_o}. \tag{12.16}$$

By virtue of the assumed general form in Eq. 12.12 we find for the potassium and proton current mediated by Nha1

$$I_K^{Nha1} = -g^{Nha1}(V_m - E_{Nha1}) \tag{12.17}$$

$$I_H^{Nha1} = -2I_K^{Nha1}. \tag{12.18}$$

The last equation follows from the antiport of protons and potassium in the stoichiometric ratio 2:1.

For Nha1 it is not *a priory* clear whether the conductivity $g^{Nha1}$ is constant or whether it is regulated by the membrane potential (Kahm et al. 2012). Both situations are compared in the second row of Fig. 12.2. For the leak currents, it is reasonable to assume constant conductivity.

These examples illustrate the idea of modelling passive transport using Eq. 12.12. First, the resting potential $E$ is determined by thermodynamic considerations. Second, the conductivity $g$ is modified to incorporate information about the direction or activity of transport. In our simple model, modulations of the conductivity by the membrane potential were considered. However, in other models it is also conceivable to take concentration dependence of the conductivity into account, e.g. by including Michaelis-Menten- or Hill-expressions (Kremling 2013). In addition, known regulatory control of transporter activity by other signalling proteins could in principle be included. This provides a principled way to integrate transporter models with models of signal transduction networks.

Modelling the active transport of protons via Pma1 requires the incorporation of the ATP dependent thermodynamic and activation effects. One principled way to deduce a model for Pma1 is described in Box 12.2. All the current voltage relationships are plotted in Fig. 12.2.

## 12.4 Predictive Models of Ion Transport in Yeast

Ionic, pH and membrane voltage regulation are fundamental processes which have been studied theoretically for years in various organisms (Armstrong 2003; Tosteson and Hoffman 1960) including plants (Gradmann 2001), mammals (Mori et al. 2008) and animals (Weinstein and Sontag 2009). Surprisingly, little has been published for yeast. The basic model described in Sect. 12.3 was taken as a starting point in Kahm et al. (2012) to answer the question: How can yeast cells manage to maintain a minimal intracellular concentration of potassium under conditions of low external potassium? The data basis for this model came from potassium starvation experiments (Navarrete et al. 2010), where yeast cells grown in media with sufficiently high potassium concentrations (50 mM) were transferred to potassium free medium. Intracellular potassium concentrations and cellular volume was determined and compared to the prediction of model simulations. Initially, these comparisons produced disappointing results. The measurements showed that the intracellular potassium concentration rapidly fell during the first 60 min after starvation, but that the cells were able to maintain a minimal intracellular concentration for 4–5 h (Herrera et al. 2014; Kahm et al. 2012; Navarrete et al. 2010). In contrast, the *in silico* simulations predicted that potassium would be lost completely or almost completely within 10–20 min. These differences indicated that an important biological process is not covered by this model and that additional regulatory responses must be present in reality.

**Box 12.2 Modelling active transport via Pma1**

The Gibbs free energy of ATP-driven proton pumping

$$\Delta G_{Pma1} = q_e \left( V_m - \left( E_H + \frac{\Delta G_{ATP}}{F} \right) \right)$$

against the membrane potential $V_m$ depends on the equilibrium potential $E_H$ for protons (see Eq. 12.7 with $j = H^+$) and on the Gibbs free energy $\Delta G_{ATP} \approx -40\,\mathrm{kJ/mol}$ of ATP hydrolysis (Rodriguez-Navarro 2000). Following (Endresen et al. 2000) we assume that the ratio of forward and backward transport rates $k_+$ and $k_-$ is given by the Boltzmann-distribution

$$\frac{k_+}{k_-} = \exp\left( \frac{-\Delta G_{Pma1}}{RT} \right).$$

This reflects the thermodynamic requirement that $k_+ = k_-$ for $\Delta G_{Pma1} = 0$. A further assumption is that the sum of the forward and backward rate

$$\lambda = k_+ + k_-$$

is a constant. The electrical current carried by the protons driven by Pma1 is proportional to the difference between the forward and the backward rates

$$I_H^{Pma1} = I_{max}^{Pma1} (k_+ - k_-).$$

The last three equations can be used to derive the expression.

$$I_H^{Pma1} = I_{max}^{Pma1} \tanh\left( \frac{F}{2RT} \left( V_m - E_H - \frac{\Delta G_{ATP}}{F} \right) \right) \tag{12.19}$$

The parameter $I_{max}^{Pma1}$ is maximum pump current of $Pma1$.

Initial suggestions came from the experimental side. It was known for a long time that both the apparent $V_{max}$ and the apparent affinity of the main potassium uptake system Trk1 increase strongly with decreasing potassium concentrations (Rodríguez-Navarro and Ramos 1984). However, even when this additional knowledge was built into the model, the predictions did not change. The dynamics of potassium loss was largely insensitive to increased Trk1,2 transport activity (increased conductivity $g_K^{Trk1,2}$) and the very strong potassium gradient could not be compensated by any conceivable manipulation of the Trk1,2 system in the model. An initial analysis of the intracellular charge balance suggested, that additional proton charges are required to compensate the potassium gradient. For this reason, the model was complemented by a model for the bicarbonate system which is the major intracellular buffer for protons. But even with this amendment, the predicted loss of potassium was much faster than in reality.

Progress was made, when the model was systematically searched for dynamic external inputs which were able to generate a sufficient fit to the experimental time course data. The basic idea was to identify the points of application of these dynamic

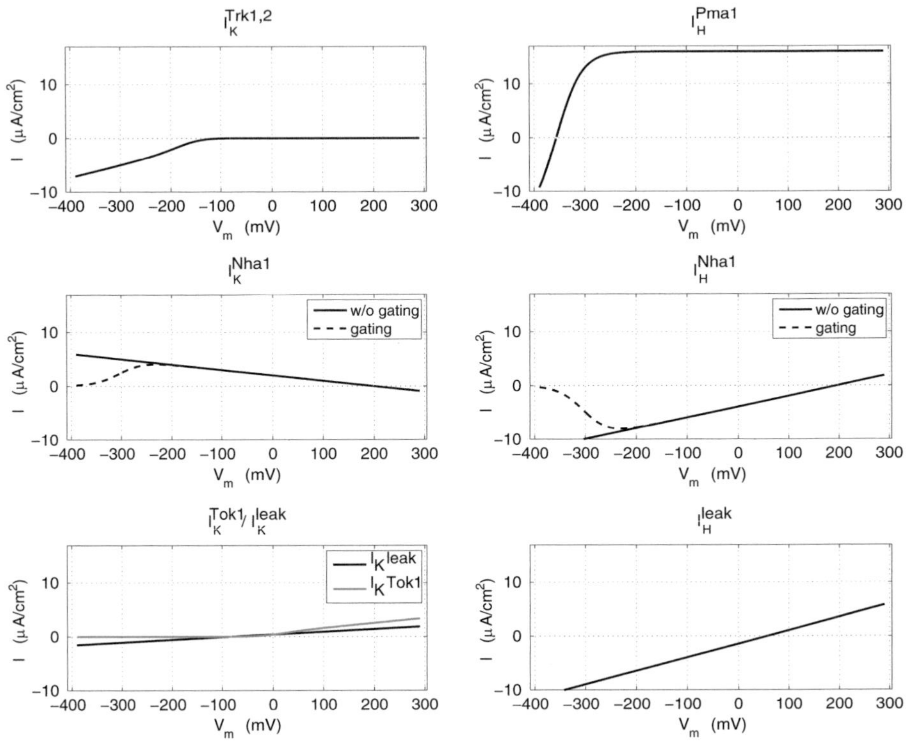


**Fig. 12.2** Current voltage relationships for the transport systems of the minimal potassium model. The proton and potassium leaks are modelled by constant conductivities. Voltage gating was assumed for Trk1,2 and Tok1. For Nha1, the current voltage relationship for a model with constant conductivity (*solid line*) and with a voltage gated conductivity (*dashed line*) is compared. Parameter values can be found in Kahm et al. (2012)

model inputs which were sufficient to explain the data (see Fig. 12.1b). This was done with a control theoretic approach called the reverse tracking algorithm. Interestingly, this approach predicted a transient activation of the bicarbonate reaction system and an activation of proton pumping via Pma1 in response to potassium starvation. Careful experiments confirmed that the ATPase activity of Pma1 and the mRNA expression of the gene NCE103 coding for carbonic anhydrase are elevated in response to shortage of potassium. An immediate question for future research arises from these findings: How is this response triggered and what senses external potassium?

The second published model (Ke et al. 2013) integrates transporters, regulatory molecules, signaling pathways and changes in cell volume. For the ion regulation part, the authors employ similar expressions as derived in Sect. 12.3. However, the model contains many more components and incorporates also sodium and calcium ions and signalling effects from Hog1p, calcineurin (CN), Ppz proteins and Nrg1p.

The authors considered four stress conditions: NaCl, osmotic, KCl and alkaline pH stresses.

The model predicts, that phosphorylation of Hog1p in response to osmotic stress inhibits the activity of Tok1 to depolarise the plasma membrane. This prevents the influx of toxic sodium. Based on their simulations the authors speculate, that Hog mediated depolarisation of the cell membrane is a protective short term response to prevent uptake of toxic compounds.

This short term response in not specific for the type of osmotic stress. However, the long term adaptation of the model to NaCl stress was found to be different from sorbitol stress. In the simulations, initial responses to NaCl stress are by Nha1 upregulation, which is then replaced by upregulation of Ena1. Thus, the simulations suggest that upregulation of Nha1 acts as a short term response to NaCl stress. This high Nha1 activity can not be sustained for long, because it leads to an acidification of the intracellular medium. Instead, Ena1 is transcriptionally activated and replaces Nha1 activity. According to Ke et al. (2013), Ena1 plays also an important role in the response to KCl stress and low external pH by establishing the membrane potential to maintain normal uptake of nutrients through secondary transporters. These prediction are very interesting and plausible, but are still awaiting experimental validation.

## 12.5 Conclusion

Constructing a useful mathematical model is not simple. The model has to capture the key processes and to omit unimportant details. This difficulty is most severe, when fundamental cell biological states (e.g. ph and membrane potential) and processes are involved, which influence a variety of cell biology functions. Thus, there are many important processes which can not all be modelled. One principle way to overcome this problem is to restrict the model to the system of primary interest. This small model might possibly underfit the data, but this might be informative in and of itself. The next step would be to consider all the unmodelled dynamics as external inputs and try to estimate these from the data. This approach was taken in Kahm et al. (2012), where the stimulation of proton production and pump activity was identified as one important response to potassium shortage. Other approaches aim to generate several models and to automatically select a subset of them which robustly explain the data. For both, advanced computational algorithms are currently under development (Babtie et al. 2014).

Modelling ion transport requires combining biological knowledge with biophysical, mathematical and computational skills. It is hoped, that this article will help modellers and biologists understand some principles of mechanistic modelling of ion transport and to pursue new exciting research in the future.

**Acknowledgements** Maik Kschischo and Matthias Kahm were supported by BMBF grant 0315786C (SysMo2/Translucent 2). Both authors thank the members of the TRANSLUCENT consortium for helpful discussions.

## References

Arino J, Ramos J, Sychrova H (2010) Alkali-metal-cation transport and homeostasis in yeasts. Microbiol Mol Biol Rev 74:95–120

Armstrong CM (2003) The na/k pump, cl ion, and osmotic stabilization of cells. Proc Natl Acad Sci 100:6257–6262

Babtie AC, Kirk P, Stumpf MPH (2014) Topological sensitivity analysis for systems biology. Proc Natl Acad Sci 111:18507–18512

Barreto L, Canadell D, Petrezsélyová S, Navarrete C, Maresová L et al (2011) A genomewide screen for tolerance to cationic drugs reveals genes important for potassium homeostasis in saccharomyces cerevisiae. Eukaryot Cell 10:1241–1250

Bertl A, Slayman CL, Gradmann D (1993) Gating and conductance in an outward-rectifying $k^+$ channel from the plasma membrane of saccharomyces cerevisiae. J Membr Biol 132:183–199

Bihler H, Slayman CL, Bertl A (1998) NSC1: a novel high-current inward rectifier for cations in the plasma membrane of saccharomyces cerevisiae. FEBS Lett 432:59–64

Bihler H, Slayman CL, Bertl A (2002) Low-affinity potassium uptake by saccharomyces cerevisiae is mediated by NSC1, a calcium-blocked non-specific cation channel. Biochim Biophys Acta 1558:109–118

Bordbar A, Monk JM, King ZA, Palsson BO (2014) Constraint-based models predict metabolic and associated cellular functions. Nat Rev Genet 15:107–120

Boron WF (2004) Regulation of intracellular pH. Adv Physiol Educ 28:160–179

Canadell D, González A, Casado C, Ariño J (2015) Functional interactions between potassium and phosphate homeostasis in *S accharomyces cerevisiae*: functional $k^+$ and phosphate interactions in yeast. Mol Microbiol 95:555–572

Cyert MS, Philpott CC (2013) Regulation of cation balance in saccharomyces cerevisiae. Genetics 193:677–713

Endresen LP, Hall K, Høye JS, Myrheim J (2000) A theory for the membrane potential of living cells. Eur Biophys J 29:90–103

Gaber RF, Styles CA, Fink GR (1988) TRK1 encodes a plasma membrane protein required for high-affinity potassium transport in saccharomyces cerevisiae. Mol Cell Biol 8:2848–2859

Gennemark P, Nordlander B, Hohmann S, Wedelin D (2006) A simple mathematical model of adaptation to high osmolarity in yeast. Silico Biol (Gedrukt) 6:193–214

Grabe M, Oster G (2001) Regulation of organelle acidity. J Gen Physiol 117:329–344

Gradmann DD (2001) Models for oscillations in plants. Aust J Plant Physiol 28:577–590

Haro R, Garciadeblas B, Rodríguez-Navarro A (1991) A novel p-type ATPase from yeast involved in sodium transport. FEBS Lett 291:189–191

Herrera R, Álvarez MC, Gelis S, Kodedová M, Sychrová H et al (2014) Role of saccharomyces cerevisiae trk1 in stabilization of intracellular potassium content upon changes in external potassium levels. Biochim Biophys Acta – Biomembranes 1838:127–133

Hodgkin AL, Huxley AF (1952) A quantitative description of membrane current and its application to conduction and excitation in nerve. J Physiol (Lond) 117:500–544

Kahm M, Navarrete C, Llopis-Torregrosa V, Herrera R, Barreto L et al (2012) Potassium starvation in yeast: mechanisms of homeostasis revealed by mathematical modeling. PLoS Comput Biol 8:e1002548

Katchalsky A, Curran PF (1967) Nonequilibrium thermodynamics in Biophysics. Harvard University Press, Cambridge

Ke R, Ingram PJ, Haynes K (2013) An integrative model of ion regulation in yeast. PLoS Comput Biol 9:e1002879

Keener J, Sneyd J (2004) Mathematical physiology. Springer, New York

Ko CH, Buckley AM, Gaber RF (1990) TRK2 is required for low affinity $k^+$ transport in saccharomyces cerevisiae. Genetics 125:305–312

Kremling A (2013) Systems biology: mathematical modeling and model analysis. Chapman and Hall/CRC Mathematical and Computational Biology. https://www.crcpress.com/Systems-Biology-Mathematical-Modeling-and-Model-Analysis/Kremling/9781466567894

Kuroda T, Bihler H, Bashi E, Slayman C, Rivetta A (2004) Chloride channel function in the yeast trk-potassium transporters. J Membr Biol 198:177–192

Lam FH, Ghaderi A, Fink GR, Stephanopoulos G (2014) Engineering alcohol tolerance in yeast. Science 346:71–75

Martinez-Muñoz GA, Peña A (2005) In situ study of $k^+$ transport into the vacuole of saccharomyces cerevisiae. Yeast 22:689–704

Merchan S, Pedelini L, Hueso G, Calzada A, Serrano R et al (2011) Genetic alterations leading to increases in internal potassium concentrations are detrimental for DNA integrity in saccharomyces cerevisiae. Genes Cells 16:152–165

Mo ML, Palsson B, Herrgård MJ (2009) Connecting extracellular metabolomic measurements to intracellular flux states in yeast. BMC Syst Biol 3:37

Mori Y, Fishman GI, Peskin CS (2008) Ephaptic conduction in a cardiac strand model with 3d electrodiffusion. Proc Natl Acad Sci 105:6463–6468

Navarrete C, Petrezsélyová S, Barreto L, Martínez JL, Zahrádka J et al (2010) Lack of main k+ uptake systems in saccharomyces cerevisiae cells affects yeast performance in both potassium-sufficient and potassium-limiting conditions: potassium transport and yeast physiology. FEMS Yeast Res 10:508–517

Rodriguez-Navarro A (2000) Potassium transport in fungi and plants. Biochim Biophys Acta 1469:1–30

Rodríguez-Navarro A, Ramos J (1984) Dual system for potassium transport in saccharomyces cerevisiae. J Bacteriol 159:940–945

Schmidt G, Hecht L, Thannhauser SJ (1949) The effect of potassium ions on the absorption of orthophosphate and the formation of metaphosphate by bakers' yeast. J Biol Chem 178:733–742

Serrano R (1978) Characterization of the plasma membrane ATPase of saccharomyces cerevisiae. Mol Cell Biochem 22:51–63

Snoep JL, Westerhoff HV (2005) From isolation to integration, a systems biology approach for building the silicon cell. In: Alberghina L, Westerhoff H (eds) Systems biology, vol 13. Springer, Berlin/Heidelberg, pp 13–30. http://www.springerlink.com/index/10.1007/b106456.00000

Tosteson DC, Hoffman JF (1960) Regulation of cell volume by active cation transport in high and low potassium sheep red cells. J Gen Physiol 44:169–194

Weinstein AM, Sontag ED (2009) Modeling proximal tubule cell homeostasis: tracking changes in luminal flow. Bull Math Biol 71:1285–1322

Weiss TF (1996) Cellular biophysics – transport. Bradford, Cambridge

# Chapter 13
# Transport Systems in Halophilic Fungi

**Ana Plemenitaš, Tilen Konte*, Cene Gostinčar*, and Nina Gunde Cimerman**

**Abstract** Fungi that tolerate very high environmental NaCl concentrations are good model systems to study mechanisms that enable them to endure osmotic and salinity stress. The whole genome sequences of six such fungal species have been analysed: *Hortaea werneckii, Wallemia ichthyophaga* and four *Aureobasidium* spp.: *A. pullulans, A. subglaciale, A. melanogenum* and *A. namibiae*. These fungi show different levels of halotolerance, with the presence of numerous membrane transport systems uncovered here that are believed to maintain physiological intracellular concentrations of alkali metal cations. Despite some differences, the intracellular cation contents of *H. werneckii, A. pullulans* and *W. ichthyophaga* remain low even under extreme extracellular salinities, which suggests that these species have efficient cation transport systems. We speculate that cation transporters prevent intracellular accumulation of $Na^+$, and thus avoid the toxic effects that such $Na^+$ accumulation would have, while also maintaining the high $K^+/Na^+$ ratio that is required for the full functioning of the cell – another crucial task in high-$Na^+$ environments. This chapter primarily summarises the cation transport systems of these selected fungi, and it also describes other membrane transporters that might be involved in their mechanisms of halotolerance.

**Keywords** Extremely halotolerant *H. werneckii* • Halotolerant *Aurobasidium* spp. • Halophilic *W. ichthyophaga* • Cation transporters • Mechanisms of halotolerance

---

*Author contributed equally with all other contributors.

A. Plemenitaš (✉) • T. Konte
Institute of Biochemistry, Faculty of Medicine, University of Ljubljana, Ljubljana, Slovenia
e-mail: ana.plemenitas@mf.uni-lj.si

C. Gostinčar
National Institute of Biology, Večna pot 111, SI-1000 Ljubljana, Slovenia

N.G. Cimerman
Biology Department, Biotechnical Faculty, University of Ljubljana, Ljubljana, Slovenia

Centre of Excellence for Integrated Approaches in Chemistry and Biology of Proteins (CIPKeBiP), Jamova 39, SI-1000 Ljubljana, Slovenia

 
J. Ramos et al. (eds.), *Yeast Membrane Transport*, Advances in Experimental Medicine and Biology 892, DOI 10.1007/978-3-319-25304-6_13

## 13.1 Introduction

Exposure to high salinity represents osmotic stress as well as ionic stress for cells. Osmotic stress triggers water efflux from the cell, which results in reduction of turgor pressure and dehydration of the cytoplasm, thereby increasing the solute concentrations in the cell cytosol. Ionic stress on the other hand is caused by the entry of $Na^+$ into the cell cytoplasm, and this increased intracellular $Na^+$ concentration can damage membrane systems and cytosolic proteins.

While in eukaryotes a large part of our knowledge on adaptation to high salinity originates from studies of *Saccharomyces cerevisiae*, it has to be kept in mind that this species is relatively salt-sensitive and thus cannot adapt to hypersaline conditions. The opposite is true for some fungi that are found in natural and anthropogenic hypersaline habitats, such as coastal evaporation ponds or solar salterns, and thus these are much more suitable model organisms to study halotolerance in eukaryotes than *S. cerevisiae*. Some of these fungal species, such as the black ascomycetous yeast *Aureubasidium pullulans* and *Hortaea werneckii*, and the basidiomycetous *Wallemia ichthyophaga*, thrive under conditions characterised by extremes of salinity, UV irradiation, and pH that prohibit the growth of most other eukaryotes (as well as the majority of prokaryotes). Furthermore, these fungi are separated by large evolutionary distances, and they appear to have developed different strategies to cope with the same problems of ion toxicity and loss of water (Gunde-Cimerman et al. 2009; Gostinčar et al. 2011; Plemenitas et al. 2014).

In hypersaline waters worldwide, melanised black yeast are the most consistently occurring and most abundant fungal group (Gunde-Cimerman et al. 2000, 2004). Among these black yeast (indeed, also considering all fungi), *H. werneckii* is the most successful in adapting to the whole salinity range. *H. werneckii* has adapted to fluctuating salt concentrations to the extent that it can grow without salt as well as in almost saturated NaCl solutions. In the upper salinity range, *H. werneckii* is outcompeted by the obligatory halophilic fungus *W. ichthyophaga*, a rare case of a basidiomycetous halophile that can grow even in media saturated with NaCl. However, *W. ichthyophaga* does not grow in media with less than ~1.5 M salt, a trait usually seen in halophilic Archaea, and not in fungi. The black yeast *A. pullulans* is extreme in its own way – while *A. pullulans* can grow in media with up to only 3 M NaCl, it can tolerate many other types of stress as well, and hence it is considered to be polyextremotolerant (Gostinčar et al. 2011).

Three varieties that were previously described as *A. pullulans* were recently defined as separate species that can be distinguished by differences in their growth limits (among other factors): *Aureubasidium subglaciale*, *Aureubasidium melanogenum* and *Aureubasidium namibiae*. In contrast to *A. pullulans*, other *Aureubasidium* spp. have a growth limit slightly above 1.5 M NaCl. *A. melanogenum* has been isolated mainly from oligotrophic, aqueous environments, and it grows at 37 °C, while these other *Aureubasidium* spp. cannot. On the other hand, *A. subglaciale* is unique in terms of its psychrotolerant nature (i.e., it can grow at 4 °C) and its exclusive occurrence in glacial habitats. Finally, *A. namibiae* was named based on a single isolate from Namib Desert marble (Zalar et al. 2008). These species use the strategy of compatible solutes for osmoadaptation, and they accumulate glycerol as the main compatible solute at increased salinities (Gunde-Cimerman et al. 2009; Zajc et al. 2014).

Although these fungi all thrive in similar hypersaline environments, they show different levels of halotolerance: as indicated, *A. pullulans* can grow only up to 3 M NaCl; *H. werneckii* is interesting due to its extreme halotolerance, as it can withstand a wide range of salt concentrations, almost up to saturation point; and *W. ichthyophaga*, on the other hand, is extreme in a different way, as it is an obligate halophilic specialist, because as well as growing in media saturated with NaCl, *W. ichthyophaga* requires at least 1.5 M NaCl, and is therefore a true halophile. *W. ichthyophaga* also shows greater growth under high concentrations of NaCl, compared to non-ionic solutes (Zajc et al. 2014).

In growing in high-salt media, these fungi need to maintain an intracellular osmotic pressure that matches or exceeds that of the environment, which is accompanied by the constant entry of alkali metal cations, such as $Na^+$. These alkali metal cations can be toxic to several cell components, including intracellular membranes and enzymes, and they also replace the intracellular $K^+$, which is crucial for full functioning of the cell machinery. Although these fungal species can all maintain low intracellular cation levels, they show differences in the mechanisms behind the management of this ion homeostasis. When compared to the extremely halotolerant *H. werneckii* and the moderately halotolerant *A. pullulans*, in the halophilic *W. ichthyophaga* the intracellular $K^+/Na^+$ ratio is higher across the whole salinity range of the external media. Moreover, the growth of *W. ichthyophaga* is greatest when the $Na^+$ content exceeds that of $K^+$ (Zajc et al. 2014), which suggests that $Na^+$ ions are less toxic to *W. ichthyophaga* cells.

Recently the whole genomes of *H. werneckii*, *W. ichthyophaga* and these four *Aureobasidium* spp. (*A. pullulans*, *A. subglaciale*, *A. melanogenum*, *A. namibiae*) were sequenced (Lenassi et al. 2013; Zajc et al. 2013; Gostinčar et al. 2014). Analysis of these fungal genomes has provided better insights into the mechanisms behind this halotolerance. It should also be stressed that the large genetic redundancy for *H. werneckii* that appears to have resulted from whole genome duplication is an important characteristic here (Lenassi et al. 2013). As cation homeostasis and maintenance of an appropriate intracellular $K^+/Na^+$ ratio is essential for survival, analysis of the genes that code for cation transporters in these fungi has been of particular interest.

## 13.2 Alkali Metal Cation Transporters in Halophilic Fungi

### *13.2.1 Alkali Metal Cation Transporters in H. werneckii*

The extremely halotolerant *H. werneckii* thrives in environments that have fluctuating $K^+$ and $Na^+$ concentrations (Gunde-Cimerman et al. 2000). Under such conditions, maintenance of a low intracellular $Na^+$ concentration and a high intracellular $K^+/Na^+$ ratio is crucial for survival. Regulated $K^+$ uptake and efflux across the cell membrane is important for maintenance of the plasma membrane potential and for basic physiological functions, including protein synthesis, enzyme activation, and osmotic regulation. On the other hand, high intracellular concentrations of $Na^+$ interfere with cell growth through inhibition of many important enzymatic functions. As well as causing osmotic stress, high extracellular $Na^+$ concentrations disturb $K^+$ homeostasis, because this $Na^+$ entry into cells can also displace $K^+$ (Rodriguez-Navarro 2000; Arino et al. 2010).

Although in its hypersaline habitats *H. werneckii* is exposed to very high $Na^+$ concentrations (Gostinčar et al. 2011), these cells maintain a stable and relatively high $K^+$ concentration. Even when the $K^+/Na^+$ ratio drops at very high $Na^+$ concentrations, the intracellular $Na^+$ in *H. werneckii* is very low (Kogej et al. 2005). These data suggested efficient transport systems for the elimination of these toxic $Na^+$ ions. Indeed, according to the recently sequenced whole genome of *H. werneckii*, a large proportion of the *H. werneckii* predicted proteins can be assigned to transport systems (Lenassi et al. 2013). In particular, homologues of alkali metal cation transporters to those known from *S. cerevisiae* were abundant, which appears to have arisen from several gene duplications.

In the *H. werneckii* genome, we identified homologues of known *S. cerevisiae* plasma-membrane transporters, including Trk1, Trk2, Tok1, Nha1, Pho89, and the Ena proteins, and also the intracellular cation transporters Kha1, Mrs7, Vnx1 and Nhx1as. We also identified homologues of the $H^+$-ATPases, Pma and the V-ATPase complex. On the other hand, no homologues were identified in the *H. werneckii* genome for the less common Hak1 $H^+/K^+$ ($Na^+/K^+$) symporters and the Acu1 P-type $K^+$ ATPases that are present in some non-conventional yeast (Benito et al. 2011; Ramos et al. 2011), and also in the below-described *Aureobasidium* spp.

#### 13.2.1.1 Plasma-Membrane Cation Transporters in *H. werneckii*

In the mesophilic *S. cerevisiae*, the Trk1 and Trk2 high-affinity $K^+$ transporters are involved in $K^+$ uptake (Ko and Gaber 1991). Also, the Tok1 $K^+$ channel (Ketchum et al. 1995), the Ena1-5 ATPases (Haro et al. 1991; Garciadeblas et al. 1993; Wieland et al. 1995), and the Nha1 $Na^+/K^+$ antiporter, which was first identified as a $Na^+$ efflux system (Prior et al. 1996), are involved in $K^+$ efflux in *S. cerevisiae*. The Ena P-type ATPases are invoved in $Na^+$ (or $K^+$) export from these cells at alkaline pH, and they are coupled to ATP hydrolysis. With an acidic external pH, the Nha1 antiporter can promote $Na^+$ (or $K^+$) efflux down this $H^+$ gradient (Arino et al. 2010).

Multiple copies of the genes encoding these transporters are present in the genome of the extremely halotolerant *H. werneckii*. We identified eight homologues of the Trk1 and Trk2 $K^+$ transporters, with the conserved Trk H domain that is essential for the activity of these cation transport proteins. While these Trk homologues show relatively low homology to *S. cerevisiae* Trk1, their amino-acid sequence identities increase in the Trk H domain (Lenassi et al. 2013). As well as these Trk $K^+$ transport systems, four homologues of the Tok1 $K^+$ channels were identified in the genome of *H. werneckii*, which show the two conserved transmembrane helices that are typical for the Tok1 ion-channel family.

The genome of *H. werneckii* is also enriched in proteins that are homologues of the Nha1 $Na^+/K^+$, $H^+$ antiporter. Here, eight protein homologues were identified, which show the conserved PF00999 transmembrane region at the N-terminal. However, only two of these homologues contain the C-terminal cytoplasmic PF08619 region, with little similarity across the family.

In contrast, only four homologues of the *S. cerevisiae* Ena1, Ena 2 and Ena 5 $Na^+$ P-type ATPases were identified in the genome of *H. werneckii* (Lenassi et al. 2013). These homologues contain the conserved N-terminal and C-terminal regions found in the P-type ATPases that transport $H^+$, $Na^+$, $Ca^{2+}$, $Na^+/K^+$ and $H^+/K^+$. These differences in the numbers of *H. werneckii* Nha and Ena homologues are of interest. While both Nha and Ena can export $Na^+$ from cells, it is known that Nha is active under slightly acidic conditions; the Ena proteins on the other hand, function better in alkaline environments (Banuelos et al. 1998).

Regulation of the transcript numbers of the *ENA* genes is known to occur at the transcriptional level in *S. cerevisiae* (Garciadeblas et al. 1993; Mendoza et al. 1994; Wieland et al. 1995). In *H. werneckii*, increased salinity and alkaline pH have also been shown to induce transcription of the *ENA* genes (Gorjan and Plemenitas 2006). On the other hand, *NHA* gene expression is generally constitutive and low. Nha1 also has a role in regulation of the cell cycle, cell volume, membrane potential, and intracellular pH (Banuelos et al. 1998; Kinclova et al. 2001; Simon et al. 2001; Kinclova-Zimmermannova et al. 2006). It is therefore possible that this Nha enrichment in the halotolerant *H. werneckii* provides specialisation for other functions, as well as material for the evolution of some paralogues towards changed specificity (for $Na^+$ or $K^+$).

Alkaline pH also induces transcription of the *PHO89* gene, even when cells are grown under normal phosphate concentrations (Serrano et al. 2002). The genome of *H. werneckii* has six homologues of the Pho89 $Na^+$,$P_i$ symporter, which has resulted from several duplication steps that occurred at various times after the separation of its ancestor from the *S. cerevisiae* lineage. We have proposed that under conditions of high salinity and alkaline pH in *H. werneckii*, where the $H^+$ gradient cannot efficiently promote $P_i$ import, the $Na^+$ gradient represents an alternative energy source for $P_i$ transport (Lenassi et al. 2013).

Unlike in the closely related plant pathogen *Mycosphaerella graminicola*, which has homologues of both the Acu P-type ATPase, which mediates high affinity $K^+$ or $Na^+$ uptake, and the Hak1 $K^+$, $H^+$ symporter (Benito et al. 2011), *H. werneckii* does not have any homologues of Acu or Hak1. It thus appears that *H. werneckii* regulates the $K^+/Na^+$ ratio through mechanisms that are distinct from those of *M. graminicola*, which has only one genome copy of Trk and the Tok channel, compared to the eight and four copies, respectively, in *H. werneckii*. It appears that *M. graminicola* requires active import of $K^+$, while in *H. werneckii* the passive transport is more pronounced. These two closely related fungi with such different cation transport systems might well reflect the environments in which they are found, with *H. werneckii* as an extremely halotolerant fungi and *M. graminicola* as a plant pathogen.

#### 13.2.1.2 Plasma Membrane $H^+$ ATPase in *H. werneckii*

The number of P-type ATPases varies among fungal species (Palmgren and Axelsen 1998). In the search of the predicted proteome of *H. werneckii,* 40 members of the P-type ATPases were identified. When compared to *S. cerevisiae,* transporters from different groups of the P-type ATPases are enriched in *H. werneckii*. Among these, there are four members of group IIA P-type ATPase transporters, which are involved in $Ca^{2+}$ transport, six members of group IIIA proton transporters, and 14 members of group IV, which are involved in phospholipid transport (Lenassi et al. 2013).

These different P-type ATPases have a common mechanism of action: hydrolysis of ATP to provide the necessary energy for the transport of different ions and other substrates across membranes (Kuhlbrandt 2004). Among these different P-type ATPases, the export of protons is of major importance, as the activities of many other transporters are coupled to the gradient of protons across membranes. Proton export is mediated by the Pma1 P-type ATPase at the plasma membrane (Serrano et al. 1986; Ambesi et al. 2000) and the V-type ATPase at the vacuole membrane (Graham et al. 2000). We identified four homologues of Pma1 in *H. werneckii*, *Hw*Pma1A, *Hw*Pma1B, *Hw*Pma2A and *Hw*Pma2B. Each of these has conserved domains that are found in the *S. cerevisiae* Pma1 and Pma2 proteins: the cation_ATPase_N domain (PF00690), the E1-E2ATPase domain (PF00122), and the hydrolase_like2 domain (PF13246).

In *S. cerevisiae*, of the two copies of the *PMA* gene, only *PMA1* has an impact on ion homeostasis, as *PMA2* is expressed only at very low levels (Schlesser et al. 1988). Additionally, *PMA1* expression is not induced by salt (Yale and Bohnert 2001). In contrast, in the halotolerant *H. werneckii*, expression of both *HwPMA1* and *HwPMA2* is salt dependent. The expression profiles of these two genes during growth under different NaCl concentrations demonstrated the responsiveness to these different salinity conditions. The lowest expression of *HwPMA1* homologues was at 3 M NaCl, which is the optimal salinity for growth of *H. werneckii* (Kogej et al. 2005). This is in agreement with our previous studies that showed lower expression of the *HwPMA* homologue with the sequence that corresponds to the

*HwPMA1B* gene at 3 M NaCl, in comparison to the expression at 4 M NaCl (Vaupotic and Plemenitas 2007). The *HwPMA2* expression profile differs from *HwPMA1*, and it is low at low salt concentrations (e.g., 1.5 M) as well as at very high salt concentrations (e.g., 4 M).

#### 13.2.1.3 Intracellular Cation Transporters in *H. werneckii*

Maintenance of the appropriate $K^+/Na^+$ ratio in the cell cytosol is also connected to cation transport across the organelle membranes. These transporters are important for the regulation of organelle pH and volume (Arino et al. 2010). Endosomal Nhx1 (Nass and Rao 1999) and Kha1 in the Golgi apparatus (Maresova and Sychrova 2005) are typical $Na^+/H^+$ exchangers, and they are similar to the plasma-membrane Nha1 (Prior et al. 1996). Mitochondrial Mdm38 and Mrs7, and vacuolar Vnx1 (Nowikovsky et al. 2004; Cagnac et al. 2007; Zotova et al. 2010) have functions similar to the $Na^+/K^+$, $H^+$ exchanger.

In the *H. werneckii* genome, the homologues of Nhx1 and Kha1 and the homologues of the human LETM1 transporter, which have high homology to the Mdm38 and Mrs7 transporters from *S. cerevisiae*, have been duplicated (as have almost all of the *H. werneckii* genes, due to whole genome duplication in this species). On the other hand, vacuolar Vnx1 transporters are enriched in *H. werneckii*, with eight homologues of the Vnx1 $Na^+/K^+$,$H^+$ antiporter in the genome of *H. werneckii*. It is possible that by accumulating $Na^+$ in the vacuoles, vacuolar $Na^+/K^+$, $H^+$ antiporters are involved in detoxification of cytosolic $Na^+$ and contribute to the highly haloadaptable character of *H. werneckii* (Lenassi et al. 2013).

Vacuolar ATPase has a crucial role in the acidification of the vacuolar lumen (Arino et al. 2010). Analysis of the *H. werneckii* genome revealed homologues of all of the subunits of the V-ATP complex of *S. cerevisiae* (Lenassi et al. 2013). Yeast V-ATPase is composed of a catalytic $V_1$ sub-complex that is coded by eight genes (*VMA1, VMA2, VMA4, VMA5, VMA7, VMA8, VMA10, VMA13)* and a proton-translocating membrane $V_0$ sub-complex, which is coded by six different genes (*VPH1*, *STV1*, *VMA3*, *VMA6*, *VMA11*, *VMA16*) (Graham et al. 2000). The *H. werneckii* $V_1$ subunits share a lot of similarity with the *S. cerevisiae* V1 subunits. Interestingly, the intein sequence, which is typical for *S. cerevisiae* (Kane et al. 1990), is missing in the *H. werneckii* Vma1 homologues.

In *S. cerevisiae, VPH1* and *STV1* code for two homologues of the $V_0$ subunit. Vph1 is part of the vacuolar membrane V-ATPase, whereas Stv1 is a part of the V-ATPase on the Golgi apparatus or endosome membranes (Graham et al. 2000). In *H. werneckii*, we identified two copies of Vph1 homologues, *Hw*Vph1A and *Hw*Vph1B, although localisation of the ATPase complexes in *H. werneckii* remains to be carried out. While the transcription of *HwPMA*s is salt-regulated, expression of the *VMA* homologues is not affected by changes in NaCl concentrations (Lenassi et al. 2013). In contrast, salt stress induces the expression of vacuolar ATPase subunits in *S. cerevisiae* (Yale and Bohnert 2001).

The abundance of cation transporters in the *H. werneckii* genome indicates the importance of regulation of $K^+$ and $Na^+$ transport across the plasma membrane. Enrichment is not only observed in the transporters involved in the efflux of $Na^+$ (i.e., *Hw*Nha) and the uptake of $K^+$ (i.e., *Hw*Trk), which maintain high intracellular $K^+/Na^+$ ratios together with low intracellular concentrations of $Na^+$ in hypersaline environments with extremely high NaCl concentrations (Shabala and Cuin 2008), but also for transporters involved in $K^+$ efflux (*Hw*Tok), and possibly even with $Na^+$ intake (*Hw*Pho). We can speculate that this is advantageous for *H. werneckii* due to its life in environments with strongly fluctuating NaCl concentrations. The import of $K^+$ and export of $Na^+$ are crucial for the survival of an organism when the medium concentrations of NaCl are high, but when the concentration of NaCl suddenly drops, $K^+$ might accumulate to excessive concentrations (Benito et al. 2002). Under these conditions, the enriched Tok outward $K^+$ channels would allow this potentially harmful surplus of $K^+$ to be quickly released from *H. werneckii*.

Cell responses to hypersaline stress are energetically demanding, and they reduce the ATP pools of the cells. The $H^+$ gradient generated by the plasma membrane and vacuolar $H^+$-ATPase have important roles in providing the necessary energy for secondary cation transport (Arino et al. 2010). Studying the regulation and functioning of *H. werneckii* Pma and vacuolar $H^+$-ATPase homologues contributes to our understanding of these adaptation processes. The identification of multiple copies of Pma transporters in *H. werneckii*, together with their salt-dependent expression profiles, and identification of two copies of Vph1 homologues, *Hw*Vph1A and *Hw*Vph1B, supports the idea that these transporters have important roles in cation homeostasis in *H. werneckii* exposed to changing environmental salinities.

### 13.2.2 Alkali Metal Cation Transporters in Aureobasidum spp.

An in-depth analysis of fungal transporters in 13 black fungi with sequenced genomes revealed the great diversity of alkali metal ion transporters also in other species, besides *H. werneckii* (Gostinčar et al. 2014). Separate phylogenetic clusters were identified in the families of all of the major transporter types, which have resulted from evolutionary old duplications. The species that have retained representatives in most of these clusters were the polyextremotolerant yeast from the genus *Aureobasidium*: *A. pullulans, A. subglaciale, A. melanogenum* and *A. namibiae* (Gostinčar et al. 2014).

While gene duplications were observed in the phylogeny of plasma-membrane transporters, this was not the case for the transporters of the intracellular organelles (Gostinčar et al. 2014). This can be explained because the transporters of the intracellular organelles are not exposed to the large extracellular fluctuations of inorganic salt ions, as they are located in the relatively stable intracellular environment (Kogej et al. 2005).

The most striking among the features of alkali metal cation transporters observed in *Aureobasidium* spp. is the abundance of the $K^+$ transport systems (Gostinčar et al. 2014). Both the Trk uptake systems and the Tok efflux channels are duplicated in all four *Aureobasidium* spp.. Also, *A. pullulans* and *A. subglaciale* contain an additional (i.e., third) homologue of Trk. Both of these proteins contain a domain that is typical of Trk, but while the copy in *A. pullulans* is very similar to its homologues, the predicted protein of *A. subglaciale* is uncharacteristically short. Among nine other black fungi analysed, the phylogenetic cluster of this third Trk copy also contains predicted proteins from *H. werneckii* and *Cochliobolus heterostrophus*, but not from other species. A fourth Trk cluster contains representatives from *Aspergillus nidulans*, *C. heterostrophus* and *H. werneckii* (but none from the *Aureobasidium* spp.) (Gostinčar et al. 2014).

While the numbers of predicted Trk and Tok transporters are higher in *H. werneckii* compared to *Aureobasidum* spp., these latter contain two additional and less-common $K^+$ uptake systems that are not present in *H. werneckii* (Gostinčar et al. 2014). One homologue of the Hak $H^+/K^+$ ($Na^+/K^+$) symporters (Ramos et al. 2011) was found in each of these four *Aureobasidum* spp. genomes (Gostinčar et al. 2014). Additionally, each of these genomes contains two copies of genes encoding the otherwise relatively rarely found Acu transporters (Gostinčar et al. 2014), which function as P-type $K^+$ ATPases (Ramos et al. 2011). These two genes originate from an ancient duplication, and were also found in some of the other species analysed: *M. graminicola*, *Cladosporim fulvum*, *Sclerotinia sclerotiorum*, *Baudoinia compniacensis* and *C. heterostrophus*. However, apart from the *Aureobasidium* spp., only *B. compniacensis* and *C. heterostrophus* contain representatives of both of the phylogenetic clusters that arose from the duplication. As it appears likely that both the Hak and Acu $K^+$ uptake systems were already present in the ancestor of *H. werneckii* and were later lost despite the extremely halotolerant nature of this yeast, the role of these two systems in adaptation to salt should be further investigated.

Among the transporters with $Na^+$-exporting function, two duplications have been reported for the Nha $Na^+/H^+$ antiporter, and one duplication in the case of the Ena P-type $Na^+$ ATPase. This has resulted in three copies of the Nha gene and two copies of the Ena gene in these *Aureobasidium* spp., with the notable exception of *A. pullulans*, which is the most halotolerant of these four *Aureobasidium* spp., where one of the Ena copies has been lost. As the number of Ena-encoding genes is less than the number of Nha-encoding genes also in *H. werneckii* (Lenassi et al. 2013), this might indicate a relatively more important role of Nha transporters in the natural habitat of the fungi investigated. In light of this, it needs to be borne in mind that the characteristics of the Nha and Ena transporter types are not the same: Nha proteins are more active under slightly acidic conditions, whereas the optimal pH for the Ena proteins is alkaline, as under these conditions the proton motive force (which is needed to drive Nha transporters) is a less favourable source of energy (Banuelos et al. 1998).

While in *H. werneckii* an enrichment has been reported for the Pho $Na^+/P_i$ symporters, the situation in *Aureobasidium* spp. is less clear. For this gene, *A. namibiae* contains one copy, *A. melanogenum* contains two copies, and *A. subglaciale* and *A. pullulans* each contain three copies (Gostinčar et al. 2014). The presence of three copies of the Pho $Na^+/P_i$ symporters in the most halotolerant of these four

*Aureobasidium* spp. appears to support the explanation that was suggested from the example of *H. werneckii*: that in hypersaline environments, the $Na^+$ gradient can provide a good alternative energy source for influx of $P_i$ (Lenassi et al. 2013).

On the basis of most of the data that have been presented to date, it would appear reasonable to associate the high numbers of alkali metal cation transporters with the well-known halotolerance of *H. werneckii* and *Aureobasidium* spp. However, other data have indicated that the relationship between the number of transporters and halotolerance might not be this simple, and that any conclusions should be drawn with great caution. Even taking into account the whole genome duplication of *H. werneckii,* which will also have doubled the number of transporters, the largest number of transporter genes among these black fungi that have been analysed was actually in *C. heterostrophus* (Gostinčar et al. 2014) which has, at least to the best of our knowledge, never been studied or reported as a halotolerant species, nor has *C. heterostrophus* been found in habitats with elevated concentrations of salt. Additionally, the most halophilic fungus known, which is the basidiomycete *W. ichthyophaga*, manages to grow in solutions saturated with NaCl despite having very few metal cation transporters in its genome (Zajc et al. 2013), as is described below in more detail. And finally, while *A. pullulans* shows greater halotolerance than the other three *Aureobasidium* spp. studied here by several percent of NaCl (w/v) (Zalar et al. 2008), it does not have the highest number, nor the greatest diversity, of transporters among these four *Aureobasidium* spp.. Indeed, *A. pullulans* has one less copy of the Ena genes, even though in *S. cerevisiae* the Ena pumps are believed to be a major determinant of salt tolerance (Arino et al. 2010).

Another intriguing question that arises from the large diversity of alkali metal cation transporter genes in *Aureobasidium* spp. (and also in some other related fungi), is why has this diversity been retained in the genomes throughout the considerable evolutionary time since it was generated through gene duplications? Is this only because of the beneficial effects of this gene redundancy, which led to the extinction of all of the ancestors that had lost the functionality of one of the gene copies, or is it because of other reasons, such as functional differences between the genome duplicates that have remained undiscovered to date?

The electrochemical gradient of protons across the plasma membrane that is needed for the function of many of the above-described transporters has to be generated by active transport. This is mostly achieved by the Pma ATPases, which are encoded by only one gene in each of the *Aureobasidium* spp., a number that is in stark contrast with the four such genes in *H. werneckii.*

However, the action of Pma ATPases is not the only way to build an electrochemical gradient of protons, although it has been the most thoroughly studied to date. For example, in the dothideomycetous fungus *Leptosphaeria maculans*, it has been demonstrated that bacteriorhodopsins can function as light-driven proton pumps (Waschuk et al. 2005). The genome of *A. melanogenum* contains three copies of these bacteriorhodopsin genes, while the other three *Aureobasidum* spp. contain two copies each. These proteins form two well-defined phylogenetic clusters, and differ in their intron-exon patterns (Gostinčar et al. 2014). It was also argued previously that such an alternative way to build up a proton motive force by harvesting light instead of using ATP would be extremely useful in the hypersaline habitats of *H. werneckii*. This is due to the energetically demanding maintenance of an acceptable intracellular ion composition and the osmotic pressure under these hypersaline conditions (Gostinčar et al. 2011). This will also apply to the frequently oligotrophic habitats of *Aureobasidium* spp. (Gostinčar et al. 2014). Indeed, *H. werneckii* and these *Aureobasidium* spp. are commonly found in habitats with an abundance of light.

### 13.2.3 Alkali Metal Cation Transporters in *W. ichthyophaga*

Essential alkali metal cation transporters have also been reported in the genome of the halophilic *W. ichthyophaga*. When compared to *H. werneckii* with its multiple copies of genes that code for the various cation transporters, as described above, the number of transporter-encoding genes is smaller in *W. ichthyophaga*. This small number of genes is also accompanied by their relatively low expression levels, and their relative unresponsiveness to different salinities (Zajc et al. 2013).

*W. ichthyophaga* is a halophilic basidiomycetous fungus. In addition to phylogenetic differences, *W. ichthyophaga* is distinguished from the other representatives of the *Wallemia* genus by its characteristic morphology and halophilic physiology (Kralj Kunčič et al. 2010). It is an extraordinary example of an obligate halophilic fungal specialist (Gostinčar et al. 2010). The growth optimum of *W. ichthyophaga* is between 2.6 M and 3.5 M NaCl, although it can even survive in saturated NaCl solution. As *W. ichthyophaga* is considered the most halophilic eukaryote known to date (Zajc et al. 2014), its unusual physiology makes it an interesting model organism for ion homeostasis and ion transport studies.

In *W. ichthyophaga,* the protein family of cation-transporting ATPases includes three $H^+$ P-type ATPases (Pma) and two $Na^+$ P-type ATPases (Ena) in the plasma membrane, one $Ca^{2+}$ P-type ATPase in the vacuole (Pmc) and one in the Golgi apparatus (Pmr), and a protein with unknown specificity.

In contrast to the Pma P-type ATPases that are represented by three putative $H^+$ pumps in the *W. ichthyophaga* genome, in the closely related *Wallemia sebi*, there are only two Pma ATPases (Padamsee et al. 2012), which show over 91 % identity to the *W. ichthyophaga* Pma ATPases. The enrichment of Pma ATPases in *W. ichthyophaga*, as compared to *W. sebi*, might thus implicate the halophilic ecotype of *W. ichthyophaga*. In addition to the Pma ATPases, the V-type $H^+$ ATPases are also involved in the regulation of pH homeostasis. Such vacuolar V-type $H^+$ ATPase multiprotein complex is also present in the genome of *W. ichthyophaga*.

Two types of $Na^+$ extrusion systems are involved in the regulation of $Na^+$ homeostasis. The export of $Na^+$ by Nha antiporters is coupled to the transmembrane $H^+$ gradient that is maintained by the Pma pumps, while the export of $Na^+$ (or $K^+$) by the Ena P-type ATPases is driven by the hydrolysis of ATP. Despite their complementary roles, the function of Ena ATPases is more pronounced at high pH, where Nha antiporters cannot function correctly. The two Ena ATPases of

*W. ichthyophaga* share 92 % identical residues with each other, 83 % identical residues with the single Ena ATPase from *W. sebi*, and 36 % identical residues with the Ena ATPases from *S. cerevisiae*.

As indicated above, the P-type ATPase family of *W. ichthyophaga* also includes one Pmc $Ca^{2+}$ ATPase of the vacuole and one Pmr $Ca^{2+}$ ATPase of the Golgi apparatus. In *S. cerevisiae*, Pmc1 transports $Ca^{2+}$ from the cytoplasm into the vacuole (Marchi et al. 1999), while Pmr1 transports $Ca^{2+}$ and $Mn^{2+}$ ions into the Golgi apparatus, which is necessary for $Ca^{2+}$-dependent protein sorting and processing (Vashist et al. 2002). Orthologues of both Pmc1 and Pmr1 are also found in the genome of *W. sebi*.

Beside these P-type ATPases, a plethora of other transporters maintain the cellular ion homeostasis, to keep the intracellular amounts of the highly toxic $Na^+$ low, and those of $K^+$ constant. In environments with high concentrations of $Na^+$, this is not an easy task. Orthologues of the known secondary active transporters from *S. cerevisiae* (Arino et al. 2010) and unconventional yeast (Ramos et al. 2011) have also been searched for in the genome of *W. ichthyophaga*. Here, several plasma-membrane (Nha1, Trk1, Pho89) and intracellular (Kha1, Nhx1, Vnx1, Mrs7/Mdm37) transporters were identified. Considering the halophilic nature of *W. ichthyophaga*, the representation of these transporters is relatively weak in comparison to the extremely halotolerant ascomycete *H. werneckii*, where a significant increase in the numbers of alkali cation transporters has taken place due to gene duplications (Lenassi et al. 2013). Along with the Ena P-type ATPases, Nha antiporters are also involved in the export of toxic $Na^+$ from the cytosol. The genome of *W. ichthyophaga* contains two Nha1 orthologues (Zajc et al. 2013).

In *S. cerevisiae*, the major part of the $K^+$ influx is mediated by two high-affinity uniporters, Trk1 and Trk2 (Arino et al. 2010). Only one orthologue of these Trk transporters was found in *W. ichthyophaga* (Zajc et al. 2013), which is highly similar to the TrkH-domain-containing protein of *W. sebi*. However, the Tok channel that mediates the efflux of $K^+$ was not found in the genome of *W. ichthyophaga* (Zajc et al. 2013).

Under some conditions, the passive transport of $K^+$ might not be sufficient, such as in with acidic environmental pH and low $K^+$ levels, or under hypersaline conditions. Here, the ATP-dependent import of $K^+$ is necessary. In such a scenario, the Hak $K^+$-$H^+$ symporters or the Acu $K^+(Na^+)$-ATPase transporters are involved (Benito et al. 2004; Ramos et al. 2011). However, there are no Hak orthologues in the genome of *W. ichthyophaga*, although it contains two putative Acu ATPases. As both of these Acu transporters were also found in the genome of *W. sebi*, this might represent the $K^+$-influx system of the genus *Wallemia* (Padamsee et al. 2012; Zajc et al. 2013).

Intracellular membrane transporters include the alkali metal cation/$H^+$ antiporters of the Golgi apparatus (Kha1), endosomes (Nhx1) and vacuoles (Vnx1), and the mitochondrial $K^+/H^+$ antiport exchange system (Mdm38, Mrs7) (Arino et al. 2010). Two orthologues of Kha1 and one orthologue of these other intracellular ion transporters have been identified in *W. ichthyophaga* (Zajc et al. 2013).

## 13.3 Other Transporters

### *13.3.1 Major Facilitator Superfamily Sugar Transporters*

The major facilitator superfamily (MFS) transporters are the largest family of secondary transporters, and they show wide substrate specificity that ranges from ions to carbohydrates, lipids, amino acids, peptides, nucleosides and other molecules (Yan et al. 2013). The following paragraphs provide a more detailed look at these transporters in *Aureobasidium* spp. and *W. ichthyophaga*.

A significant enrichment of proteins containing the Pfam domain (http://pfam.xfam.org/) of sugar transporter proteins has been reported for the genomes of these *Aureobasidium* spp. (Gostinčar et al. 2014). Significant differences have also been reported for these proteins between the individual *Aureobasidium* spp., with a relative enrichment of the genes in *A. pullulans*, and a relative depletion in *A. melanogenum*. Thus, while *A. melanogenum* contains "only" 73 predicted members of the sugar porter family, *A. pullulans* contains 92. The largest difference here is in the number of putative maltose or α-glucoside:$H^+$ symporters, whereby *A. melanogenum* contains only half as many as the other *Aureobasidium* spp. (Gostinčar et al. 2014).

These observations have been interpreted in the light of the ecology of *Aureobasidium* spp. (Gostinčar et al. 2014). Nutritionally, these *Aureobasidium* spp. are very diverse, a trait that is reflected not only in the number of their MFS transporters, but also in the high numbers of predicted extracellular enzymes, of which more than 300 in each *Aureobasidium* spp. have been predicted to act on carbohydrates (according to the Carbohydrate-Active Enzyme database), again with the highest number in *A. pullulans* (Gostinčar et al. 2014). Furthermore, *A. pullulans* is a well-known epiphyte (Andrews et al. 2002; Olstorpe et al. 2010; Grube et al. 2011), and therefore it has access to the plant biomass, which will include various carbohydrate substrates. On the other hand, *A. melanogenum* is more frequently found in freshwater habitats, and of these *Aureobasidium* spp., it is phylogenetically the furthest away from *A. pullulans* (Zalar et al. 2008). Thus, the access of *A. melanogenum* to various sugars will be comparably poorer, and therefore significantly fewer sugar transporters (and carbohydrate-active enzymes) in the genome of *A. melanogenum* might not be unexpected.

The apparent numbers of MFS sugar transporters in these *Aureobasidium* spp. are high also in comparison with some other fungi. They have similar numbers to the moulds *Aspergillus oryzae* (119) and *A. nidulans* (100) (Gostinčar et al. 2014), but are much higher than in *S. cerevisiae* (31) (Leandro et al. 2009) and *Neurospora crassa* (32) (Galagan et al. 2003). More than three quarters of the transporters in *Aureobasidium* spp. belong to the group of sugar:$H^+$ symporters (Gostinčar et al. 2014).

In contrast to *Aureobasidium* spp., the genome of *W. ichthyophaga* contains statistically significantly fewer proteins that are involved in the transport of various molecules (Zajc et al. 2013). Within the MFS transporters, in the genome of

*W. ichthyophaga* the MFS_1 domain (PF07690) is represented in only 51 members, which is significantly less than in its closest relative *W. sebi* (69) and in *Cryptococcus neoformans* (114). On the other hand, the Sugar_tr domain (PF00083) of MFS transporters is present in 19 *W. ichthyophaga* proteins (15 in *W. sebi*, 64 in *C. neoformans*). The glycerol/ proton symporter Slt1 is a member of the MFS sugar transporter family and is important under conditions of hyperosmotic shock in *S. cerevisiae*; in *W. ichthyophaga*, there are four orthologues of Slt1.

The fewer MFS transporters in *W. ichthyophaga*, which also has fewer ABC transporters, might be the result of the limited competition with other species for this obligate halophile. However, the loss of proteins in both of these superfamilies might be an effect of other factors as well. For example, *W. ichthyophaga* will also have reduced needs for extrusion of heavy metals and secondary metabolites, nutritional specialisation, and prevention of compatible solutes leaking via broad specificity transporters. In addition, the diffusion of toxic compounds might be limited by the exceptionally thick cell wall of *W. ichthyophaga* (Kralj Kunčič et al. 2010).

### *13.3.2 Aquaporins*

Aquaporins are selective channels for the transport of water across membranes that operate at extremely fast rates (i.e., $10^9$-fold faster than ion channels). The aquaglyceroporins are selective not only for water, but also for glycerol and other small uncharged molecules, such as other polyols, urea, or arsenite (Borgnia and Agre 2001). While the management of cellular water is crucial for all life, it is an even greater challenge in osmotically stressful environments, and thus the water- and solute-permeable aquaporins can also have important osmotic implications (Borgnia et al. 1999; Hohmann et al. 2000; Alleva et al. 2012).

A search for aquaporin-like and aquaglyceroporin-like genes identified 11 genes in *A. pullulans, A. subglaciale* and *A. melanogenum,* and 12 in *A. namibiae* (Gostinčar et al. 2014). Such an abundance of aquaporins is not unusual in mammals or plants, but this greatly exceeds the numbers observed in other fungal species, which can contain up to seven aquaporins, although the majority have only one or two (Xu et al. 2013). It has been suggested that the high number of aquaporins in *Aureobasidium* spp. is crucial for their survival in the water-challenged habitats in which they are frequently found (Gostinčar et al. 2014), however this remains to be studied in more detail.

In *W. ichthyophaga*, a total of three aquaporin-like and aquaglyceroporin-like genes were identified (Zajc et al. 2013). The same numbers of highly similar proteins are also present in *W. sebi* (Padamsee et al. 2012). As *W. ichthyophaga* thrives in environments with very low water activity, this number is relatively low compared to the *Aureobasidium* spp., which contain more than 11 (Gostinčar et al. 2014). Thus, *W. ichthyophaga* appears to use other mechanisms to combat the detrimental effects of high osmolarity.

## 13.4 Conclusions

In conclusion, in natural hypersaline environments where the concentrations of toxic $Na^+$ ions exceed $K^+$ ions, the mechanisms that maintain the stable and high intracellular $K^+/Na^+$ ratio are crucial for the survival of organisms. Major alkali cation transporters of selected halophilic fungi, described in this chapter, are summarized in Table 13.1. It has been demonstrated that the halophilic/ halotolerant fungal species studied can maintain high $K^+/Na^+$ ratios over a wide range of environmental $Na^+$ concentrations (Kogej et al. 2005; Zajc et al. 2014). In *H. werneckii*, this homeostasis is maintained by regulated transport of $K^+$ and $Na^+$ across the plasma membrane, as the cation transporters are diverse and highly enriched in this fungus. However, in *W. ichthyophaga*, it appears that as these transporters are present in lower numbers and are expressed at lower levels, they do not represent the main system for the regulation of the intracellular cation concentrations. Studies on the cell-wall structure of *W. ichthyophaga* indicate that it can prevent the entry of toxic sodium ions by dynamic cell-wall restructuring (Kralj Kunčič et al. 2010; Plemenitas et al. 2014). An explanation for these differences in the salt-exclusion strategies of *H. werneckii* and *W. ichthyophaga* might be the need for *H. werneckii* to rapidly adapt to highly dynamic concentrations of NaCl (and other salts), which will be typical of its natural environment; *W. ichthyophaga* instead thrives under continuous high salinity. On the other hand, the genomes

**Table 13.1** Major alkali cation transporters in halophilic fungi

| Description | *S. cerevisiae** | *H. werneckii** | *Aureobasidium* spp.* | *W. ichthyophaga** |
|---|---|---|---|---|
| $K^+$ efflux channel, plasma membrane | Tok1 | 4 | 2 | *none* |
| $K^+$ uptake uniporter, plasma membrane | Trk1,2 | 8 | 2–3 | 1 |
| $K^+(Na^+)$ efflux antiporter (Hak), plasma membrane | *none* | *none* | 1 | *none* |
| $K^+$ efflux P-type ATPase (Acu), plasma membrane | *none* | *none* | 2 | 2 |
| $Na^+$ efflux antiporter, plasma membrane | Nha1 | 8 | 3 | 2 |
| $Na^+$ (and $Li^+$) efflux P-type ATPase, plasma membrane | Ena1–5 | 4 | 1–2 | 2 |
| $Na^+/P_i$ symporter, plasma membrane | Pho89 | 6 | 1–3 | 1 |
| $H^+$ exporter P-type ATPase, plasma membrane | Pma1 | 4 | 1 | 3 |

*The columns contain protein names or number of homologues of each protein. For *Aureobasidium* spp. the range found in the genomes of *A. pullulans, A. melanogenum, A. subglaciale* and *A. namibiae* is presented of the moderately halotolerant *Aureobasidium* spp. include additional transporters: the Hak $H^+/K^+$ symporters that can act as $Na^+/K^+$ symporters; and the Acu P-type $K^+$ importing ATPases that are encoded by two relatively divergent genes. Additionally, the high abundance of aquaporins in *Aureobasidium* spp. might be vital for the survival and adaptability of these polyextremotolerant species to osmotically stressful environments. These water channels and osmoregulators will allow the passage of sufficient water and/or compatible solute fluxes into and out of the cells under the various osmotic conditions.

As has been revealed by the analysis of the recently sequenced genomes of these six halotolerant fungi, they have at their disposal different transport systems and other mechanisms. This indicates that these different phylogenetic lineages of fungi have evolved different strategies to survive and thrive under hypersaline conditions that prohibit the growth of most other species.

## References


Alleva K, Chara O, Amodeo G (2012) Aquaporins: another piece in the osmotic puzzle. FEBS Lett 586(19):2991–2999. doi:10.1016/j.febslet.2012.06.013

Ambesi A, Miranda M, Petrov VV, Slayman CW (2000) Biogenesis and function of the yeast plasma-membrane H+-ATPase. J Exp Biol 203(1):155–160

Andrews JH, Spear RN, Nordheim EV (2002) Population biology of *Aureobasidium pullulans* on apple leaf surfaces. Can J Microbiol 48(6):500–513

Arino J, Ramos J, Sychrova H (2010) Alkali metal cation transport and homeostasis in yeasts. Microbiol Mol Biol Rev 74(1):95–120. doi:10.1128/MMBR.00042-09

Banuelos MA, Sychrova H, Bleykasten-Grosshans C, Souciet JL, Potier S (1998) The Nha1 antiporter of *Saccharomyces cerevisiae* mediates sodium and potassium efflux. Microbiology 144:2749–2758

Benito B, Garciadeblas B, Rodriguez-Navarro A (2002) Potassium- or sodium-efflux ATPase, a key enzyme in the evolution of fungi. Microbiology 148:933–941

Benito B, Garciadeblas B, Schreier P, Rodriguez-Navarro A (2004) Novel P-type ATPases mediate high-affinity potassium or sodium uptake in fungi. Eukaryot Cell 3(2):359–368. doi:10.1128/Ec.3.2.359-368.2004

Benito B, Garciadeblas B, Fraile-Escanciano A, Rodriguez-Navarro A (2011) Potassium and sodium uptake systems in fungi. The transporter diversity of *Magnaporthe oryzae*. Fungal Genet Biol 48(8):812–822. doi:10.1016/j.fgb.2011.03.002

Borgnia MJ, Agre P (2001) Reconstitution and functional comparison of purified GlpF and AqpZ, the glycerol and water channels from *Escherichia coli*. Proc Natl Acad Sci U S A 98(5):2888–2893. doi:10.1073/pnas.051628098

Borgnia M, Nielsen S, Engel A, Agre P (1999) Cellular and molecular biology of the aquaporin water channels. Annu Rev Biochem 68:425–458. doi:10.1146/annurev.biochem.68.1.425

Cagnac O, Leterrier M, Yeager M, Blumwald E (2007) Identification and characterization of Vnx1p, a novel type of vacuolar monovalent Cation/H+ antiporter of *Saccharomyces cerevisiae*. J Biol Chem 282(33):24284–24293. doi:10.1074/jbc.M703116200

Galagan JE, Calvo SE, Borkovich KA, Selker EU, Read ND, Jaffe D, FitzHugh W, Ma LJ, Smirnov S, Purcell S, Rehman B, Elkins T, Engels R, Wang S, Nielsen CB, Butler J, Endrizzi M, Qui D, Ianakiev P, Bell-Pedersen D, Nelson MA, Werner-Washburne M, Selitrennikoff CP, Kinsey JA, Braun EL, Zelter A, Schulte U, Kothe GO, Jedd G, Mewes W, Staben C, Marcotte E, Greenberg D, Roy A, Foley K, Naylor J, Stange-Thomann N, Barrett R, Gnerre S, Kamal

M, Kamvysselis M, Mauceli E, Bielke C, Rudd S, Frishman D, Krystofova S, Rasmussen C, Metzenberg RL, Perkins DD, Kroken S, Cogoni C, Macino G, Catcheside D, Li W, Pratt RJ, Osmani SA, DeSouza CP, Glass L, Orbach MJ, Berglund JA, Voelker R, Yarden O, Plamann M, Seiler S, Dunlap J, Radford A, Aramayo R, Natvig DO, Alex LA, Mannhaupt G, Ebbole DJ, Freitag M, Paulsen I, Sachs MS, Lander ES, Nusbaum C, Birren B (2003) The genome sequence of the filamentous fungus *Neurospora crassa*. Nature 422(6934):859–868

Garciadeblas B, Rubio F, Quintero FJ, Banuelos MA, Haro R, Rodriguez-Navarro A (1993) Differential expression of two genes encoding isoforms of the ATPase involved in sodium efflux in *Saccharomyces cerevisiae*. Mol Gen Genet 236(2–3):363–368

Gorjan A, Plemenitas A (2006) Identification and characterization of ENA ATPases HwENA1 and HwENA2 from the halophilic black yeast *Hortaea werneckii*. FEMS Microbiol Lett 265(1):41–50. doi:10.1111/j.1574-6968.2006.00473.x

Gostinčar C, Grube M, de Hoog S, Zalar P, Gunde-Cimerman N (2010) Extremotolerance in fungi: evolution on the edge. FEMS Microbiol Ecol 71(1):2–11. doi:10.1111/j.1574-6941.2009.00794.x

Gostinčar C, Lenassi M, Gunde-Cimerman N, Plemenitas A (2011) Fungal adaptation to extremely high salt concentrations. Adv Appl Microbiol 77:71–96. doi:10.1016/B978-0-12-387044-5.00003-0

Gostinčar C, Ohm RA, Kogej T, Sonjak S, Turk M, Zajc J, Zalar P, Grube M, Sun H, Han J, Sharma A, Chiniquy J, Ngan CY, Lipzen A, Barry K, Grigoriev IV, Gunde-Cimerman N (2014) Genome sequencing of four *Aureobasidium pullulans* varieties: biotechnological potential, stress tolerance, and description of new species. BMC Genomics 15:549. doi:10.1186/1471-2164-15-549

Graham LA, Powell B, Stevens TH (2000) Composition and assembly of the yeast vacuolar H(+)-ATPase complex. J Exp Biol 203(Pt 1):61–70

Grube M, Schmid F, Berg G (2011) Black fungi and associated bacterial communities in the phyllosphere of grapevine. Fungal Biol 115(10):978–986. doi:10.1016/j.funbio.2011.04.004

Gunde-Cimerman N, Zalar P, de Hoog S, Plemenitas A (2000) Hypersaline waters in salterns – natural ecological niches for halophilic black yeasts. FEMS Microbiol Ecol 32(3):235–240

Gunde-Cimerman N, Zalar P, Petrovič U, Turk M, Kogej T, De Hoog GS, Plemenitaš A (2004) Fungi in the salterns. In: Ventosa A (ed) Halophilic microorganisms. Springer, Berlin, pp 103–111

Gunde-Cimerman N, Ramos J, Plemenitaš A (2009) Halotolerant and halophilic fungi. Mycol Res 113(Pt 11):1231–1241. doi:10.1016/j.mycres.2009.09.002

Haro R, Garciadeblas B, Rodriguez-Navarro A (1991) A novel P-type ATPase from yeast involved in sodium-transport. FEBS Lett 291(2):189–191. doi:10.1016/0014-5793(91)81280-L

Hohmann S, Bill RM, Kayingo G, Prior BA (2000) Microbial MIP channels. Trends Microbiol 8(1):33–38. doi:10.1016/S0966-842x(99)01645-5

Kane PM, Yamashiro CT, Wolczyk DF, Neff N, Goebl M, Stevens TH (1990) Protein splicing converts the yeast Tfp1 gene-product to the 69-Kd subunit of the vacuolar $H^+$-adenosine triphosphatase. Science 250(4981):651–657. doi:10.1126/science.2146742

Ketchum KA, Joiner WJ, Sellers AJ, Kaczmarek LK, Goldstein SA (1995) A new family of outwardly rectifying potassium channel proteins with two pore domains in tandem. Nature 376(6542):690–695. doi:10.1038/376690a0

Kinclova O, Poitier S, Sychrova H (2001) The *Zygosaccharomyces rouxii* strain CBS732 contains only one copy of the *HOG1* and the *SOD2* genes. J Biotechnol 88(2):151–158

Kinclova-Zimmermannova O, Gaskova D, Sychrova H (2006) The Na+, K+/H+-antiporter Nha1 influences the plasma membrane potential of Saccharomyces crevisiae. FEMS Yeast Res 6(5):792–800. doi:10.1111/j.1567-1364.2006.00062.x

Ko CH, Gaber RF (1991) Trk1 and Trk2 encode structurally related $K^+$ transporters in *Saccharomyces cerevisiae*. Mol Cell Biol 11(8):4266–4273

Kogej T, Ramos J, Plemenitas A, Gunde-Cimerman N (2005) The halophilic fungus *Hortaea werneckii* and the halotolerant fungus *Aureobasidium pullulans* maintain low intracellular cation concentrations in hypersaline environments. Appl Environ Microbiol 71(11):6600–6605. doi:10.1128/aem.71.11.6600-6605.2005

Kralj Kunčič M, Kogej T, Drobne D, Gunde-Cimerman N (2010) Morphological response of the halophilic fungal genus *Wallemia* to high salinity. Appl Environ Microbiol 76(1):329–337. doi:10.1128/AEM.02318-09

Kuhlbrandt W (2004) Biology, structure and mechanism of P-type ATPases. Nat Rev Mol Cell Biol 5(4):282–295. doi:10.1038/nrm1354

Leandro MJ, Fonseca C, Goncalves P (2009) Hexose and pentose transport in ascomycetous yeasts: an overview. FEMS Yeast Res 9(4):511–525. doi:10.1111/j.1567-1364.2009.00509.x

Lenassi M, Gostinčar C, Jackman S, Turk M, Sadowski I, Nislow C, Jones S, Birol I, Cimerman NG, Plemenitas A (2013) Whole genome duplication and enrichment of metal cation transporters revealed by de novo genome sequencing of extremely halotolerant black yeast *Hortaea werneckii*. PLoS ONE 8(8):e71328. doi:10.1371/journal.pone.0071328

Marchi V, Sorin A, Wei Y, Rao R (1999) Induction of vacuolar Ca2+-ATPase and H+/Ca2+ exchange activity in yeast mutants lacking Pmr1, the Golgi Ca2+-ATPase. FEBS Lett 454(3):181–186

Maresova L, Sychrova H (2005) Physiological characterization of *Saccharomyces cerevisiae kha1* deletion mutants. Mol Microbiol 55(2):588–600. doi:10.1111/j.1365-2958.2004.04410.x

Mendoza I, Rubio F, Rodrigueznavarro A, Pardo JM (1994) The protein phosphatase calcineurin is essential for NaCl tolerance of *Saccharomyces cerevisiae*. J Biol Chem 269(12):8792–8796

Nass R, Rao R (1999) The yeast endosomal Na+/H+ exchanger, Nhx1, confers osmotolerance following acute hypertonic shock. Microbiology 145:3221–3228

Nowikovsky K, Froschauer EM, Zsurka G, Samaj J, Reipert S, Kolisek M, Wiesenberger G, Schweyen RJ (2004) The LETM1/YOL027 gene family encodes a factor of the mitochondrial K+ homeostasis with a potential role in the Wolf-Hirschhorn syndrome. J Biol Chem 279(29):30307–30315. doi:10.1074/jbc.M403607200

Olstorpe M, Axelsson L, Schnurer J, Passoth V (2010) Effect of starter culture inoculation on feed hygiene and microbial population development in fermented pig feed composed of a cereal grain mix with wet wheat distillers' grain. J Appl Microbiol 108(1):129–138. doi:10.1111/j.1365-2672.2009.04399.x

Padamsee M, Kumar TK, Riley R, Binder M, Boyd A, Calvo AM, Furukawa K, Hesse C, Hohmann S, James TY, LaButti K, Lapidus A, Lindquist E, Lucas S, Miller K, Shantappa S, Grigoriev IV, Hibbett DS, McLaughlin DJ, Spatafora JW, Aime MC (2012) The genome of the xerotolerant mold *Wallemia sebi* reveals adaptations to osmotic stress and suggests cryptic sexual reproduction. Fungal Genet Biol 49(3):217–226. doi:10.1016/j.fgb.2012.01.007

Palmgren MG, Axelsen KB (1998) Evolution of P-type ATPases. Biochim Biophys Acta 1365(1–2):37–45

Plemenitas A, Lenassi M, Konte T, Kejzar A, Zajc J, Gostinčar C, Gunde-Cimerman N (2014) Adaptation to high salt concentrations in halotolerant/halophilic fungi: a molecular perspective. Front Microbiol 5:199. doi:10.3389/fmicb.2014.00199

Prior C, Potier S, Souciet JL, Sychrova H (1996) Characterization of the NHA1 gene encoding a Na+/H+-antiporter of the yeast *Saccharomyces cerevisiae*. FEBS Lett 387(1):89–93. doi:10.1016/0014-5793(96)00470-X

Ramos J, Arino J, Sychrova H (2011) Alkali-metal-cation influx and efflux systems in nonconventional yeast species. FEMS Microbiol Lett 317(1):1–8. doi:10.1111/j.1574-6968.2011.02214.x

Rodriguez-Navarro A (2000) Potassium transport in fungi and plants. Biochim Biophys Acta Rev Biomembr 1469(1):1–30. doi:10.1016/S0304-4157(99)00013-1

Schlesser A, Ulaszewski S, Ghislain M, Goffeau A (1988) A second transport ATPase gene in *Saccharomyces cerevisiae*. J Biol Chem 263(36):19480–19487

Serrano R, Kiellandbrandt MC, Fink GR (1986) Yeast plasma-membrane ATPase is essential for growth and has homology with ($Na^+$+$K^+$), $K^+$- and $Ca^{2+}$-ATPases. Nature 319(6055):689–693. doi:10.1038/319689a0

Serrano R, Ruiz A, Bernal D, Chambers JR, Arino J (2002) The transcriptional response to alkaline pH in *Saccharomyces cerevisiae*: evidence for calcium-mediated signalling. Mol Microbiol 46(5):1319–1333. doi:10.1046/j.1365-2958.2002.03246.x

Shabala S, Cuin TA (2008) Potassium transport and plant salt tolerance. Physiol Plant 133(4):651–669. doi:10.1111/j.1399-3054.2007.01008.x

Simon E, Clotet J, Calero F, Ramos J, Arino J (2001) A screening for high copy suppressors of the sit4 hal3 synthetically lethal phenotype reveals a role for the yeast Nha1 antiporter in cell cycle regulation. J Biol Chem 276(32):29740–29747. doi:10.1074/jbc.M101992200

Vashist S, Frank CG, Jakob CA, Ng DT (2002) Two distinctly localized p-type ATPases collaborate to maintain organelle homeostasis required for glycoprotein processing and quality control. Mol Biol Cell 13(11):3955–3966. doi:10.1091/mbc.02-06-0090

Vaupotic T, Plemenitas A (2007) Differential gene expression and Hog1 interaction with osmoresponsive genes in the extremely halotolerant black yeast *Hortaea werneckii*. BMC Genomics 8:280. doi:10.1186/1471-2164-8-280

Waschuk SA, Bezerra AG Jr, Shi L, Brown LS (2005) *Leptosphaeria* rhodopsin: bacteriorhodopsin-like proton pump from a eukaryote. Proc Natl Acad Sci U S A 102(19):6879–6883. doi:10.1073/pnas.0409659102

Wieland J, Nitsche AM, Strayle J, Steiner H, Rudolph HK (1995) The Pmr2 gene-cluster encodes functionally distinct isoforms of a putative $Na^+$ pump in the yeast plasma-membrane. EMBO J 14(16):3870–3882

Xu H, Cooke JEK, Zwiazek JJ (2013) Phylogenetic analysis of fungal aquaporins provides insight into their possible role in water transport of mycorrhizal associations. Botany-Botanique 91(8):495–504. doi:10.1139/cjb-2013-0041

Yale J, Bohnert HJ (2001) Transcript expression in *Saccharomyces cerevisiae* at high salinity. J Biol Chem 276(19):15996–16007. doi:10.1074/jbc.M008209200

Yan A, Hu X, Wang K, Sun J (2013) Discriminating of ATP competitive Src kinase inhibitors and decoys using self-organizing map and support vector machine. Mol Divers 17(1):75–83. doi:10.1007/s11030-012-9411-0

Zajc J, Liu Y, Dai W, Yang Z, Hu J, Gostinčar C, Gunde-Cimerman N (2013) Genome and transcriptome sequencing of the halophilic fungus *Wallemia ichthyophaga*: haloadaptations present and absent. BMC Genomics 14(1):617. doi:10.1186/1471-2164-14-617

Zajc J, Kogej T, Galinski EA, Ramos J, Gunde-Cimerman N (2014) Osmoadaptation strategy of the most halophilic fungus, Wallemia ichthyophaga, growing optimally at salinities above 15% NaCl. Appl Environ Microbiol 80(1):247–256. doi:10.1128/AEM.02702-13

Zalar P, Gostinčar C, de Hoog GS, Ursic V, Sudhadham M, Gunde-Cimerman N (2008) Redefinition of Aureobasidium pullulans and its varieties. Stud Mycol 61:21–38. doi:10.3114/sim.2008.61.02

Zotova L, Aleschko M, Sponder G, Baumgartner R, Reipert S, Prinz M, Schweyen RJ, Nowikovsky K (2010) Novel components of an active mitochondrial $K+/H+$ exchange. J Biol Chem 285(19):14399–14414. doi:10.1074/jbc.M109.059956

# Chapter 14
# Antifungals: Mechanism of Action and Drug Resistance

**Rajendra Prasad, Abdul Haseeb Shah, and Manpreet Kaur Rawal**

**Abstract** There are currently few antifungals in use which show efficacy against fungal diseases. These antifungals mostly target specific components of fungal plasma membrane or its biosynthetic pathways. However, more recent class of antifungals in use is echinocandins which target the fungal cell wall components. The availability of mostly fungistatic antifungals in clinical use, often led to the development of tolerance to these very drugs by the pathogenic fungal species. Thus, the development of clinical multidrug resistance (MDR) leads to higher tolerance to drugs and its emergence is helped by multiple mechanisms. MDR is indeed a multifactorial phenomenon wherein a resistant organism possesses several mechanisms which contribute to display reduced susceptibility to not only single drug in use but also show collateral resistance to several drugs. Considering the limited availability of antifungals in use and the emergence of MDR in fungal infections, there is a continuous need for the development of novel broad spectrum antifungal drugs with better efficacy. Here, we briefly present an overview of the current understanding of the antifungal drugs in use, their mechanism of action and the emerging possible novel antifungal drugs with great promise.

**Keywords** Multidrug resistance • Antifungal agents • Azoles • Combination therapy • Drug efflux • Erg11p

R. Prasad (✉)
Membrane Biology Laboratory, School of Life Sciences, Jawaharlal Nehru University, New Delhi 110067, India

AMITY Institute of Integrative Sciences and Health (AIISH), Amity University Haryana, Manesar, Gurgaon, Haryana, India
e-mail: rp47jnu@gmail.com; rprasad@ggn.amity.edu

A.H. Shah • M.K. Rawal
Membrane Biology Laboratory, School of Life Sciences, Jawaharlal Nehru University, New Delhi 110067, India
e-mail: hasb789biotech@gmail.com; manpreet.rawal@gmail.com



J. Ramos et al. (eds.), *Yeast Membrane Transport*, Advances in Experimental Medicine and Biology 892, DOI 10.1007/978-3-319-25304-6_14

## 14.1 Introduction

Fungal infections have emerged as one of the major causes of human disease, especially in immunocompromised individuals (Shapiro et al. 2011). Fungal infections which are generally superficial can also turn into systemic infections as the disease incidence prolongs (Cannon et al. 2009; Brown et al. 2012). Among the different mycotic infections caused by these opportunistic fungi, candidiasis, an infection caused by *Candida*, is the most threatening due to severity of the disease and higher worldwide occurrence. Other fungal diseases like cryptococcal meningitis and invasive aspergillosis are also life threatening (López-Martínez 2010). The pathogenicity of fungal infections proceeds in well-organized steps. For example, *Candida* cell surface adhesion factors first promote its adherence to host surface, followed by an invasion and damage of the host tissues due to release of various virulence factors (Gow and Hube 2012). Eukaryotic fungal pathogens pose an additional therapeutic challenge since they show close evolutionary relationship with the human hosts, thus minimizing the choice of novel drug targets that can be exploited to selectively kill the pathogen (Heitman 2011; Shapiro et al. 2011). Nonetheless, there are many drug categories currently in use against fungal infections which exploit exclusive novel fungal targets. For example, most antifungals are directed against ergosterol which is a typical sterol of fungal cells. The sterol component of cell membranes of fungi is targeted either by blocking the enzymes important for its synthesis or by directly depleting the ergosterol from the plasma membrane (PM) (White et al. 1998). In addition, several other drugs, now in use, also target unique components of cell wall (CW) in fungal cells (White et al. 1998). The fact that the increase in incidence of worldwide fungal infections and emergence of antifungal drug resistance which outcompetes the development of novel antifungal compounds, it becomes important to understand the various facets of infections and to understand the basic mechanisms that govern the development of resistance. This enforces the widening of the hunt for the development of new antifungals targeting novel pathways. This chapter focuses on some of the aspects of antifungals, their mechanisms of action and development of resistance against them.

## 14.2 Antifungal Drugs

The limited availability of antifungals is a major impediment for the effective treatment of fungal infections (Vandeputte et al. 2012). This is further compounded by the fact that the generation of newer antifungals has lagged behind when compared to the pace of emergence of fungal infections. The components of the fungal CW such as mannans, glucans and chitins; and a few of the enzymes of the ergosterol biosynthetic pathways which are unique to fungal cells are commonly targeted for the development of antifungal agents (St Georgiev 2000; Munro et al. 2001). Among the enzymes of the ergosterol biosynthetic pathway, squalene epoxidase (*ERG1*), 14α-lanosterol demethylase or CYP51 (*ERG11*), $\Delta^{14}$–reductase (*ERG24*) and $\Delta^8$-$\Delta^7$-isomerase (*ERG2*) have been the targets of most antifungal agents (Fig. 14.1) (Sanglard et al. 2003). Some of the commonly used antifungal drugs and their mechanisms of action are discussed below:

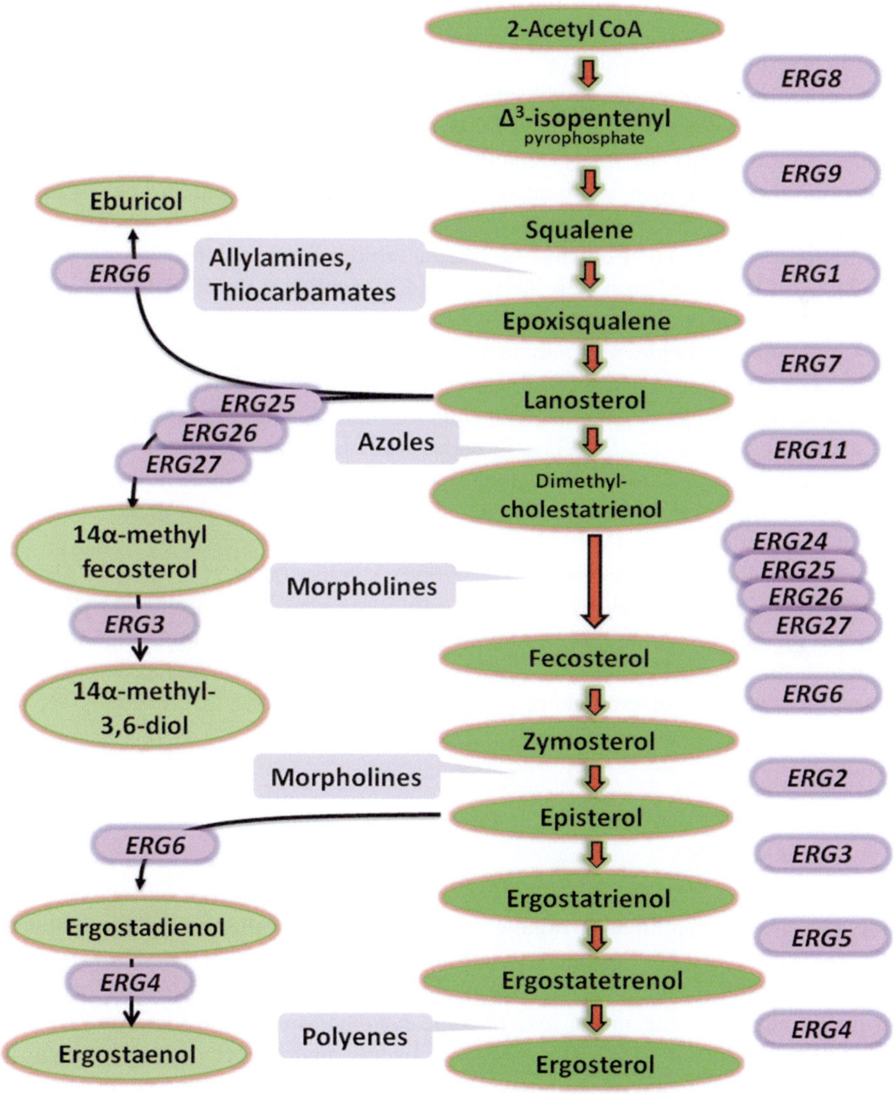


**Fig. 14.1** Ergosterol biosynthesis pathway showing specific point of action of select antifungal drugs. Different classes of antifungal drugs are shown on *left* against the steps of their action in pathway with corresponding enzymes catalyzing the reaction steps shown on *right*

### 14.2.1 Azoles

The fungistatic azoles primarily act on ergosterol biosynthesis by targeting 14α-lanosterol demethylase encoded by *ERG11* gene resulting in the inhibition of cytochrome P450-dependent conversion of lanosterol to ergosterol (Fig. 14.1). The resulting ergosterol depletion interferes with the bulk functions of ergosterol as a membrane component, but more importantly, severe ergosterol depletion may also interfere with the "sparking" functions of ergosterol, affecting cell growth and proliferation (White et al. 1998; Sanglard et al. 2009; Shapiro et al. 2011). The blocking of 14α-demethylase results in the accumulation of toxic methylated sterols leading to the membrane stress (Shapiro et al. 2011). In case of yeast *Cryptococcus neoformans,* azoles such as fluconazole (FLC) and itraconazole also result in the accumulation of obtusifolione in the ergosterol biosynthetic pathway, mainly due to the inhibition of NADPH-dependent 3-ketosteroid reductase (*ERG 27*), catalyzing the last C-4 demethylation step in ergosterol biosynthesis (Vanden Bossche et al. 1993; Ghannoum et al. 1994).

Azoles mainly include two subclasses based on the number of nitrogen atoms in a ring; The first class includes imidazoles which consist of miconazole, oxiconazole, econazole, ketoconazole, tioconazole, and clotrimazole with two nitrogen atoms in an azole ring, while another class includes triazoles such as FLC, posaconazole, itraconazole, terconazole, and voriconazole which contain three nitrogen atoms in a cyclic ring (Fig. 14.2). Imidazoles are mainly used for the mucosal fungal infections while triazoles are administered both for the systemic as well as for the mucosal infections (Sanglard et al. 2009; Vandeputte et al. 2012). Depletion of membrane ergosterol due to the use of azoles are also shown to disrupt vacuolar ATPase functions resulting in an impairment of the vacuolar acidification and ion homeostasis (Zhang et al. 2010). Since azoles are fungistatic, their prolonged use poses greater threat of emergence of drug resistance among the surviving fungal population (Shapiro et al. 2011).

### 14.2.2 Polyenes

Polyenes are the amphipathic organic natural molecules called macrolides and are generally produced by Streptomyces (Vandeputte et al. 2012). Polyenes directly bind to ergosterol of fungal cell membranes leading to the formation of pores in membrane, resulting in the loss of ionic balance, membrane integrity and cell death (Sanglard et al. 2009) (Fig. 14.1). Polyenes mainly include amphotericin B (AmpB), natamycin and nystatin (Fig. 14.3). AmpB is mostly effective in systemic invasive fungal infections and is used generally against *Cryptococcus*, *Candida* and *Aspergillus* species (Lemke et al. 2005; Sanglard et al. 2009) while nystatin and natamycin are preferred for topical infections due to their low absorption (Vandeputte et al. 2012). Although polyenes are fungicidal in nature and have been

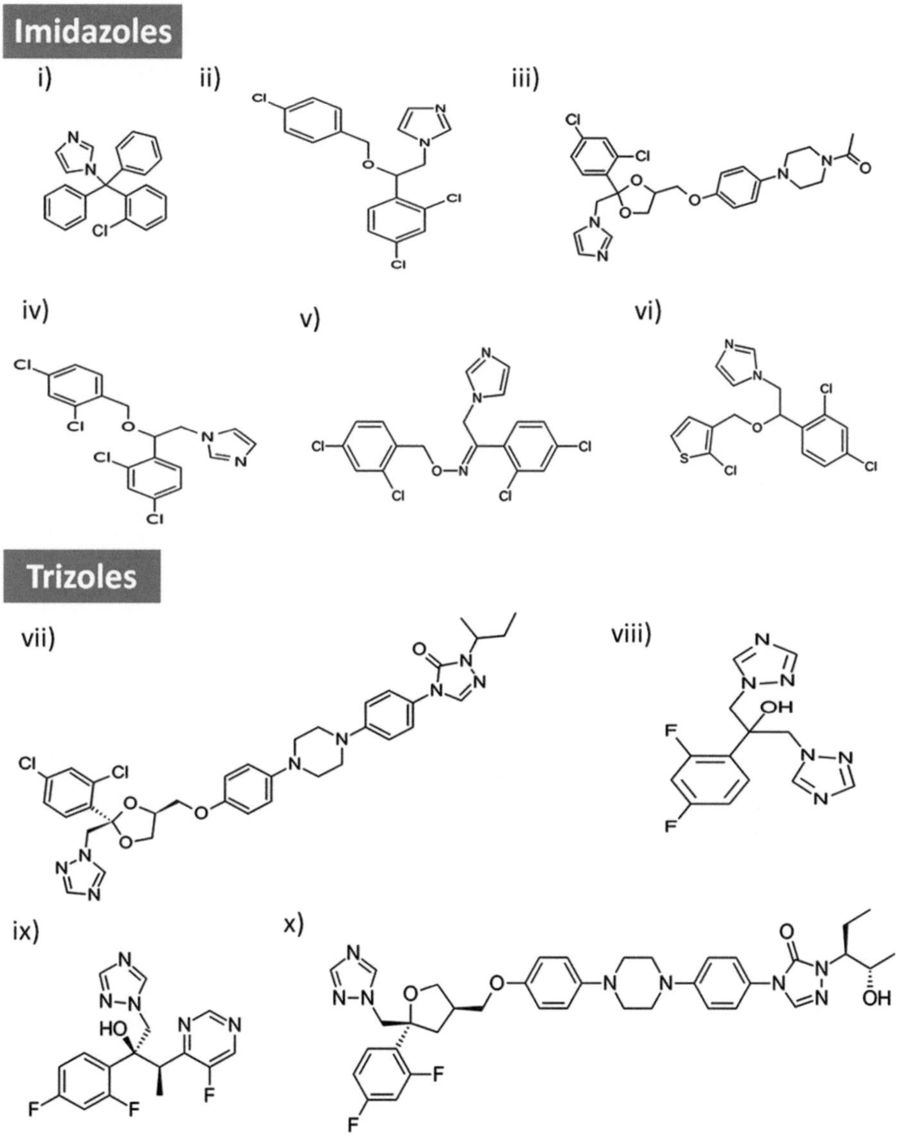


**Fig. 14.2** Structure of various azole antifungal compounds. These include imidazoles with two nitrogen atoms in a ring (*i*) Clotrimazole, (*ii*) Econazole, (*iii*) Ketoconazole (*iv*) Miconazole, (*v*) Oxiconazole, (*vi*) Tioconazole or triazoles containing three nitrogen atoms in a ring, (*vii*) Itraconazole (*viii*) Fluconazole (*ix*) Voriconazole (*x*) Posaconazole

in use for a long time but they show many side effects in humans which limits their use. However, lipid formulations of AmpB are less toxic and are relatively better for the treatment of fungal infections (Shapiro et al. 2011).

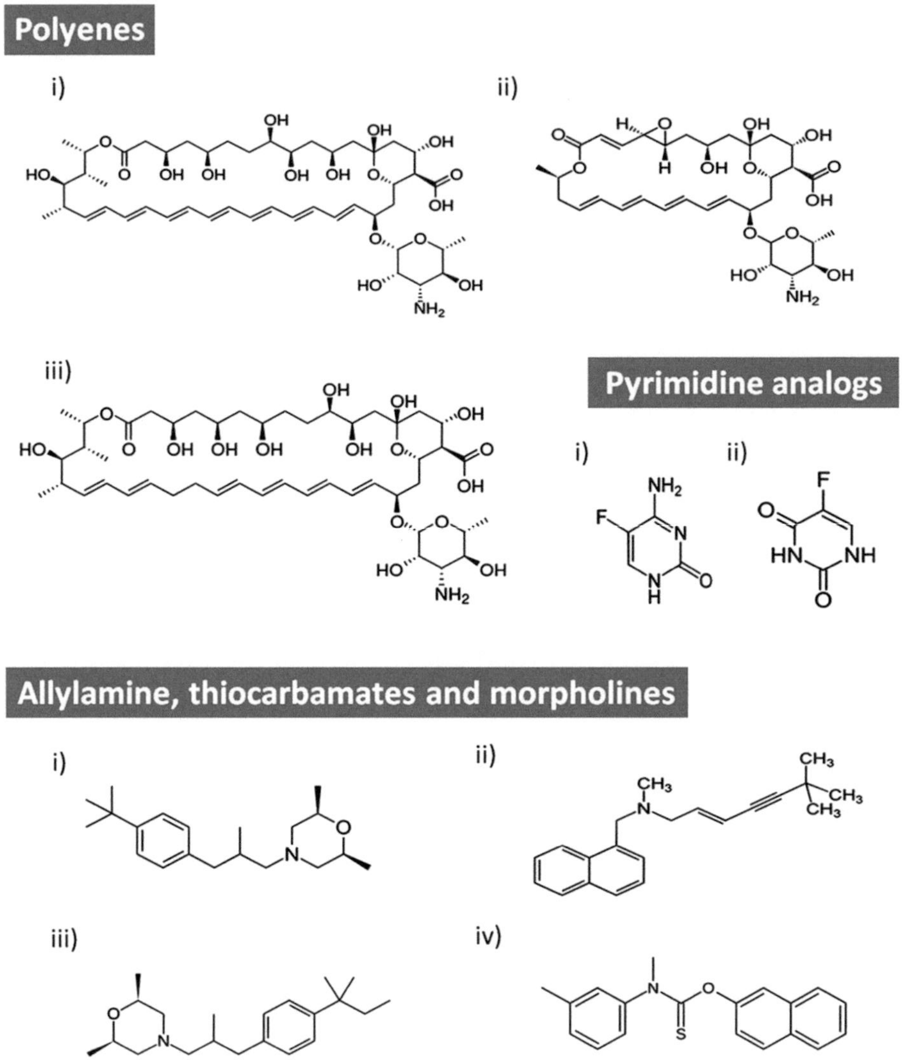


**Fig. 14.3** Structure of various antifungal compounds. These include polyenes (*i*) AmpB (*ii*) Natamycin (*iii*) Nystatin; pyrimidine analogs (*i*) 5 Fluorocytosine (*ii*) 5 Fluorouracil and Allylamine, thiocarbamates and morpholines antifungals as (*i*) Fenpropimorph (*ii*) Terbinafine (*iii*) Amorolfine (*iv*) Tolnaftate

### *14.2.3 Pyrimidine Analogs*

Pyrimidine analogs which include 5-fluorocytosine (5-FC) and 5-fluorouracil (5-FU) are the synthetic structural analogs of nucleotide cytosine (Fig. 14.3). Pyrimidine analog 5-FC is converted to 5-FU by cytosine deaminase which after conversion to downstream products gets incorporated into DNA and RNA during the synthesis of these biomolecules where it inhibits cellular functioning by blocking protein synthesis or inhibiting DNA replication. These drug analogs show activity against different *Candida* and *Cryptococcus* species (Lemke et al. 2005; Sanglard et al. 2009). 5-FC is rapidly absorbed and thus gives good bioavailability, however; it also shows many side effects (Lemke et al. 2005; Vandeputte et al. 2012). 5-FC is comparatively less effective antifungal drug because the fungal cells frequently develop tolerance to it. For this reason, it is generally preferred in combination therapy (Sanglard et al. 2009).

### 14.2.4 Allylamine, Thiocarbamates and Morpholines

Allylamines and thiocarbamates inhibit the *ERG1* gene of ergosterol biosynthesis while morpholines which include fenpropimorph and amorolfine (Fig. 14.3) inhibit the *ERG24* and *ERG2* genes of ergosterol biosynthesis (Fig. 14.1). Allylamines include terbinafine while thiocarbamates include tolnaftate (Fig. 14.3). All of these drugs are mostly used for the control of dermatophyte fungal infections (Gubbins and Anaissie 2006; Sanglard et al. 2009).

### 14.2.5 Echinocandins

The lipopeptide echinocandins which include caspofungin, micafungin and anidulafungin (Fig. 14.4) are comparatively recent class of antifungal drugs which target the synthesis of CW components by acting as non-competitive inhibitors of β-1,3 glucan synthase required for β-glucan synthesis (Fig. 14.5) (Perlin 2011; Shapiro et al. 2011). Defects in the synthesis of CW components affect the integrity of fungal cells resulting in CW stress. As a result, echinocandin treated cells become osmotically sensitive, form pseudohyphae, show separation defects, reduced sterol contents and thickened CW. Echinocandins are generally non-toxic to mammalian cells because they act on specific CW synthesis pathway unique to fungal cells (Sanglard et al. 2009; Perlin 2011; Shapiro et al. 2011).

### 14.2.6 Emerging Novel Antifungals

Keeping in view the facts that there are limited antifungal strategies and paucity of effective drugs, there is continuous hunt for the development of novel and effective antifungals to combat the fungal infections. Many different drug categories which show promising antifungal activity are at various stages of the development and some of them are listed in Table 14.1 and few potential antifungal strategies are discussed in the following section.

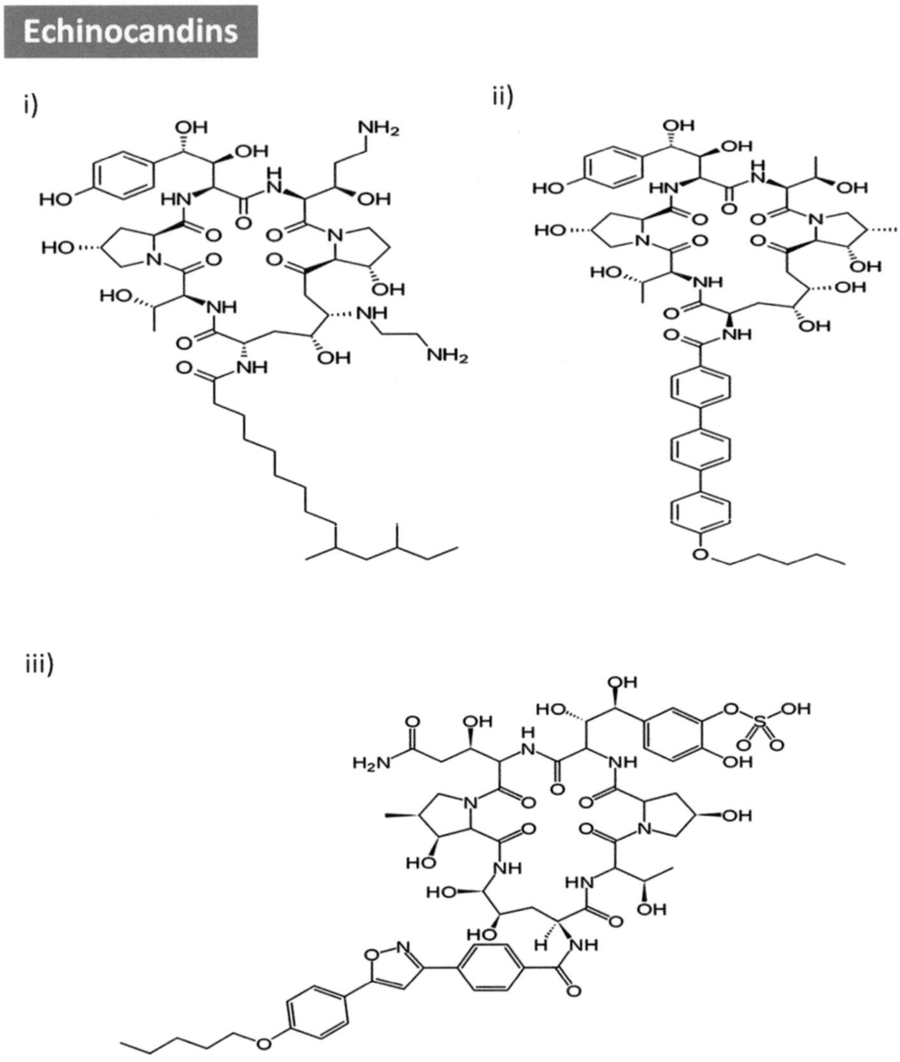


**Fig. 14.4** Structure of various echinocandin antifungal compounds targeting unique components in fungal CW. These include (*i*) Caspofungin (*ii*) Anidulafungin and (*iii*) Micafungin

## *14.2.7 Emerging Antifungal Strategies*

As mentioned above, the development of novel drug candidates have not kept pace with the frequency of the development of tolerance to available drugs, it is imperative to search for the different strategies to combat fungal infections. For instance, the transcription factor Upc2 which regulates the expression of ERG genes has been exploited as a potential target for the development of antifungal drugs. Many

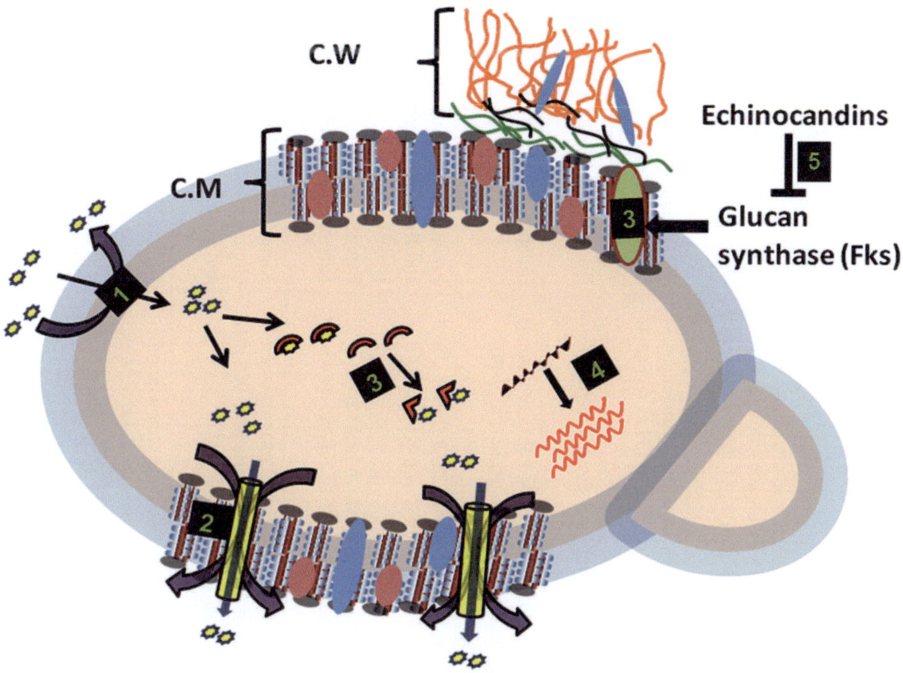


**Fig. 14.5** Different mechanisms of multidrug resistance adopted by fungal cells. The commonly observed mechanisms of drug resistance particularly against azoles, polyenes and echinocandins include (*1*) changes in membrane property/ lipid composition affecting normal drug import (*2*) over expression of drug efflux proteins leading to rapid drug extrusion (*3*) an alteration of the drug target (genes encoding ergosterol biosynthetic pathway enzymes or glucan synthases) leading to poor binding of toxic drugs to its target (*4*) overproduction of genes synthesizing the drug target proteins. (*5*) Echinocandins block the activity of glucan synthase important for synthesis of CW components affecting CW integrity leading to cell stress

small molecules screened from the commercially available compounds collection library have already been shown to inhibit the azole mediated up regulation of Upc2 and its target genes in *S. cerevisiae* and *C. glabrata*. However, the full potential of Upc2 remains to be explored before it could be successfully employed to improve antifungal strategies (Gallo-Ebert et al. 2014). Several new targets which include glucan synthesis, 26S proteasome, cAMP homeostasis, microtubule dynamics, and translational elongation also show great promise as antifungal targets (Roemer and Krysan 2014).

Combination therapy wherein different drugs are given in combination to treat fungal infections is among most favorite strategies. Some of the benefits of combination therapies include broad spectrum of treatment, synergy of effects between different drugs, lower doses of drug usage and lesser chances of the development of drug resistance. Synergistic effect shown by the drugs mainly occurs due to the additive effects on both CW and cell membrane components of

**Table 14.1** Novel antifungal agents showing promise for antifungal drug development either against particular fungal species or showing broad range antifungal effects

| S. No. | Drug | Category | Features | References |
|---|---|---|---|---|
| 1. | SPK-843 | Polyene | Compound is under clinical trials and shows better activity than AmpB and micafungin formulations against *Candida* spp., *C. neoformans*, and *Aspergillus* spp. | Kantarcioglu et al. (2003), Kakeya et al. (2008a, b) |
| 2 | T-2307 | Arylamide | Shows treatment potential against candidiasis, cryptococcosis and aspergillosis, is effective even against echinocandin and FLC resistant *Candida* species. | Mitsuyama et al. 2008, Shibata et al. (2012), and Wiederhold et al. (2015) |
| 3 | Isavuconazole (BAL4815) | Triazole | Compound is under late stage of clinical trials and shows activity against *Candida* and *Aspergillus* Spp. | Espinel-Ingroff (2009) |
| 4 | CPA18 and CPA109 | Azoles | Effective against drug resistant strains of *C. albicans*. CPA compounds in addition to decreasing the hyphae formation in *Candida*, leads to change in membrane polarity & lipid composition | Calabrese et al. (2013) |
| 5 | Enfumifungin and its derivative MK-3118 | Natural acidic terpenoid | Inhibits 1,3- β -glucan synthesis in fungal CW. MK-3118 shows enhanced efficacy against echinocandin resistant *C. albicans* and *C. glabrata* isolates | Roemer and Krysan (2014) and Jiménez-Ortigosa et al. (2014) |
| 6 | D75-4590 | Pyridobenzimidazole derivative | Inhibits 1,3- β -glucan synthase, Krep6 enzyme which lacks any mammalian homolog. Shows potent activity against *C. albicans* where it also inhibits hyphae formation | Espinel-Ingroff (2009) and Kitamura et al. (2009) |
| 7 | Papulacandin, Corynecandin and Mer-WF3010 | Glycolipid and Glycolipid analogs | Papulacandin leads to an inhibition of cell growth and promotes cell lysis. Corynecandin inhibits growth of *C. albicans* while Mer-WF3010 inhibits growth of *C.albicans* and *C. kufy*r. | Kaneto et al. (1993), Gunawardana et al. (1997), and Fostel and Lartey (2000) |
| 8 | Sordarins and related compounds GR135402 and BE31405 | Natural sugar moieties | Inhibits protein synthesis specifically in fungi. Sordarins show good activity against *C. tropicalis, C. albicans, A. flavus* and *C. kefyr*, while it is less effective against *C. parapsilosis, C. glabrata, A. fumigatus* and *C. krusei* | Kinsman et al. (1998), Okada et al. (1998), Domínguez et al. (1998), Domínguez and Martín (1998), and Herreros et al. (1998) |

| | | | | |
|---|---|---|---|---|
| 9 | Cispentacin, FR109615 and PDL118 (Bay10-8888) | Cyclic β-amino acids | These compounds get accumulated inside the fungal cells and inhibit the aminoacyl tRNA synthases, thus deregulating cellular amino acid metabolism and inhibiting cell growth. Cispentacin is effective against *Candida* and *Crypotococcus* species but not against *Aspergillus* species | Konishi et al. (1989), Iwamoto et al. (1990), Capobianco et al. (1993), Fostel and Lartey (2000), and Petraitis et al. (2004) |
| 10 | Aureobasidin A (AbA) | cyclic depsipeptide | AbA inhibits Inositol phosphoceramide (IPC) synthase activity. AbA also interacts with fungal ABC transporters like *YOR1* and *PDR5* and inhibits their transport activity | Nagiec et al. (1997) and Ogawa et al. (1998) |
| 11 | Rustmicin (galbonolide A) and galbonolide B | Macrolides | Inhibit sphingolipid biosynthesis | Mandala et al. (1998), Harris et al. (1998), and Fostel and Lartey (2000) |
| 12 | Khafrefungin | Aldonic acid linked via an ester to a C22 modified alkyl chain | Inhibits IPC synthase activity, leading to an accumulation of ceramide and depletion of downstream complex sphingolipids | Mandala et al. (1997) |
| 13 | Parnafungins | Isoxazolidinone-containing natural products | Inhibits the poly A polymerase activity and show broad spectrum of action. Parnafungins are active against *Candida* and *Aspergillus* species. | Jiang et al. (2008), Parish et al. (2008), Bills et al. (2009), and Roemer and Krysan (2014), |
| 14 | Curcumin | Natural polyphenol | It acts as an antifungal agent via generation of oxidative stress and inhibits hyphae development. Modulates efflux activity of ABC transporters of *Candida*. | Sharma et al. (2009,2010) |
| 15 | Malachite Green | Triarylmethane dye | Shows potent antifungal activity against different species of *Candida* by shifting the metabolic flux towards fermentation, ROS generation and by cell necrosis. | Dhamgaye et al. (2012) |
| 16 | Berberine | Isoquinoline plant alkaloid | *Candida* cells treated with berberine compromises CW integrity via the calcineurin pathway leading to cell death. | Dhamgaye et al. (2014) |

fungal cell. For instance, the damaged CW due to one of the antifungal components potentiates effective action of drugs directed against cell membrane components. The compromised CW integrity could also facilitate permeability of drugs across cell membranes to intracellular targets. Combination of azoles and allylamines displays synergistic effects due to the inhibition of the same pathway at different steps (Tobudic et al. 2010a; Rodrigues et al. 2014). Notably, combination therapy requires critical evaluation of the possible antagonistic and agonist property of different drugs when administered in combinations (Lewis and Kontoyiannis 2001). Table 14.2 summarizes some of the drugs which show better efficacy when given in combination.

## 14.3 Resistance Against Antifungal Drugs

Fungal cells have developed several strategies to deal with the antifungals. They have learnt to modify the antifungal drug targets or most commonly increase the efflux of the incoming drugs. Some of the known mechanisms of MDR are depicted in Fig. 14.5 and are briefly discussed below.

### *14.3.1 Azole Target Protein (Erg11p) Is Modified in Resistant Isolates of Candida*

The modification of the target protein represents one of the commonest mechanisms of MDR where the target protein of azoles, Erg11p, is modified by the chromosomal mutations leading to the replacement of native amino acids. This is evident from the fact that several point mutations in *ERG11* gene which encodes Erg11p have been identified in clinical drug resistant isolates of *Candida*. Interestingly, these mutations appear to be predominantly restricted to certain hot spot regions of Erg11p. The exact placement of all the identified mutations in a 3D model of the protein confirms that these mutations are not randomly distributed but rather are clustered in select hot spot regions (Marichal et al. 1999; Wang et al. 2009). These point mutations either individually or in combination, invariably prevent normal binding of FLC to target protein by reducing the affinity of the drug towards Erg11p (Wang et al. 2009; Morio et al. 2010).

### *14.3.2 Azole Resistance Leads to an Overexpression of ERG11*

Apart from spontaneous point mutations in *ERG11* (discussed above), in many FLC resistant clinical isolates, very often an over expression of *ERG11* is also observed (Hoot et al. 2011; Flowers et al. 2012; Sasse et al. 2012). The zinc

**Table 14.2** Various regimes of combinatorial antifungal therapy. Various antifungal and/or non-antifungal drug combinations which show better efficacy in combination compared to the effects shown by independent drugs

| | Drug 1 | Drug 2 | Target | References |
|---|---|---|---|---|
| 1 | Amp B (polyene) | Posaconazole (azole) | *Candida* biofilms | Tobudic et al. (2010a, b), Bink et al. (2011), and Rodrigues et al. (2014) |
| | | Caspofungin (echinocandin) | | |
| 2 | Micafungin (echinocandin) | Fluconazole (azole) | *Candida* and *Cryptococcus* infections | Serena et al. (2005), Nishi et al. (2009), and Espinel-Ingroff et al. (2009) |
| | | Voriconazole (azole) | | |
| | | Amp B (polyene) | | |
| 3 | Caspofungin (echinocandin) | Liposomal AmpB | *Candida glabrata* | Olson et al. (2005), Hodgetts et al. (2008), and Espinel-Ingroff et al. (2009) |
| | | Human antibody fragment against HSP90 (efungumab) | *Candida* infections | |
| 4 | Flucytosine (pyrimidine analog) | Voriconazole (azole) | *Candida* infections | Pai et al. (2008) and Bink et al. (2011) |
| 5 | Fluconazole (azole) | FK506 (calcineurin pathway inhibitor) | *Candida* biofilm | Uppuluri et al. (2008) and Bink et al. (2011) |
| | | Cyclosporine A (CsA) (calcineurin pathway inhibitor) | | |
| 6 | Minocycline (tetracycline antibiotic) | Fluconazole (azole) | *C.albicans* biofilms | Shi et al. (2010) and Bink et al. (2011) |
| 7 | Doxycycline (antibiotic/antibacterial agent) | AmpB (polyene) | *Candida albicans* biofilm | Miceli et al. (2009) |
| | | Caspofungin (Echinocandin) | | |
| | | Fluconazole (azole) | | |
| 8 | AmpB (polyene) | Itraconazole (azole) | Invasive aspergillosis | Garbati et al. (2012) |
| | | Caspofungin (echinocandin) | | |

cluster transcription factor Upc2p regulates the expression of *ERG11* and other genes involved in ergosterol biosynthesis (White and Silver 2005). Several studies have confirmed that an overexpression of *UPC2* increases resistance of *Candida* cells towards azole drugs, while its disruption results in cells hypersusceptibility to azoles. A comparison of sequence of *UPC2* between genetically matched pair azole susceptible and resistant isolates led to the identification of point mutations in the encoded protein. These gain of function (GOF) mutations results in an over expression of *ERG11* and hyper-resistance to azoles (Heilmann et al. 2010; Hoot et al. 2011; Flowers et al. 2012).

### 14.3.3 Azole Resistance and Ergosterol Biosynthetic Pathway

In clinical azole resistance, several mutations have also been detected in other ERG genes. For example, the point mutations in *ERG3* also occur which could be present either alone or in combination with *ERG11* mutations, resulting in a change in the ratios of various cell sterol biosynthetic intermediates, and increased tolerance to azoles and polyenes. The cytochrome P450 spectral studies performed in a system reconstituted with purified *ERG5* ($\Delta^{22}$-desaturase or CYP61) of *C. glabrata* revealed an interactions between azoles and the heme-protein, implying that *ERG5* could also be a target of azoles and may contribute in the development of antifungal resistance (Lamb et al. 1999). Indeed, *C. albicans* drug resistant clinical isolate with a combination of a single mutation in the *ERG5* gene along with a stretch of amino acid duplication in the *ERG11* gene has been identified (Martel et al. 2010). This mutant was not only resistant to azoles but also displayed collateral resistance to AmpB due to the depletion of membrane ergosterol (Martel et al. 2010). *ERG6* in *C. glabrata* is involved in azole resistance due to various base pair alterations leading to missense mutations (Vandeputte et al. 2007). Similarly, an *erg6* disruptant strain of *C. lusitaniae* was susceptible to AmpB due to decreased membrane ergosterol levels. Coinciding with this, several clinical isolates of *C. lusitaniae* show increased expression of *ERG6* along with a decrease in *ERG3* expression and enhanced resistance to AmpB (Young et al. 2003).

Together, the azole-induced upregulation of *ERG11,* along with other genes of the ergosterol biosynthetic pathway, suggests the existence of a common mechanism of upregulation in *C. albicans* (Henry et al. 2000). Transcript profiling of azole treated *Candida* show that almost 15 % of genes differentially expressed upon drug treatment fall under the category of sterol metabolism in a wild type strain of *C. albicans* (Liu et al. 2005). Notably, while a global regulation of *ERG* genes was evident from the transcript profiling, several genes of diverse functions as well as of unknown functions were also differentially regulated by the drug treatment (De Backer et al. 2001; Liu et al. 2005). This reinforces that azole resistance could be the result of many factors which remains to be identified. The dissection of the mechanisms mediating these phenotypes could provide newer insights into the phenomenon of MDR.

### 14.3.4 Drug Import Impacts Tolerance

The hydrophobic nature of drugs facilitates their easy import by passive diffusion. However, the contribution of drug import in the overall scenario of MDR is not well established. Nonetheless, there are a few instances to suggest that passive diffusion of drugs could be an important determinant of MDR. For example, fluctuations in membrane fluidity are shown to affect passive diffusion leading to an increase in susceptibility to drugs. The *erg* mutants of *C. albicans* possess high membrane fluidity, which led to an enhanced diffusion and susceptibility to azoles (Kohli et al. 2002; Prasad et al. 2010). In another study, permeability constrains imposed by *Candida* cells have been reemphasized in the development of MDR. It is shown that azoles can enter in *C. albicans*, *C. kruesi* and *C. neoformans* cells by diffusion (Mansfield et al. 2010). The kinetics of import in de-energized cells suggests that FLC import proceeds via facilitated diffusion (FD) mediated through a transporter rather than by passive diffusion. Other azoles compete for FLC import, suggesting that all the azoles utilize the same FD mechanism. FLC import was also shown to vary among *C. albicans* resistant clinical isolates, suggesting that altered FD may be a previously uncharacterized mechanism of resistance to azole drugs (Mansfield et al. 2010). However, the identification of a membrane transporter protein involved in FD of azoles remains elusive (Mansfield et al. 2010). Interestingly, drug inactivation which is a common mechanism in bacteria has not been observed in *Candida* cells.

### 14.3.5 Drug Efflux as a Common Strategy of Drug Tolerance

Increased efflux, which results in reduced intracellular accumulation of the incoming drugs, is another prominent mechanism of MDR in fungi (Prasad et al. 1995; Prasad and Kapoor 2005). In *C. albicans,* for example, this is achieved by increasing the efflux of drugs from cells by overproducing the PM efflux pump proteins. An over expression of genes encoding efflux pump proteins, particularly ABC (ATP Binding Cassette) multidrug transporter proteins Cdr1 and Cdr2 or MFS (Major Facilitator Superfamily) efflux pump protein Mdr1, have been commonly observed in azole resistant clinical isolates of *C. albicans* (White T et al. 2002; Karababa et al. 2004; Kusch et al. 2004; Prasad and Kapoor 2005). Invariably, MDR *Candida* cells, which show enhanced expression of efflux pump encoding genes, also show simultaneous increase in the efflux of drugs, thus implying a causal relationship between efflux pump encoding gene expression levels and intracellular concentration of the drug (Cannon et al. 2009). A brief description of these transporters is included, however, for more details, the reader is recommended to see the accompanying chapter "Efflux Pump Proteins of *Candida* in Clinical Drug Resistance".

#### 14.3.5.1 ABC Transporters

The inventory of ABC transporters of *C. albicans* revealed that there are twenty-eight putative ABC superfamily members, including twelve half transporters that largely remain uncharacterized (Gaur et al. 2005). ABC transport proteins are classified into nine families (A to I) according to the nomenclature adopted by the Human Genome Organization (HUGO) (Dean et al. 2001; Verrier et al. 2008). Of these, yeast proteins belonging to ABCB (MDR) (Thornewell et al. 1997; Sanguinetti et al. 2006; Lamping et al. 2010), ABCC (MRP) (Decottignies et al. 1998; Pagant et al. 2010) and ABCG (PDR) (Golin et al. 2007; Prasad and Goffeau 2012) transporters are most often associated with the antifungal resistance.

Full ABC proteins are made up of two (or three) transmembrane domains (TMDs) and two cytoplasmic nucleotide-binding domains (NBDs). NBDs are the nucleotide binding sites, which bind and hydrolyze ATP required to power the efflux of substrates bound within TMDs drug binding sites. Each TMD is usually comprised of six transmembrane segments (TMS), which generally are continuous alpha helices arranged to form drug binding sites (Prasad and Goffeau 2012).

The PDR protein subfamily of *C. albicans* comprises seven full-size members: Cdr1p, Cdr2p, Cdr3p, Cdr4p, Cdr11p, CaSnq2p and Ca4531. The *C. albicans* Cdr1p and Cdr2p proteins are active multidrug transporters, while Cdr3p and Cdr4p do not efflux drugs and play no apparent role in the development of antifungal resistance (Prasad and Goffeau 2012). Other transporters in related fungi, including *CgCDR1* (Sanglard et al. 1999), *CgCDR2 (PDH1)* (Miyazaki et al. 1998) and *SNQ2* (Torelli et al. 2008) in *C. glabrata*, *ABC1* in *C. krusei* (Katiyar and Edlind 2001) and *AFR1* in *C. neoformans* (Sanguinetti et al. 2006), are multidrug transporters and play a role in the development of MDR in these pathogenic species.

#### 14.3.5.2 MFS Transporters

MFS transporters are the second major superfamily of transporters (Saier et al. 1999). A phylogenetic analysis identified 95 putative MFS transporters in *C. albicans* (Gaur et al. 2008). Most MFS transporters consist of two domains of six-TMSs within a single polypeptide chain with few exceptions (Stephanie et al. 1998). On the basis of hydropathy and phylogenetic analysis, the drug efflux MFS proteins can be divided into two distinct types; Drug: $H^+$ Antiporter-1 (DHA1), consisting of 12 TMSs and Drug: $H^+$ Antiporter-2 (DHA2) that contains 14 TMSs. MDR1 of DHA1 subfamily is a major multidrug transporter of *C. albicans*. Homologues of *CaMDR1* have been identified from *C. dubliniensis* and *C. glabrata*, which are designated as *CdMDR1* and *CgMDR1*, respectively (Moran et al. 1998; Sanglard et al. 1999). It appears that an increased expression of *CdMDR1* is one of the main mechanisms of FLC resistance in clinical isolates of *C. dubliniensis* (Moran et al. 1998). Since *CgMDR1* confers specific resistance to FLC, its constitutive expression in *C. glabrata* may be responsible for the intrinsically low susceptibility of this yeast species to triazoles (Sanglard et al. 1999).

Among all the MFS proteins, only one member, *MDR1,* has been implicated clinically to be involved in azole resistance in *S. cerevisiae. FLU1*, a close homologue of *MDR1* has also been implicated in FLC resistance in *S. cerevisiae*. However, an over expression of *FLU1* has not been detected in FLC resistant clinical isolates of *C. albicans*. None of the other 95 members of this superfamily are implicated in MDR (Gaur et al. 2008).

As an important MDR gene of the MFS family, *MDR1* of *C. albicans* has been extensively studied for its role in drug resistance. The functional evaluation of critical amino acid residues of the Mdr1 protein revealed that the residues of TMS5 which harbor antiporter motifs are potentially significant for their functionality and contribute to drug:$H^+$ transport. Independent of the substrate specificity of the antiporter, the antiporter motif in the predicted TMS5 is well conserved in all of the functionally related subgroups in bacteria and plants (Pasrija et al. 2007).

### *14.3.6 Echinocandin Resistance*

Echinocandins inhibit the synthesis of β-1,3-glucans which is one of the major component of fungal CW (Fig. 14.5). Mutation in the FKS genes encoding echinocandin drug target glucan synthase enzyme results in its decreased sensitivity towards drug and development of resistance (Fig. 14.5) (Perlin 2007). Point mutations in FKS genes are the only known mechanism by which fungi develop resistance to echinocandin antifungal drugs (Park et al. 2005; Balashov et al. 2006). Drug resistant mutations developed in FKS genes generally fall in two "hot spots" regions of *FKS1*, essential for enzyme activity (Perlin 2007). Garcia-Effron et al. has also reported that mutations in Fks1 protein (Fks1p) lower the activity of β-glucan synthase without altering its affinity for the drug as is also the case with azole drug target Erg11p (Garcia-Effron et al. 2009). Fks1p mutations in "hot-spot" regions have been characterized in *C. albicans* as well as in non-albicans spp. (Perlin 2011). A paralog of *FKS1* in *C. glabrata*, *FKS2*, is also responsible for echinocandin resistance (Perlin 2007).

## 14.4 Concluding Remarks

The available arsenals of antifungals targeting mostly sterols or its synthesis machinery or CW components of fungal cells are reasonably successful in combating fungal infections. However, the fungistatic nature of many of these antifungals limits their success. The synergy among different drugs is being also projected as an alternate strategy. The limited availability of antifungals and emergence of clinical drug resistance necessitate search of newer compounds and new targets. The current research does promise for novel targets and better drugs.

# References

Balashov SV, Park S, Perlin DS (2006) Assessing resistance to the echinocandin antifungal drug caspofungin in *Candida albicans* by profiling mutations in *FKS1*. Antimicrob Agents Chemother 50:2058–2063

Bills GF, Platas G, Overy DP, Collado J, Fillola A, Jiménez MR et al (2009) Discovery of the parnafungins, antifungal metabolites that inhibit mRNA polyadenylation, from the Fusarium larvarum complex and other Hypocrealean fungi. Mycologia 101:449–472

Bink A, Pellens K, Cammue BPA, Thevissen K (2011) Anti-Biofilm Strategies: how to Eradicate *Candida* Biofilms? Open Mycol J 5:29–38

Brown GD, Denning DW, Gow NA, Levitz SM, Netea MG, White TC (2012) Hidden killers: human fungal infections. Sci Transl Med 4:165rv13

Calabrese EC, Castellano S, Santoriello M, Sgherri C, Quartacci MF, Calucci L et al (2013) Antifungal activity of azole compounds CPA18 and CPA109 against azole-susceptible and -resistant strains of *Candida albicans*. J Antimicrob Chemother 68:1111–1119

Cannon RD, Lamping E, Holmes AR, Niimi K, Baret PV, Keniya MV et al (2009) Efflux-mediated antifungal drug resistance. Clin Microbiol Rev 22:291–321

Capobianco JO, Zakula D, Coen ML, Goldman RC (1993) Anti-Candida activity of cispentacin: the active transport by amino acid permeases and possible mechanisms of action. Biochem Biophys Res Commun 190:1037–1044

De Backer MD, Ilyina T, Ma XJ, Vandoninck S, Luyten WH, Vanden Bossche H (2001) Genomic profiling of the response of Candida albicans to itraconazole treatment using a DNA microarray. Antimicrob Agents Chemother 45:1660–1670

Dean M, Andrey R, Rando A (2001) The human ATP-binding cassette (ABC) transporter superfamily. Genome Res 11:1156–1166

Decottignies A, Grant AM, Nichols JW, De Wet H, McIntosh DB, Goffeau A (1998) ATPase and multidrug transport activities of the overexpressed yeast ABC protein Yor1p. J Biol Chem 273:12612–12622

Dhamgaye S, Devaux F, Manoharlal R, Vandeputte P, Shah AH, Singh A et al (2012) In vitro effect of malachite green on *Candida albicans* involves multiple pathways and transcriptional regulators *UPC2* and *STP2*. Antimicrob Agents Chemother 56:495–506

Dhamgaye S, Devaux F, Vandeputte P, Khandelwal NK, Sanglard D, Mukhopadhyay G et al (2014) Molecular mechanisms of action of herbal antifungal alkaloid berberine, in *Candida albicans*. PLoS One 9:e104554

Domínguez JM, Martín JJ (1998) Identification of elongation factor 2 as the essential protein targeted by sordarins in *Candida albicans*. Antimicrob Agents Chemother 42:2279–2283

Domínguez JM, Kelly VA, Kinsman OS, Marriott MS, Gómez de las Heras F, Martín JJ (1998) Sordarins: a new class of antifungals with selective inhibition of the protein synthesis elongation cycle in yeasts. Antimicrob Agents Chemother 42:2274–2278

Espinel-Ingroff A (2009) Novel antifungal agents, targets or therapeutic strategies for the treatment of invasive fungal diseases: a review of the literature (2005–2009). Rev Iberoam Micol 26:15–22

Espinel-Ingroff A, Canton E, Martin-Mazuelos E, Pemán J (2009) Pharmacotherapy of *Candida* Infections with Echinocandins. Clin Med Ther 1:889–897

Flowers SA, Barker KS, Berkow EL, Toner G, Chadwick SG, Gygax SE et al (2012) Gain-of function mutations in *UPC2* are a frequent cause of *ERG11* upregulation in azole-resistant clinical isolates of *Candida albicans*. Eukaryot Cell 11:1289–1299

Fostel JM, Lartey PA (2000) Emerging novel antifungal agents. Drug Discov Today 5:25–32

Gallo-Ebert C, Donigan M, Stroke IL, Swanson RN, Manners MT, Francisco J et al (2014) Novel antifungal drug discovery based on targeting pathways regulating the fungus-conserved Upc2 transcription factor. Antimicrob Agents Chemother 58:258–266

Garbati MA, Alasmari FA, Al-Tannir MA, Tleyjeh IM (2012) The role of combination antifungal therapy in the treatment of invasive aspergillosis: a systematic review. Int J Infect Dis 16:e76–e81

Garcia-Effron G, Park S, Perlin DS (2009) Correlating echinocandin MIC and kinetic inhibition of fks1 mutant glucan synthases for *Candida albicans*: implications for interpretive breakpoints. Antimicrob Agents Chemother 53:112–122
Gaur M, Choudhury D, Prasad R (2005) Complete inventory of ABC proteins in human pathogenic yeast, *Candida albicans*. J Mol Microbiol Biotechnol 9:3–15
Gaur M, Puri N, Manoharlal R, Rai V, Mukhopadhayay G, Choudhury D et al (2008) MFS transportome of the human pathogenic yeast *Candida albicans*. BMC Genomics 9:579
Ghannoum MA, Spellberg BJ, Ibrahim AS, Ritchie JA, Currie B, Spitzer ED et al (1994) Sterol composition of *Cryptococcus neoformans* in the presence and absence of fluconazole. Antimicrob Agents Chemother 38:2029–2033
Golin J, Ambudkar SV, May L (2007) The yeast Pdr5p multidrug transporter: how does it recognize so many substrates? Biochem Biophys Res Commun 356:1–5
Gow NA, Hube B (2012) Importance of the *Candida albicans* cell wall during commensalism and infection. Curr Opin Microbiol 15:406–412
Gubbins PO, Anaissie E (2006) Overview of antifungal agents. Pharmacy practice news special edition 59–64
Gunawardana G, Rasmussen RR, Scherr M, Frost D, Brandt KD, Choi W et al (1997) Corynecandin: a novel antifungal glycolipid from Coryneum modonium. J Antibiot (Tokyo) 50:884–886
Harris GH, Shafiee A, Cabello MA, Curotto JE, Genilloud O, Göklen KE et al (1998) Inhibition of fungal sphingolipid biosynthesis by rustmicin, galbonolide B and their new 21-hydroxy analogs. J Antibiot (Tokyo) 51:837–844
Heilmann CJ, Schneider S, Barker KS, Rogers PD, Morschhäuser J (2010) An A643T mutation in the transcription factor Upc2p causes constitutive *ERG11* upregulation and increased fluconazole resistance in *Candida albicans*. Antimicrob Agents Chemother 54:353–359
Heitman J (2011) Microbial pathogens in the fungal kingdom. Fungal Biol Rev 25:48–60
Henry KW, Nickels JT, Edlind TD (2000) Upregulation of ERG genes in *Candida* species by azoles and other sterol biosynthesis inhibitors. Antimicrob Agents Chemother 44:2693–2700
Herreros E, Martinez CM, Almela MJ, Marriott MS, De Las Heras FG, Gargallo-Viola D (1998) Sordarins: in vitro activities of new antifungal derivatives against pathogenic yeasts, *Pneumocystis carinii*, and filamentous fungi. Antimicrob Agents Chemother 42:2863–2869
Hodgetts S, Nooney L, Al-Akeel R, Curry A, Awad S, Matthews R et al (2008) Efungumab and caspofungin: pre-clinical data supporting synergy. J Antimicrob Chemother 61:1132–1139
Hoot SJ, Smith AR, Brown RP, White TC (2011) An A643V amino acid substitution in Upc2p contributes to azole resistance in well-characterized clinical isolates of *Candida albicans*. Antimicrob Agents Chemother 55:940–942
Iwamoto T, Tsujii E, Ezaki M, Fujie A, Hashimoto S, Okuhara M et al (1990) FR109615, a new antifungal antibiotic from *Streptomyces setonii*. Taxonomy, fermentation, isolation, physico-chemical properties and biological activity. J Antibiot (Tokyo) 43:1–7
Jiang B, Xu D, Allocco J, Parish C, Davison J, Veillette K et al (2008) PAP inhibitor with in vivo efficacy identified by *Candida albicans* genetic profiling of natural products. Chem Biol 15:363–374
Jiménez-Ortigosa C, Paderu P, Motyl MR, Perlin DS (2014) Enfumafungin derivative MK-3118 shows increased in vitro potency against clinical echinocandin-resistant *Candida* Species and *Aspergillus* species isolates. Antimicrob Agents Chemother 58:1248–1251
Kakeya H, Miyazaki Y, Senda H, Kobayashi T, Seki M, Izumikawa K et al (2008a) Efficacy of SPK-843, a novel polyene antifungal, in comparison with amphotericin B, liposomal amphotericin B, and micafungin against murine pulmonary aspergillosis. Antimicrob Agents Chemother 52:1868–1870
Kakeya H, Miyazaki Y, Senda H, Kobayashi T, Seki M, Izumikawa K et al (2008b) Efficacy of SPK-843, a novel polyene antifungal, in a murine model of systemic cryptococcosis. Antimicrob Agents Chemother 52:1871–1872
Kaneto R, Chiba H, Agematu H, Shibamoto N, Yoshioka T, Nishida H et al (1993) Mer-WF3010, a new member of the papulacandin family. I. Fermentation, isolation and characterization. J Antibiot (Tokyo) 46:247–250

Kantarcioglu AS, Yucel A, Vidotto V (2003) In vitro activity of a new polyene SPK-843 against *Candida* spp, *Cryptococcus neoformans* and *Aspergillus* spp. clinical isolates. J Chemother 15:296–298

Karababa M, Coste AT, Rognon B, Bille J, Sanglard D (2004) Comparison of gene expression profiles of *Candida albicans* azole-resistant clinical isolates and laboratory strains exposed to drugs inducing multidrug transporters. Antimicrob Agents Chemother 48:3064–3079

Katiyar SK, Edlind TD (2001) Identification and expression of multidrug resistance related ABC transporter genes in *Candida krusei*. Med Mycol 39:109–116

Kinsman OS, Chalk PA, Jackson HC, Middleton RF, Shuttleworth A, Rudd BA et al (1998) Isolation and characterisation of an antifungal antibiotic (GR135402) with protein synthesis inhibition. J Antibiot (Tokyo) 51:41–49

Kitamura A, Someya K, Hata M, Nakajima R, Takemura M (2009) Discovery of a small-molecule inhibitor of {beta}-1,6-glucan synthesis. Antimicrob Agents Chemother 53:670–677

Kohli A, Smirti, Mukhopadhyay K, Rattan A, Prasad R (2002) In vitro low-level resistance to azole in *Candida albicans* is associated with changes in membrane fluidity and asymmetry. Antimicrob Agents Chemother 46:1046–1052

Konishi M, Nishio M, Saitoh K, Miyaki T, Oki T, Kawaguchi H (1989) Cispentacin, a new antifungal antibiotic. I. Production, isolation, physico-chemical properties and structure. J Antibiot (Tokyo) 42:1749–1755

Kusch H, Biswas K, Schwanfelder S, Engelmann S, Rogers PD, Hecker M et al (2004) A proteomic approach to understanding the development of multidrug-resistant Candida albicans strains. Mol Genet Genomics 271:554–565

Lamb DC, Maspahy S, Kelly DE, Manning NJ, Geber A, Bennett JE et al (1999) Purification, reconstitution, and inhibition of cytochrome P-450 sterol delta22-desaturase from the pathogenic fungus *Candida glabrata*. Antimicrob Agents Chemother 43:1725–1728

Lamping E, Baret PV, Holmes AR, Monk BC, Goffeau A, Cannon RD (2010) Fungal PDR transporters: phylogeny, topology, motifs and function. Fungal Genet Biol 47:127–142

Lemke A, Kiderlen AF, Kayser O (2005) Amphotericin B. Appl Microbiol Biotechnol 68:151–162

Lewis RE, Kontoyiannis DP (2001) Rationale for combination antifungal therapy. Pharmacotherapy 21:149S–164S

Liu TT, Lee REB, Barker KS, Lee RE, Wei L, Homayouni R et al (2005) Genome-wide expression profiling of the response to azole, polyene, echinocandin, and pyrimidine antifungal agents in *Candida albicans*. Antimicrob Agents Chemother 49:2226–2236

López-Martínez R (2010) Candidosis, a new challenge. Clin Dermatol 28:178–184

Mandala SM, Thornton RA, Rosenbach M, Milligan J, Garcia-Calvo M, Bull HG et al (1997) Khafrefungin, a novel inhibitor of sphingolipid synthesis. J Biol Chem 272:32709–32714

Mandala SM, Thornton RA, Milligan J, Rosenbach M, Garcia-Calvo M, Bull HG et al (1998) Rustmicin, a potent antifungal agent, inhibits sphingolipid synthesis at inositol phosphoceramide synthase. J Biol Chem 273:14942–14949

Mansfield BE, Oltean HN, Oliver BG, Hoot SJ, Leyde SE, Hedstrom L et al (2010) Azole drugs are imported by facilitated diffusion in *Candida albicans* and other pathogenic fungi. PLoS Pathog 6(9):e1001126

Marichal P, Koymans L, Willemsens S, Bellens D, Verhasselt P, Luyten W et al (1999) Contribution of mutations in the cytochrome P450 14alpha-demethylase (Erg11p, Cyp51p) to azole resistance in *Candida albicans*. Microbiology 145:2701–2713

Martel CM, Parker JE, Bader O, Weig M, Gross U, Warrilow AG et al (2010) A clinical isolate of *Candida albicans* with mutations in *ERG11* (encoding sterol 14alpha-demethylase) and *ERG5* (encoding C22 desaturase) is cross resistant to azoles and amphotericin B. Antimicrob Agents Chemother 54:3578–3583

Miceli MH, Bernardo SM, Lee SA (2009) In vitro analyses of the combination of high-dose doxycycline and antifungal agents against *Candida albicans* biofilms. Int J Antimicrob Agents 34:326–332

Mitsuyama J, Nomura N, Hashimoto K, Yamada E, Nishikawa H, Kaeriyama M et al (2008) In vitro and in vivo antifungal activities of T-2307, a novel arylamidine. Antimicrob Agents Chemother 52:1318–1324

Miyazaki H, Miyazaki Y, Geber A, Parkinson T, Hitchcock C et al (1998) Fluconazole resistance associated with drug efflux and increased transcription of a drug transporter gene, PDH1, in *Candida glabrata*. Antimicrob Agents Chemother 42:1695–1701

Moran GP, Sanglard D, Donnelly SM, Shanley DB, Sullivan DJ, Coleman DC (1998) Identification and expression of multidrug transporters responsible for fluconazole resistance in *Candida dubliniensis*. Antimicrob Agents Chemother 42:1819–1830

Morio F, Loge C, Besse B, Hennequin C, Le Pape P (2010) Screening for amino acid substitutions in the *Candida albicans* Erg11 protein of azole-susceptible and azole-resistant clinical isolates: new substitutions and a review of the literature. Diagn Microbiol Infect Dis 66:373–384

Munro CA, Winter K, Buchan A, Henry K, Becker JM, Brown AJ et al (2001) Chs1 of *Candida albicans* is an essential chitin synthase required for synthesis of the septum and for cell integrity. Mol Microbiol 39:1414–1426

Nagiec MM, Nagiec EE, Baltisberger JA, Wells GB, Lester RL, Dickson RC (1997) Sphingolipid synthesis as a target for antifungal drugs. Complementation of the inositol phosphorylceramide synthase defect in a mutant strain of *Saccharomyces cerevisiae* by the *AUR1* gene. J Biol Chem 272:9809–9817

Nishi I, Sunada A, Toyokawa M, Asari S, Iwatani Y (2009) In vitro antifungal combination effects of micafungin with fluconazole, voriconazole, amphotericin B, and flucytosine against clinical isolates of *Candida* species. J Infect Chemother 15:1–5

Ogawa A, Hashida-Okado T, Endo M, Yoshioka H, Tsuruo T, Takesako K et al (1998) Role of ABC transporters in aureobasidin A resistance. Antimicrob Agents Chemother 42:755–761

Okada H, Kamiya S, Shiina Y, Suwa H, Nagashima M, Nakajima S et al (1998) BE-31405, a new antifungal antibiotic produced by Penicillium minioluteum. I. Description of producing organism, fermentation, isolation, physico-chemical and biological properties. J Antibiot (Tokyo) 51:1081–1086

Olson JA, Adler-Moore JP, Smith PJ, Proffitt RT (2005) Treatment of *Candida glabrata* infection in immunosuppressed mice by using a combination of liposomal amphotericin B with caspofungin or micafungin. Antimicrob Agents Chemother 49:4895–4902

Pagant S, Halliday JJ, Kougentakis C, Miller EA (2010) Intragenic suppressing mutations correct the folding and intracellular traffic of misfolded mutants of Yor1p, a eukaryotic drug transporter. J Biol Chem 285:36304–36314

Pai MP, Samples ML, Mercier RC, Spilde MN (2008) Activities and ultrastructural effects of antifungal combinations against simulated *Candida* endocardial vegetations. Antimicrob Agents Chemother 52:2367–2376

Parish CA, Smith SK, Calati K, Zink D, Wilson K, Roemer T et al (2008) Isolation and structure elucidation of parnafungins, antifungal natural products that inhibit mRNA polyadenylation. J Am Chem Soc 130:7060–7066

Park S, Kelly R, Kahn JN, Robles J, Hsu MJ, Register E et al (2005) Specific substitutions in the echinocandin target Fks1p account for reduced susceptibility of rare laboratory and clinical *Candida* sp. isolates. Antimicrob Agents Chemother 49:3264–3273

Pasrija R, Banerjee D, Prasad R (2007) Structure and function analysis of CaMdr1p, a major facilitator superfamily antifungal efflux transporter protein of *Candida albicans*: identification of amino acid residues critical for drug/H+ transport. Eukaryot Cell 6:443–453

Perlin DS (2007) Resistance to echinocandin-class antifungal drugs. Drug Resist Updat 10:121–130

Perlin DS (2011) Current perspectives on echinocandin class drugs. Future Microbiol 6:441–457

Petraitis V, Petraitiene R, Kelaher AM, Sarafandi AA, Sein T, Mickiene D et al (2004) Efficacy of PLD-118, a novel inhibitor of candida isoleucyl-tRNA synthetase, against experimental oropharyngeal and esophageal candidiasis caused by fluconazole-resistant *C. albicans*. Antimicrob Agents Chemother 48:3959–3967

Prasad R, Goffeau A (2012) Yeast ATP-binding cassette transporters conferring multidrug resistance. Annu Rev Microbiol 66:39–63

Prasad R, Kapoor K (2005) Multidrug resistance in yeast *Candida*. Int Rev Cytol 242:215–248

Prasad R, DeWergifosse P, Goffeau A, Balzi E (1995) Molecular cloning and characterization of a novel gene of Candida albicans, CDR1, conferring multiple resistance to drugs and antifungals. Curr Genet 27:320–329

Prasad T, Hameed S, Manoharlal R, Biswas S, Mukhopadhyay CK, Goswami SK et al (2010) Morphogenic regulator EFG1 affects the drug susceptibilities of pathogenic *Candida albicans*. FEMS Yeast Res 10:587–596

Rodrigues ME, Silva S, Azeredo J, Henriques M (2014) Novel strategies to fight *Candida* species infection. Crit Rev Microbiol 10:1–13

Roemer T, Krysan DJ (2014) Antifungal drug development: challenges, unmet clinical needs, and new approaches. Cold Spring Harb Perspect Med 4 pii:a019703

Saier MH Jr, Beatty JT, Goffeau A, Harley KT, Heijne WHM, Huang SC et al (1999) The major facilitator superfamily. J Mol Microbiol Biotechnol 1:257–279

Sanglard D, Ischer F, Calabrese D, Majcherczyk PA, Bille J (1999) The ATP binding cassette transporter gene *CgCDR1* from *Candida glabrata* is involved in the resistance of clinical isolates to azole antifungal agents. Antimicrob Agents Chemother 43:2753–2765

Sanglard D, Ischer F, Parkinson T, Falconer D, Bille J (2003) *Candida albicans* mutations in the ergosterol biosynthetic pathway and resistance to several antifungal agents. Antimicrob Agents Chemother 47:2404–2412

Sanglard D, Coste A, Ferrari S (2009) Antifungal drug resistance mechanisms in fungal pathogens from the perspective of transcriptional gene regulation. FEMS Yeast Res 9:1029–1050

Sanguinetti M, Posteraro B, La Sorda M, Torelli R, Fiori B, Santangelo R et al (2006) Role of AFR1, an ABC transporter-encoding gene, in the in vivo response to fluconazole and virulence of *Cryptococcus neoformans*. Infect Immun 74:1352–1359

Sasse C, Schillig R, Reimund A, Merk J, Morschhäuser J (2012) Inducible and constitutive activation of two polymorphic promoter alleles of the *Candida albicans* multidrug efflux pump *MDR1*. Antimicrob Agents Chemother 56:4490–4494

Serena C, Fernández-Torres B, Pastor FJ, Trilles L, Lazéra Mdos S, Nolard N et al (2005) In vitro interactions of micafungin with other antifungal drugs against clinical isolates of four species of *Cryptococcus*. Antimicrob Agents Chemother 49:2994–2996

Shapiro RS, Robbins N, Cowen LE (2011) Regulatory circuitry governing fungal development, drug resistance, and disease. Microbiol Mol Biol Rev 75:213–267

Sharma M, Manoharlal R, Shukla S, Puri N, Prasad T, Ambudkar SV et al (2009) Curcumin modulates efflux mediated by yeast ABC multidrug transporters and is synergistic with antifungals. Antimicrob Agents Chemother 53:3256–3265

Sharma M, Manoharlal R, Puri N, Prasad R (2010) Antifungal curcumin induces reactive oxygen species and triggers an early apoptosis but prevents hyphae development by targeting the global repressor *TUP1* in *Candida albicans*. Biosci Rep 30:391–404

Shi W, Chen Z, Chen X, Cao L, Liu P, Sun S (2010) The combination of minocycline and fluconazole causes synergistic growth inhibition against *Candida albicans*: an in vitro interaction of antifungal and antibacterial agents. FEMS Yeast Res 10:885–893

Shibata T, Takahashi T, Yamada E, Kimura A, Nishikawa H, Hayakawa H et al (2012) T-2307 causes collapse of mitochondrial membrane potential in yeast. Antimicrob Agents Chemother 56:5892–5897

St Georgiev V (2000) Membrane transporters and antifungal drug resistance. Curr Drug Targets 1:261–284

Stephanie SP, Paulsen IT, Saier MH Jr (1998) Major facilitator superfamily. Microbiol Mol Biol Rev 62:1–34

Thornewell SJ, Peery RB, Skatrud PL (1997) Cloning and characterization of CneMDR1: a *Cryptococcus neoformans* gene encoding a protein related to multidrug resistance proteins. Gene 201:21–29

Tobudic S, Kratzer C, Lassnigg A, Graninger W, Presterl E (2010a) In vitro activity of antifungal combinations against *Candida albicans* biofilms. J Antimicrob Chemother 65:271–274

Tobudic S, Lassnigg A, Kratzer C, Graninger W, Presterl E (2010b) Antifungal activity of amphotericin B, caspofungin and posaconazole on *Candida albicans* biofilms in intermediate and mature development phases. Mycoses 53:208–214

Torelli R, Posteraro B, Ferrari S, La Sorda M, Fadda G, Sanglard D et al (2008) The ATP-binding cassette transporter-encoding gene *CgSNQ2* is contributing to the CgPDR1-dependent azole resistance of *Candida glabrata*. Mol Microbiol 68:186–201

Uppuluri P, Nett J, Heitman J, Andes D (2008) Synergistic effect of calcineurin inhibitors and fluconazole against *Candida albicans* biofilms. Antimicrob Agents Chemother 52:1127–1132

Vanden Bossche H, Marichal P, Le Jeune L, Coene MC, Gorrens J, Cools W (1993) Effects of itraconazole on cytochrome P-450-dependent sterol 14 alpha-demethylation and reduction of 3-ketosteroids in *Cryptococcus neoformans*. Antimicrob Agents Chemother 37:2101–2105

Vandeputte P, Tronchin G, Berge's T, Hennequin C, Chabasse D, Bouchara JP (2007) Reduced susceptibility to polyenes associated with a missense mutation in the *ERG6* gene in a clinical isolate of *Candida glabrata* with pseudohyphal growth. Antimicrob Agents Chemother 51:982–990

Vandeputte P, Ferrari S, Coste AT (2012) Antifungal resistance and new strategies to control fungal infections. Int J Microbiol 2012:713687

Verrier PJ, Bird D, Burla B, Dassa E, Forestier C, Geisler M et al (2008) Plant ABC proteins—a unified nomenclature and updated inventory. Trends Plant Sci 13:151–159

Wang H, Kong F, Sorrell TC, Wang B, McNicholas P, Pantarat N et al (2009) Rapid detection of *ERG11* gene mutations in clinical *Candida albicans* isolates with reduced susceptibility to fluconazole by rolling circle amplification and DNA sequencing. BMC Microbiol 14:167

White TC, Silver PM (2005) Regulation of sterol metabolism in *Candida albicans* by the *UPC2* gene. Biochem Soc Trans 33:1215–1218

White TC, Marr KA, Bowden RA (1998) Clinical, cellular, and molecular factors that contribute to antifungal drug resistance. Clin Microbiol Rev 11:382–402

White TC, Holleman S, Dy F, Mirels LF, Stevens DA (2002) Resistance mechanisms in clinical isolates of *Candida albicans*. Antimicrob Agents Chemother 46:1704–1713

Wiederhold NP, Najvar LK, Fothergill AW, Bocanegra R, Olivo M, McCarthy DI et al (2015) The Novel Arylamidine T-2307 Maintains In Vitro and In Vivo Activity against Echinocandin-Resistant *Candida albicans*. Antimicrob Agents Chemother 59:1341–1343

Young LY, Hull CM, Heitman J (2003) Disruption of ergosterol biosynthesis confers resistance to amphotericin B in *Candida lusitaniae*. Antimicrob Agents Chemother 47:2717–2724

Zhang YQ, Gamarra S, Garcia-Effron G, Park S, Perlin DS, Rao R (2010) Requirement for ergosterol in V-ATPase function underlies antifungal activity of azole drugs. PLoS Pathog 6:e1000939

# Chapter 15
# Candida Efflux ATPases and Antiporters in Clinical Drug Resistance

**Rajendra Prasad, Manpreet Kaur Rawal, and Abdul Haseeb Shah**

**Abstract** An enhanced expression of genes encoding ATP binding cassette (ABC) and major facilitator superfamily (MFS) transport proteins are known to contribute to the development of tolerance to antifungals in pathogenic yeasts. For example, the azole resistant (AR) clinical isolates of the opportunistic human fungal pathogen *Candida albicans* show an overexpression of *CDR1* and/or *CaMDR1* belonging to ABC and MFS, superfamilies, respectively. The reduced accumulation (due to rapid efflux) of drugs in AR isolates confirms the role of efflux pump proteins in the development of drug tolerance. Considering the importance of major multidrug transporters, the focus of recent research has been to understand the structure and function of these proteins which could help to design inhibitors/modulators of these pump proteins. This chapter focuses on some aspects of the structure and function of yeast transporter proteins particularly in relation to MDR in *Candida*.

**Keywords** Drug resistance • ABC transporters • MFS transporters • Cdr1p • CaMdr1p

## 15.1 Introduction

The occurrence of fungal infections has risen dramatically over the past few decades because of the increase in number of immunocompromised patients undergoing, transplantation surgery, cancer chemotherapy and HIV infection etc. (Richardson 2005). The severity of the fungal infections varies from the superficial infections

R. Prasad (✉)
Membrane Biology Laboratory, School of Life Sciences, Jawaharlal Nehru University, New Delhi 110067, India

AMITY Institute of Integrative Sciences and Health (AIISH), Amity University Haryana, Manesar, Gurgaon, Haryana, India
e-mail: rp47jnu@gmail.com; rprasad@ggn.amity.edu

M.K. Rawal • A.H. Shah
Membrane Biology Laboratory, School of Life Sciences, Jawaharlal Nehru University, New Delhi 110067, India
e-mail: manpreet.rawal@gmail.com; hasb789biotech@gmail.com



J. Ramos et al. (eds.), *Yeast Membrane Transport*, Advances in Experimental Medicine and Biology 892, DOI 10.1007/978-3-319-25304-6_15

to life-threatening systemic infections. The fungal genera most often associated with the invasive fungal infections include species of *Aspergillus, Candida, and Cryptococcus* (Pfaller and Diekema 2007). The opportunistic *Candida albicans* accounts for approximately 50–60 % causes of candidiasis particularly in immunocompromised patients. But the infections caused by non-albicans species such as by *C. glabrata, C. parapsilosis, C. tropicalis, and C. krusei* are also common (Pfaller et al. 2014).

Fungal infections are commonly treated with the azoles or echinocandin drugs (Perlin 2011). However, the widespread and prolonged use of antifungals results in the development of tolerance to not only drug in use but also results in collateral increased resistance to other drugs. This development of resistance against a spectrum of drugs that share neither a common target nor a common structure is termed as multidrug resistance (MDR). Intrinsic and acquired antifungal resistance is a growing clinical problem, which poses a hurdle in antifungal therapy. Combating MDR poses a major challenge to clinicians since it is a multi-factorial phenomenon where a combination of mechanisms could contribute in the development of drug tolerance.

One of the most important mechanisms of drug resistance is the failure of drug accumulation inside the cell resulting from an up regulation of membrane-associated transporters and antiporters acting as multidrug efflux pumps. Several azole resistant (AR) clinical isolates of *C. albicans* as well as of other fungal pathogens like *A. fumigates* and *C. neoformans* display transcriptional activation of genes encoding these drug efflux pump proteins which coincides with the reduced intracellular accumulation of the drugs, thus confirming the role of efflux proteins in drug extrusion and tolerance (Prasad et al. 2002). These drug transporters act like cell bouncers controlling the permeability of the cells. Therefore, the focus of research has been to understand the structure and function of multidrug transporter proteins and to design inhibitors/modulators which can jam the efflux activity of these transporters.

## 15.2 Efflux Pumps

There are two main classes of efflux pump proteins responsible for fungal drug resistance, involving different mechanistic strategies to power efflux of drugs. For example, while superfamily of ATP binding cassette (ABC) proteins which are primary active transporters utilize energy derived directly from ATP hydrolysis to efflux drugs. The major facilitator superfamily (MFS) proteins are secondary active transporters that utilize energy derived from the proton motive force to efflux different substrates (Cannon et al. 2009). Both classes of transporters are integral membrane proteins with distinctive functional domains. ABC transporters contain transmembrane domains (TMDs) as well as nucleotide binding domains (NBDs) while MFS transporters only possess TMDs (Figs. 15.1 and 15.2).

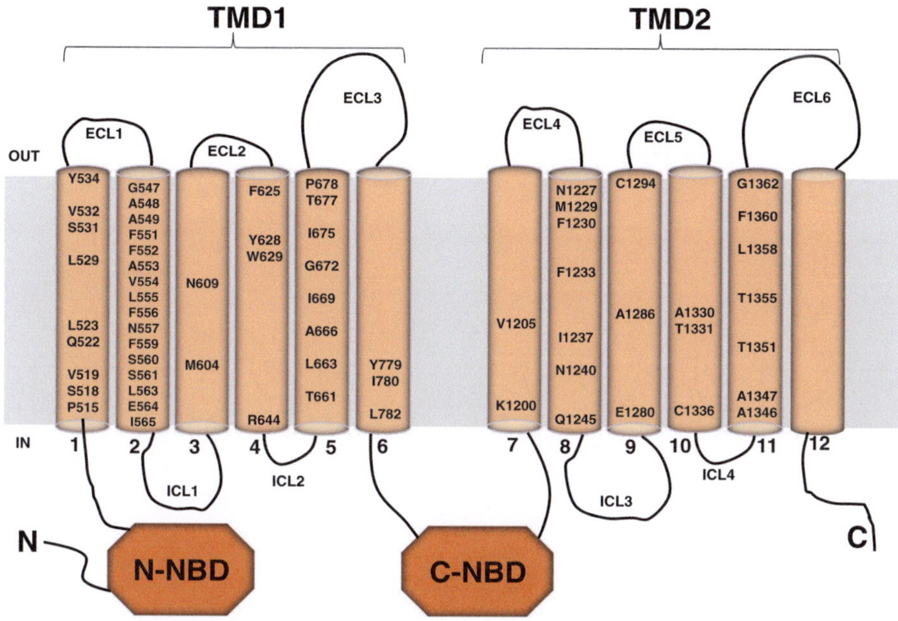


**Fig. 15.1** Schematic representation of Cdr1p predicted topology. Cdr1p is predicted to have two transmembrane domains (TMD1 and TMD2) proceeded by two NBDs, (N- and C-terminal NBDs). Each TMD consists of six TMSs, which are numbered *1–6* in TMD1 and *7–12* in TMD2. Interconnecting extracellular loops (ECL1-6) and intracellular loops (ICL1-4) are indicated. The amino and carboxyl terminals of protein are indicated. TMDs residues that result in susceptible phenotype upon mutation are shown within respective TMSs with TMS1, TMS2, TMS5, TMS8 and TMS11 contributing higher number of critical residues as compared to TMS3, TMS4, TMS6, TMS7, TMS9 and TMS10

### *15.2.1 ABC Transporters*

Sequencing of fungal genomes has revealed the repertoire of ABC proteins in a range of species (Dassa and Bouige 2001). For example, both *Saccharomyces cerevisiae and C. albicans* have over 25 putative ABC proteins in its genome (Decottignies and Goffeau 1997; Gaur et al. 2005). The putative ABC transporters of yeasts are listed in Table 15.1. Yeast ABC transporters are localized in the outer plasma membrane or in the membranes of the intracellular organelles and are capable of transporting a wide variety of agents which includes, ions, steroids, sugars, amino acids, vitamins, peptides, membrane phospholipid translocation (Rawal et al. 2013; Prasad and Rawal 2014). ABC transport proteins are classified into nine families (A to I) according to the nomenclature adopted by the human genome organization (HUGO) (Dean et al. 2001). Of these, yeast proteins belonging to ABCB (MDR), ABCC (MRP) and ABCG (PDR) transporters are most often associated with the antifungal resistance (Thornewell et al. 1997; Decottignies et al. 1998; Sanguinetti et al. 2006; Lamping et al. 2010; Pagant et al. 2010; Prasad and Goffeau 2012).

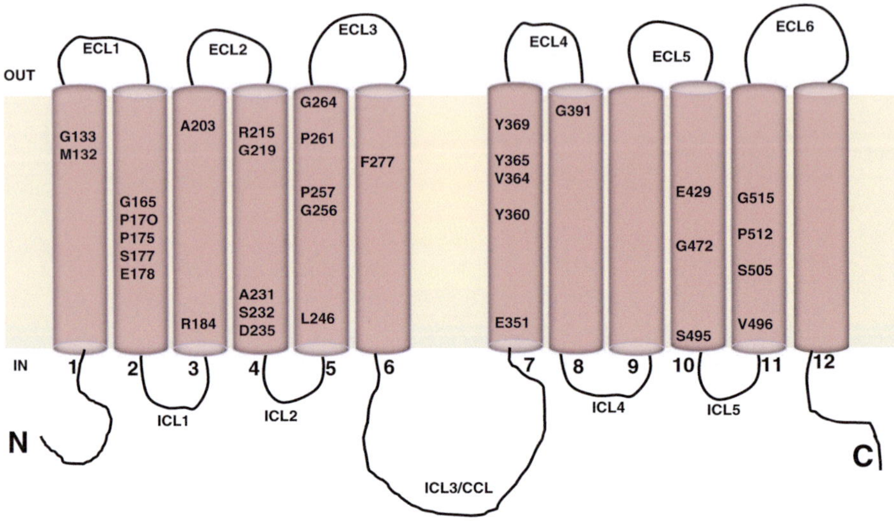


**Fig. 15.2** Schematic representation of *Ca*Mdr1p. *Ca*Mdr1p is characterized by two structural units of six TMSs, linked by a central cytoplasmic loop (ICL3/CCL). All the TMSs (numbered *1–12*) are interconnected by ECLs (ECL1-6) and ICLs (ICL1-5). The amino and carboxyl terminals of protein are indicated. TMDs residues that result in susceptible phenotype upon mutation are shown within respective TMSs

### *15.2.2 ABC Proteins of Candida as Multidrug Transporters*

Among 28 putative ABC transporters which exist in the genome of *C. albicans*, *Candida* Drug Resistance Protein (Cdr1p) is a major drug transporter (Prasad et al. 1995). An overexpression of *CDR1* gene encoding Cdr1p is associated with an increased efflux of incoming drug in AR clinical isolates recovered from patients receiving long term antifungal therapy (Sanglard et al. 1995). It is observed that the high level of expression of *CDR1* results in an increased efflux of fluconazole (FLC) and enhanced resistance to drugs (Krishnamurthy et al. 1998; Sanglard et al. 1999). Reduced intracellular accumulation of drugs mediated by efflux pump proteins in AR clinical isolates has not only been observed in *C. albicans* but has also been the case with other MDR pathogenic species such as *C. tropicalis, C. glabrata, C. krusei and C. dubliniensis*. In *C. glabrata* azole resistance is attributed to an overexpression of the ABC proteins, CgCdr1p and CgCdr2p (Miyazaki et al. 1998; Moran et al. 1998). A potential role of two putative ABC transporters *CkABC1* and *CkABC2* in drug resistance has also been suggested for *C. krusei* (Katiyar and Edlind 2001). In *C. neoformans*, the ABC proteins CneAfr1p and CneMdr1p are the only efflux pump proteins that are linked to antifungal resistance (Thornewell et al. 1997; Sanguinetti et al. 2006). Some of the major efflux pump proteins involved in the azole resistance belonging to PDR subfamily of ABC transporters are listed in Table 15.1.

**Table 15.1** ABC transporters of yeast

| Organism | [a]Subfamily | Gene/ORF name | [b]Size | [c]Location | Functions | References |
|---|---|---|---|---|---|---|
| *Saccharomyces cerevisiae* | PDR | *PDR5* | 1511 | PM | Multidrug efflux pump, phospholipid translocator | Balzi et al. (1994) and Decottignies and Goffeau (1997) |
| | | *PDR10* | 1564 | PM | Drug efflux pump | Wolfger et al. (1997) |
| | | *PDR11* | 1411 | PM | Sterol uptake | Wilcox et al. (2002) |
| | | *PDR12* | 1511 | PM | Weak acid efflux | Piper et al. (1998) |
| | | *PDR15* | 1529 | PM | Potential drug efflux pump | Wolfger et al. (1997) |
| | | *PDR18* | 1333 | PM | Drug efflux pump, sterol incorporation in PM | Cabrito et al. (2011) |
| | | *SNQ2* | 1501 | PM | Drug efflux pump | Decottignies et al. (1995) |
| | | *AUS1* | 1394 | PM | Sterol uptake | Wilcox et al. (2002) |
| | MRP | *YOR1/YRS1* | 1477 | PM | Oligomycin resistance, organic anions & multidrug resistance | Katzmann et al. (1995) |
| | | *YCF1* | 1515 | Vacuole | Cadmium resistance, $Cd^{2+}$ & glutathione-S conjugate pump | Li et al. (1996) |
| | | *BPT1* | 1559 | Vacuole | Phosphatidylcholine translocator, $Ca^{2+}$ homeostasis | Sasser et al. (2012) |
| | | *YBT1* | 1661 | Vacuole | Bile acid transporter | Ortiz et al. (1997) |
| | | *VMR1* | 1592 | Vacuole | Multidrug resistance | Wawrzycka et al. (2010) |
| | MDR | *STE6* | 1290 | PM | a-factor export | McGrath and Vershavsky (1989) |
| | | *ATM1* | 694 | iMT | Mitochondrial DNA maintenance, iron homeostasis, Fe-S proteins export | Kispal et al. (1997) |
| | | *MDL1* | 695 | iMT | Transport of peptides, regulates resistance to oxidative stress | Young et al. (2001) |
| | | *MDL2/SSH1* | 773 | iMT | Transport of peptides | Dean et al. (1994) |

(continued)

**Table 15.1** (continued)

| Organism | [a]Subfamily | Gene/ORF name | [b]Size | [c]Location | Functions | References |
|---|---|---|---|---|---|---|
| | ALDP | *PXA1* | 758 | PX | Fatty acid transport | Shani et al. (1995) |
| | | *PXA2* | 853 | PX | Interaction with PXA1b & oxidation and transport of long chain fatty acids | Shani and Valle (1996) |
| | YEF3/RL1 | *YEF3* | 1044 | Cytosol | Involved in translation | Sandbaken et al. (1990) |
| | | *GCN20* | 752 | Polysomes | Involved in translation | Vazquez de Aldana et al. (1995) |
| | Others | *CAF16* | 289 | ND | Involved in transcription | Liu et al. (2001) |
| | Half-size PDR | *ADP1* | ND | ND | Putative ATP dependent permease, membrane transport | Purnelle et al. (1991) |
| *Schizosaccharomyces pombe* | PDR | *BFR1/HBA2* | 1530 | PM | Brefeldin A transport, tributyltin resistance | Nagao et al. (1995) |
| | MDR | *ABC1* | 1427 | ER | Tolerance to cerulenin | Christensen et al. (1997a) |
| | | *ABC2* | 1463 | Vacuole | Vacuolar sequestration and detoxification of pigments, tolerance to heavy metals | Mendoza-Cózatl et al. (2010) |
| | | *ABC3* | 1465 | Vacuole | Iron transport, tolerance to cerulenin | Iwaki et al. (2006) |
| | | *ABC4* | 1469 | Vacuole | Vacuolar sequestration and detoxification of pigments | Iwaki et al. (2006) |
| | | *MAM1* | 1336 | Endosome and golgi | M-factor, mating pheromone transport | Christensen et al. (1997b) |
| | | *PMD1* | 1362 | PM | Drug efflux pump | Nishi et al. (1993) |
| | ALDP | *HMT1* | 830 | Vacuole | Phytochelatins and glutathione- cadmium complex transport, vacuolar transport | Huang et al. (2012) |
| | | *MDL1* | 726 | MIT | Mitochondrial peptide transport | Iwaki et al. (2006) |
| | | *ATM1* | 693 | MIT | Cellular iron homeostasis | Iwaki et al. (2005) |

| | | | | | | |
|---|---|---|---|---|---|---|
| *Candida albicans* | PDR | *CACDR1* | 1501 | PM | Drug efflux pump, phospholipid translocator | Prasad et al. (1995) and Smriti et al. (2002) |
| | | *CACDR2* | 1499 | PM | Drug efflux pump, phospholipid translocator | Sanglard et al. (1997) and Smriti et al. (2002) |
| | | *CACDR3* | 949 | PM | Opaque- phase specific transporter, Phospholipid translocator | Smriti et al. (2002) |
| | | *CACDR4* | 1490 | ND | Phospholipid translocator | Franz et al. (1998) |
| | | *CaO19.5759* | 1493 | ND | ND | Gaur et al. (2005) |
| | | *CaO19.4531* | 1245 | ND | ND | Gaur et al. (2005) |
| | | *CACDR99* | 683 | ND | Drug efflux pump | Gaur et al. (2005) |
| | | *CaO19.919* | 819 | ND | ND | Gaur et al. (2005) |
| | | *CaO19.3120* | 584 | ND | ND | Gaur et al. (2005) |
| | | *CaO19.459* | 719 | ND | ND | Gaur et al. (2005) |
| | MDR | *CaHST6* | 1328 | ND | Transport of α-mating factor | Raymond et al. (1998) |
| | | *CaMDL1* | 659 | ND | ATP-dependent permease | Gaur et al. (2005) |
| | | *CaOI9.13043* | 726 | ND | ND | Gaur et al. (2005) |
| | | *CaOI9.1077* | 730 | ND | ND | Gaur et al. (2005) |
| | MRP | *CaYOR1* | 1606 | PM | Drug efflux pump | Ogawa et al. (1998) |
| | | *CaYCF1* | 1512 | Vacuole | Cadmium resistance | Wang et al. (2006) |
| | | *CaMLT1* | 1544 | Vacuole | Involved in virulence | Theiss et al (2002) |
| | | *CaO19.6382* | 1254 | ND | ND | Gaur et al. (2005) |
| | | CaO19.1784 | 763 | ND | ND | Gaur et al. (2005) |
| | ALDP | CaO19.7500 | 652 | ND | ND | Gaur et al. (2005) |
| | | CaO19.5255 | 615 | ND | ND | Gaur et al. (2005) |
| | EF3 | CaO19.6060 | 751 | ND | ND | Gaur et al. (2005) |
| | | CaO19.2183 | 585 | ND | ND | Gaur et al. (2005) |
| | | CaELF1 | 1097 | ND | m-RNA export factor | Sturtevant et al. (1998) |
| | | CaCEF3 | 978 | ND | ND | Gaur et al. (2005) |

(continued)

**Table 15.1** (continued)

| Organism | [a]Subfamily | Gene/ORF name | [b]Size | [c]Location | Functions | References |
|---|---|---|---|---|---|---|
| | RL1 | CaO19.3034 | 602 | ND | ND | Gaur et al. (2005) |
| | Others | CaO19.5029 | 545 | ND | ND | Gaur et al. (2005) |
| | | CaO19.388 | 276 | ND | ND | Gaur et al. (2005) |
| *Candida glabrata* | PDR | *CgCDR1* | 1499 | PM | Drug efflux pump | Sanglard et al. (1999) |
| | | *PDH1 (CgCDR2)* | 1542 | ND | Drug efflux pump | Miyazaki et al. (1998) |
| | | *CgSNQ2* | 1507 | ND | Drug efflux pump | Torelli et al. (2008) |
| | | *Cg AUS1* | 1398 | ND | Involved in sterol uptake, drug resistance | Nakayama et al. (2007) |
| *Candida dubliensis* | PDR | *CdCDR1* | 1501 | ND | Drug efflux pump | Moran et al. (1998) |
| | | *CdCDR2* | 1500 | ND | Drug efflux pump | Moran et al. (1998) |
| *Candida krusei* | ND | *CkABC1* | ND | ND | Drug efflux pump | Katiyar and Edlind (2001) |
| | | *CkABC2* | ND | ND | Drug efflux pump | Katiyar and Edlind (2001) |
| *Candida lusitaniae* | ND | *ABC9* | ND | ND | Role in drug resistance | Reboutier et al. (2009) |
| | | *ABC12* | ND | ND | Role in drug resistance | Reboutier et al. (2009) |
| | | *ABC15* | ND | ND | Role in drug resistance | Reboutier et al. (2009) |
| | | *ABC16* | ND | ND | Role in drug resistance | Reboutier et al. (2009) |
| *Candida tropicalis* | PDR | *CtCDR1* | 1472 | ND | Drug efflux pump | Yong et al. (2006) |
| *Cryptococcus neoformans* | MDR | *CneMDR1* | 1408 | ND | Azole drug resistance | Thornewell et al. (1997) |
| | PDR | *CnAFR1* | ND | ND | Azole drug resistance, virulence | Sanguinetti et al. (2006) |

[a]Names of subfamilies based on sequence similarity with human ALDP, MRP/CFTR, MDR and yeast (PDR, YEF3) ABC transporters
[b]Number of amino acids residues
[c]*iMT* inner mitochondrial membrane, *PM* plasma membrane, *PX* peroxisomes, *ER* endoplasmic reticulum, *MIT* mitochondria
*ND* signifies not defined

### 15.2.3 *Cdr1p, a Major ABC Multidrug Transporter of C. albicans*

ABC transporter Cdr1p transports a huge array of structurally unrelated compounds and this promiscuity toward substrates is a characteristic feature of all the multidrug transporters and hence makes their functionality all the more complex to understand (Prasad et al. 2002; Puri et al. 2010; Prasad and Rawal 2014). Cdr1p is an integral plasma membrane protein, with a predicted molecular weight of ~170 kDa (Prasad and Goffeau 2012). It is organized as two homologous halves wherein each half begins with a cytoplasmic hydrophilic NBD followed by a highly hydrophobic TMD (Fig. 15.1). Each TMD comprised of α-helices of six transmembrane segments (TMS), which harbours substrate binding sites and imparts promiscuity to the protein. While the twelve TMSs of Cdr1p are variable structure, the NBDs which bind and hydrolyse the ATP to drive drug efflux have conserved sub domains. Conformational changes induced by ATP binding and/or hydrolysis are transmitted from the NBDs to the TMDs, which can result in the drug efflux or resetting of the reaction cycle (Cannon et al. 2009). There have been intensive efforts to understand the molecular basis of the drug binding and transport. Expectedly, predicting the substrate binding pocket without high-resolution structural data remains a challenge. Yet by employing site directed mutagenesis, followed by phenotyping of the mutant variants, it has been possible to map the substrate binding pockets of Cdr1p with some success (Shukla et al. 2003; Puri et al. 2009; Rawal et al. 2013).

#### 15.2.3.1 Substrate Binding Site of Cdr1p

For most structural and functional related studies, Cdr1p was over expressed in *S. cerevisiae* AD1-8u$^-$ strain wherein its expression was driven by the *ScPDR5* promoter (Nakamura et al. 2001). The expression system of AD1-8u$^-$ strain lacks the seven major ABC transporters ScYor1p, ScSnq2p, ScPdr5p, ScPdr10p, ScPdr11p, ScYcf1p, ScPdr15p thereby rendering it hyper susceptible to various drugs (Decottignies et al. 1998). The constitutive hyper-expression of heterologous proteins in this strain is facilitated by the presence of a hyperactive allele of Pdr1p (pdr1-3), one of the transcription factors that regulates the expression of ScPdr5p (Shukla et al. 2003). By employing this expression system, it has been shown that the photoaffinity analogue of ABC multidrug transporter human P-glycoprotein (P-gp) substrate iodoarylazido prazosin (IAAP) and azidopine, a dihydropyridine photoaffinity analogue of its modulator, verapamil could bind to Cdr1p. Interestingly, while polyene nystatin could compete with IAAP binding, the binding of azidopine could only be competed out by miconazole, implying different drug interacting sites of the two analogues (Shukla et al. 2003). This was reinforced by an extensive site directed mutational analysis of Cdr1p which identified several critical residues involved in drug transport, however, these studies could provide only limited insight into the drug binding cavity. For example, contributions of some conserved amino acids of the NBDs and TMDs to Cdr1p localization, stability and substrate specificity could be demonstrated (Shukla et al. 2003). Similarly, the alanine scanning mutagenesis of TMS11 where all the residues were individually replaced with alanine highlighted functionally relevant residues of the helix of Cdr1p (Saini et al. 2005). Further experiments revealed that similar to P-gp, Cdr1p also harbours multiple binding sites specific to a drug or to a group of drugs (Puri et al. 2009). A more recent systematic study probed the nature of drug binding pocket of Cdr1p by performing alanine scanning mutagenesis of the entire primary sequences of both the TMDs, wherein, each of the residues of 12 TMSs was replaced with an alanine (Rawal et al. 2013). This approach enabled the identification of residues whose replacement completely, or selectively, eliminated resistance to the drugs tested. These critical residues were found to be distributed in almost all the TMSs but the number of critical residue in each TMS was extremely variable (Fig. 15.1). The behaviour of all the TMS residues provided an opportunity to conceptualize the entire drug binding cavity of Cdr1p. A comparison of critical residues for azoles revealed that while there are several residues which are important for the transport of all the azoles, however, each drug also has exclusive residues which may be necessary for its binding and transport (Fig. 15.3). Biochemical data and homology modeling of Cdr1p, based on the structure for *S. cerevisiae,* ScPdr5p revealed multiple overlapping drug-binding sites within a large centrally located polyspecific drug binding pocket, wherein, the residues critical for drug susceptibility were predominantly clustered on TMS1, TMS2 and TMS11 on one side and TMS4, TMS5 and TMS8 on other side of the Cdr1p (Rawal et al. 2013). Together, the mutagenesis of both TMDs of Cdr1p, combined with phenotypic, biochemical and biophysical mapping, could identify amino acid residues that are critical structural/functional determinants of the drug transport. The data of drug susceptibility profiles and the deduced 3D model supports multiple distinct mini-binding sites within a common large drug-binding pocket, a finding that may help resolve the substrate promiscuity of Cdr1p and enable rational design of its inhibitors.

#### 15.2.3.2 ATP Catalysis Sites

All ABC drug transporters are primary transporters and directly utilize the energy derived from ATP hydrolysis to power drug extrusion. The NBDs located at the cytoplasmic periphery are the hub of such an activity. As compared to TMDs which comprised of 12 TMSs, the NBDs are highly conserved structures (Prasad and Goffeau 2012). The NBDs of all ABC transporters, irrespective of their origin and nature of transport substrate, share extensive amino acid sequence identity and typical motifs, which are considered to be critical for the domain's functionality (Higgins and Linton 2004). These include, the Walker A motif with a consensus sequence GxxGxGKS/T, where 'x' represents any amino acid, the Walker B motif i.e. hhhhD, where 'h' represents any aliphatic residue, and an ABC signature, LSGGQQ/R/KQR. Interestingly, though N-NBD and C-NBD of Cdr1p contains the conserved Walker A, Walker B and conserved ABC signature sequence, they pos-

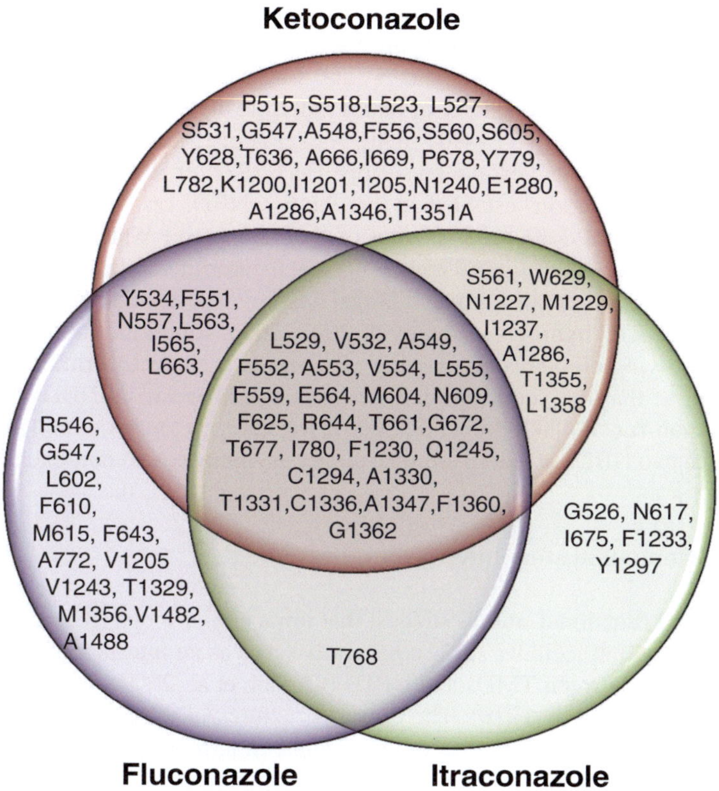


**Fig. 15.3** Venn diagram indicating overlapping drug binding sites. Venn diagram depicting common as well as drug specific residues involved in fluconazole, ketoconazole and itraconazole drug transport

sess a few unique and typical replacements of residues. Such sequence degeneracy which exist in otherwise conserved sub-domains of NBDs appears to be a common feature among fungal ABC transporters (Prasad and Goffeau 2012). For example, while, N-NBD of Cdr1p has degenerated Walker A (GRPGAGCST) and Walker B (IQCWD) but a well conserved ABC signature sequence (VSGGERKRVS), the C-NBD has conserved Walker A (GASGAGKTT) and B (LLFLD) but a degenerated ABC signature sequence (LNVEQRKRLT) (degenerated residues are underlined). In the Walker A motif of Cdr1p, the commonly occurring lysine residue is replaced by a cysetine. Of note, Walker A in the C-NBD of most of yeast ABC transporters contains the evolutionary conserved lysine. Structural and biochemical analyses of NBDs show that the typical lysine residue of Walker A motif binds to the β- and γ-phosphates of ribonucleotides and plays a critical role in ATP hydrolysis (Schneider and Hunke 1998). The role of a typical cysteine in N-NBD of Cdr1p was explored by over-expressing and purifying the domain as soluble protein independent of its C-terminal counterpart and both TMDs. It was demonstrated that the purified domain elicits ATPase activity and an evolutionarily divergent Cys193 of Walker A of N-NBD is critical for ATP hydrolysis (Jha et al. 2003). A comparison of the relative contribution of both the N- and C-terminal NBDs in ATP binding, hydrolysis and transporter activity of native Cdr1p revealed that Cys193 of Walker A of N-NBD and conserved Lys901 in the Walker A of C-NBD are essential for the functioning of protein and cannot be swapped as C193K or K901C (Jha et al. 2003, 2004). These observations established that the two NBDs respond asymmetrically to the substitution of conserved residues of their respective Walker A motifs. The conserved substitutions of typical residues reveal their critical roles in ATP catalysis. Since these substitutions are unique to fungal ABC transporters, they represent novel mechanism of ATP hydrolysis (Jha et al. 2003, 2004; Rai et al. 2005, 2006). The studies employing chimeras comprising of either two N- or C-terminal NBDs of Cdr1p suggest that the variant protein comprising either two N-NBDs or two C-NBDs are non-functional (Saini et al. 2006). Together, the overwhelming data suggest that the two NBDs of Cdr1p are structurally asymmetric and non-exchangeable.

#### 15.2.3.3 Communicating Interface

Structural and functional studies suggest that intracellular loops (ICLs) connecting various TMS on intracellular surface generally serve as an inter-domain communication interface between TMDs and NBDs (Cannon et al. 2009; Ananthaswamy et al. 2010; Lamping et al. 2010). The crystal structures of ABC transporters reveal that NBDs contact the TMDs via the communication interface provided by ICLs (Dawson and Locher 2006; Aller et al. 2009; Lamping et al. 2010). Fungal ABC transporters belonging to PDR subfamily possess a specific ICL arrangement which is architecturally different from the MDR transporters. In MDR transporters, each TMD contains two ICLs that mediate communication between NBDs and TMDs through specific coupling helices (Cannon et al. 2009; Ananthaswamy et al. 2010; Lamping et al. 2010). The ICLs from both TMDs of MDR transporters are large and comparable in size, while PDR transporters show a peculiar arrangement of ICLs and possess one large and one small ICL in each TMD (Fig. 15.1). The primary sequence and secondary structural elements of ICL1 and ICL3 of PDR transporters are generally conserved (Lamping et al. 2010). ICL1 and ICL3 of Cdr1p possess helical secondary structures with ICL1 containing a centrally located coupling helix reminiscent of the arrangement found in MDR transporters (Dawson and Locher 2006; Aller et al. 2009).

The precise role of ICLs of Cdr1p is not known. However, biochemical studies from other similar transporters have provided evidence that ICLs function as communication link between NBDs and TMDs (Sauna et al. 2008; Ananthaswamy et al. 2010). There are many instances where the drug-susceptibility phenotype resulting due to a point mutation in one of the TMSs could be repressed by chromosomal suppressor with a mutation in NBD (Sauna et al. 2008; Pagant et al. 2010). Considering the importance of inter-domain communication in yeast ABC transporters, detailed knowledge of the communication interface is also essential in designing inhibitors/modulators for similar transporters.

### 15.2.4 Physiological Roles of ABC Transporters

The rapidly growing super family of ABC transporter is involved in performing diverse functions. Besides, playing an important role in drug transport, additional roles of ABC transporters in absorption, excretion, signal transduction and bacterial pathogenesis reflect their diverse functions (Mendez and Salas 2001; Goldman and Kranz 2001; Schneider 2001). Considering the existence of large numbers of ABC transporters in yeasts genome, there has been a focus on research to explore their physiological relevance. Table 15.2 lists some of the functions of ABC transporters.

**Table 15.2** Physiological roles of ABC transporters

| Physiological roles | Transporters involved | References |
|---|---|---|
| Transport of sterols | ScPdr5p, ScSnq2p, ScAus1p and ScPdr11p, CaCdr1p | Krishnamurthy et al. (1998) and Wilcox et al. (2002) |
| Translocation of membrane phospholipids | ScPdr5p and ScYor1p | Dogra et al. (1999) |
| | CaCdr1p, CaCdr2p and CaCdr4p elicit in-to-out translocation (floppase activity) of phospholipids, CaCdr3p acts as a flippase (out –to –in translocation) | |
| Peptide transport | ScMdl1p, ScMdl2p and ScSte6p | McGrath and Vershavsky (1989) |
| Transport of weak acids | ScPdr12p | Piper et al. (1998) |
| Heavy metal detoxification | ScYcf1p, ScBpt1p and ScYbt1p, SpABC2, SpAtm1p | Iwaki et al. (2006) |
| Mitochondrial iron homeostasis | ScAtm1p | Leighton and Schatz (1995) |
| Protection against cellular toxins/metabolites | ScPdr5p, ScPdr15p and CaCdr1p | Krishnamurthy et al. (1998) |
| Ion transport | Human P-gp act as a $H^+/K^+$ pump | Fritz et al. (1999) |
| Brefeldin A transport | ScBfr1p | Nagao et al. (1995) |
| Fatty acid transport | ScPxa1p | Shani et al. (1995, 1996) |
| Involved in translation | ScYef3p, Gcn20 | Sandbaken et al. (1990) |
| Involved in transcription | Caf16 | Liu et al. (2001) |
| Pheromone transport MATα | SpMam1p, CaHst6p | Christensen et al. (1997b) |
| Oligomycin resistance | ScYor1p | Katzmann et al. (1995) |
| Cadmium resistance | ScYcf1p, CaYcf1p | Li et al. (1996) |
| Bile acid transporter | ScYbt1p | Ortiz et al. (1997) |
| α-factor export | ScSte6p | McGrath and Vershavsky (1989) |
| Iron homeostasis | Atm1p | Kispal et al. (1997) |
| mRNA export | CaEfl1p | Sturtevant et al. (1998) |

*Sp-Schizosaccharomyces pombe; Sc-Saccharomyces cerevisiae; Ca-Candida albicans*

### 15.2.5 Modulators of Cdr1p

MDR in cancer cells is an obstacle to effective chemotherapy. ABC transporters including ABCB1, ABCC1 and ABCG2 have important roles in the development of frequently encountered MDR in cancer cells. Here again, among different approaches employed to overcome MDR, inhibition of the drug extrusion pump activity represents an attractive approach. Many clinically relevant anticancer drugs such as vinca alkaloids (vinblastine and vincristine), anthracyclines (doxorubicin and daunorubicin), taxenes (paclitaxel and docetaxel), epipodophyyltoxins (etoposide and teniposide), camptothecins (topotecan) and anthracenes are identified as modulators of human ABC transporters which offer great hope in successful cancer chemotherapy (Wu et al. 2011). In comparison, modulators of MDR pump proteins in pathogenic yeasts are only beginning to be characterized. There are already examples of compounds such as enniatins, milbemycins, synthetic D-octapeptides, isonitrile and unnarmicins which modulate drug efflux by inhibiting the fungal multidrug transporters (Table 15.3). An alcohol antabuse drug disulfiram, acts as a modulator of Cdr1p by inhibiting oligomycin-sensitive ATP hydrolysis and affects drug binding (Shukla et al. 2004). The polyphenol curcumin and quorum sensing molecule farnesol are also specific modulator of Cdr1p (Sharma et al. 2009; Sharma and Prasad 2011).

**Table 15.3** Modulators of Cdr1p

| Modulators | References |
|---|---|
| FK506 | Schuetzer-Muehlbauer et al. (2003) |
| Enniatin | Hiraga et al. (2005) |
| Milbemycin | Lee et al. (2001) |
| Synthetic D-octapeptides | Niimi et al. (2004) |
| Cyclosporine | Puri et al. (2010) |
| Isonitrile | Yamamoto et al. (2005) |
| Disulfiram | Shukla et al. (2004) |
| Ibuprofen | Pina-Vaz et al. (2005) |
| Quinazolinone derivatives | Watkins et al. (2007) |
| Unnarmicins | Tanabe et al. (2007) |
| FK-520 | Nim et al. (2014) |
| Curcumin | Sharma et al. (2009) |
| Farnesol | Sharma et al. (2011) |
| Trichostatin A | Smith and Edlind (2002) |

## 15.3 MFS Transporters

MFS superfamily of transporters belongs to a large family of secondary active transporters which use the proton motive force for the transport of molecules (Ben-Yaacov et al. 1994). On the basis of hydropathy and phylogenetic analysis, these transporters are divided into two distinct types; first group includes, Drug: $H^+$ Antiporter-1 (DHA1), consisting of 12 TMSs and the second is called Drug: $H^+$ Antiporter-2 (DHA2) that contains 14 TMSs (Pao et al. 1998). DHA1 transporters are comprised of two halves; each typically made of 6 TMSs which form a central pore for drug transport. TMSs are connected by ICLs and ECLs, with both their N- and C-termini located in the cytoplasm (Pao et al. 1998). There are few members of MFS with only 6 TMSs and even with as many 24 TMSs (Pao et al. 1998). Similar to ABC transporters, the promiscuous MFS pump proteins hold an enormous diversity of substrate which can be expelled by these proteins. For example, organic and inorganic ions, sugars, sugar phosphates, drugs, neurotransmitters, nucleosides, amino acids and polyols are well known substrates of MFS proteins (Ben-Yaacov et al. 1994; Gupta et al. 1998; Calabrese et al. 2000).

### *15.3.1 MFS Proteins of Candida as Multidrug Transporters*

The genome wide inventory of *C. albicans* revealed the presence of 95 putative MFS proteins that clustered into 17 families (Gaur et al. 2008). Out of all MFS proteins, only *CaMDR1* and *FLU1* encoding proteins are involved in drug efflux in *C. albicans* cells (Sanglard et al. 1999; Gaur et al. 2008). *FLU1* was shown to confer resistance to fluconazole, however, no clinical relevance of this drug transporter has been reported.

Homologues of *CaMDR1* have been identified in *C. dubliniensis* and *C. glabrata* which are designated as *CdMDR1* and *CgMDR1*, respectively (Moran et al. 1998; Sanglard et al. 1999). Increased expression of *CdMDR1* appears to be the main mechanism involved in fluconazole resistance in *C. dubliniensis* clinical isolates (Moran et al. 1998). Similarly, constitutive expression of *CgMDR1* in *C. glabrata* is responsible for the intrinsically low susceptibility of this yeast species to fluconazole (Sanglard et al. 1999). Table 15.4 lists MFS transporters which have roles in drug resistance and transport.

### *15.3.2 CaMdr1p, a Major MFS Multidrug Transporter of C. albicans*

All the MFS proteins which exist in *Candida* genome, only CaMdr1p is shown to extrude drugs and its overexpression has been linked to clinical resistance to azoles (Braun et al. 2005). The CaMdr1p belongs to the DHA1 family, which

**Table 15.4** MFS transporters of yeast

| Organism | Gene name | Family | Function/resistance to/substrate | References |
|---|---|---|---|---|
| *Saccharomyces cerevisiae* | *AQR1* | DHA1 | Ketoconazole, crystal violet, quinine, quinidine, short-chain monocarboxylic acids | Tenreiro et al. (2002) |
| | *DTR1* | DHA1 | Quinine, quinidine, propionic acid, butyric acid, benzoic acid | Felder et al. (2002) |
| | *FLR1* | DHA1 | Fluconazole, cycloheximide, 4-nitroquinoline 1-oxide, diethylmaleate, methotrexate, mancozeb, benomyl, diazoborine, cerulenin, diamide, menadione, alkaloid antifungal Canthin-6-one | Dejos et al. (2013) |
| | *HOL1* | DHA1 | Mutations in the pump encoding gene allows the uptake of histidinol and other cations non selectively | Wright et al. (1996) |
| | *QDR1* | DHA1 | Quinidine, fluconazole, barban, ketoconazole | Nunes et al. (2001) |
| | *QDR2* | DHA1 | Quinidine, barban, ketoconazole, cisplatin, bleomycin, metal ion homeostasis | Vargas et al. (2007) |
| | *QDR3* | DHA1 | Quinidine, barban, cisplatin, bleomycin, polyamine tolerance | Tenreiro et al. (2005) |
| | *TPO1* | DHA1 | Quinidine, Polyamines spermine, spermidine, ibuprofen, putrescene; paraquat, artesunate, caspofungin | Tomitori et al. (2001) |
| | *TPO2* | DHA1 | Spermine, weak organic acids (acetic, propionic, benzoic and octanoic acids) | Tomitori et al. (2001) |
| | *TPO3* | DHA1 | Spermine, weak organic acids (acetic, propionic, benzoic and octanoic acids) | Tomitori et al. (2001) |
| | *TPO4* | DHA1 | Spermine, quinidine, cycloheximide | Tomitori et al. (2001) |
| | *YHK8* | DHA1 | Itraconazole, fluconazole | Barker et al. (2003) |
| | *Atg22* | DHA1 | Vacuolar amino acid efflux | Yang et al. (2006) |
| | *ATR1* | DHA2 | Main boron exporter in yeast, aminotriazole and $Sn^{2+}$ ions transport, exports L-cysteine | Kaya et al. (2009) |
| | *AZR1* | DHA2 | Acetic acid, propionic acid, crystal violet, ketoconazole, fluconazole, polymyxin B | Tenreiro et al. (2000) |
| | *SGE1* | DHA2 | Crystal violet, ethidium bromide, methylmethane sulfonate, malachite green, 10-N-nonyl acridine orange | Jacquot et al. (1997) |

| | | | | |
|---|---|---|---|---|
| | *VBA1* | DHA2 | Vacuolar basic amino acid transporter 1 | Shimazu et al. (2005) |
| | *VBA2* | DHA2 | Transporter required for vacuolar uptake of histidine, arginine and lysine and to a lesser extent tyrosine | Shimazu et al. (2005) |
| | *VBA3* | DHA2 | Vacuolar basic amino acid (arginine, lysine, histidine) transporter | Shimazu et al. (2005) |
| | *VBA4* | DHA2 | Amino-acid transport | Shimazu et al. (2005) |
| | *VBA5* | DHA2 | Amino-acid transport | Shimazu et al. (2005) |
| *Schizosaccharomyces pombe* | *CAR1* | DHA1 | Pyridoxine, pyridoxal, pyridoxamine, amiloride:, $H^+$ cotransporter | Stolz et al. (2005) |
| *Candida maltosa* | *cyh$^R$* | DHA1 | Cycloheximide | Sasnauskas et al. (1992) |
| *Candida albicans* | *CaMDR1* | DHA1 | Benomyl, cycloheximide, methotrexate, fluconazole, sulfomthuron methyl, 4-nitroquinoline 1-oxide | Pasrija et al. (2007) |
| | *FLU1* | DHA1 | Mycophenolic acid | Calabrese et al. (2000) |
| *Candida glabrata* | *CgMDR1* | DHA1 | Fluconazole, benomyl, cycloheximide, 4-nitroquinoline 1-oxide | Sanglard et al. (1999) |
| | *CgQDR2* | DHA1 | Quinidine, miconazole, tioconazole, clotrimazole, Ketoconazole | Costa et al. (2013) |
| *Candida dublinensis* | *CdMDR1* | DHA1 | Fluconazole, amorolfine, benomyl, brefeldin A, cerulenin, cucloheximide, fluphenazine, 4-nitroquinoline 1-oxide | Moran et al. (1998) |

efflux drugs in exchange with one or more $H^+$ (Paulsen and Skurray 1993). It is characterized by two structural units of six TMSs, linked by a central cytoplasmic loop (CCL). Six TMSs are interconnected by ECLs and ICLs (Fig. 15.2). An over expression of *CaMDR1* is also linked to an emergence of colateral resistance to several unrelated drugs e.g. fluconazole, 4-nitroquinoline–N-oxide, cycloheximide, benomyl, methotrexate and cerulenin (Ben-Yaacov et al. 1994; Gupta et al. 1998). In addition to pronounced expression of *CaMDR1* in some of the azole resistant clinical isolates of *C. albicans*, its expression can also be induced in vitro by the exposure of cells to benomyl, methotrexate and several other unrelated drugs (Gupta et al. 1998).

### *15.3.3 Drug/H⁺ Binding Sites in CaMdr1p*

Multiple-sequence analysis of the MFS transporters reveals that proteins within this family share greater similarity between their N terminal halves than in their C-terminal halves, and it is assumed that the later half is responsible for substrate recognition and N-terminal regions are involved in proton translocation (Saier and Reizer 1991). Additionally, the MFS drug antiporter proteins possess many conserved residues scattered throughout the length of the protein. For example, motifs A and B are conserved throughout the MFS, while motif 'C' (also known as antiporter motif) is conserved in only 12 and 14-TMS subfamilies (Paulsen et al. 1996). Independent of the substrate specificity of the antiporter which is majorly govern by TMSs of the protein, the antiporter motif [G(X8)G(X3)GP(X2)GG] of TMS5 is well conserved in all of the functionally related subgroups in bacteria and plants (Ginn et al. 2000). Based on the sequence homology, CaMdr1p also has a conserved antiporter motif [G(X6)G(X3)GP(X2)GP(X2)G] in TMS5 (Pasrija et al. 2007).

For a detail structure and function analysis, Ca*MDR1* gene was overexpressed as a GFP-tagged protein in a heterologous host *S. cerevisiae* (Pasrija et al. 2007). This recombinant protein was well folded and properly localized to the plasma membrane. The study established the importance of TMS5 where several residues including those of antiporter motif are potentially significant for its functionality and contribute to drug: $H^+$ transport (Pasrija et al. 2007). This led to computational based analysis of CaMdr1p where membrane environment based approach was employed to rationally predict the critical residues of the protein (Kapoort et al. 2009, 2010). When rationally selected important residues were subjected to site-directed mutational analysis followed by phenotyping, the data proved that the predicted residues were indeed important in maintaining inter-helical interactions and function of the transporter (Kapoor et al. 2009). The study could show that by employing similar rational approach, critical residues in other transporters can also be predicted (Kapoor et al. 2010).

Unlike ICLs of ABC transporters, which acts as communicating link between TMDs and NBDs; ICLs in MFS transporters are essential structural element of protein. For example, a recent study on ICL3/CCL connecting the two TMDs of CaMdr1p not only highlighted the critical role of this central cytoplasmic loop in fungal MFS belonging to DHA1 family in drug transport but also showed it as an essential structural element of CaMdr1p (Mandal et al. 2012).

## 15.4 Concluding Remarks

The ABC and MFS multidrug transporters of *Candida* are very promiscuous proteins which is evident from the range of compounds which they can export. This unique feature poses a major challenge in dissecting their structure and function relationship. How and where drug binds and released and how the entire process is coupled with the energy, remain key questions. The structure and function activity of transporter proteins have so far failed to predict common minimum descriptors of substrate structures. However, the studies conducted so far are beginning to address some of the aspects of drug efflux. For example, recent studies involving directed mutagenesis of these transporter proteins have revealed that these proteins harbour a large binding cavity which can accommodate diverse compounds to export. Elucidating the location and biophysical properties of the drug-binding sites of these multidrug transporter proteins would provide deeper insights into the molecular interaction between substrates/inhibitors and the drug transporter, which would facilitate the development of potent inhibitors of drug efflux pump.

## References

Aller SG, Yu J, Ward A, Weng Y, Chittaboina S, Zhuo R, Harrell PM, Trinh YT, Zhang Q, Urbatsch IL, Chang G (2009) Structure of P-glycoprotein reveals a molecular basis for poly-specific drug binding. Science 323:1718–1722

Ananthaswamy N, Rutledge R, Sauna ZE, Ambudkar SV, Dine E, Nelson E, Xia D, Gollin J (2010) The signaling interface of the yeast multidrug transporter Pdr5 adopts a cis conformation, and there are functional overlap and equivalence of the deviant and canonical Q-loop residues. Biochemistry 49:4440–4449

Balzi E, Wang M, Leterme S, Van Dyck L, Goffeau A (1994) PDR5, a novel yeast multidrug resistance conferring transporter controlled by the transcription regulator PDR1. J Biol Chem 269:2206–2214

Barker KS, Pearson MM, Rogers PD (2003) Identification of genes differentially expressed in association with reduced azole susceptibility in *Saccharomyces cerevisiae*. J Antimicrob Chemother 51:1131–1140

Ben-Yaacov R, Knoller S, Caldwell GA, Becker JM, Koltin Y (1994) *Candida albicans* gene encoding resistance to benomyl and methotrexate is a multidrug resistance gene. Antimicrob Agents Chemother 38:648–652

Braun BR, Vanhet HM, d'Enfert C, Martchenko M, Dungan J, Kuo A, Inglis DO et al (2005) A human-curated annotation of the *Candida albicans* genome. PLoS Genet 1:36–57

Cabrito TR, Teixeira MC, Singh A, Prasad R, Sa-Correia I (2011) The yeast ABC transporter Pdr18 (ORF YNR070w) controls plasma membrane sterol composition, playing a role in multidrug resistance. Biochem J 440:195–202

Calabrese D, Bille J, Sanglard D (2000) A novel multidrug efflux transporter gene of the major facilitator superfamily from *Candida albicans* (FLU1) conferring resistance to fluconazole. Microbiology 146:2743–2754

Cannon RD, Lamping E, Holmes AR, Niimi K, Baret PV, Keniya MV, Tanabe K, Niimi M, Goffeau A, Monk BC (2009) Efflux-mediated antifungal drug resistance. Clin Microbiol Rev 22:291–321

Christensen PU, Davis K, Nielsen O, Davey J (1997a) Abc1: a new ABC transporter from the fission yeast *Schizosaccharomyces pombe*. FEMS Microbiol Lett 147:97–102

Christensen PU, Davey J, Nielsen O (1997b) The Schizosaccharomyces pombe mam1 gene encodes an ABC transporter mediating secretion of M-factor. Mol Gen Genet 255:226–236

Costa C, Pires C, Cabrito TR, Renaudin A, Ohno M, Chibana H, Sa-Correia I, Teixeira MC (2013) *Candida glabrata* drug: $H^+$ antiporter CgQdr2 confers imidazole drug resistance, being activated by transcription factor CgPdr1. Antimicrob Agents Chemother 57:3159–3167. doi:10.1128/AAC.00811-12

Dassa E, Bouige P (2001) The ABC of ABCS: a phylogenetic and functional classification of ABC systems in living organisms. Res Microbiol 152:211–229

Dawson RJ, Locher KP (2006) Structure of a bacterial multidrug ABC transporter. Nature 443:180–185

Dean M, Allikmets R, Gerrard B, Stewart C, Kistler A, Shafer B, Michaelis S, Strathern J (1994) Mapping and sequencing of two yeast genes belonging to the ATP- binding cassete superfamily. Yeast 10:377–383

Dean M, Andrey R, Rando A (2001) The human ATP-binding cassette (ABC) transporter superfamily. Genome Res 11:1156–1166

Decottignies A, Goffeau A (1997) Complete inventory of the yeast ABC proteins. Nat Genet 15:137–145

Decottignies A, Lambert L, Catty P, Degand H, Epping EA, Moye-Rowley WS, Balzi E, Goffeau A (1995) Identification and characterization of *SNQ2*, a new multidrug ATP binding cassette transporter of the yeast plasma membrane. J Biol Chem 270:18150–18157

Decottignies A, Grant AM, Nichols JW, De Wet H, McIntosh DB, Goffeau A (1998) ATPase and multidrug transport activities of the overexpressed yeast ABC protein Yor1p. J Biol Chem 273:12612–12622

Dejos C, Régnacq M, Bernard M, Voisin P, Bergès T (2013) The MFS-type efflux pump Flr1 induced by Yap1 promotes canthin-6-one resistance in yeast. FEBS Lett 587:3045–3051

Dogra S, Krishnamurthy S, Gupta V, Dixit BL, Gupta CM, Sanglard D, Prasad R (1999) Asymmetric distribution of phosphatidylethanolamine in C. albicans: possible mediation by CDR1, a multidrug transporter belonging to ATP binding cassette (ABC) superfamily. Yeast 15:111–121

Felder T, Bogengruber E, Tenreiro S, Ellinger A, Sá-Correia I, Briza P (2002) Dtrlp, a multidrug resistance transporter of the major facilitator superfamily, plays an essential role in spore wall maturation in *Saccharomyces cerevisiae*. Eukaryot Cell 1:799–810

Franz R, Michel S, Morschhauser J (1998) A fourth gene from the *Candida albicans* CDR family of ABC transporters. Gene 220:91–98

Fritz F, Howard EM, Hoffman MM, Roepe P (1999) Evidence for altered ion transport in *Saccharomyces cerevisiae* overexpressing human MDR 1 protein. Biochemistry 38:4214–4226

Gaur M, Choudhury D, Prasad R (2005) Complete inventory of ABC proteins in human pathogenic yeast, *Candida albicans*. J Mol Microbiol Biotechnol 9:3–15

Gaur M, Puri N, Manoharlal R, Rai V, Mukhopadhayay G, Choudhary D, Prasad R (2008) MFS transportome of the human pathogenic yeast *Candida albicans*. BMC Genomics 9:579. doi:10.1186/1471-2164-9-579

Ginn SL, Brown MH, Skurray RA (2000) The TetA(K) tetracycline/H(+) antiporter from Staphylococcus aureus: mutagenesis and functional analysis of motif C. J Bacteriol 182:1492–1498. doi:10.1128/JB.182.6.1492-1498.2000

Goldman BS, Kranz RG (2001) ABC transporters associated with cytochrome c biogenesis. Res Microbiol 152:323–329

Gupta V, Kohli AK, Krishnamurthy S, Puri N, Aalamgeer SA, Panwar SL, Prasad R (1998) Identification of mutant alleles of CaMDR1, a major facilitator of Candida albicans which confers multidrug resistance and its in vitro transcriptional activation. Curr Genet 3:192–199

Higgins CF, Linton KJ (2004) The ATP switch model for ABC transporters. Nat Struct Mol Biol 11:918–926

Hiraga K, Yamamoto S, Fukuda H, Hamanaka N, Oda K (2005) Enniatin has a new function as an inhibitor of Pdr5p, one of the ABC transporters in *Saccharomyces cerevisiae*. Biochem Biophys Res Commun 328:1119–1125

Huang J, Zhang Y, Peng JS, Zhong C, Yi HY, Ow DW, Gong JM (2012) Fission yeast HMT1 lowers seed cadmium through phytochelatin-dependent vacuolar sequestration in Arabidopsis. Plant Physiol 158:1779–1788. doi:10.1104/pp.111.192872

Iwaki T, Fujita Y, Tanaka N, Giga-Hama Y, Takegawa K (2005) Mitochondrial ABC transporter Atm1p is required for protection against oxidative stress and vacuolar functions in *Schizosaccharomyces pombe*. Biosci Biotechnol Biochem 69:2109–2116

Iwaki T, Giga-Hama Y, Takegawa K (2006) A survey of all 11 ABC transporters in fission yeast: two novel ABC transporters are required for red pigment accumulation in a *Schizosaccharomyces pombe* adenine biosynthetic mutant. Microbiology 152:2309–2321

Jacquot C, Julien R, Guilloton M (1997) The *Saccharomyces cerevisiae* MFS superfamily SGE1 gene confers resistance to cationic dyes. Yeast 13:891–902

Jha S, Karnani N, Dhar SK, Mukhopadhayay K, Shukla S, Saini P, Mukhopadhayay G, Prasad R (2003) Purification and characterization of the N-terminal nucleotide binding domain of an ABC drug transporter of *Candida albicans*: uncommon cysteine 193 of Walker A is critical for ATP hydrolysis. Biochemistry 42:10822–10832

Jha S, Dabas N, Karnani N, Saini P, Prasad R (2004) ABC multidrug transporter Cdr1p of *Candida albicans* has divergent nucleotide-binding domains which display functional asymmetry. FEMS Yeast Res 5:63–72

Kapoor K, Rehan M, Kaushiki A, Pasrija R, Lynn AM, Prasad R (2009) Rational mutational analysis of a multidrug MFS transporter CaMdr1p of *Candida albicans* by employing a membrane environment based computational approach. PLoS Comput Biol 5:1–11

Kapoor K, Rehan M, Lynn AM, Prasad R (2010) Employing information theoretic measures and mutagenesis to identify residues critical for drug-proton antiport function in Mdr1p of *Candida albicans*. PLoS One 5(6):e11041. doi:10.1371/journal.pone.0011041

Katiyar SK, Edlind TD (2001) Identification and expression of multidrug resistance-related ABC transporter gene in *Candida krusei*. Med Mycol 39:109–116

Katzmann DJ, Hallstrom TC, Voet M, Wysock W, Golin J, Volckert G, Moye-Rowley WS (1995) Expression of an ATP-binding cassette transporter-encoding gene (YOR1) is required for oligomycin resistance in *Saccharomyces cerevisiae*. Mol Cell Biol 15:6875–6883

Kaya A, Karakaya HC, Fomenko DE, Gladyshev VN, Koc A (2009) Identification of a novel system for boron transport: Atr1 is a main boron exporter in yeast. Mol Cell Biol 29:3665–3674

Kispal G, Csere P, Guiard B, Lill R (1997) The ABC transporter Atm1p is required for mitochondrial iron homeostasis. FEBS Lett 418:346–350

Krishnamurthy S, Gupta V, Prasad R, Panwar SL, Prasad R (1998) Expression of CDR1, a multidrug resistance gene of *Candida albicans*: In vitro transcriptional activation by heat shock, drugs and human steroid hormones. FEMS Microbiol Lett 160:191–197

Lamping E, Baret PV, Holmes AR, Monk BC, Goffeau A, Cannon RD (2010) Fungal PDR transporters: phylogeny, topology, motifs and function. Fungal Genet Biol 47:117–142

Lee MD, Galazzo JL, Staley AL, Lee JC, Warren W, Fuernkranz H, Chamberland S, Lomovskaya O, Miller GH (2001) Microbial fermentation-derived inhibitors of efflux-pump-mediated drug resistance. Farmaco 56:81–85

Leighton J, Schatz G (1995) An ABC transporter in the mitochondrial inner membrane is required for normal growth of yeast. EMBO J 14:188–195

Li ZS, Szczypka M, Lu YP, Thiele DJ, Rea PA (1996) The yeast cadmium factor protein (YCF1) is a vacuolar glutathione S-conjugate pump. J Biol Chem 271:6509–6517

Liu HY, Chiang YC, Pan J, Chen J, Salvadore C, Audino DC, Badarinarayana V, Palaniswamy V, Anderson B, Denis CL (2001) Characterization of CAF4 and CAF16 reveals a functional connection between the CCR4-NOT complex and a subset of SRB proteins of the RNA polymerase II holoenzyme. J Biol Chem 276:7541–7548

Mandal A, Kumar A, Singh A, Lynn AM, Kapoor K, Prasad R (2012) A key structural domain of the *Candida albicans* Mdr1 protein. Biochem J 445:313–322

McGrath JP, Vershavsky A (1989) The yeast STE6 gene encodes a homologue of the mammalian multidrug resistance P-glycoprotein. Nature 340:400–404

Mendez C, Salas JA (2001) The role of ABC transporters in antibiotic producing organisms: drug secretion and resistance mechanisms. Res Microbiol 152:341–350

Mendoza-Cózatl DG, Zhai Z, Jobe TO, Akmakjian GZ, Song WY et al (2010) Tonoplast-localized Abc2 transporter mediates phytochelatin accumulation in vacuoles and confers cadmium tolerance. J Biol Chem 285:40416–40426

Miyazaki H, Miyazaki Y, Geber A, Parkinson T, Hitchcock CA, Falconer DJ, Ward DJ, Marsden K, Bennett JE (1998) Fluconazole resistance associated with drug efflux and increased transcription of a drug transporter gene, PDH1, in *Candida glabrata*. Antimicrob Agents Chemother 42:1695–1701

Moran GP, Sanglard D, Donnelly SM, Shanley DB, Sullivan DJ, Coleman DC (1998) Identification and expression of multidrug transporters responsible for fluconazole resistance in *Candida dubliniensis*. Antimicrob Agents Chemother 42:1819–1830

Nagao K, Taguchi Y, Arioka M, Kadokura H, Takatsuki A, Yoda K, Yamasaki M (1995) bfr1+, a novel gene of *Schizosaccharomyces pombe* which confers brefeldin A resistance, is structurally related to the ATP-binding cassette superfamily. J Bacteriol 177:1536–1543

Nakamura K, Niimi M, Niimi K, Holmes AR, Yates JE, Decottignies A, Monk BC, Goffeau A, Cannon RD (2001) Functional expression of *Candida albicans* drug efflux pump Cdr1p in a *Saccharomyces cerevisiae* strain deficient in membrane transporters. Antimicrob Agents Chemother 45:3366–3374

Nakayama H, Tanabe K, Bard M, Hodgson W, Wu S, Takemori D, Aoyama T, Kumaraswami NS, Metzler L, Takano Y, Chibana H, Niimi M (2007) The *Candida glabrata* putative sterol transporter gene CgAUS1 protects cells against azoles in the presence of serum. J Antimicrob Chemother 60:1264–1272

Niimi K, Harding DR, Parshot R, King A, Lun DJ, Decottignies A, Niimi M, Lin S, Cannon RD, Goffeau A, Monk BC (2004) Chemosensitization of fluconazole resistance in *Saccharomyces cerevisiae* and pathogenic fungi by a D-octapeptide derivative. Antimicrob Agents Chemother 48:1256–1271

Nim S, Rawal MK, Prasad R (2014) FK520 interacts with the discrete intrahelical amino acids of multidrug transporter Cdr1 protein and acts as antagonist to selectively chemosensitize azole-resistant clinical isolates of *Candida albicans*. FEMS Yeast Res 14:624–632. doi:10.1111/1567-1364 12149

Nishi K, Yoshida M, Horinouchi S, Beppu T (1993) Mating of the fission yeast independently of pmd1$^+$ gene product, a structural homologue of budding yeast STE6 and mammalian P-glycoproteins. Arch Microbiol 160:162–165

Nunes PA, Tenreiro S, Sá-Correia I (2001) Resistance and adaptation to quinidine in *Saccharomyces cerevisiae*: role of QDR1 (YIL120w), encoding a plasma membrane transporter of the major facilitator superfamily required for multidrug resistance. Antimicrob Agents Chemother 45:1528–1534

Ogawa A, Hashida Okado T, Endo M, Yoshioka H, Tsuruo T, Takesako K, Kato I (1998) Role of ABC transporters in aureobasidin A resistance. Antimicrob Agents Chemother 42:755–761

Ortiz DF, Pierre MV, Abdulmessih A, Arias IM (1997) A yeast ATP binding cassette-type protein mediating ATP-dependent bile acid transport. J Biol Chem 272:15358–15365

Pagant S, Halliday JJ, Kougentakis C, Miller EA (2010) Intragenic suppressing mutations correct the folding and intracellular traffic of misfolded mutants of Yor1p, a eukaryotic drug transporter. J Biol Chem 285:36304–36314

Pao SS, Paulsen IT, Saier MH Jr (1998) Major facilitator superfamily. Microbiol Mol Biol Rev 62:1–34

Pasrija R, Banerjee D, Prasad R (2007) Structure and function analysis of CaMdr1p, a MFS antifungal efflux transporter protein of Candida albicans: identification of amino acid residues critical for drug/H+ transport. Eukaryot Cell 6:443–453

Paulsen IT, Skurray RA (1993) Topology, structure and evolution of two families of proteins involved in antibiotic and antiseptic resistance in eukaryotes and prokaryotes- an analysis. Gene 124:1–11

Paulsen IT, Brown MH, Skurray RA (1996) Proton-dependent multidrug efflux systems. Microbiol Rev 60:575–608

Perlin DS (2011) Current perspectives on echinocandin class drugs. Future Microbiol 6:41–57

Pfaller MA, Diekema DJ (2007) Epidemiology of invasive candidiasis: a persistent public health problem. Clin Microbiol Rev 20:133–163

Pfaller MA, Andes DR, Diekema DJ, Horn DL, Reboli AC, Rotstein C, Franks B, Azie NE (2014) Epidemiology and outcomes of invasive candidiasis due to non-albicans species of Candida in 2,496 patients: data from the Prospective Antifungal Therapy (PATH) registry 2004–2008. PLoS One 9(7):e101510. doi:10.1371/journal.pone.0101510. eCollection 2014

Pina-Vaz C, Rodrigues AG, Costa-de-Oliveira S, Ricardo E, Mardh PA (2005) Potent synergic effect between ibuprofen and azoles on *Candida* resulting from blockade of efflux pumps as determined by FUN-1 staining and flow cytometry. J Antimicrob Chemother 56:678–685

Piper P, Mahe Y, Thompson S, Pandjaitan R, Holyoak C, Egner R, Muhlbauer M, Coote P, Kuchler K (1998) The Pdr12 ABC transporter is required for the development of weak organic acid resistance in yeast. EMBO J 17:4257–4265

Prasad R, Goffeau A (2012) Yeast ATP binding cassette transporters conferring multidrug resistance. Annu Rev Microbiol 66:39–63

Prasad R, Rawal MK (2014) Efflux pump proteins in antifungal resistance. Front Pharmacol 5:202. doi:10.3389/fphar.2014.00202. eCollection 2014. Review

Prasad R, Worgifosse PD, Goffeau A, Balzi E (1995) Molecular cloning and characterisation of a novel gene of *C. albicans,* CDR1, conferring multiple resistance to drugs and antifungals. Curr Genet 27:320–329

Prasad R, Panwar SL, Smriti (2002) Drug resistance in yeasts as an emerging scenario. Adv Microb Physiol 46:155–201

Puri N, Gaur M, Sharma M, Shukla S, Ambudkar SV, Prasad R (2009) The amino acid residues of transmembrane helix 5 of multidrug resistance protein CaCdr1p of *Candida albicans* are involved in substrate specificity and drug transport. Biochim Biophys Acta 1788:1752–1761

Puri N, Prakash O, Manoharlal R, Sharma M, Ghosh I, Prasad R (2010) Analysis of physico-chemical properties of substrates of ABC and MFS multidrug transporters of pathogenic Candida albicans. Eur J Med Chem 45:4813–4826

Purnelle B, Skala J, Goffeau A (1991) The product of the YCR105 gene located on the chromosome III from *Saccharomyces cerevisiae* presents homologies to ATP-dependent permeases. Yeast 7:867–872

Rai V, Shukla S, Jha S, Komath SS, Prasad R (2005) Functional characterization of N-terminal nucleotide binding domain (NBD-1) of a major ABC drug transporter Cdr1p of Candida albicans: uncommon but conserved Trp326 of Walker B is important for ATP binding. Biochemistry 44:6650–6661

Rai V, Gaur M, Shukla S, Shukla S, Ambudkar SV, Komath SS, Prasad R (2006) Conserved Asp327 of Walker B motif in the N-terminal nucleotide binding domain (NBD-1) of Cdr1p of *Candida albicans* has acquired a new role in ATP hydrolysis. Biochemistry 45:14726–14739

Rawal MK, Khan MF, Kapoor K, Goyal N, Sen S, Saxena AK, Lynn AM, Tyndal JD, Monk BC, Cannon RD, Komath SS, Prasad R (2013) Insight into pleiotropic drug resistance ATP-binding cassette pump drug transport through mutagenesis of Cdr1p transmembrane domains. J Biol Chem 288:24480–24493. doi:10.1074/jbc.M113.488353

Raymond M, Dignard D, Alarco AM, Mainville N, Magee BB, Thomas DY (1998) A Ste6p/P-glycoprotein homologue from the asexual yeast *Candida albicans* transports the a-factor mating pheromone in *Saccharomyces cerevisiae*. Mol Microbiol 27:587–598

Reboutier D, Piednoel M, Boisnard S, Conti A, Chevalier V, Florent M, Gibot-Leclerc S, Da Silva B, Chastin C, Fallague K, Favel A, Noël T, Ruprich-Robert G, Chapeland-Leclerc F, Papon N (2009) Combination of different molecular mechanisms leading to fluconazole resistance in a *Candida lusitaniae* clinical isolate. Diagn Microbiol Infect Dis 63:188–193. doi:10.1016/j.diagmicrobio.2008.10.019

Richardson DM (2005) Changing patterns and trends in systemic function infections. J Antimicrob Chemother 56:5–112

Saier MH Jr, Reizer J (1991) Families and super families of transport proteins common to prokaryotes and eukaryotes. Curr Opinion Struct Biol 1:362–368

Saini P, Prasad T, Gaur NA, Shukla S, Jha S, Komath SS, Khan LA, Haq QMR, Prasad R (2005) Alanine scanning of transmembrane helix 11 of Cdr1p ABC antifungal efflux pump of *Candida albicans*: identification of amino acid residues critical for drug efflux. J Antimicrob Chemother 56:77–86

Saini P, Gaur NA, Prasad R (2006) Chimeras of the ABC drug transporter Cdr1p reveal functional indispensability of transmembrane domains and nucleotide-binding domains, but transmembrane segment 12 is replaceable with the corresponding homologous region of the non-drug transporter Cdr3p. Microbiology 152:1559–1573

Sandbaken MG, Lupisella JA, DiDomenico B, Chakraburtty K (1990) Protein synthesis in yeast. Structural and functional analysis of the gene encoding elongation factor 3. J Biol Chem 265:15838–15844

Sanglard D, Kuchler K, Ischer F, Pagani JL, Monod M, Bille J (1995) Mechanisms of resistance to azole antifungal agents in *Candida albicans* isolates from AIDS patients involve specific multidrug transporters. Antimicrob Agents Chemother 39:2378–2386

Sanglard D, Ischer F, Monod M, Bille J (1997) Cloning of *Candida albicans* genes conferring resistance to azole antifungal agents: characterisation of CDR2, a new multidrug ABC transporter gene. Microbiology 143:405–416

Sanglard D, Ischer F, Calabrese D, Majcherczyk PA, Bille J (1999) The ATP binding cassette transporter gene CgCDR1 from *Candida glabrata* is involved in the resistance of clinical isolates to azole antifungal agents. Antimicrob Agents Chemother 43:2753–2765

Sanguinetti M, Posteraro B, La Sorda M, Torelli R, Fiori B, Santangelo R, Delogu G, Fadda G (2006) Role of AFR1, an ABC transporter-encoding gene, in the in vivo response to fluconazole and virulence of *Cryptococcus neoformans*. Infect Immun 74:1352–1359

Sasnauskas K, Jomantiene R, Lebediene E, Lebedys J, Januska A, Janulaitis A (1992) Cloning and sequence analysis of a *Candida maltosa* gene which confers resistance to cycloheximide. Gene 116:105–108

Sasser TL, Padolina M, Fratti RA (2012) The yeast vacuolar ABC transporter Ybt1p regulates membrane fusion through $Ca2+$ transport modulation. Biochem J 448:365–372. doi:10.1042/BJ20120847

Sauna ZE, Bohn SS, Rutledge R, Dougherty MP, Cronin S, May L, Xia D, Ambudkar SV, Golin J (2008) Mutations define cross-talk between the N-terminal nucleotide-binding domain and transmembrane helix-2 of the yeast multidrug transporter Pdr5: possible conservation of a signaling interface for coupling ATP hydrolysis to drug transport. J Biol Chem 283:35010–35022

Schneider E (2001) ABC transporters catalyzing carbohydrate uptake. Res Microbiol 152:303–310

Schneider E, Hunke S (1998) ATP-binding-cassette (ABC) transport systems: functional and structural aspects of the ATP-hydrolyzing subunits/domains. FEMS Microbiol Rev 22:1–20

Schuetzer-Muehlbauer M, Willinger B, Egner R, Ecker G, Kuchler K (2003) Reversal of antifungal resistance mediated by ABC efflux pumps from *Candida albicans* functionally expressed in yeast. Int J Antimicrob Agents 22:291–300

Shani N, Valle D (1996) A *Saccharomyces cerevisiae* homolog of the human adrenoleukodystrophy transporter is a heterodimer of two half ATP-binding cassette transporters. Proc Natl Acad Sci U S A 93:11901–11906

Shani N, Watkins PA, Valle D (1995) PXA1, a possible *Saccharomyces cerevisiae* ortholog of the human adrenoleukodystrophy gene. Proc Natl Acad Sci U S A 92:6012–6016

Sharma M, Prasad R (2011) The quorum-sensing molecule farnesol is a modulator of drug efflux mediated by ABC multidrug transporters and synergizes with drugs in *Candida albicans*. Antimicrob Agents Chemother 55:4834–4843

Sharma M, Manoharlal R, Shukla S, Puri N, Prasad T, Ambudkar SV, Prasad R (2009) Curcumin modulates efflux mediated by yeast ABC multidrug transporters and is synergistic with antifungals. Antimicrob Agents Chemother 53:3256–3265

Shimazu M, Sekito T, Akiyama K, Ohsumi Y, Kakinuma Y (2005) A family of basic amino acid transporters of the vacuolar membrane from *Saccharomyces cerevisiae*. J Biol Chem 280:4851–4857

Shukla S, Saini P, Smriti JS, Ambudkar SV, Prasad R (2003) Functional characterization of *Candida albicans* ABC transporter Cdr1p. Eukaryot Cell 2:1361–1375

Shukla S, Sauna ZE, Prasad R, Ambudkar SV (2004) Disulfiram is a potent modulator of multidrug transporter Cdr1p of *Candida albicans*. Biochem Biophys Res Commun 322:520–525

Smith WL, Edlind TD (2002) Histone deacetylase inhibitors enhance *Candida albicans* sensitivity to azoles and related antifungals: correlation with reduction in CDR and ERG upregulation. Antimicrob Agents Chemother 46:3532–3539

Smriti Krishnamurthy S, Dixit BL, Gupta CM, Milewski S, Prasad R (2002) ABC transporters Cdr1p, Cdr2p and Cdr3p of a human pathogen *Candida albicans* are general phospholipid translocators. Yeast 19:303–318

Stolz J, Wöhrmann HJ, Vogl C (2005) Amiloride uptake and toxicity in fission yeast are caused by the pyridoxine transporter encoded by bsu1+ (car1+). Eukaryot Cell 4:319–326

Sturtevant J, Cihlar R, Calderone R (1998) Disruption studies of a *Candida albicans* gene, ELF1: a member of the ATP-binding cassette family. Microbiology 144:2311–2321

Tanabe K, Lamping E, Adachi K, Takano Y, Kawabata K, Shizuri Y, Niimi M, Uehara Y (2007) Inhibition of fungal ABC transporters by unnarmicin A and unnarmicin C, novel cyclic peptides from marine bacterium. Biochem Biophys Res Commun 364:990–995

Tenreiro S, Rosa PC, Viegas CA, Sa-Correia I (2000) Expression of the AZR1 gene (ORF YGR224w), encoding a plasma membrane transporter of the major facilitator superfamily, is required for adaptation to acetic acid and resistance to azoles in *Saccharomyces cerevisiae*. Yeast 16:1469–1481

Tenreiro S, Nunes PA, Viegas CA, Neves MS, Teixeira MC, Cabral MG, Sa-Correia I (2002) *AQR1* gene (ORF YNL065w) encodes a plasma membrane transporter of the major facilitator superfamily that confers resistance to short-chain monocarboxylic acids and quinidine in *Saccharomyces cerevisiae*. Biochem Biophys Res Commun 292:741–748

Tenreiro S, Vargas RC, Teixeira MC, Magnani C, Sa-Correia I (2005) The yeast multidrug transporter Qdr3 (Ybr043c): localization and role as a determinant of resistance to quinidine, barban, cisplatin, and bleomycin. Biochem Biophys Res Commun 327:952–959

Theiss S, Kretschmar M, Nichterlein T, Hof H, Agabian N, Hacker J, Kohler GA (2002) Functional analysis of a vacuolar ABC transporter in wild-type *Candida albicans* reveals its involvement in virulence. Mol Microbiol 43:571–584

Thornewell SJ, Peery RB, Skatrud PL (1997) Cloning and characterization of CneMDR1, a *Cryptococcus neoformans* gene encoding a protein related to multidrug resistance proteins. Gene 201:21–29

Tomitori H, Kashiwagi K, Asakawa T, Kakinuma Y, Michael AJ, Igarashi K (2001) Multiple polyamine transport systems on the vacuolar membrane in yeast. Biochem J 353:681–688

Torelli R, Posteraro B, Ferrari S, La Sorda M, Fadda G, Sanglard D, Sanguinetti M (2008) The ATP-binding cassette transporter-encoding gene CgSNQ2 is contributing to the CgPDR1-dependent azole resistance of *Candida glabrata*. Mol Microbiol 68:186–201. doi:10.1111/j.1365-2958.2008.06143.x

Vargas RC, Garcia-Salcedo R, Tenreiro S, Teixeira MC, Fernandes AR, Ramos J, Sa-Correia I (2007) *Saccharomyces cerevisiae* multidrug resistance transporter Qdr2 is implicated in potassium uptake, providing a physiological advantage to quinidine-stressed cells. Eukaryot Cell 6:134–142

Vazquez de Aldana CR, Marton MJ, Hinnebusch AG (1995) GCN20, a novel ATP binding cassette protein, and GCN1 reside in a complex that mediates activation of the eIF-2 alpha kinase GCN2 in amino acid-starved cells. EMBO J 14:3184–3199

Wang Y, Cao YY, Jia XM, Cao YB, Gao PH, Fu XP, Ying K, Chen WS, Jiang YY (2006) Cap1p is involved in multiple pathways of oxidative stress response in *Candida albicans*. Free Radic Biol Med 40:1201–1209

Watkins WJ, Chong L, Cho A, Hilgenkamp R, Ludwikow M, Garizi N, Iqbal N, Barnard J, Singh R, Madsen D, Lolans K, Lomovskaya O, Oza U, Kumaraswamy P, Blecken A, Bai S, Loury DJ, Griffith DC, Dudley MN (2007) Quinazolinone fungal efflux pump inhibitors. Part 3: (N-Methyl)piperazine variants and pharmacokinetic optimization. Bioorg Med Chem Lett 17:2802–2806

Wawrzycka D, Sobczak I, Bartosz G, Bocer T, Ułaszewski S, Goffeau A (2010) Vmr 1p is a novel vacuolar multidrug resistance ABC transporter in *Saccharomyces cerevisiae*. FEMS Yeast Res 10:828–838. doi:10.1111/j.1567-1364.2010.00673.x

Wilcox LJ, Balderes DA, Wharton B, Tinkelenberg AH, Rao G, Sturley SL (2002) Transcriptional profiling identifies two members of the ATP-binding cassette transporter superfamily required for sterol uptake in yeast. J Biol Chem 277:32466–32472

Wolfger H, Mahe Y, Parle-McDermott A, Delahodde A, Kuchler K (1997) The yeast ATP binding cassette (ABC) protein genes PDR10 and PDR15 are novel targets for the Pdr1 and Pdr3 transcriptional regulators. FEBS Lett 418:269–274

Wright MB, Howell EA, Gaber RF (1996) Amino acid substitutions in membrane-spanning domains of Hol1, a member of the major facilitator superfamily of transporters, confer nonselective cation uptake in *Saccharomyces cerevisiae*. J Bacteriol 178:7197–7205

Wu CP, Ohnuma S, Ambudkar SV (2011) Discovering natural product modulators to overcome multidrug resistance in cancer chemotherapy. Curr Pharm Biotechnol 12:609–620

Yamamoto S, Hiraga K, Abiko A, Hamanaka N, Oda K (2005) A new function of isonitrile as an inhibitor of the Pdr5p multidrug ABC transporter in *Saccharomyces cerevisiae*. Biochem Biophys Res Commun 330:622–628

Yang Z, Huang J, Geng J, Nair U, Klionsky DJ (2006) Atg22 recycles amino acids to link the degradative and recycling functions of autophagy. Mol Biol Cell 17:5094–5104

Yong VC, Chew LM, Kwan CL, Ng KP, Seow HF et al (2006) Transcriptional and sequencing analysis of CtCDR1, CtMDR1 and CtERG11 genes in fluconazole-resistant *Candida tropicalis* isolates from Candidemia patients. J Med Sci 6:713–723

Young L, Leonhard K, Tatsuta T, Trowsdale J, Langer T (2001) Role of the ABC transporter Mdl1 in peptide export from mitochondria. Science 291:2135–2138

# Index





J. Ramos et al. (eds.), *Yeast Membrane Transport*, Advances in Experimental Medicine and Biology 892, DOI 10.1007/978-3-319-25304-6

MIX
Papier aus verantwortungsvollen Quellen
Paper from responsible sources
FSC® C105338

If you have any concerns about our products,
you can contact us on
**ProductSafety@springernature.com**

In case Publisher is established outside the EU,
the EU authorized representative is:
**Springer Nature Customer Service Center GmbH**
**Europaplatz 3, 69115 Heidelberg, Germany**

Printed by Libri Plureos GmbH
in Hamburg, Germany